Bob Scholl

73165129

2A MECH ENG

# APPLIED STATISTICAL METHODS

## OPERATIONS RESEARCH
## AND INDUSTRIAL ENGINEERING

Consulting Editor: *J. William Schmidt*

Virginia Polytechnic Institute and State University, Blacksburg, Virginia

---

Applied Statistical Methods
*I. W. Burr*

*IN PREPARATION:*

Mathematical Foundations of Management Science
and Systems Analysis
*J. William Schmidt*

# APPLIED STATISTICAL METHODS

*Irving W. Burr*

PURDUE UNIVERSITY

ACADEMIC PRESS   New York and London

*A Subsidiary of Harcourt Brace Jovanovich, Publishers*

ACADEMIC PRESS, INC.
111 Fifth Avenue, New York, New York 10003

*United Kingdom Edition published by*
ACADEMIC PRESS, INC. (LONDON) LTD.
24/28 Oval Road, London NW1

Library of Congress Cataloging in Publication Data

Burr, Irving Wingate, Date
     Applied statistical methods.

     (Operations research and industrial engineering)
     Includes bibliographical references.
     1.  Mathematical statistics.    2.    Probabilities.
I.    Title
QA276.B863            519.5            73-2077
ISBN 0-12-146150-5

AMS (MOS) 1970 Subject Classifications: 62N99, 62P99, 62N10

*To my wife Elsie*

# TABLE OF CONTENTS

*Preface*                                                                      xvii

*Acknowledgments*                                                              xix

Chapter 1  **Introduction**

1.1  Why Statistical Methods?                                                   1
1.2  Advice to the Student                                                      2

Chapter 2  **The Frequency Distribution——
A Tool and a Concept**

2.1  Introduction                                                              4
2.2  An Example of a Frequency Distribution                                    4
2.3  Frequency Class Nomenclature and Tabulation                              6
2.4  Discrete versus Continuous Data                                          8
2.5  Graphical Representation of a Frequency Distribution                     8
2.6  The Cumulative Frequency Graph                                          10
2.7  What a Frequency Distribution Shows                                     12
2.8  Some Examples of Use of Frequency Tables and Graphs                     13
2.9  Sample versus Population                                                16
2.10 Summary                                                                19
     Problems                                                                19

Chapter 3  **Summarization of Data by Objective Measures**

3.1  Introduction                                                           33
3.2  Some Averages                                                          33
3.3  Some Measures of Variability                                           35
3.4  Efficient Calculation of Averages and Standard Deviations             38

     3.4.1  *Calculations for Frequency Data*                              41

3.5* Further Descriptive Measures of Frequency Distributions,
       Third and Fourth Moments                                           42
3.6  Summary                                                              45
     Problems                                                             46

vii

## Chapter 4   **Some Elementary Probability**

4.1     Introduction                                                           48
4.2     Sample Spaces of Outcomes                                              49
4.3     Events                                                                 50

    4.3.1   *Relations of Events*                                          51
    4.3.2   *Combinations of Events*                                       52

4.4     Probabilities of Events                                                55
4.5     Probabilities on Discrete Sample Spaces                                57

    4.5.1   *Countably Infinite Spaces*                                    58
    4.5.2   *Events over Discrete Spaces*                                  58

4.6     Independent and Dependent Events                                       59

    4.6.1   *Conditional Probabilities*                                    62
    4.6.2   *Repeated Trials*                                              63

4.7     Discrete Probabilities                                                 66

    4.7.1   *Permutations and Combinations*                                66
    4.7.2   *Discrete Probability Examples*                                69

4.8     Probabilities on Continuous Spaces                                     77
4.9     Applied Bayes' Probabilities—Posterior Probabilities                   80
4.10    Interpretation of a Probability                                        82
4.11    Random Variables                                                       83
4.12    Summary                                                                83
       Problems                                                        84

## Chapter 5   **Some Discrete Probability Distributions**

5.1     Theoretical Populations                                                90
5.2     Discrete Probability Distributions in General                          91

    5.2.1   *Expected Values for Y and Functions of Y*                     92
    5.2.2*  *Population Curve-Shape Characteristics*                      93
    5.2.3   *Algebra of Expectations*                                     93
    5.2.4*  *Further on Population Moments*                               95

5.3     The Binomial Distribution                                              95

    5.3.1   *Examples of the Binomial Distribution*                        97
    5.3.2   *Population Moments for Binomial Distributions*                99
    5.3.3   *Use of Binomial Tables*                                      101
    5.3.4   *Calculation of a Binomial Distribution*                      101
    5.3.5   *Approximations to the Binomial Distribution*                 103
    5.3.6   *Conditions of Applicability of the Binomial Distribution*    103

5.4     The Poisson Distribution                                               103

    5.4.1*  *A Derivation of the Poisson Probability Function*           106
    5.4.2   *Examples of the Poisson Distribution*                        107

5.4.3   *Tables of the Poisson Distribution*                            109
5.4.4   *Using the Poisson Distribution to Approximate the Binomial*    109
5.4.5*  *Derivation of the Poisson as a Limit of the Binomial*          109
5.4.6   *Conditions of Applicability of the Poisson Distribution*       110

5.5     The Hypergeometric Distribution                                 111

5.5.1   *Tables for the Hypergeometric Distribution*                    113
5.5.2   *Examples of the Hypergeometric Distribution*                   114
5.5.3*  *Moments for the Hypergeometric Distribution*                   115
5.5.4*  *Binomial Approximations to the Hypergeometric Distribution*    116
5.5.5*  *Poisson Approximations to the Hypergeometric Distribution*     117
5.5.6   *Approximations to Sums of Terms of the Hypergeometric*
        *Distribution*                                                  118
5.5.7   *Conditions of Applicability of the Hypergeometric Distribution* 118
5.5.8   *Applications of the Hypergeometric Distribution*               119

5.6     The Uniform Distribution                                        119
5.7*    The Geometric Distribution                                      120
5.8*    The Negative Binomial Distribution                              122
5.9     Generating Samples from Discrete Distributions                  124
5.10    Summary                                                         125
        References                                                      125
        Problems                                                        126

## Chapter 6   **Some Continuous Probability Distributions**

6.1     Continuous Probability Distributions                            131
6.2     Some General Properties of Continuous Distributions             131

6.2.1   *Moments for a Continuous Distribution*                         134

6.3     The Normal Curve                                                135

6.3.1   *Properties of the Normal Distribution*                         136
6.3.2   *The General Normal Curve*                                      138
6.3.3   *Sketching a Normal Curve*                                      138
6.3.4   *Approximating Probabilities by a Normal Distribution*          139

6.4     The Rectangular Distribution                                    141
6.5     The Exponential Distribution                                    142
6.6*    The Gamma Distribution                                          144

6.6.1*  *Tables of the Gamma Distribution*                              146
6.6.2*  *Relation to the Normal Distribution*                           146
6.6.3*  *Use of the Gamma Distribution to Approximate Discrete*
        *Distributions*                                                 146

6.7*    The Beta Distribution                                           148
6.8*    The Weibull Distribution                                        150
6.9*    The Pearson System of Distributions                             151
6.10*   An Easily Fitted General System of Frequency Curves             152
6.11    Sums and Averages and a Central Limit Theorem                   152

6.12* Tchebycheff's Theorem                                                     155
6.13  Summary                                                                    156
6.14* Proofs of Some Relations in Section 6.11                                   157
      References                                                                 160
      Problems                                                                   161


Chapter 7  **Some Sampling Distributions**

7.1   Distribution of Sample Statistics from Populations                         166
7.2   Choice of Sample                                                           167

      7.2.1* *Sampling from a Probability Distribution*                          168
      7.2.2* *Machine Generation of Random Samples*                              169

7.3   Sampling Distributions of a Sample Statistic                              169
7.4   Distribution of Sample Means                                              171

      7.4.1  *Standardized Distribution for Means*                              171
      7.4.2  *Distribution of Means when Standard Deviation Is Unknown*         171
      7.4.3  *Areas for the t Distribution*                                     173
      7.4.4* *Interpolation Note*                                              173
      7.4.5  *Distribution of Means from Nonnormal Populations*                174

7.5   Distribution of Sample Variances                                          174

      7.5.1  *Distribution of Sample Standard Deviation*                        176
      7.5.2* *Population of y's Nonnormal*                                      176
      7.5.3  *Tables of Chi-Square*                                            177

7.6*  Joint Distribution of $\bar{y}$ and $s$ from a Normal Population          177
7.7   Two Normal Populations, Independent Samples                               177

      7.7.1  *Sum and Difference of Two Means, Standard Deviations Known*       177
      7.7.2  *Sums and Differences of Two Means, Standard Deviations*
             *Unknown but Equal*                                               178
      7.7.3  *Two Variances, F Distribution*                                   179
      7.7.4* *Two Variances, Large Samples*                                    180

7.8   Sampling Aspects of the Binomial and Poisson Distributions               181
7.9*  Sum of Two Independent Chi-Square Variables                              182
7.10* Noncentral Distributions                                                 182
7.11  Summary                                                                   183
      References                                                                183
      Problems                                                                  184


Chapter 8  **Statistical Tests of Hypotheses——**
           **General and One Sample**

8.1   Introduction                                                             187
8.2   An Example                                                               188

      8.2.1  *Approach* 1  *Given n, Set Significance Level* $\alpha$          189
      8.2.2  *Approach* 2  *Set Two Risks:* $\alpha$ *and a* $\beta_\mu$ *, and Find n*   192

8.3 Summary of the Elements in Tests of Hypotheses on One Parameter     194
8.4 Summary of Significance Testing for One Mean with $\sigma$ Unknown     196
8.5 Interpretation of Decisions in Hypothesis Testing     197
8.6 Nonnormal Populations of $y$'s     198
8.7 Significance Testing for Mean $\mu$, with $\sigma$ Unknown     198

    8.7.1 *Example*     200
    8.7.2 *Operating Characteristic Curve*     201

8.8 Significance Tests for Variability     202

    8.8.1 *An Example of First Approach*     203
    8.8.2 *The Second Approach of 4, in Section 8.8*     204
    8.8.3\* *Operating Characteristic Curves for Variability Tests*     206
    8.8.4\* *Large Samples*     206

8.9 Significance Testing for Attributes     207

    8.9.1 *The Binomial Tests*     207
    8.9.2 *The Poisson Tests*     209
    8.9.3\* *Other Attribute Distributions*     210
    8.9.4 *Operating Characteristic Curves*     211

8.10 Relation of Significance Testing to Decision Theory     212
8.11 Summary     214
    References     217
    Problems     218

## Chapter 9    Significance Tests——Two Samples

9.1 The General Problem     223
9.2 Tests on Two Variances——The $F$ Test     225

    9.2.1 *An Example*     226
    9.2.2 *Large Sample Tests*     227
    9.2.3 *A Large Sample Example*     227

9.3 Differences between Means     228

    9.3.1 *Standard Deviations Known*     228
    9.3.2 *Standard Deviations Equal but Unknown*     230
    9.3.3\* *Standard Deviations Unknown and Possibly Unequal*     232

9.4 Significance of Differences——Binomial Data     234
9.5 Significance of Differences——Poisson Data     236

    9.5.1 *Unequal Areas of Opportunity*     237

9.6 Matched Pair Data. Importance of Experimental Design     239

    9.6.1\* *Matched-Pair Model*     241

9.7 Sample Sizes Needed for Tests of Two Means     242
9.8 Summary     246
    References     246
    Problems     247

## Chapter 10  **Estimation of Population Characteristics**

10.1   Point Estimates——General Idea                                      253
10.2   Which Estimator to Use——Characteristics of Estimation             254

    10.2.1* *Consistency and Sufficiency*                            256

10.3*  How to Find a Desirable Estimator                                  256
10.4   Point Estimates——Common Cases                                      256
10.5   Interval Estimation in General                                     257

    10.5.1* *Geometrical Argument for Confidence Intervals*         258

10.6   Confidence Intervals for $\mu$                                     259
10.7   Confidence Intervals for $\sigma$                                  261

    10.7.1* *Large Confidence Limits for $\sigma$*                   262

10.8   How to Have Narrower Confidence Intervals                          263
10.9   Confidence Intervals for Functions of Two Parameters——
     Two Samples                                                   263

    10.9.1  *Confidence Limits on the Difference of Means*           264
    10.9.2* *Confidence Limits on the Ratio of $\sigma$'s*          266
    10.9.3  *Paired Differences*                                     267

10.10  Confidence Limits for Attribute Data                              267

    10.10.1  *Exact Method for Binomial Population*                  268
    10.10.2  *Normal Approximation for Confidence Limits for the Binomial*  269
    10.10.3  *Exact Method for Poisson Population*                   270
    10.10.4  *Normal Approximation for Confidence Limits for the Poisson*  271
    10.10.5  *Tables of Confidence Limits*                           271
    10.10.6* *Confidence Limits for Two Samples of Attribute Data*  272

10.11  Relation between Interval Estimation and Significance Testing      273
10.12  Summary                                                           275
    References                                                     275
    Problems                                                       278

## Chapter 11  **Simple Regression**

11.1   Regression, A Study of Relationship                                285
11.2   The Scatter Diagram                                                286
11.3   Line of Best Fit to "Linear" Data                                  287

    11.3.1  *Least Squares Fitting*                                  287
    11.3.2  *Calculational Aspects*                                  290
    11.3.3  *The Linear Model and Its Parameters*                     292

11.4   Sampling Distributions for Estimates                               293
11.5   Significance Tests and Confidence Intervals for Parameters in
     Linear Regression                                             295

    11.5.1  *Slope*                                                  296

11.5.2  *Intercept*                                                        297
11.5.3  *Mean of Y's:* $\mu_Y$                                             297
11.5.4  *Error Variance* $\sigma_\epsilon{}^2$, *and* $\sigma_\epsilon$     297
11.5.5  *Regression Line Mean:* $\mu_{Y \cdot X} = \mu_Y + \beta(X - \bar{X})$  298

11.6     Correlational Aspects                                             298
11.7*    Grouped Bivariate Data                                            300
11.8     Special Case $\mu_{Y \cdot X} = \beta_1 X$                         304
11.9*    Significance of Differences between Two Slopes                    305
11.10    Nonlinearity Test                                                 305
11.11*   Use of Least Squares Fitting for Other Trends                     306

11.11.1*  *Functions Linear in the Parameters*                             306
11.11.2*  *Least Squares after a Transformation*                           307
11.11.3*  *Intrinsically Nonlinear Cases*                                  310

11.12    Applications to Industry and the Laboratory                       310
11.13    Summary                                                           312
         References                                                        312
         Problems                                                          313

## Chapter 12  Simple Analysis of Variance

12.1     General Concept of Analysis of Variance                          322
12.2     One-Factor Analysis of Variance                                  323

12.2.1  *The Model*                                                        323
12.2.2  *The Formulas and Test*                                            325
12.2.3  *The Case of Unequal Sample Sizes*                                 329
12.2.4*  *Orthogonal Contrasts*                                            331

12.3     Orthogonal Polynomials and Tests                                 334
12.4     A Method of Multiple Contrasts                                    341

12.4.1  *The Newman–Keuls Multiple Range Test*                             341
12.4.2  *Example*                                                          343
12.4.3  *Interpretation of Risk* $\alpha$                                  344

12.5*    Testing Homogeneity of Variances                                 345

12.5.1  *An Example*                                                       346
12.5.2  *Q Test with Unequal Degrees of Freedom*                          347
12.5.3  *Q Test for Ranges*                                                347

12.6     Types of Factors                                                  348
12.7     Analysis of Variance for Two Factors                             349

12.7.1  *An Example*                                                       349
12.7.2  *The Models and Assumptions*                                       355
12.7.3  *Expected Mean Squares and Significance Tests*                     356
12.7.4  *Interpretation of Significant Factors*                            348
12.7.5  *Case of Unreplicated Two-Factor Experiments*                      348
12.7.6  *Example of an Unreplicated Two-Factor Completely
         Randomized Design*                                               360

12.8   Other Models                                                         362
12.9   Summary                                                              363
       References                                                           363
       Problems                                                             365


Chapter 13   **Multiple Regression**

13.1   Introduction                                                         372
13.2   First Approach                                                       374

       13.2.1   *Data Table Format*                                         374
       13.2.2   *Fitting the Equation*                                      374
       13.2.3   *Alternative Forms of Normal Equations and Regression*      376
       13.2.4   *Describing Goodness of Fit*                                379
       13.2.5   *Systematic Solution of the Normal Equations*              380
       13.2.6   *Significance Tests on the Explained Variation*             384
       13.2.7   *Simple Example*                                            385
       13.2.8   *Second Example*                                            387

13.3   Second Approach                                                      388

       13.3.1   *Vectors and Matrices*                                      388
       13.3.2   *The Matrix Approach*                                       391
       13.3.3   *Selection of a Set of Predictors*                          396
       13.3.4   *Calculation of an Inverse*                                 397

13.4   Summary of Approach                                                  397
13.5   Adequacy of Regression Model                                         398
13.6   Comments and Precautions                                             399
       References                                                           400
       Problems                                                             401


Chapter 14   **Goodness of Fit Tests, Contingency Tables**

14.1   Introduction                                                         407
14.2   The Chi-square Test for Cell Frequencies, Observed versus
          Theoretical                                                       408

       14.2.1   *An Example*                                                409

14.3   Testing Goodness of Fit of Theoretical Distributions                409

       14.3.1   *A Binomial Example*                                        411
       14.3.2   *Examples of Tests of Normality*                           412
       14.3.3*  *Example of a Gamma Distribution Fit*                       414
       14.3.4   *Example of a Poisson Distribution*                         415

14.4*  Other Goodness of Fit Tests                                          416
14.5   Contingency Tables                                                   417

       14.5.1   *An Example*                                                417
       14.5.2   *The General Setup of a Contingency Table*                  418
       14.5.3   *A $2 \cdot 2$ Contingency Table*                           420
       14.5.4   *Case of a $2 \cdot b$ Contingency Table*                   421

14.6    The Sign Test                                                                    422
14.7    Summary                                                                          423
        References                                                                       423
        Problems                                                                         424


**Appendix**                                                                             429


**Answers to Odd-Numbered Problems**                                                     461


*Index*                                                                                  469

# PREFACE

The purpose of this book is to provide the student with the fundamental understanding of statistical methods necessary to deal with a wide variety of practical problems. Two points have been emphasized in developing the text. First, the selection of topics for inclusion was based upon breadth of applicability to practical problems. Second, the topics covered are presented in a manner which stresses clarity of understanding, interpretation, and method of application.

The text is intended primarily for upper division undergraduate and graduate students in the mathematical, physical, and engineering sciences, and in economics, business, and related areas. In addition, researchers and line personnel in industry and government will find this book useful in self-study. The background required for complete understanding of the entire text is limited to an elementary knowledge of integral and differential calculus. However, much of the material covered does not require calculus and may therefore be pursued by students with a background limited to college algebra.

In general, results given are presented without proof or derivation. In some cases, however, proper comprehension requires a proof or derivation of a particular result. The development and discussion of the statistical methods treated is accompanied by an extensive set of examples to demonstrate the application of the material to practical problems. In many instances these examples have been taken directly from industrial problems encountered by the author. In addition, each chapter includes a liberal set of problems for student assignment. Answers for odd-numbered problems are provided; solutions for even-numbered problems are available upon request from the Publisher.

The text is intended for use in a one-semester course or a sequence of two-quarter courses in statistical methods. The variety of material presented should provide the instructor with a reasonable degree of flexibility in the choice of topics to be presented. Several sections within a number of chapters are starred. While these sections are of significant importance in specific areas of application, they may be omitted without loss of continuity.

In Chapter 1 the importance of statistical analysis is illustrated, and this provides motivation for the study of the material in the remainder of the text. Chapters 2 and 3 introduce the student to methods of data summarization, including frequency distributions, cumulative frequency distributions, and measures of central tendency and variability. In Chapter 4 the fundamental principles of probability are discussed. The concepts of sample spaces, outcomes, events, probability, independence of events, and random variables are introduced. Chapters 5 and 6 treat discrete and continuous random variables and their characterization. The distributions of several important statistics such as the sample mean, the sample variance, the difference of sample means, and the ratio of sample variances are presented in Chapter 7.

In Chapter 8 the student is introduced to statistical tests of hypotheses, stressing the significance of sample size, type I and type II errors, and the design of statistical tests. Tests of hypotheses concerning a single mean and a single variance are then presented with an interpretation of the results of these tests. Two sample tests of hypotheses are presented in Chapter 9, including tests of hypotheses for the comparison of two means and two variances. Point and interval estimation as well as a discussion of the desirable properties of point estimates are treated in Chapter 10. Simple linear regression is the topic of Chapter 11. Here the method of least squares is discussed and illustrated. In addition, tests of hypotheses about and confidence intervals for sample estimates of the parameters of a simple linear regression model are presented. The elements of single- and two-factor analysis of variance are discussed in Chapter 12. The design of analysis of variance experiments is discussed along with tests for homogeneity of variance and multiple contrasts. The results of Chapter 11 are extended to multivariate regression analysis in Chapter 13. Relevant significance tests and matrix methods in multiple regression analysis are also presented. Finally, in Chapter 14, chi-square tests for goodness of fit and data categorized in contingency tables are treated and illustrated.

The author would much appreciate receipt of any errors or inaccuracies found in the text, or suggestions for future improvement.

# ACKNOWLEDGMENTS

The author is much indebted to his colleague Professor Virgil L. Anderson who read the original manuscript and made helpful suggestions. He is also indebted to Professor J. W. Schmidt of Virginia Polytechnic Institute who reviewed the manuscript, making useful suggestions. Thanks are also due Mrs. Dorothy Penner, whose accurate typing was a great help.

The tables at the back of the book have been either calculated for this text (and checked against other sources) or else copied or abstracted, by kind permission, from Professor Donald B. Owen's "Handbook of Statistical Tables," published by Addison-Wesley Publishing Co., to whom thanks are tendered. In using rounding off in the few cases where the best last digit could not be ascertained, the larger digit has been used. Table VII has been reproduced by the kind permission of Dr. H. L. Harter of the Aerospace Research Laboratories, Wright-Patterson Air Force Base, from his "Order Statistics and Their Use in Testing and Estimation," Volume 1.

Special thanks go to my wife Elsie, ever a source of encouragement.

# Chapter 1

# INTRODUCTION

## 1.1. WHY STATISTICAL METHODS?

The word "statistics" has two meanings, both of which are of importance to us. The first of these is as a plural word, meaning facts or observations, especially numerical results or data. The second meaning is as a singular word, denoting a branch of knowledge.

Whenever outcomes or results vary unpredictably or in random fashion, we have statistical data, whether we like it or not, whether we know it or not. If we have statistical data, we ought to be using appropriate statistical methods to analyze them, so as to be objective and to avoid the many traps, pitfalls, and misinterpretations which lie in wait for the unwary. "Statistics" may in fact be defined as the science of proper collection, condensation, analysis, and interpretation of data which *vary* (from time to time, place to place, trial to trial, person to person, material to material, etc.). The objects of statistics are estimation, prediction, and decision-making from analysis of properly collected data which vary. In doing so some risks of wrong decisions must be accepted, but statistical methods enable us to measure such risks and to control them within economic limits. Basically there are two errors we can make from data: (1) fail to learn something we should; (2) "learn" something or take some action not justified by the data. Both errors are being made continually.

Now why is "statistics" an important discipline? The first reason is, as we have just seen, that of making sound use of data for decision-making, prediction, estimation, and so on, in our job or profession. The second reason is because of the universality of variation in all areas.

1

It is not at all difficult to show the need for statistical reasoning in each and every field of concentration in a large university, not only for those experimenting in the field but also for anyone reading about the field in question. Professional books and journals carry much statistical material. The same thing is true of our lives as citizens. We are continually bombarded by data and statements therefrom, in newspapers, magazines, radio and television, public speeches, conversation, advertisements, and in politics. We need statistical sense to be good citizens.

In essence statistics studies the relationships between populations and samples. Of course, if we have the entire population at hand and it is not too large a collection, we have a purely descriptive job at hand. (Statistical methods of condensation, averaging, etc., can help in this description.) But in general we can only obtain a sample, and from this we will wish to draw inferences about what the population is like. Or we may specify a hypothetical population and ask whether our observed sample could readily have come from such a population. In both cases we have the problem of the relation between population and sample.

Statistics provides us with a set of mathematical models of patterns of variation, which we "hold up to nature." The more accurately our model reflects nature, the safer are our conclusions.

In this book is presented a kit of statistical tools of remarkable versatility and utility for the analysis of data. These tools can either be applied directly or with a little ingenuity or imagination be adapted to the problem at hand. Regarding the latter it may even be said that there is always some approximation involved in attributing a mathematical or probabilistic model to nature. In fact one friend of the author's said he had never seen any *perfectly* pure application of statistical methods, while another used to give a course entitled "Messy Data." Nevertheless our applications are in general "good enough," if not absolutely perfect.

## 1.2. ADVICE TO THE STUDENT

Statistical reasoning may well seem rather foreign to the student if this book is his first contact with the field of statistics. But as the relations between population and sample become clear, the approaches will begin to make sense and to seem entirely natural.

The student is urged to work many numerical problems, for this helps him get the "feel" of the methods. Indeed the great early leader in statistics, Sir Ronald Fisher, is reputed to have said that all he ever learned was through the calculating machine. Nowadays, with such fantastic calculational potential in digital computers, the student may wonder why he should ever calculate any results, especially those in the latter chapters of this book. There are three reasons: (1) to aid in interpretation of results and familiarity with the methods, as just pointed out; (2) the desirability of becoming adept with a desk calculator or possibly a slide rule; and (3) because it is often quicker and more convenient to calculate a result via a desk calculator than to dig out the necessary program, put the data onto cards, and then enter the deck and wait for the results.

One thing against which the student should be constantly on guard, no matter how he calculates or with what equipment, is errors due to rounding off. In fact good advice is to carry many more places than seem necessary and then to round off at the end. This is especially true when "heavy cancellation" occurs. For example, $2927.3386 - 2927.3144 = .0242$, where the difference between these two eight-significant digit numbers is only good for three. A number of shortcut methods have this potential danger.

Another suggestion is that the student make a guess as to the numerical result prior to calculating, for example, an average or a standard deviation. This has two virtues: (1) it tends to sharpen the student's numerical sense, and (2) it will provide warning, if a gross error has been committed in the calculation. Further it can be regarded as a game to liven up the problem!

Finally the student may or may not find numerical examples and problems from his *own* particular field. He is urged to use his *imagination* and think up, from his own field of interest, as many illustrations as he can, for each new technique or tool as it comes along in the book. The author cannot possibly give examples from every field for each technique. Certainly there are differences in the various fields. Everyone can see such differences. What we are here emphasizing is not the differences, but the similarities—notably variation and patterns thereof and inference-making in the presence of variation.

In the examples and problems the student will find many names for objects or things. If they are meaningless to you, just pass right over the name and forget it, since it will not bother you any. The author would not be able to describe each object, industrial piece or chemical, and so on. These problems and examples have virtually all come from actual experience in science and industry and are a small part of some 30 notebooks of cases.

# Chapter 2

# THE FREQUENCY DISTRIBUTION—
# A TOOL AND A CONCEPT

## 2.1. INTRODUCTION

As has been pointed out in Chapter 1, variation in numerical results occurs almost universally. For example, we may measure some characteristic for each of a series of manufactured parts, or we may measure the outcome, perhaps yield, for a series of laboratory runs. If our measurements are precise enough, there will always be some variation even when all conditions are closely controlled.

Whenever we have a large number of numerical results, say 25 or more, we will find it desirable to try to condense them to give the general pattern. An excellent way to do this is to tabulate the results into a frequency table, which will give the number of results in each of a set of numerical classes. We may then construct a graph to give further picturization of the pattern of variation of the results. Or we may calculate from the table certain objective statistical measures such as an average and a measure of variability. The idea is to condense and describe the distribution of the results.

In this chapter we are concerned with frequency tables and graphs. Moreover we introduce the basic concepts of sample and population. Chapter 3 discusses the statistical measures.

## 2.2. AN EXAMPLE OF A FREQUENCY DISTRIBUTION

Table 2.1 shows the percentage of manganese in iron from a blast furnace. A single analysis was made on each of the five casts per day.

4

By looking through the 125 analyses in the table one can obtain a rough idea of the general average level, say 1.4%, and by careful inspection find the minimum 1.06% and maximum 1.80%. But little else is obtained by inspection (other than eyestrain).

TABLE 2.1

Percent Manganese in 125 Casts of Iron from a Blast Furnace[a]

| April | | | | | % Manganese | | | | | |
|-------|------|------|------|------|------|------|------|------|------|------|
| 1, 2 | 1.40 | 1.28 | 1.36 | 1.38 | 1.44 | 1.40 | 1.34 | 1.54 | 1.44 | 1.46 |
| 3, 4 | 1.80 | 1.44 | 1.46 | 1.50 | 1.38 | 1.54 | 1.50 | 1.48 | 1.52 | 1.58 |
| 5, 6 | 1.52 | 1.46 | 1.42 | 1.58 | 1.70 | 1.62 | 1.58 | 1.62 | 1.76 | 1.68 |
| 7, 8 | 1.68 | 1.66 | 1.62 | 1.72 | 1.60 | 1.62 | 1.46 | 1.38 | 1.42 | 1.38 |
| 9, 10 | 1.60 | 1.44 | 1.46 | 1.38 | 1.34 | 1.38 | 1.34 | 1.36 | 1.58 | 1.38 |
| 11, 12 | 1.34 | 1.28 | 1.08 | 1.08 | 1.36 | 1.50 | 1.46 | 1.28 | 1.18 | 1.28 |
| 13, 14 | 1.26 | 1.50 | 1.52 | 1.38 | 1.50 | 1.52 | 1.50 | 1.46 | 1.34 | 1.40 |
| 15, 16 | 1.50 | 1.42 | 1.38 | 1.36 | 1.38 | 1.42 | 1.34 | 1.48 | 1.36 | 1.38 |
| 17, 18 | 1.32 | 1.40 | 1.40 | 1.26 | 1.26 | 1.16 | 1.34 | 1.40 | 1.16 | 1.54 |
| 19, 20 | 1.24 | 1.22 | 1.20 | 1.30 | 1.36 | 1.30 | 1.48 | 1.28 | 1.18 | 1.28 |
| 21, 22 | 1.30 | 1.52 | 1.76 | 1.16 | 1.28 | 1.48 | 1.46 | 1.48 | 1.42 | 1.36 |
| 23, 24 | 1.32 | 1.22 | 1.72 | 1.18 | 1.36 | 1.44 | 1.28 | 1.10 | 1.06 | 1.10 |
| 25 | 1.16 | 1.22 | 1.24 | 1.22 | 1.34 | | | | | |

[a] Five casts per day.

By constructing a frequency table we can picture much better the general pattern of variation of the analyses. We first must choose numerical classes for the data, perhaps eight to ten, so as to cover the range 1.06–1.80%. In this choice we note that the data were measured to the nearest .02%, and thus the limits of the classes are listed only to even hundredths of a percent. Convenient classes are 1.00–1.08%, 1.10–1.18%, and so on. (Note that there can be no question as to which class a 1.09% reading would go into, because such a number will never occur.)

To tabulate the data, we write all the classes in a column. Then the first number, 1.40, is tallied opposite the class 1.40–1.48, the next, 1.28, opposite 1.20–1.28, and so forth, until all 125 have been tallied. (The tally marks themselves give us a rough graph of the distribution.) Thus we obtain the class frequencies as shown in the last column of Table 2.2. The second and third columns give additional information about the corresponding class; see the next section.

Table 2.2 shows the class frequencies rising steadily to a maximum of 32, and then falling away perhaps a trifle more slowly than they rose.

The average level of manganese is quite apparently about 1.40%. One can also find the proportion of cases in various ranges. For example, 98/125 or 78.4% of the cases lie between 1.19 and 1.59%.

TABLE 2.2

Frequency Table for Percents Manganese in Iron from 125 Casts from a Blast Furnace

| Class limits (%) | Class boundaries (%) | Midvalue (%) | Class frequency |
|---|---|---|---|
| 1.00–1.08 | .99–1.09 | 1.04 | 3 |
| 1.10–1.18 | 1.09–1.19 | 1.14 | 9 |
| 1.20–1.28 | 1.19–1.29 | 1.24 | 18 |
| 1.30–1.38 | 1.29–1.39 | 1.34 | 32 |
| 1.40–1.48 | 1.39–1.49 | 1.44 | 29 |
| 1.50–1.58 | 1.49–1.59 | 1.54 | 19 |
| 1.60–1.68 | 1.59–1.69 | 1.64 | 9 |
| 1.70–1.78 | 1.69–1.79 | 1.74 | 5 |
| 1.80–1.88 | 1.79–1.89 | 1.84 | 1 |
| | | | —— |
| | | | 125 |

## 2.3. FREQUENCY CLASS NOMENCLATURE AND TABULATION

We now define some of the wording used in Table 2.2. By "class limits" we mean the smallest and largest observed numbers which are tallied within the class. These are written to the precision in which the data were obtained. Thus, since in Table 2.1 all data were to even hundredths of a percent, we use limits *to this precision*, for example, 1.00–1.08%, not 1.00–1.09%.

Now what would a recorded value of 1.08% mean? Since it is to the nearest .02%, such a recorded value can represent any *true value* from 1.07 to 1.09%. Likewise a value listed as 1.00% can be for any true value from .99 to 1.01%. Thus all *true values* from .99 to 1.09% will be recorded in the first class. Such extremes for the true values recorded within a class are called the "class boundaries." Note that if the original measurements had been made to the nearest .01%, then a 1.09% could occur and now our *class limits* would have been taken as 1.00–1.09%. Then the *class boundaries* would be .995–1.095%, since these are now the extremes possible within the class.

The "midvalue" or "class mark" for any class is defined to be half-way between the two *class boundaries* of that class. Thus we find the midvalues as shown in Table 2.2. Note that in this example the point half-way between the class limits is identical to that between the boundaries for each class. But such is not always the case. For example, consider a class of conventional ages in years: class limits 20–29 years. What is the meaning of "29"? It means that the person is at least 29 and may be just barely less than 30. Thus the class boundaries are 20–30 for this class. The midvalue is $(20 + 30) 2 = 25$, not $(20 + 29) 2 = 24.5$. As another example, data may be on a life test. At the end of each hour a check is made to see whether any of the devices under test have failed. If they have, the elapsed time till that hour is recorded. Now, if a class has *limits* 20–29 hours, the *class boundaries* are 19–29, since a recorded life of 20 may be for a device failing just a moment after the nineteenth-hour check. But a "29" means 28 to 29. Here the midvalue is $(19 + 29) 2 = 24$ (hours). Thus a class with class limits 20–29 could have a midvalue of 24, 24.5, or 25. It all depends on the meaning of the recorded data.

Finally we have as "class interval" the difference between two con-secutive midvalues. For Table 2.2, the class interval is $1.14 - 1.04\% = .10\%$. It is good practice to have a constant class interval throughout, that is, uniform width classes.

In order to tabulate a list of data into a frequency table the following steps may be taken:

1. Find the maximum and minimum numbers occurring in the list, and the difference between these, that is, the range.
2. Decide about how many classes to set up to cover the range. Somewhere between 7 and 20 classes are usually used, the former for small samples, say 25 to 50 numbers, the latter for large samples.
3. Divide the range by the desired number of classes to find the approximate size of the class interval. Take a convenient number near the quotient obtained.
4. Set up convenient class limits for a class to include the minimum number. (In most cases the difference between the two class limits will be one measurement unit less than the class interval. For example, in percent manganese $1.08 - 1.00\% = .08\%$, while the class interval was $.10\%$.)
5. After writing the limits for all classes, in a column, make sure that there is one, and just one, class for each possible number in the data. Then take the data one at a time and tally opposite the class containing the number. Add the frequencies to make sure their total is the sample size.

## 2.4. DISCRETE VERSUS CONTINUOUS DATA

In any statistical analysis it is desirable to make clear the distinction between two different types of data. In the first place we have measurements. Such data are obtained in general by comparison with a scale of some kind, for example, length, time, or weight. Commonly there will be decimal places. But even if all recorded data are whole numbers, such as, 658 lb for a block to be pressed into a railroad car wheel, decimals could occur by greater precision of measurement. "Measurement" or "continuous data" are theoretically indefinitely subdivisible. (In fact the striving for higher and higher precision of measurement is one of the fascinating stories of technology. Extreme precision of measurement was vital to reach the moon safely.) In general, analysis of continuous data is called the "method of variables."

On the other hand, we have "discrete data." Such data always arise from some sort of counting rather than measurement. Some examples are the number of "pinholes" in a painted area, the number of typographical errors on a page, the number of alpha particles noted in a time interval, and the number of defective castings in a lot of 1000. All these data would be whole numbers, that is, integers, and no subdivision is possible. Most discrete data are of integers, but prices of common stock, for example, $28\frac{1}{8}$, are discrete since no finer precision is ever used. Likewise, the fraction defective is a decimal, but discrete. For example, the fraction defective of lots of 100 carburetors at the end of production is .02 if two are defective. The only possible "fractions defective" are .00, .01, .02, ..., 1.00, and nothing else can occur. No subdivision to more decimals is possible. These are both examples of discrete data, with fractions.

In constructing frequency tables we should keep clear the distinction between discrete and continuous data, since it can make a difference. (See Problems 2.1 and 2.16.) Moreover we have quite different probability models to describe patterns of variability for discrete data and continuous data (see Chapters 5 and 6, respectively). Analysis of discrete data is called the "method of attributes."

## 2.5. GRAPHICAL REPRESENTATION OF A FREQUENCY DISTRIBUTION

Although a table such as Table 2.2 gives the frequency distribution for the classes, several kinds of graphs can aid in visualizing the distribution. The two most common kinds are now described. A *histogram* pictures the distribution by a series of rectangles, the bases of which run from the lower to the upper class *boundary* for the class, and the height is proportional to the corresponding frequency (see Fig. 2.1). Such a histogram thus shows blocks and emphasizes class boundaries by intervals.

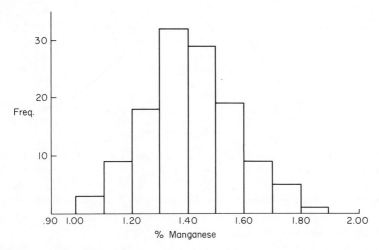

**Fig. 2.1.**   Frequency histogram for percents manganese in iron from 125 casts from a blast furnace. From Table 2.2.

A *frequency polygon* is made by plotting the frequency versus the corresponding midvalue and connecting points for consecutive midvalues. The point with the lowest nonzero frequency is connected to the immediately lower midvalue having zero frequency and similarly at the highest nonzero frequency, to complete the polygon. We thus have a broken line "curve" picturing the distribution. It emphasizes midvalues and curve shape. (See Fig. 2.2.)

Variations can be made on the foregoing types of graphs. One such is to use a percentage of the sample size on the vertical scale, instead of frequencies. A variation on the histogram is to use a more or less thick bar centered on the midvalue, instead of a rectangle with the base covering the range of the boundaries. This latter variation is often effective for discrete data, especially so if there is but one possible count per class, which is therefore the midvalue.

Another useful variation is used in industry, where measurements are computerized and stored in the memory. Then when it is time to print out the distribution, the computer automatically picks out an appropriate multiple of a case to use for each "X" in the printout. If the multiple

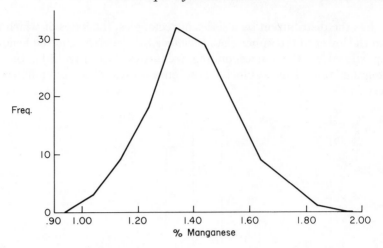

**Fig. 2.2.** Frequency polygon for percents manganese in iron from 125 casts from a blast furnace. From Table 2.2.

is .5, then a frequency of 12 would have six X's. By choosing such a multiple, the size of the printed distribution can be kept within bounds, regardless of the sample size.

## 2.6. THE CUMULATIVE FREQUENCY GRAPH

A third type of graphical presentation, the cumulative graph, is often used. The cumulative graph shows the number of cases (data) with values less than the various *class boundaries*. It is constructed from a cumulative table. See Table 2.3, which has been made from Table 2.2.

**TABLE 2.3**

Cumulative Frequency Table for Percents Manganese in 125 Casts of Iron from a Blast Furnace

| | | | | | |
|---|---|---|---|---|---|
| Below | .99 % | are | 0 | cases or | 0 % |
| | 1.09 % | are | 3 | cases or | 2.4 % |
| | 1.19 % | are | 12 | cases or | 9.6 % |
| | 1.29 % | are | 30 | cases or | 24.0 % |
| | 1.39 % | are | 62 | cases or | 49.6 % |
| | 1.49 % | are | 91 | cases or | 72.8 % |
| | 1.59 % | are | 110 | cases or | 88.0 % |
| | 1.69 % | are | 119 | cases or | 95.2 % |
| | 1.79 % | are | 124 | cases or | 99.2 % |
| Below | 1.89 % | are | 125 | cases or | 100.0 % |

Notice particularly that in constructing a cumulative frequency table such as Table 2.3 from an ordinary frequency table like Table 2.2, we *cannot* find the cumulative frequency below any *midvalue*. Thus we know that there are at least $3 + 9 = 12$ cases below the midvalue 1.24%. Moreover undoubtedly some of the next 18 are below the midvalue, while others of the 18 are above. We can only guess that about half, or 9 of the 18, are below 1.24%. But we *can* be sure that all of the 18 do lie below the upper boundary 1.29%, and hence that 30 altogether lie below this boundary, as shown in Table 2.3.

While a cumulative frequency column could be added to Table 2.2, it is much safer and clearer to make a new table, such as Table 2.3.

Figure 2.3 shows the cumulative frequency graph for the data of Table 2.3. The various points are connected by straight lines. Such

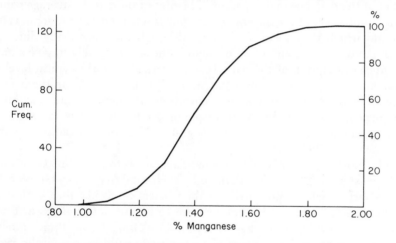

**Fig. 2.3.** Cumulative frequency graph for percents manganese in iron from 125 casts from a blast furnace. From Table 2.3.

connecting assumes a uniform distribution of cases between each pair of the class boundaries, which is an approximation. This graph may be used to answer the following two kinds of questions: (1) About how many of the 125 analyses on the percentage of manganese were below 1.65%? (2) About what percentage of manganese has 25 of the 125 cases below it? Either reading Fig. 2.3 or linearly interpolating in Table 2.3, one obtains the estimates 115 cases and 1.26% Mn, respectively. We also note that a vertical scale giving the percentage of cases could be shown, such as that on the right side of Fig. 2.3, which comes from the last column of Table 2.3.

The author suggests that in the case of discrete data with class boundaries, such as 5–9, 10–14, 15–19 (say, defects in large electronic assemblies), one plot the cumulative frequency below 9.5, 14.5, and so on, rather than either boundary 9 or 10. Why?

A final point is that if one wishes to avoid the assumption of uniform distribution between the boundaries of a class, he can connect the cumulative frequency points with smooth curves rather than straight lines.

## 2.7. WHAT A FREQUENCY DISTRIBUTION SHOWS

Upon condensing a listing of data into a frequency table and constructing a histogram or frequency polygon, three important characteristics of the data become apparent. The first of these is the average value. Commonly a set of measurements or counts will tend to have some value around which they seem to cluster. Although there are many different averages in use, the aim of all of them is the same, namely to give a one-number description of the data. It is concerned with the typical level of the data. One good average is the "mode." It is the figure in the data at which there is the greatest concentration. If there is one maximum frequency, then it will be the corresponding midvalue. This will show up clearly on either table or a frequency graph. Thus in the example we find the mode is 1.34% Mn.

The most important and widely used average is the "arithmetic mean." It is the sum of the numbers divided by their total frequency. Graphically it is a sort of balance point. It is at the vertical centroidal line, such that, if the rectangles of the histogram were to be cut out of uniform cardboard, they would exactly balance around this line. The arithmetic mean for the data on the percentage of manganese would be practically right at 1.40% Mn.

A third average, the "median," can be found readily from either the cumulative frequency table or graph. It is the value in the data such that half of the total frequency lies above and half below. Thus in our example we seek the percentage of manganese corresponding to a cumulative frequency of 125/2 = 62.5. The median is then 1.392% Mn.

A second characteristic of data is the tendency to vary. This shows clearly in the frequency table or graph. One measure of variability is the range. We saw that this was $1.80 - 1.06\% = .74\%$ in Table 2.1. But it also shows up in either Table 2.2 or Fig. 2.2 by the difference between the highest and lowest midvalues for which there are cases, namely, $1.84 - 1.04\% = .80\%$. Why is there a difference in the two ranges?

As we shall see in Chapter 3 there are other more useful measures of variability. But they cannot be found by inspection.

The third characteristic of a frequency distribution is the curve shape. Some distributions rise steadily to a maximum frequency, then the frequencies fall away in a manner such as to give quite a symmetrical histogram or frequency polygon. Others have frequencies which rise much more steeply on one side than they fall away on the opposite side. See the tables in the problem section of this chapter.

Besides the foregoing general characteristics of data, a frequency table or graph may show up one or more irregularities such as the following: There may be two or even more high frequencies separated by one or more considerably lower frequencies. Such an effect may well indicate nonhomogeneity of conditions. For example, two high frequencies were noted in times for an employee to perform a certain task. It turned out that the longer times were occurring when the employee suspected she was being timed on the operation, whereas the shorter times were when she thought she was not being timed. Such "bimodal" data often occur from two different machine settings, two different production heads, two different mold cavities, and so on.

Another irregularity is the tendency of alternating ups and downs in frequencies, resulting from measuring bias. Thus whole thousandths of an inch may be much more often recorded than half-thousandths. There may also be a tendency for a very high clustering of measurements to be right at the specified minimum and maximum limits for the characteristic, with few or none just outside the limits. Such a tendency is called "flinching," and is not uncommon. Then there can well be a few isolated cases very far out from the main body of data, due to some known or unknown cause. Such irregularities as these are well worth studying, when they appear, and then taking appropriate action.

## 2.8. SOME EXAMPLES OF USE OF FREQUENCY TABLES AND GRAPHS

The examples we next discuss show that use of a frequency tabulation alone can provide a useful basis for action and may prove a decisive aid in working with those who have little background in statistics.

The first example is concerned with the manufacture of railroad car wheels. The process consists basically of (1) cutting blocks weighing six or seven hundred pounds from a rotating round ingot, (2) reheating the blocks, and (3) pressing between two dies by an enormously powerful press. The percentage of defective wheels from the process was intolerably high. After investigation it was suspected that part of the trouble was caused by excessive variability in the weights of the blocks before pressing. Accordingly a number were weighed, and the resulting frequency distribution, showing a 70-lb range in weight, was shown to the man in charge of the cutting. He showed much surprise at the amount of variation in the weights, and said he thought he could do better. He was immediately able to reduce the range to about 20 lb, and, accompanying the improvement in uniformity of the blocks, there was an immediate and dramatic decrease in rejection after pressing. They continued weighing the blocks. The latest distribution at the time the author visited the plant is shown in Table 2.4.

TABLE 2.4

Weights in Pounds of Railroad Car Wheel Blocks

| Weights (lb) | Frequency |
|:---:|:---:|
| 648–649 | 1 |
| 650–651 | 19 |
| 652–653 | 30 |
| 654–655 | 36 |
| 656–657 | 26 |
| 658–659 | 9 |
| 660–661 | 5 |
| 662–663 | 2 |
| Total | 128 |

Our second example is concerned with the thickness of rubber for caps. Extruded rubber tubing is cut into 30-in. lengths and cured on mandrils (cylindrical steel bars). Each such tube is then cut into about 300 "gaskets" which are assembled and locked into the caps to provide the seal against the glass jar to be packed with food products. If the gasket is too thin, the seal may leak and the vacuum is lost; or if too thick, the cap may come off from the jar. Hence specifications are set for the thickness, which is actually measured doubled. At the time of this example, every one of the 30-in. tubes was measured for doubled thickness at several positions and placed in one of several piles representing desired specification ranges, such as .096–.100 in. and .100–.106 in.

Table 2.5 shows the distributions of random samples of gaskets cut from the two piles of tubes, as sorted according to the two specifications, from a single shipment. One can immediately see in the table the large amount of overlapping of the two distributions. It is even more striking

**TABLE 2.5**

Doubled Thickness of Gaskets Cut from Rubber Tubes Sorted into Two Piles According to Specifications

| Doubled thickness (in .001 in.) | Frequencies for specifications | |
|---|---|---|
| | 96–100 | 100–106 |
| 94 | 3 | 1 |
| 95 | 7 | 0 |
| 96 | 15 | 2 |
| 97 | 40 | 9 |
| 98 | 51 | 28 |
| 99 | 61 | 21 |
| 100 | 67 | 36 |
| 101 | 66 | 62 |
| 102 | 62 | 63 |
| 103 | 46 | 81 |
| 104 | 27 | 56 |
| 105 | 14 | 42 |
| 106 | 10 | 21 |
| 107 | 5 | 11 |
| 108 | 0 | 5 |
| Totals | 474 | 438 |
| Averages | 100.43 | 102.26 |

in Fig. 2.4 where the frequency polygons are shown. The differences in everages .10226 in. — .10043 in. = .00183 in. is much less than the difference between the middles of the specification ranges, that is, .103 in. — .098 in. = .005 in. Also one can easily see the large number of gaskets outside of each specified range.

The reason for the trouble lay in the curing of the tubes on the mandrils. The rubber could pull together somewhat at each end, but less so in the middle. This gave a large variation in the gaskets from just one tube, so much so that gaskets from only one tube could never fully meet the specifications no matter what average the tube had. This study led to much more realistic specifications and to the use of sampling of tubes to replace the expensive 100% sorting of them into piles.

Our third example is simply a recasting of the data of Table 2.1 into a different form to show how much improvement might be obtained by maintaining better control of the process. Consider the five percentages of manganese for April 1—1.40, 1.28, 1.36, 1.38, and 1.44. The average of these is 1.372. For simplicity we use 1.37 and subtract this from each

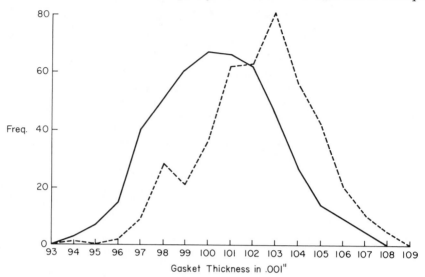

**Fig. 2.4.** Comparative frequency polygons for gaskets from two piles into which rubber tubes were sorted before being cut into gaskets. Polygons show great overlapping instead of distinct separated distributions as was desired.

of the five percentages, obtaining $+.03$, $-.09$, $-.01$, $+.01$, and $+.07$. These are "deviations" from that day's average. Now if all of the daily averages were constant, such deviations would give a frequency polygon centering around zero but of the same shape, and same (or very slightly less) variability as that for the original data. But in fact when we plot the deviations from their daily averages we find the frequency polygon, as shown in Fig. 2.5, to have much less spread than that for the original data. This is due to lack of control of the blast furnace process relative to percent manganese. If the process can be better controlled, the narrower polygon in Fig. 2.5 can be obtained. In Chapter 11 we consider tests for studying and improving the control of a process.

## 2.9. SAMPLE VERSUS POPULATION

It is desirable at this point to make clear the meaning of two important concepts, since sound applications of statistical methods can be made only where their distinction is appreciated. As used here a *population* is the whole collection of measurements or other numbers of the kind in which we are at the moment interested. The simplest illustration of a population is a set of chips in a bowl, each chip bearing some number. Another population is the whole aggregation of diameters of a steel rod,

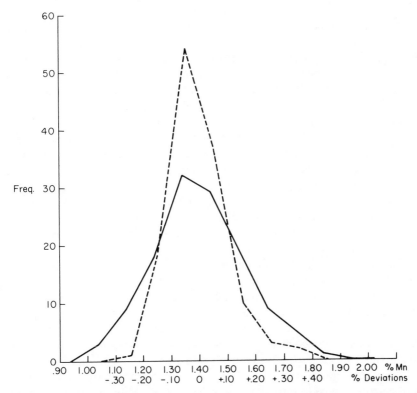

**Fig. 2.5.** Comparative frequency polygons for percents manganese in iron, blast furnace casts. The solid-line polygon shows the original as in Table 2.2 and Fig. 2.2. The dashed-line polygon shows spread which should be obtainable if the average of percent manganese were closely controlled from day to day. Obtained by finding and tabulating the deviations of the five daily percentages from their daily average, for the 25 days.

measured at all distances from one end and in all angular directions, but always measured by the same person, using the same micrometer or dial gage. We call such a population *infinite* because the number of possible repetitions is unlimited. A similar example is repeated determinations of the atomic weight of an isotope. A fourth example is the set of times till each fuse "blows" under controlled circuit conditions, in a lot of 5000 fuses. This population is *finite*, since each fuse can provide just one time, because when it blows it is destroyed. Commonly we ask that a population represent as homogeneous a set of conditions as possible. All of these are populations of numbers. The word "population" can also be used to mean a collection of objects, people, or animals. We shall be more concerned with the former meaning.

A *sample* is a set of results or numbers taken from the population. In the first example just given, we might reach into the bowl, mix up the chips, and withdraw 10 of them simultaneously. This would be "sampling without replacement." Or we could "sample with replacement" by drawing a single chip, reading it, replacing, mixing, and drawing again. The former of these would best simulate the process of testing 10 of the fuses, while the latter would best simulate the measurement of the diameter of the rod or the determination of atomic weight, since after each measurement (or drawing) the population is just what it was before.

Now naturally the main aim in taking a sample is to learn something about the population, or at least to aid one in making some kind of a decision. The whole problem of the relation of sample to population in all its ramifications is perhaps the central problem of statistical theory and practice. At this point, however, we shall concern ourselves only with two facets of the problem.

The first idea is that in order to seek an unbiased sample we let "chance do the work," by taking our sample at random. We draw the chips from the bowl first, after we have made sure that they are uniform in size, shape, and weight, and second, after thorough mixing, thus trying to ensure that any one chip has as good a change of being chosen as any other. We select the sample of fuses at random over the whole lot, instead of taking them all from one corner. The choice of position and direction on the rod is made at random instead of always measuring at the same place. The fundamental reason for taking samples at random is to minimize the chance of conscious or unconscious bias in selection.

To point up the importance of sample selection consider the following true story. A sample of 100 piston-ring castings was taken from a lot of 3000 and inspected. Twenty-five of these 100 proved to be defectives. It was thought that some 725 of the remaining 2900 castings might be defective, so they were all sorted 100%. Only four of the 2900 proved to be defectives. What do you think of the manner in which the original 100 were drawn?

Operationally we often think of "random sampling" as choosing by use of "random digits" 0 to 9, or 00 to 99 as in Table X. There are many ways in which we can associate an integer with an object in the population and then draw by using random digits.

The second idea we wish to examine is the notion that a random sample tends to resemble and represent the population. Now it is of course true that there is such a tendency, but how good a job the sample does toward picturing the population depends on (1) the makeup or characteristics of the population, (2) the size of the sample, and (3) luck or chance. In general, however, the larger the random sample, the more confidence we can place in it as a describer of the population.

Sufficiently large random samples can tell just about everything about the populations from which they are chosen, whereas a small random sample may tell almost nothing about the population from which it was chosen. Note that if conditions were held quite constant, some of the large samples, such as that shown in Table 2.18, must surely give an excellent picture of the population related to the conditions involved.

In making any application of statistics to practical problems, one must constantly ask himself the questions

1. Has the sample been chosen in a random manner?
2. Is my sample large enough for the purpose in mind?
3. Am I dealing with one or several populations? In other words, are all conditions essentially the same, or have there been different conditions?

## 2.10. SUMMARY

Construction of a frequency distribution from a series of data relating to a given set of conditions is a good way of condensing the results and describing them. The frequency graph shows (1) approximately what the average value is; (2) the total extent of variation in the data; (3) the general way in which the data cluster around the average, that is, the shape of the frequency distribution; and (4) perhaps some irregularities of importance. Graphs enable one to visualize the data. Shortcomings of this approach are (1) the information provided is not objective enough; (2) the accuracy with which the sample represents the population is not apparent; and (3) there is no adequate way to tell whether the data came from one population or from several different populations—that is, the order in the data has been lost in the tabulation.

### Problems

2.1.  For each of the following, suppose we were preparing to tabulate data into a frequency table. Choose suitable classes, then for the lowest two classes, list the class limits, class boundaries, midvalues, and class interval:

(a)  Measurements of "eccentricity" in needle valves, the

distance from a conical point to the center line of the base (it would be zero if perfect): .0002–.0133 in., measured to the nearest .0001 in.

(b)   Automobile speeds, 34–82 mph, measured to the nearest mile per hour.

(c)   Tensile strengths of iron samples running from 54,300 to 83,200 pounds per square inch, measured to the nearest 100 psi.

(d)   Ordinary ages of a group of people, running from 17 to 73.

(e)   Counts of the number of specks in test areas of paper, running from 0 to 87 per area.

(f)   Examination scores to the nearest whole percent, 45–98%.

(g)   Postal weights of parcels in ounces in which any weight from practically zero to precisely 1 ounce is recorded as 1 oz, and so on. Weights 1–13 oz.

(h)   Capacities of bottles in fluid ounces 31.58–31.81, measured to the nearest .01.

(i)   Bowling scores 113–197.

(j)   Fraction defective: the proportion of defective pens, in lots of 200, from .00 to .095.

2.2.   Collect a series of numerical data, and tabulate into a frequency table. Draw an appropriate frequency graph and comment.

For the table corresponding to each assigned problem below choose a class interval and class limits and tabulate into a frequency table. List the class boundaries and draw a frequency graph. Comment. If assigned, draw a cumulative frequency graph.

|       |            |       |            |
|-------|------------|-------|------------|
| 2.3.  | Table 2.6. | 2.5.  | Table 2.8. |
| 2.4.  | Table 2.7. | 2.6.  | Table 2.9. |

For the frequency tables corresponding to assigned problems below, draw a histogram or frequency polygon and comment on what you see. Also draw a cumulative frequency graph and find the median from it.

|        |             |        |             |        |             |
|--------|-------------|--------|-------------|--------|-------------|
| 2.7.   | Table 2.10. | 2.14.  | Table 2.17. | 2.20.  | Table 2.23. |
| 2.8.   | Table 2.11. | 2.15.  | Table 2.18. | 2.21.  | Table 2.24. |
| 2.9.   | Table 2.12. | 2.16.  | Table 2.19. | 2.22.  | Table 2.25. |
| 2.10.  | Table 2.13. | 2.17.  | Table 2.20. | 2.23.  | Table 2.26. |
| 2.11.  | Table 2.14. | 2.18.  | Table 2.21. | 2.24.  | Table 2.27. |
| 2.12.  | Table 2.15. | 2.19.  | Table 2.22. | 2.25.  | Table 2.28. |
| 2.13.  | Table 2.16. |        |             |        |             |

TABLE 2.6

Laboratory Sieve Analyses of a Molding Powder[a]

| | | | | | | | | | |
|---|---|---|---|---|---|---|---|---|---|
| 97.4 | 98.6 | 97.6 | 98.1 | 95.6 | 95.2 | 95.2 | 95.6 | 95.4 | 98.2 |
| 98.5 | 98.2 | 97.0 | 99.6 | 98.9 | 99.1 | 98.6 | 99.3 | 97.6 | 98.0 |
| 96.4 | 94.8 | 97.7 | 96.1 | 97.0 | 96.5 | 100.0 | 98.8 | 98.1 | 98.5 |
| 97.0 | 99.0 | 95.3 | 99.3 | 98.5 | 96.7 | 97.2 | 96.4 | 96.4 | 97.0 |
| 97.3 | 97.6 | 99.3 | 98.9 | 99.2 | 99.0 | 99.3 | 95.4 | 98.2 | 96.4 |
| 97.8 | 98.5 | 98.5 | 98.8 | 96.8 | 97.0 | 98.0 | 98.5 | 98.4 | 98.4 |
| 98.4 | 97.2 | 97.9 | 98.0 | 99.6 | 97.4 | 99.6 | 98.8 | 96.9 | 98.5 |
| 99.0 | 98.5 | 98.0 | 98.5 | 97.5 | 98.0 | 99.7 | 99.3 | 99.4 | 98.8 |
| 98.3 | 99.6 | 98.4 | 98.3 | 98.0 | 98.0 | 99.0 | 97.5 | 98.0 | 99.0 |
| 98.5 | 97.7 | 99.3 | 98.7 | 98.5 | 98.9 | 97.1 | 95.2 | 99.6 | 98.0 |
| 98.4 | 98.0 | 98.7 | 96.3 | 94.3 | 96.5 | 97.3 | 98.4 | 97.6 | 98.5 |
| 98.8 | 98.2 | 99.6 | 95.9 | 98.0 | 99.1 | 98.2 | 98.7 | 98.3 | 97.1 |

[a] Percentage on and above a U.S. No. 8 sieve.

TABLE 2.7

Capacities of 100 Consecutive Bottles from the Same Cavity in Fluid Ounces

| | | | | | | | | | |
|---|---|---|---|---|---|---|---|---|---|
| 31.67 | 31.68 | 31.67 | 31.64 | 31.71 | 31.64 | 31.62 | 31.72 | 31.72 | 31.65 |
| 31.70 | 31.78 | 31.69 | 31.72 | 31.75 | 31.67 | 31.76 | 31.67 | 31.69 | 31.68 |
| 31.58 | 31.68 | 31.62 | 31.66 | 31.68 | 31.71 | 31.73 | 31.74 | 31.76 | 31.74 |
| 31.76 | 31.68 | 31.70 | 31.67 | 31.72 | 31.71 | 31.69 | 31.70 | 31.72 | 31.67 |
| 31.71 | 31.70 | 31.71 | 31.76 | 31.75 | 31.70 | 31.67 | 31.79 | 31.66 | 31.74 |
| 31.69 | 31.68 | 31.73 | 31.79 | 31.76 | 31.73 | 31.77 | 31.75 | 31.74 | 31.76 |
| 31.73 | 31.75 | 31.79 | 31.71 | 31.76 | 31.63 | 31.74 | 31.73 | 31.79 | 31.80 |
| 31.69 | 31.70 | 31.75 | 31.71 | 31.80 | 31.70 | 31.78 | 31.73 | 31.77 | 31.78 |
| 31.74 | 31.72 | 31.65 | 31.66 | 31.73 | 31.76 | 31.75 | 31.68 | 31.67 | 31.70 |
| 31.69 | 31.79 | 31.68 | 31.72 | 31.71 | 31.72 | 31.72 | 31.66 | 31.74 | 31.79 |

TABLE 2.8

Tensile Strength of Paper Samples in Grams

| | | | | | | | | | |
|---|---|---|---|---|---|---|---|---|---|
| 242.8 | 242.9 | 255.7 | 260.0 | 239.8 | 259.2 | 263.8 | 261.8 | 256.7 | 243.1 |
| 250.2 | 237.3 | 251.9 | 275.4 | 258.9 | 280.0 | 276.4 | 255.9 | 255.2 | 266.5 |
| 270.9 | 262.4 | 267.8 | 277.8 | 259.3 | 281.4 | 259.7 | 264.0 | 269.0 | 254.5 |
| 257.1 | 271.8 | 262.4 | 262.8 | 253.3 | 255.1 | 272.4 | 260.5 | 248.0 | 264.6 |
| 260.1 | 269.6 | 256.0 | 261.6 | 267.5 | 257.0 | 248.5 | 270.5 | 268.2 | 263.7 |
| 262.4 | 260.6 | 276.7 | 264.4 | 249.2 | 257.4 | 271.2 | 273.5 | 266.8 | 252.7 |
| 250.0 | 256.4 | 265.1 | 243.3 | 292.0 | 290.4 | 262.3 | 255.5 | 255.1 | 245.5 |
| 259.3 | 274.5 | 291.9 | 271.3 | 276.2 | 267.1 | 243.7 | 265.2 | 273.2 | 254.3 |
| 288.6 | 256.3 | 278.8 | 266.2 | 236.4 | 236.8 | 265.5 | 261.4 | 258.0 | 260.5 |
| 277.3 | 282.0 | 262.1 | 281.3 | 276.2 | 266.9 | 265.1 | 265.7 | 249.8 | 227.0 |
| 281.3 | 277.0 | 272.6 | 278.8 | 272.7 | 260.5 | 258.1 | 257.2 | 276.1 | 233.4 |
| 255.3 | 245.5 | 243.7 | 252.4 | 247.5 | 249.6 | 253.3 | 258.7 | 251.2 | 274.4 |
| 257.8 | 258.6 | 271.2 | 265.4 | 276.5 | 293.1 | 280.0 | 263.5 | 273.0 | 265.1 |
| 263.4 | 270.3 | 251.1 | 244.0 | 241.4 | 237.5 | 255.2 | 251.3 | 253.8 | 253.4 |
| 241.0 | 247.6 | 260.2 | 272.7 | 275.1 | 246.0 | 244.0 | 264.0 | 259.1 | 253.8 |
| 281.6 | 280.2 | 276.4 | 267.8 | 280.3 | 241.0 | 284.4 | 262.1 | 276.3 | 291.7 |
| 277.1 | 292.5 | 270.8 | 276.5 | 292.3 | 254.2 | 245.5 | 238.3 | 263.2 | 245.5 |
| 253.4 | 249.3 | 260.7 | 255.5 | 285.5 | 244.0 | 263.0 | 273.2 | 290.2 | 251.1 |
| 263.4 | 250.7 | 276.2 | 272.8 | 280.6 | 290.7 | 256.6 | 260.3 | 279.3 | 235.5 |
| 265.2 | 276.2 | 257.0 | 271.8 | 278.5 | 277.0 | 273.7 | 282.0 | 275.5 | 260.0 |
| 265.8 | 263.2 | 263.5 | 257.3 | | | | | | |

TABLE 2.9

Thicknesses of Cork Disks for Crowns for Sealing Bottles[a]

| | | | | | | | | | | | | | | |
|---|---|---|---|---|---|---|---|---|---|---|---|---|---|---|
| 69 | 66 | 66 | 76 | 62 | 61 | 63 | 59 | 65 | 66 | 63 | 67 | 65 | 72 | 64 |
| 64 | 66 | 74 | 67 | 62 | 61 | 68 | 69 | 67 | 68 | 60 | 65 | 65 | 67 | 66 |
| 63 | 63 | 63 | 64 | 72 | 69 | 68 | 67 | 62 | 61 | 65 | 69 | 66 | 65 | 65 |
| 70 | 61 | 67 | 67 | 69 | 65 | 68 | 65 | 63 | 71 | 68 | 70 | 73 | 68 | 69 |
| 59 | 62 | 64 | 62 | 63 | 65 | 67 | 64 | 65 | 59 | 69 | 69 | 67 | 68 | 62 |
| 66 | 66 | 64 | 64 | 69 | 70 | 64 | 68 | 65 | 66 | 64 | 60 | 66 | 64 | 60 |
| 66 | 70 | 65 | 72 | 63 | 70 | 65 | 65 | 66 | 68 | 64 | 66 | 63 | 64 | 64 |
| 64 | 68 | 67 | 63 | 69 | 64 | 68 | 68 | 63 | 62 | 65 | 67 | 65 | 67 | 69 |
| 65 | 65 | 63 | 67 | 66 | 67 | 67 | 65 | 67 | 63 | 67 | 63 | 66 | 60 | 61 |
| 62 | 68 | 63 | 61 | 67 | 66 | 67 | 69 | 63 | 69 | 65 | 71 | 66 | 64 | 64 |
| 72 | 65 | 65 | 66 | 69 | 67 | 63 | 65 | 70 | 65 | 61 | 68 | 69 | 66 | 60 |
| 66 | 69 | 65 | 65 | 63 | 64 | 65 | 63 | 62 | 66 | 66 | 66 | 67 | 65 | 65 |
| 66 | 66 | 64 | 62 | 67 | 66 | 66 | 67 | 66 | 66 | 64 | 64 | 66 | 65 | 65 |
| 65 | 66 | 64 | 60 | 63 | 63 | 62 | 68 | 67 | 67 | 68 | 65 | 68 | 65 | 65 |
| 68 | 66 | 67 | 62 | 66 | 61 | 67 | 63 | 64 | 66 | 64 | 64 | 64 | 65 | 67 |
| 66 | 66 | 68 | 66 | 62 | 66 | 64 | 68 | 68 | 63 | 68 | 63 | 65 | 63 | 62 |
| 68 | 66 | 57 | 70 | 66 | 65 | 61 | 62 | 62 | 65 | 60 | 64 | 63 | 71 | 70 |
| 65 | 65 | 69 | 66 | 64 | 62 | 64 | 61 | 64 | 63 | 68 | 64 | 67 | 70 | 61 |
| 69 | 64 | 62 | 61 | 69 | 58 | 60 | 65 | 66 | 65 | 70 | 57 | 69 | 61 | 67 |
| 66 | 64 | 66 | 64 | 61 | 66 | 67 | 66 | 66 | 67 | 62 | 65 | 67 | 65 | 65 |

[a] In .001 in. units. Nominal $\frac{1}{16}$ in.

**TABLE 2.10**

Photogrammetric Measurements of Final Pay
Quantities in Highway Construction[a]

| Error (in ft) | Frequency |
|:---:|:---:|
| +.7 | 1 |
| +.6 | 4 |
| +.5 | 6 |
| +.4 | 8 |
| +.3 | 17 |
| +.2 | 29 |
| +.1 | 34 |
| 0 | 42 |
| −.1 | 34 |
| −.2 | 32 |
| −.3 | 14 |
| −.4 | 7 |
| −.5 | 6 |
| −.6 | 5 |
| Total | 239 |

[a] Elevations.

**TABLE 2.11**

Dimensions for an Electrical Contact[a]

| Class limits | Frequency |
|:---:|:---:|
| 17, 18 | 1 |
| 19, 20 | 21 |
| 21, 22 | 107 |
| 23, 24 | 167 |
| 25, 26 | 170 |
| 27, 28 | 134 |
| 29, 30 | 120 |
| 31, 32 | 84 |
| 33, 34 | 57 |
| 35, 36 | 48 |
| 37, 38 | 34 |
| 39, 40 | 34 |
| 41, 42 | 14 |
| 43, 44 | 9 |
| 45, 46 | 5 |
| Total | 1005 |

[a] In .001 in.

**TABLE 2.12**

Hardness of Motor Clutch Cams

| Rockwell-C | Frequency |
|:---:|:---:|
| 47 | 1 |
| 48 | 0 |
| 49 | 3 |
| 50 | 8 |
| 51 | 12 |
| 52 | 19 |
| 53 | 29 |
| 54 | 36 |
| 55 | 35 |
| 56 | 38 |
| 57 | 29 |
| 58 | 25 |
| 59 | 17 |
| 60 | 12 |
| 61 | 4 |
| 62 | 2 |
| 63 | 3 |
| Total | 273 |

**TABLE 2.13**

Thicknesses of Cork Disks for Crowns for Bottles[a]

| Thickness | Frequencies | |
| | 6/4/58 | 6/5/58 |
|:---:|:---:|:---:|
| 56, 57 | 6 | 0 |
| 58, 59 | 5 | 12 |
| 60, 61 | 34 | 33 |
| 62, 63 | 82 | 86 |
| 64, 65 | 153 | 164 |
| 66, 67 | 137 | 130 |
| 68, 69 | 77 | 65 |
| 70, 71 | 33 | 37 |
| 72, 73 | 11 | 23 |
| 74, 75 | 1 | |
| 76, 77 | 1 | |
| Total | 540 | 550 |

[a] In .001 in.

**TABLE 2.14**

Gasket Space in Caps, Distance from Top to Rubber Gasket Edge

| Space (in.) | Frequency |
|:---:|:---:|
| .105 | 1 |
| .110 | 2 |
| .115 | 32 |
| .120 | 47 |
| .125 | 88 |
| .130 | 31 |
| | Total 201 |

**TABLE 2.15**

Warpage within Piston Rings, Maximum – Minimum for Each Ring[a]

| Midvalue | Frequency |
|:---:|:---:|
| 2.5 | 16 |
| 12.5 | 155 |
| 22.5 | 308 |
| 32.5 | 258 |
| 42.5 | 132 |
| 52.5 | 60 |
| 62.5 | 33 |
| 72.5 | 15 |
| 82.5 | 13 |
| 92.5 | 7 |
| 102.5 | 4 |
| 112.5 | 1 |
| | Total 1002 |

[a] In .0001 in.

**TABLE 2.16**

Lengths of Cut of Rubber Gaskets in .001 in. from the Nominal Length

| Midvalue | Frequency |
|---|---|
| +4.5 | 1 |
| +3.0 | 13 |
| +1.5 | 144 |
| 0 | 281 |
| −1.5 | 284 |
| −3.0 | 109 |
| −4.5 | 26 |
| −6.0 | 2 |
| Total | 860 |

**TABLE 2.17**

Tensile Strength of Iron Castings[a]

| Classes | Frequency |
|---|---|
| 54,000–55,900 | 1 |
| 56,000–57,900 | 5 |
| 58,000–59,900 | 13 |
| 60,000–61,900 | 14 |
| 62,000–63,900 | 26 |
| 64,000–65,900 | 28 |
| 66,000–67,900 | 38 |
| 68,000–69,900 | 42 |
| 70,000–71,900 | 32 |
| 72,000–73,900 | 20 |
| 74,000–75,900 | 14 |
| 76,000–77,900 | 13 |
| 78,000–79,900 | 2 |
| 80,000–81,900 | 2 |
| Total | 250 |

[a] In pounds per square inch.

**TABLE 2.18**

Eccentricities in Needle Valves (Distance from Conical Point to Center of Base)[a]

| Class limits | Frequency |
|:---:|:---:|
| 0–9 | 171 |
| 10–19 | 477 |
| 20–29 | 561 |
| 30–39 | 505 |
| 40–49 | 379 |
| 50–59 | 289 |
| 60–69 | 150 |
| 70–79 | 75 |
| 80–89 | 43 |
| 90–99 | 30 |
| 100–109 | 10 |
| 110–119 | 1 |
| 120–129 | 5 |
| 130–139 | 3 |
| 140–149 | 1 |
| | Total 2700 |

[a] Measurements to the nearest .0001 in.

**TABLE 2.19**

Golf Scores for a Company Golf Day

| Class | Frequency |
|:---:|:---:|
| 70–79 | 2 |
| 80–89 | 15 |
| 90–99 | 34 |
| 100–109 | 48 |
| 110–119 | 41 |
| 120–129 | 28 |
| 130–139 | 23 |
| 140–149 | 10 |
| 150–159 | 10 |
| 160–169 | 3 |
| 170–179 | 3 |
| 180–189 | 1 |
| | Total 218 |

**TABLE 2.20**

Total Number of Letters in People's First
and Last Names

| Number | Frequency |
|--------|-----------|
| 8 | 3 |
| 9 | 5 |
| 10 | 11 |
| 11 | 23 |
| 12 | 22 |
| 13 | 17 |
| 14 | 13 |
| 15 | 7 |
| 16 | 4 |
| 17 | 0 |
| 18 | 2 |
| | —— |
| Total | 107 |

**TABLE 2.21**

Solenoid Overtravel[a]

| Midvalue (in.) | Frequency |
|----------------|-----------|
| .030 | 1 |
| .035 | 5 |
| .040 | 19 |
| .045 | 43 |
| .050 | 89 |
| .055 | 141 |
| .060 | 123 |
| .065 | 86 |
| .070 | 30 |
| .075 | 6 |
| .080 | 3 |
| | —— |
| Total | 546 |

[a] Measured to the nearest .001 in.

**TABLE 2.22**

Pressure in Pounds per Square Inch to Operate an Hydraulic Switch

| Pressure (psi) | Frequency |
|:---:|:---:|
| 50 | 1 |
| 55 | 0 |
| 60 | 5 |
| 65 | 11 |
| 70 | 13 |
| 75 | 33 |
| 80 | 21 |
| 85 | 8 |
| 90 | 4 |
| 95 | 3 |
| 100 | 1 |
| | — |
| Total | 100 |

**TABLE 2.23**

Lengths in Inches of 12 Boxes of Drawing Pencils

| Length (in.) | Frequency |
|:---:|:---:|
| 6.996 | 3 |
| 6.997 | 6 |
| 6.998 | 17 |
| 6.999 | 30 |
| 7.000 | 35 |
| 7.001 | 36 |
| 7.002 | 12 |
| 7.003 | 4 |
| 7.004 | 1 |
| | — |
| Total | 144 |

TABLE 2.24

Governor-Type Distributors[a]

| Vacuum (in. Hg) | Frequency |
|---|---|
| 3.0 | 24 |
| 3.1 | 52 |
| 3.2 | 36 |
| 3.3 | 24 |
| 3.4 | 20 |
| 3.5 | 13 |
| 3.6 | 4 |
| 3.7 | 0 |
| 3.8 | 0 |
| 3.9 | 1 |
| 4.0 | 1 |
| Total | 175 |

[a] Vacuum at 1675 rpm (revolutions per minute).

TABLE 2.25

Out-of-Roundness of Individual Bottles Old and New Molds[a]

| Classes (in .001 in.) | Frequency | |
|---|---|---|
| | Old molds | New molds |
| 0–4 | 0 | 2 |
| 5–9 | 35 | 105 |
| 10–14 | 23 | 279 |
| 15–19 | 17 | 243 |
| 20–24 | 112 | 150 |
| 25–29 | 112 | 38 |
| 30–34 | 151 | 19 |
| 35–39 | 99 | 20 |
| 40–44 | 121 | 5 |
| 45–49 | 90 | 1 |
| 50–54 | 50 | — |
| 55–59 | 32 | — |
| 60–64 | 13 | — |
| 65–69 | 2 | — |
| Totals | 857 | 862 |

[a] Measured to .001 in.

**TABLE 2.26**

Combined Capacity Distribution of 100 Bottles Each from Eight Cavities

| Midvalues (fl oz) | Frequency |
|:---:|:---:|
| 31.47 | 1 |
| 31.52 | 5 |
| 31.57 | 32 |
| 31.62 | 97 |
| 31.67 | 216 |
| 31.72 | 243 |
| 31.77 | 154 |
| 31.82 | 43 |
| 31.87 | 8 |
| 31.92 | 1 |
| | — |
| Total | 800 |

**TABLE 2.27**

Thermostat Opening Time [a] at 42 A for 6-V Light Switches

| Class limits | Frequency |
|:---:|:---:|
| 38–57 | 97 |
| 58–77 | 320 |
| 78–97 | 259 |
| 98–117 | 127 |
| 118–137 | 78 |
| 138–157 | 55 |
| 158–177 | 26 |
| 178–197 | 20 |
| 198–217 | 11 |
| 218–237 | 12 |
| 238–257 | 5 |
| 258–277 | 2 |
| 278–297 | 1 |
| 298–317 | 2 |
| 318–337 | 1 |
| | — |
| Total | 1016 |

[a] In seconds.

**TABLE** 2.28

Thicknesses of Gaskets Cut from 30-in. Rubber Tubes
Which Were Sorted into Two Piles Aimed at Two
Specifications[a]

| Midvalue (in .001 in.) | Frequencies | |
| --- | --- | --- |
| | Lower specification | Upper specification |
| 95.0 | 2 | 2 |
| 96.5 | 8 | 15 |
| 98.0 | 29 | 66 |
| 99.5 | 44 | 129 |
| 101.0 | 92 | 254 |
| 102.5 | 80 | 389 |
| 104.0 | 27 | 255 |
| 105.5 | 12 | 89 |
| 107.0 | | 5 |
| 108.5 | | 2 |
| Total | 294 | 1206 |

[a] Specifications: .096–100 and .100–.106 in.

# Chapter 3

# SUMMARIZATION OF DATA BY OBJECTIVE MEASURES

## 3.1. INTRODUCTION

Although a frequency table and a graph such as a histogram will give one quite a bit of information about the data, most of it is rather subjective. In order to describe the data more objectively and to provide tools for analysis, it is desirable to use what are called "sample statistics." Such statistics are of three kinds: averages, measures of variability, and measures of curve shape. The first two are used with most samples, the last only with large samples.

## 3.2. SOME AVERAGES

The aim of an average is to give a one-number description of the sample of numbers. If all the numbers were identical, the average would be this number. If the numbers vary, but still show some "central tendency," then it is the object of the average to locate this level. Different kinds of averages have various characteristics: advantages and disadvantages, and each kind has its special use and interpretation. Finally one may say that if the data are so scattered as to exhibit no noticeable central tendency or pattern, then it is doubtful whether any average is worth much.

Certainly the most commonly used and widely understood average is the so-called arithmetic mean. It is simply the sum of the set of numbers, divided by how many there are. If we call the individual numbers $y_1, y_2, y_3, ..., y_n$, there being $n$ of them, then their arithmetic mean is defined by

$$\bar{y} = \frac{y_1 + y_2 + \cdots + y_n}{n} = \frac{\sum y}{n}, \tag{3.1}$$

where the symbol $\sum$ means "sum of." The *sample size* is $n$. If desirable, we could have shown the limits on the summation as follows:

$$\sum_{i=1}^{n} y_i = y_1 + y_2 + \cdots + y_n. \tag{3.2}$$

The summation symbol is much used in statistics. Three simple laws are as follows ($c$ is a constant)*:

$$\sum_{i=1}^{n} c = nc \quad \text{or} \quad \sum_{i=a}^{b} c = (b - a + 1)c \tag{3.3}$$

$$\sum_{i=a}^{b} cy_i = c \sum_{i=a}^{b} y_i \tag{3.4}$$

$$\sum_{i=a}^{b} (y_i \pm x_i) = \left( \sum_{i=a}^{b} y_i \right) \pm \left( \sum_{i=a}^{b} x_i \right). \tag{3.5}$$

* The proofs of these laws are almost self-evident, if one merely writes out the expansion as in (3.2) for each $\sum$, and compares.

Some of the desirable properties of the arithmetic mean are as follows:

1.  It takes account of the value of each of the numbers and is affected by each.
2.  If each $y_i$ is multiplied by the same number $k$, then the arithmetic mean of the resulting numbers is $k\bar{y}$.
3.  If each $y_i$ is changed by the addition of $A$ to it, the arithmetic mean of the resulting numbers is $\bar{y} + A$.
4.  If $\bar{y}_1$ is the arithmetic mean of $n_1$ numbers, and $\bar{y}_2$ of $n_2$, then the arithmetic mean of all $n_1 + n_2$ numbers is the "weighted mean"

$$\frac{n_1 \bar{y}_1 + n_2 \bar{y}_2}{n_1 + n_2}.$$

5.  The mean has many desirable properties in the theory and practice of sampling, that is, the relation between sample and population.

As an example consider the following data on amounts of fuel metered by a nozzle in five trials under uniform test condition: $y$: 96.6, 97.2, 96.4, 97.4, 97.8, $\sum y = 485.4$, $\bar{y} = 97.08$ (all in cubic centimeters).

A second kind of average, the "median," we have already mentioned in Chapter 2. There it was described as the $y$ value for which the cumulative frequency is $n/2$. If $n$ is even and there are no tied values of $y$ in the middle, there will be half of the cases below the median and half above it. Or if $n$ is odd, the median is the midmost number after all the numbers have been arranged in order, that is, in an "array." As many numbers $y$ lie below the median as above it, barring ties in the middle of the array. For the small sample of amounts metered, we first rank in order in an array: 96.4, 96.6, 97.2, 97.4, 97.8 cc, and the median is then the middle number, 97.2 cc. If $n$ is even, then when we rank the numbers into an array, the median is half-way between the *two* middle numbers. Tie scores may make interpretation a little more difficult, but this does not complicate the method of finding the median.

The median has a number of desirable properties. Among those given for the arithmetic mean, it possesses 2 and 3, which should in fact be possessed by nearly any average worth consideration. In regard to property 1, the median is only indirectly affected by the value of each of the $n$ numbers. Thus if the lowest amount metered were 86.4 instead of 96.4, the median would still be 97.2 cc, but the sum would be reduced by 10 cc, and $\bar{y}$ by 2 cc to 95.08 cc. Property 4 definitely does not hold for the median, while regarding 5, the median has less desirable sampling properties than the mean.

A third type of average, the "mode," we have also introduced in Chapter 2. It is the point of greatest concentration of the numbers in the data. It is usually found as the midvalue of the class with greatest frequency. Obviously the mode then depends on whatever classes we have chosen for the tabulation. The mode possesses properties 2 and 3, but not 1 or 4, and it has little use in statistical analysis. Its prime use is in the description of a frequency distribution. For example, a distribution with two isolated high frequencies is described as "bimodal," and we could list two modes.

Since the arithmetic mean is so commonly used, we shall frequently call it simply the "average."

## 3.3. SOME MEASURES OF VARIABILITY

There are three commonly used measures of the variability in a set of numbers, namely the "range," the "mean deviation," and the "standard deviation." Of these the last is by far the most important.

The range we have already met in Chapter 2. It is defined for a sample of numbers by

$$\text{range} = R = \text{maximum } y - \text{minimum } y. \qquad (3.6)$$

Thus for the amounts metered we have

$$R = 97.8 - 96.4 = 1.4 \qquad \text{(in cc)}.$$

This measure of variability gives us the extreme amount of variation among the $n$ sample numbers. Hence it is easily interpreted. But it can be greatly affected by just one number far above or below the rest of the numbers. Nevertheless the range is a useful measure, especially for small samples.

The range is of little direct use in the problem of interpreting an individual number in relation to the average. Frequently we wish to know whether one of the numbers is exceptionally far below the average or only an ordinary amount below, and so on. For this we need to know what is a typical or average discrepancy of the numbers from their average. Thus we want to find the discrepancy or deviation of each number from the mean and then in some fashion find an average of these deviations. Consider again the amounts metered as shown below:

| Amounts metered $y$ (in cc) | Deviations $y - \bar{y}$ (in cc) | Absolute deviations $\lvert y - \bar{y} \rvert$ (in cc) | Squared deviations $(y - \bar{y})^2$ (in cc²) |
|---|---|---|---|
| 96.6 | $-.48$ | .48 | .2304 |
| 97.2 | $+.12$ | .12 | .0144 |
| 96.4 | $-.68$ | .68 | .4624 |
| 97.4 | $+.32$ | .32 | .1024 |
| 97.8 | $+.72$ | .72 | .5184 |
| $\bar{y} = 97.08$ cc | 0 | $\Sigma \lvert y - \bar{y} \rvert = 2.32$ cc | $\Sigma (y - \bar{y})^2 = 1.3280$ cc² |

Suppose we try to find the arithmetic mean of the deviations $y - \bar{y}$. Their sum is zero, and thus their mean is also zero. This does not tell us much, especially since in *all* cases $\Sigma(y - \bar{y}) = 0$ (except for round-off errors). Hence we must do something else to find a typical deviation from the mean.

One solution is to neglect signs, and sum the absolute deviations $\lvert y - \bar{y} \rvert$, then divide by $n$. This gives the "mean deviation"

$$\text{mean deviation} = \frac{\Sigma \lvert y - \bar{y} \rvert}{n}. \qquad (3.7)$$

For the illustrative data we thus have

$$\text{mean deviation} = \frac{2.32 \text{ cc}}{5} = .464 \text{ cc.}$$

Note that three of the deviations $|y - \bar{y}|$ exceed this value, and two are less.

A solution used far more often, however, leads to the "standard deviation." For it we square all of the deviations as in the last column of the table, and add. An average of the squared deviations could now be found by dividing the sum of the squared deviations, $\Sigma(y - \bar{y})^2 = 1.3280 \text{ cc}^2$, by $n = 5$. Indeed, this used to be the common practice. However, for certain theoretical reasons, which appear in Chapter 7, the present practice is to divide $\Sigma(y - \bar{y})^2$ by $n - 1$ instead of $n$. This gives what is called the "sample variance," its symbol being $s^2$.

$$\text{sample variance} = s^2 = \frac{\Sigma(y - \bar{y})^2}{n - 1}. \tag{3.8}$$

For our illustrative data this gives

$$\text{sample variance} = s^2 = \frac{1.3280 \text{ cc}^2}{5 - 1} = .332 \text{ cc}^2.$$

Note especially that the variance $s^2$ is a sort of average of squared deviations, $(y - \bar{y})^2$, and that it carries the square of the original measured unit, $\text{cc}^2$. Although the variance is of much use in various applications, as in Chapters 8–10, it is not of direct help in indicating how much of a deviation is "only to be expected" in our data, that is, of about how far from $\bar{y}$ we can expect an observed $y$ to lie.

For such a purpose we use the "sample standard deviation," its symbol being $s$. Thus

$$\text{sample standard deviation} = s = \sqrt{\frac{\Sigma(y - \bar{y})^2}{n - 1}}. \tag{3.9}$$

Note that we thus have

$$\text{sample standard deviation} = \sqrt{\text{sample variance}} \tag{3.10}$$

For the illustrative data

$$s = \sqrt{\frac{1.3280 \text{ cc}^2}{5 - 1}} = \sqrt{.332 \text{ cc}^2} = .576 \text{ cc}$$

We now have a measure in the original unit against which we can compare the deviations, $y - \bar{y}$. Two of the deviations exceed $s$ in size, and three are less than $s$ (about the usual proportion).

Thus we have two different types of "average discrepancy" of $y$'s from $\bar{y}$, $s = .576$ cc, mean deviation $= .464$ cc. The former is the larger, as is virtually always the case. They are simply two different measures like yard and meter, and are interpreted differently. In large samples with frequencies rising steadily to the maximum then falling away symmetrically, we can expect that about 57% of the cases will lie within $\pm 1$ mean deviation of the mean, while about 68% will lie within $\pm 1s$ of the mean.

A word may well be said at this point as to the comparative usage of the three measures of variability. Sampling theory may be used to determine which of them is the best for estimating the variability in the population. It may be shown that for most populations in use, the standard deviation provides the most reliable estimate, the mean deviation the next best, and the range the least reliable for all sample sizes above $n = 2$. Therefore in general we use the standard deviation, especially if the sample is at all sizable. Moreover, the standard deviation can be easily calculated by using the shortcuts discussed in the next section. On the other hand, if the sample is of only four or five cases, there is but little difference in the reliability of the three measures. Hence, although not quite as reliable, the range is often used because it is so easily calculated. This is especially true in the application of statistical quality control to industrial problems, where we may have a large number of small samples. The mean deviation is used comparatively rarely, and we shall have little further to say about it.

## 3.4. EFFICIENT CALCULATION OF AVERAGES AND STANDARD DEVIATIONS

Since the mean and the standard deviation are used so often in statistical work, it is highly desirable that one be able to calculate them in the easiest possible manner. Why settle for less efficient statistical measures when the best ones can be easily calculated, without sophisticated computers? The easiest method not only saves time, but also facilitates accuracy.

There are two algebraic aids to the computation which we now describe. The first of these simplifies the calculation of the standard deviation by making it unnecessary to subtract $\bar{y}$ from each $y$, by using instead sums and sums of squares. Thus we have the squared deviations

$$(y_1 - \bar{y})^2 = y_1{}^2 - 2\bar{y}y_1 + \bar{y}^2$$
$$(y_2 - \bar{y})^2 = y_2{}^2 - 2\bar{y}y_2 + \bar{y}^2$$
$$\vdots$$
$$\underline{(y_n - \bar{y})^2 = y_n{}^2 - 2\bar{y}y_n + \bar{y}^2}$$

$$\sum (y - \bar{y})^2 = \sum y^2 - 2\bar{y}\sum y + n\bar{y}^2$$
$$= \sum y^2 - 2\left(\sum y/n\right)\sum y + n\left(\sum y/n\right)^2$$
$$= \sum y^2 - \left(\sum y\right)^2/n$$
$$\sum (y - \bar{y})^2 = \frac{n\sum y^2 - (\sum y)^2}{n}.$$

Now dividing the left side by $n - 1$ and taking the square root will give $s$, so we do this also to the right-hand side, finding

$$s = \sqrt{\frac{n\sum y^2 - (\sum y)^2}{n(n - 1)}}. \tag{3.11}$$

The last expression looks more complicated than does (3.9), but often it is not in practice, eliminating as it does all of the subtractions for $y_i - \bar{y}$. Now, however, when the original numbers have several significant numbers, finding the sum of the squares will be cumbersome, as, for example, in the data on amounts metered: $(96.6 \text{ cc})^2$, $(97.2 \text{ cc})^2$, and so on. Moreover there is often "heavy cancellation" in finding $n\sum y^2 - (\sum y)^2$, so that we have to retain full precision in each to obtain accuracy in the difference and in $s$. For the illustrative data we have

$$\sum y = 485.4 \text{ cc}, \qquad \sum y^2 = 47{,}123.96 \text{ cc}^2$$

$$n\sum y^2 - \left(\sum y\right)^2 = 235{,}619.80 - 235{,}613.16 \text{ cc}^2 = 6.64 \text{ cc}^2.$$

Here the difference between the two terms shows up only in the last three of the eight significant figures, so *full precision was necessary* to obtain the three. We now have the numerator for (3.11) and easily finish up the work:

$$s = \sqrt{\frac{6.64 \text{ cc}^2}{5(4)}} = .576 \text{ cc}$$

as before. Clearly this is not easy, except perhaps with a desk calculator available, with which $\sum y$ and $\sum y^2$ can be found simultaneously, and full precision carried.

The real gain through the use of (3.11) is not yet apparent to the reader! But it comes about through simplifying the data by coding, using any convenient linear transformation. Let us call the transformed variable $v$, defined by

$$v = \frac{y - A}{c}, \qquad v_i = \frac{y_i - A}{c}, \tag{3.12}$$

where $A$ is any convenient number to reduce the $y$'s and $c$ is a positive number to simplify them further. For the data on amount metered, we can use $A = 97.0$ cc, and since all $y$'s are even tenths of a cubic centimeter, we can use $c = .2$ cc. Then

$$v = \frac{y - 97.0 \text{ cc}}{.2 \text{ cc}}$$

and the five $y$'s become the pure (dimensionless) numbers $-2, +1, -3,$ $+2, +4$. For these $\sum v = +2, \sum v^2 = 34$.* We now easily apply (3.1) and (3.11) for $v$ (instead of $y$):

$$\bar{v} = \frac{+2}{5} = +.4$$

$$s_v = \sqrt{\frac{n \sum v^2 - (\sum v)^2}{n(n-1)}} = \sqrt{\frac{5(34) - (2)^2}{5(4)}} = \sqrt{\frac{166}{20}} = 2.88.$$

We must now "decode" these to find $\bar{y}$ and $s_y$. Solving (3.12) for $y$ gives $y = A + cv$, from which it is easy to show from (3.3)–(3.5)

$$\bar{y} = A + c\bar{v}. \tag{3.13}$$

Also (see Problem 3.7), it can readily be shown that

$$s_y = s_{A+cv} = cs_v. \tag{3.14}$$

Then decoding

$$\bar{y} = 97.0 \text{ cc} + (.2 \text{ cc})(+.4) = 97.08 \text{ cc}$$

$$s_y = (.2 \text{ cc}) 2.88 = .576 \text{ cc}$$

as before, where the unit comes back in, through $c$.

---

* Carefully note that $\sum v^2$ is not $(\sum v)^2$. The latter would include many cross products of $v$'s as well as the pure squares of $v$'s.

### 3.4.1. Calculations for Frequency Data

Formulas (3.12)–(3.14) are especially powerful for finding $\bar{y}$ and $s$ for a frequency table. The only new features here are that instead of original numbers we use the midvalues, and that instead of single numbers we now have each midvalue occurring with the corresponding frequency $f$ in the table. It would be easy enough to find $\bar{y}$ by cumulating on a desk calculator the products of frequency by midvalue, that is, $\sum fy$, then finding $\bar{y}$ by

$$\bar{y} = \frac{\sum fy}{\sum f}.$$

But for $s$ we would need $\sum f(y - \bar{y})^2$. This would involve $(y - \bar{y})$ deviations. (How many significant figures should be retained?) Then these deviations would need to be squared and the results multiplied by the corresponding frequencies $f$. This is quite tedious. But by coding around some midvalue taken as $A$, and taking $c$ as the class interval, we can have $v$ values like $-3, -2, -1, 0, +1$, and so on. Then it is easy to cumulate $\sum fv$ and $\sum fv^2$. (Both can be cumulated simultaneously on a desk calculator.) We then use the following slight modifications of (3.13) and (3.14):

$$\bar{y} = A + c\frac{\sum fv}{n}, \qquad s = c\sqrt{\frac{n\sum fv^2 - (\sum fv)^2}{n(n - 1)}}, \qquad \sum f = n. \quad (3.15)$$

Table 3.1 illustrates the details of the calculation, for the data of Table 2.2. We may choose any convenient midvalue for $A$, here 1.34%, but always use the class interval for $c$. Here $c = .1\%$. (With a desk calculator we would use $A = 1.04\%$ so as to facilitate finding $\sum f$, $\sum fv$, $\sum fv^2$ simultaneously.) Then we obtain the $fv$ column, and often the easiest way to find $fv^2$ is by multiplication of respective $fv$ by $v$. The totals are then used as follows:

$$\bar{y} = A + c\bar{v} = 1.34\% + (.1\%)(+74)/125 = 1.3992\%$$

$$s_y = (.1\ \%)\sqrt{\frac{125(372) - (74)^2}{125(124)}} = .163\%.$$

Note that the range equals $1.84\% - 1.04\% = .80\%$ and is about five times $s$, which is quite typical for such sample sizes as 125, if there is no strong tendency for a few cases to be very extreme. With larger samples the range may be more nearly six times $s$.

TABLE 3.1

Percentages of Manganese in 125 Casts of Iron from a Blast Furnace. Coding and Calculations for $\bar{y}$ and $s$

| Midvalue $y$ (%) | Frequency $f$ | Coded variable $v$ | $fv$ | $fv^2$ | $fv^3$ [a] | $fv^4$ [a] |
|---|---|---|---|---|---|---|
| 1.04 | 3 | $-3$ | $-9$ | 27 | $-81$ | 243 |
| 1.14 | 9 | $-2$ | $-18$ | 36 | $-72$ | 144 |
| 1.24 | 18 | $-1$ | $-18$ | 18 | $-18$ | 18 |
| 1.34 | 32 | 0 | 0 | 0 | 0 | 0 |
| 1.44 | 29 | $+1$ | $+29$ | 29 | $+29$ | 29 |
| 1.54 | 19 | $+2$ | $+38$ | 76 | $+152$ | 304 |
| 1.64 | 9 | $+3$ | $+27$ | 81 | $+243$ | 729 |
| 1.74 | 5 | $+4$ | $+20$ | 80 | $+320$ | 1280 |
| 1.84 | 1 | $+5$ | $+5$ | 25 | $+125$ | 625 |
| Totals | 125 | | $+74$ | 372 | $+698$ | 3372 |
| | | | $\Sigma fv$ | $\Sigma fv^2$ | $\Sigma fv^3$ | $\Sigma fv^4$ |

$$v = (y - 1.34)/.10, \qquad A = 1.34\%, \qquad c = .1\%$$

[a] $\Sigma fv^3$ and $\Sigma fv^4$ are not used for $\bar{y}$ and $s$ but are used for $a_3$ and $a_4$. See Section 3.5.

By choosing appropriate coding the calculation of $s$ is really quite easy. If a desk calculator or large electronic computer is used, coding is usually unnecessary.

## 3.5* FURTHER DESCRIPTIVE MEASURES OF FREQUENCY DISTRIBUTIONS, THIRD AND FOURTH MOMENTS

The mean and the standard deviation of a sample give a large amount of the information contained in the sample. For most purposes this is all that we need, especially if the data are well behaved or "normal," rising regularly to the maximum frequency and falling away again symmetrically. If there are irregularities such as isolated cases or evidence of nonhomogeneity of conditions, then probably no summary measures are adequate or even appropriate. Moderately often we have large samples which have regular but unsymmetrical frequency distributions, and we find it worthwhile to use measures to supplement $\bar{y}$ and $s$. See, for example, Table 2.18.

The most important such measure is that one designed to measure the extent to which the distribution is unsymmetrical around the mode.

* Sections marked with an asterisk may be omitted without loss of continuity.

This is the property called "skewness" or lack of symmetry. Although several measures of skewness have been used, the one with the best properties and consequently the one most often used is based upon the so-called third central moment, defined by

$$m_3 = \sum (y - \bar{y})^3 / n. \tag{3.16}$$

This $m_3$ moment is called "third" because of the exponent 3, and "central" because it is for deviations from the mean. The third moment around the *origin* is

$$m_3' = \sum y^3 / n. \tag{3.17}$$

In both cases division could be by other than $n$, but we use $n$ for simplicity.

The reason for taking the third central moment may be illustrated by considering the four simple numbers 4, 11, 12, 13, and finding $m_3$ for them.

| $y$ | $y - \bar{y}$ | $(y - \bar{y})^2$ | $(y - \bar{y})^3$ |
|-----|-----|-----|-----|
| 4 | $-6$ | 36 | $-216$ |
| 11 | $+1$ | 1 | $+1$ |
| 12 | $+2$ | 4 | $+8$ |
| 13 | $+3$ | 9 | $+27$ |
| 40 | 0 | 50 | $-180$ |

Thus we find $m_3 = -180/4 = -45$. The deviations $y - \bar{y}$ add to zero, as is always the case. But note that when we cube the deviations, the one large negative deviation, when cubed, far exceeds the sum of the cubes of the other three, giving a strongly negative net total. Thus by tending to carry the sign of the numerically largest deviations, $m_3$ is capable of measuring lack of symmetry.

Now there are two serious shortcomings to $m_3$ as a skewness measure: (1) It carries the physical unit cubed, and hence $m_3$ is not readily comparable from one set of data to another. (2) A related defect is that it has no absolute meaning because it does not take into account the natural variability of the data. To remove these difficulties, we form the dimensionless quantity as follows

$$a_3 = m_3 / (m_2)^{3/2}, \tag{3.18}$$

where similarly to (3.16)

$$m_2 = \sum (y - \bar{y})^2 / n. \tag{3.19}$$

It is to be noted that $m_2$ as just defined is very similar to $s^2$, the sample variance, the difference being division by $n$. Thus to obtain $a_3$ we divided by what was essentially $s^3$.

Thus $a_3$, which is our "standardized" third central moment, is our measure of skewness or lack of symmetry. It is a pure number. Its size tells us the relative extent of the skewness, a value of $+1$ or $-1$, implying quite a strongly unsymmetrical distribution. The sign of $a_3$, being the same as for $m_3$, tells in which direction the larger sized deviations lie, positive if to high value, negative if to low ones. For the four numbers given as an illustration $a_3 = -45/(50/4)^{3/2} = -1.02$.

A second measure for curve shape, supplementing $m_3$, is $a_4$ defined by

$$a_4 = m_4/m_2{}^2 \tag{3.20}$$

in which $m_4$ is defined analogously to (3.16). Since we use the fourth power of $y - \bar{y}$ in $m_4$, all contributions from the $y$'s are positive, and there is no tendency to cancel out. Thus values of $a_4$ are always positive; in fact they are at least 1. For "normal" data $a_4$ is about 3. It can be very large positively, if there are a few numbers relatively very far from the mean in either or both directions, with the bulk of the numbers close to the mean. The measure $a_4$ is sometimes called "kurtosis," not too aptly. It basically measures the relative rapidity of the frequencies to approach zero, since it gives such heavy weight to the extreme deviations through using the fourth powers.

Neither $a_3$ nor $a_4$ is worth much unless we have a substantial sample size.

Let us now show the calculations for $a_3$ and $a_4$ for the data in Table 2.2. For this we use the totals of the last four columns of Table 3.1, substituting into

$$a_3 = \frac{n^2 \sum fv^3 - 3n(\sum fv^2)(\sum fv) + 2(\sum fv)^3}{[n \sum fv^2 - (\sum fv)^2]^{3/2}} \tag{3.21}$$

$$a_4 = \frac{n^3 \sum fv^4 - 4n^2(\sum fv^3)(\sum fv) + 6n(\sum fv^2)(\sum fv)^2 - 3(\sum fv)^4}{[n \sum fv^2 - (\sum fv)^2]^2}. \tag{3.22}$$

These are logical extensions of the short method of finding $s$ by coding as in (3.15). One should notice that although in finding $v$ from $y$ we divided by $c$, we do not have any factor $c$ in (3.21) and (3.22) as it would enter equally in numerator and denominator. Put in another way we may say that $a_3$ and $a_4$ are invariant under any linear transformation. The proof of the equivalence of (3.11) and (3.14) is left for Problem 3.7.

In the example we now have

$$a_3 = \frac{125^2(698) - 3(125)\,372(74) + 2(74)^3}{[125(372) - 74^2]^{3/2}}$$

$$a_4 = \frac{125^3(3372) - 4(125)^2\,698(74) + 6(125)\,372(74)^2 - 3(74)^4}{[125(372) - 74^2]^2}.$$

These yield $a_3 = +.168$ and $a_4 = 2.849$. Thus both figures are relatively close to the "normal" distribution values of 0 and 3, respectively. The former indicates slight positive skewness (slightly greater tailing out to the right), while the latter indicates relatively rapid approach to zero frequencies. But neither is significantly different from the "normal" values. Another use for $a_3$ and $a_4$ is in deciding which theoretical curve to fit to given data, if desirable.

In the attempt to obtain more reliable values some statisticians use the so-called $k$-statistics, which take into account the sample size, rather than using the $m_i$'s directly. Thus $s^2$ is the second $k$-statistic. But the use of $k$-statistics is essentially a correction to the $m_i$ values, dependent on $n$, and for large $n$'s such as we should have to justify calculating $a_3$ and $a_4$, the correction is quite slight. Hence we ignore it.

### 3.6. SUMMARY

In this chapter we have been concerned with several objective measures for summarizing sample data, either discrete or continuous. Of the three averages, mean, median, and mode, the first is by far the most widely used. Likewise, of the three measures of variability, standard deviation, mean deviation, and range, the first is most widely used. Such objective measures derived from the sample data form the basis of statistical analyses, such as making tests of statistical hypotheses and estimating the constants (parameters) which determine the particular population.

It should be emphasized again that if there is evidence that the data do not come from homogeneous conditions, it is not desirable to use just $\bar{y}$ and $s$, say, for we do not have a sample from just one population. Irregularities or two modes in the data would indicate nonhomogeneity of conditions. There are also statistical tests of homogeneity.

When large samples, such as many of those at the end of Chapter 2, give a substantial picture of the shape of the distribution of the population, then it may well be worthwhile calculating $a_3$ and $a_4$ to go with $\bar{y}$ and $s$ in the summary.

Efficient calculational methods have been included so that the user of statistics can obtain results quickly and accurately and need not go to a digital computer.

## Problems

3.1.　Find by the shortest method $\bar{y}$, median, $s$, and $R$ for each of the following sets of data as assigned

(a)　Outside diameter of parts: .1916 in., .1925 in., .1926 in., .1928 in., and .1925 in.

(b)　Thicknesses of cork disks for bottle crowns: .064 in., .060 in., .060 in., .063 in., .059 in.

(c)　Bowling scores for one series: 157, 145, 156.

(d)　Percent molybdenum in iron samples: .61, .60, .585, .57, .61.

(e)　Strengths of lamps, in lumens per watt: 10.37, 10.42, 10.28.

(f)　Percent moisture of polysar: .36, .36, .33, .22, .39, .37, .45, .29.

(g)　Micrograms of radioactive phosphorus removed per microgram present in rat incisors by mouthwash "W": .017, .011, .021, .019, .026, .014, .017, .018, .021, .008, .006, .007, .013, .009, .010.

(h)　Postages listed as 1, 3, 2, 1, 7 oz in a postal study.

(i)　Measurement of a property $G$ for glass in units $10^{-8}$ cm³/g sec²: 6.678, 6.671, 6.675, 6.672, 6.674.

(j)　Percent rejection of steel: 2.06, 7.35, 3.03, 2.44.

(k)　Time in seconds for a large pillbug to pass consecutive $9\frac{1}{4}$-in. tiles: 35, 17, 18, 19, 23, 20, 25, 25, 24, 25, 20, 29, 19, 23, 26. (Taken while the author was ostensibly listening to a lecture.)

3.2.　Collect a series of numerical data and tabulate into a frequency table, show graphically, and find the mean and standard deviation.

3.3.　For assigned Tables 2.10–2.27 do whichever of the following parts are assigned:

(a)　Find the mean and the standard deviation.

(b)　Approximate the value of the median for the data by making a cumulative frequency table and interpolating for the $y$ value having $n/2$ cases below it.

(c)　Using the table in (b) interpolate to approximate the number of cases between $\bar{y} - s$ and $\bar{y} + s$ using $\bar{y}$ and $s$ from (a). This is found by subtracting the cumulative frequency below $\bar{y} - s$ from that below $\bar{y} + s$. Also convert the number between to a percentage of $n$.

3.4.  Prove the laws of summations (3.3), (3.4), (3.5).

3.5.  Prove that $\Sigma(y - \bar{y}) = 0$, by laws of summations.

3.6.  Prove properties 2 and 3 for $\bar{y}$ by laws of summations. Prove (3.13).

3.7.  Prove (3.14) by substituting $y = A + cv$ and $\bar{y} = A + c\bar{v}$ into (3.9) and using (3.3).

3.8.  Suppose that we have two samples of data: $y_{11}$, $y_{12}$, ..., $y_{1n}$ and $y_{21}$, $y_{22}$, ..., $y_{2m}$, for which the respective means are $\bar{y}_1$ and $\bar{y}_2$, the sample sizes being $n$ and $m$. Prove that the mean $\bar{y}$ for all $n + m$ numbers is

$$\bar{y} = (n\bar{y}_1 + m\bar{y}_2)/(n + m).$$

3.9.  For the same situation as in Problem 3.8, derive a formula for the variance $s^2$ of the combined data, in terms of the two sample variances $s_1{}^2$, $s_2{}^2$, and $\bar{y}_1$, $\bar{y}_2$, $n$, and $m$.

3.10.  Consider the numbers $y_1$, $y_2$, ..., $y_n$. Let $B$ be any number (same unit as $y$'s). Show by laws of summations and a little calculus that

$$\sum_{i=1}^{n} (y_i - B)^2$$

is a minimum when $B = \bar{y}$. This is a property of $\bar{y}$.

3.11.  Prove (3.21). This is done by cubing and squaring out the expressions being summed in (3.16) and (3.19) using $v$ to replace $y$. Then laws of summations are applied and simplifications made to prove (3.21).

3.12.  Program a computer to calculate $\bar{y}$, $s$, $a_3$, and $a_4$, from data coded with the lowest midvalue as $A$, and class interval $c$, using (3.15), (3.21), and (3.22).

# Chapter 4

# SOME ELEMENTARY PROBABILITY

## 4.1. INTRODUCTION

Nearly everyone has at least some intuitive idea of the probability of an event or outcome. Opportunities to use probability concepts in everyday life occur often. This is especially true when we are faced with a decision between two or more courses of action. Whether to continue driving an aging auto or trade it in depends in part on the likelihood of the auto needing a major repair job, or causing an accident through failure. The individual's decision on what vocation to choose, which job offer to take, or whether and whom to marry, can involve probability reasoning. In industry many top-level decisions are based on probabilities. In the shop there are decisions such as whether to accept or reject a lot, or whether to reset, retool, or leave the process alone. In a laboratory there can be the question of when to throw away an apparently extreme value which does not "seem" to be a part of the collection of results. Or how does one interpret accurately the meaning of 27 successful and 3 unsuccessful outcomes of an experiment. What is the meaning of "chance of precipitation is 20%"? All of these illustrations involve the concepts of probability.

Although an intuitive knowledge of probability is widespread, it is unfortunately true that misconceptions and inaccuracies in probability reasoning and results are just as widespread. In fact it is rare that a probability given in a newspaper or magazine is accurate. Thus we will need to build a sound background in probability. This will be needed and used constantly in our study of statistical analysis and in the interpretation of results. In fact the dividing line between statistics and probability is nebulous, if indeed it can be drawn at all.

The subject of probability can be studied at a great variety of levels, from simple arithmetic (see Whitworth [1]) to virtually purely abstract mathematics. In this book we take a middle position, concentrating on those elements of probability of most direct use in the statistics to follow. We also strive for accuracy without extreme rigor.

We give the concepts of sample spaces of outcomes, events, and combinations of events, define the probability of an event, and give some laws of probability and interpretation.

## 4.2. SAMPLE SPACES OF OUTCOMES

In general in probability we are concerned with a trial, or a series of trials, in which the outcome is uncertain. A trial might, for example, be an experiment, or the measurement of some manufactured part. We cannot be absolutely certain which of the possible outcomes will occur. (If we could, we would not need probability nor statistics!)

It is essential not only in theory but in applications to know what outcomes are possible. The whole collection of such outcomes is called a "sample space." Sample spaces of interest can consist of two or more possible outcomes. For example, an experiment either yields a desired compound or fails to, a radio set is either "good" or "defective" (contains one or more defects), a sheet of tinplate is a good one, one which can be rerun, or a scrap sheet, a golf club is a good one, a "second," or a "reject." These spaces are of two or three possible outcomes or "points." We might take a sample of 10 razor blades from 1000. Here the sample space consists of a very large number of possible collections each of 10 blades, in fact, the number of different combinations of 1000 things taken 10 at a time, that is, C(1000, 10). This treats each blade as distinguishable from all others. Another possible sample space for the 1000 blades would be to divide them into "good," say 970, and "defective," 30, all good blades being indistinguishable and similarly all defective blades. Then the possible outcomes for the sample of 10 are for it to contain 0, 1, 2, ..., 10 defectives. Thus this sample space has only 11 points.

One may have an infinite sample space too. For example, an experiment consists of manufacturing electronic parts repeatedly till the first defective occurs. The defective may occur on the first part, on the second, and so on, with no limit to how many could be produced before the first defective. Such a sample space is called "countably infinite," because it consists only of integers. On the other hand, the sample space may be for any of the possible measurements in industry or science. Usually there are some limits to the possible measurements or "points"; possibly zero is a lower limit. For example, a United States cent is supposed to weigh between 46.00 and 50.00 grains. If we suppose that some might be outside these limits through minting errors, we might still be sure that a newly minted cent will be between 44.00 and 52.00 grains. The number of true weights between the latter limits is infinite, in fact what is called "uncountably infinite." (This is just like the number of points on any line segment, which cannot be placed in a one-to-one correspondence with the numbers 1, 2, 3, ... .) Now it is true that if we measure, accurate to the nearest .01 grain, there would be just 801 possible measured outcomes or points from 44.00 to 52.00, and one might say that the sample space consists of 801 points. This *is* a feasible sample space, but we usually instead consider the uncountably infinite sample space of all points between the limits.

The foregoing paragraph points up the distinction made in Section 2.4 between continuous and discrete data. For the most part probabilities concerned with countable spaces (finite and infinite) are more easily studied and understood than those for uncountably infinite spaces. Somewhat different mathematical approaches are used, the latter involving infinitesimal calculus. One rather troublesome problem is to define probability and the laws so as to be accurate for both types of sample spaces. Of interest in this book is that we have two distinct types of theoretical distributions in statistics for the two cases, as given in the next two chapters.

It may well be emphasized that although the outcome is often a number, it need not be. We can randomly select three boxes out of 50 to be looked at, or three part numbers out of 4000 to be given a quality audit in a plant, or select five inspectors from 20 to work in an experiment. Such outcomes are not numbers, but *are* "points" in finite sample spaces.

## 4.3. EVENTS

The concept of an "event" is both convenient and necessary. An "event" is simply a defined collection of possible outcomes. Thus, if we consider the case of 10 blades selected from 1000, an event might be the collection of possible samples which contain two or more defective blades. Or in selecting three part numbers from 4000, an event might be all such selections in which at least one is a shaft. In the case of weighing a U.S. cent, an event could be a recorded weight of 48.31

grains. The true weight is somewhere between 48.305 and 48.315 grains. Or an event could be a weight between 48.00 and 48.50 grains. Both of these events include an uncountably infinite collection of outcomes. Another event from a continuous sample space is that of a solar battery having a life exceeding 5000 hours. A somewhat similar discrete event is that of the first electronic part to fail being at least the five-hundredth one tested. In a can manufacturing line, four heads on a machine are making cans. One event is a sample of 12 which contains at least one can from each head.

It is often *convenient* to classify individual outcomes into collections or events, when we are concerned with probabilities on finite or countably infinite sample spaces. But when we are concerned with uncountably infinite sample spaces, we run into contradictions if we are not careful. We then find it *necessary* to talk about probabilities of *events* rather than of individual outcomes.

For notation we shall use small Latin letters, such as $a$, $b$, $a_1$, $x$, and so on, for individual outcomes. Events will be designated by capital Latin letters such as $A$ and $B$. Also $W$ will denote the whole sample space of possible outcomes. Just as we find zero to be a useful number in the Hindu–Arabic number system, we find some uses for $\varnothing$, the empty set, that is, the "null set" which contains no outcomes at all. When the outcome of a trial or experiment is, say, $a$, which is one of the outcomes contained in event $A$, we say that "$A$ has occurred." Both $W$ and $\varnothing$ are events, the former always occurring on every trial, the latter never occurring on any trial. Summarizing:

individual outcomes: $a$, $b$, $x$, ...,
events (sets of individual outcomes): $A$, $B$, ...,
whole space event: $W$,
empty or null set: $\varnothing$,
if $x$ occurs and is in $A$, then $A$ has occurred.

### 4.3.1. Relations of Events.

Given an event $A$, the complementary event $\bar{A}$, consists of all of the outcomes of the space $W$ which are not in $A$. Together $A$ and $\bar{A}$ make up $W$, for every outcome in $W$ but not in $A$ must be in $\bar{A}$, but there is no overlapping between $A$ and $\bar{A}$. Thus, for example, in samples of 10 razor blades out of 1000, if $A$ is the event of two or more defective blades in the sample, $\bar{A}$ is the event of none or one defective blade in the sample of 10. It might be easier to find the probability of the latter, $\bar{A}$, than the former. Likewise if $B$ is the event of a life of over 5000 hours, then $\bar{B}$ is the event of a life of 5000 hours or less. Note that $\bar{W}$ is $\varnothing$ and $\bar{\varnothing}$ is $W$.

Two events are said to be equal if they contain identical collections of outcomes. Or stated in another way, $A = B$ if and only if every outcome in $A$ is also in $B$, and conversely every one in $B$ is in $A$. A nontrivial example is the following. In sampling parts one by one from a lot of $N$ parts (each part either good or defective), let $A$ consist of the event that the *first defective* occurs on the $n$th or an earlier trial. Let $B$ consist of the event of *one or more defectives* among the first $n$ sampled. It may take a little thought to convince one that $A = B$, but it is worthwhile thinking it through.

We also have for events a type of inequality. We say $A$ contains $B$, which is written $A \supset B$, if every point or outcome in $B$ is also in $A$, but not necessarily conversely. Event $A$ may or may not contain outcomes not in $B$. Thus the event $A$, consisting of diameters measured to the nearest .0001 in., .2130–.2150 in., contains event $B$ which is a diameter measured to the nearest .001 in. and is given as .214 in. But $A$ does not "contain" event $C$, a recorded diameter of .215 in. Why not?

We also note that if $A \supset B$ and $A \subset B$ ($A$ contained in $B$), then $A = B$. Moreover $W \supset \varnothing$, $W \supset A$, and $A \supset \varnothing$ for all $A$ in $W$. And, $A$ contains any single outcome in it, say $a$; that is, $A \supset a$. Further, when $A \supset B$, then if $B$ occurs, surely $A$ does, but $A$ occurring does not guarantee that $B$ does.

Summarizing:

> $\bar{A}$, event of all outcomes not in $A$;
> $W$, whole sample space of outcomes, occurs on each trial;
> $\varnothing$, empty set, no outcomes, cannot occur on any trial;
> $A$ and $\bar{A}$, between them always comprise $W$;
> $A \supset B$ if every outcome in $B$ is also in $A$;
> $A = B$ if $A$ and $B$ are the same collections of outcomes;
> $A \supset B$ and $A \subset B$ imply $A = B$.

If $A \supset B$, then occurrence of $B$ implies the occurrence of $A$.

### 4.3.2. Combinations of Events.

We now define two events associated with any two given events, say $A$ and $B$. The "union" of $A$ and $B$ is the event which consists of all outcomes in $A$ alone or in $B$ alone, or in both $A$ and $B$, that is, points in $A$ and/or in $B$. The notation is

$$\text{union of } A \text{ and } B = \text{all outcomes in } A \text{ and/or } B = A \cup B. \qquad (4.1)$$

Note that

$$(A \cup B) \supset A \quad \text{and} \quad (A \cup B) \supset B. \tag{4.2}$$

As an example, let $A$ be the event of three to six accidents in a year at an intersection and $B$ the event of one to four accidents there. Then $A \cup B$ is the event of one to six accidents in the year there.

Another important event determined by $A$ and $B$, say, is their "intersection." It consists of all outcomes which are in *both* $A$ and $B$ at the same time, or common to both $A$ and $B$. The notation is

$$\text{intersection of } A \text{ and } B = \text{outcomes common to } A \text{ and } B$$
$$= A \cap B. \tag{4.3}$$

Note that now oppositely from (4.2)

$$(A \cap B) \subset A \quad \text{and} \quad (A \cap B) \subset B. \tag{4.4}$$

In the example above $A \cap B$ is the event of three or four accidents at the intersection, since these are the only outcomes common to both $A$ and $B$.

If two events have no outcome in common, then their intersection is the empty set $\varnothing$, and the events are said to be "mutually exclusive" or "disjoint." If one occurs, the other cannot; one excludes the other:

$$A, B \text{ mutually exclusive or disjoint:} \quad A \cap B = \varnothing. \tag{4.5}$$

If $A$ and $B$ are mutually exclusive, then some authors write

$$A \cup B = A + B \quad \text{if} \quad A \cap B = \varnothing, \tag{4.6}$$

which is the set of outcomes of $B$ in addition to those of $A$, with no overlapping. We shall not use the plus notation.

In "double sampling" a lot of $N$ articles each of which is good or defective, we may accept the lot, say, if there are no defectives in a sample of $n$, or reject if there are three or more defectives, or require a second sample if there are one or two defectives. These are mutually exclusive events, say, $A_a$, $A_r$, and $A_s$. Note that here $A_a \cup A_r \cup A_s = W$. (What is the sample space $W$?)

Purely as an aid to visualization of different sets we often use so-called Venn diagrams. In a Venn diagram we draw various closed curves which *conceptually* enclose those outcomes or points contained in the set in question and only those points. The area within such a closed curve has no meaning, in general (but can occasionally be so defined). See Fig. 4.1 for Venn diagrams illustrating some of the events just discussed.

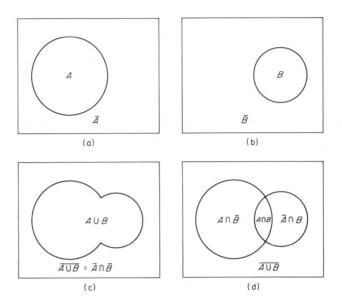

**Fig. 4.1.**   Venn diagrams illustrating certain related events and typical cases.

Some algebra of sets of outcomes or events follows:

$$W \cup \varnothing = W, \qquad W \cup A = W, \qquad W \cap \varnothing = \varnothing, \qquad A \cap \varnothing = \varnothing \quad (4.7)$$

$$A \cup \bar{A} = W, \qquad A \cap \bar{A} = \varnothing \tag{4.8}$$

$$A = (A \cap B) \cup (A \cap \bar{B}) \qquad \text{since} \quad A \cap B \quad \text{and} \quad A \cap \bar{B}$$
mutually exclusive, but together cover $A$ (4.9)

$$\overline{A \cup B} = \bar{A} \cap \bar{B} \tag{4.10}$$

$$\overline{A \cap B} = \bar{A} \cup \bar{B} \tag{4.11}$$

$$A \cup B = (A \cap \bar{B}) \cup (A \cap B) \cup (\bar{A} \cap B) \qquad \text{with three}$$
sets on the right disjoint. (4.12)

What we have been discussing here is often called the "algebra of sets of points." In the following we shall restrict ourselves to a very general class of sets, called a "Borel" class. For practical applications we can very generally say that all sets of outcomes, that is, events, are Borel sets. Hence in practice we shall not need to worry about some of the extremely clever exceptions which can be dreamed up by pure mathematicians.

## 4.4. PROBABILITIES OF EVENTS

We are now in a position to define the probability of an event. Basically the probability of an event is a numerical valued function of the event or set of points. For every such event or set there must be one and just one probability. Moreover this probability is some number from 0 to 1. It gives the "likelihood" that when a random trial or experiment is made the outcome will be among those defined for the event in question. For example, in the case in which there were 970 good razor blades and 30 defectives, the probability that a random sample of 10 contains two or fewer defectives can be calculated to be .9975, as we shall see in Chapter 5. Thus on nearly every trial we shall observe the event of two or fewer defectives in the sample of 10.

We now define the probability of an event by the following properties, using the notation $P(A)$ to denote the probability of event $A$:

Defining Properties for Probability $P(A)$

1. For every event $A$ in sample space $W$,    $P(A) \geqslant 0$.    (4.13)

2. $P(W) = 1$.    (4.14)

3. For any collection of events $A_1$, $A_2$, $A_3$, ... (finite or countably infinite), such that no two events have any outcome in common,
$$P(A_1 \cup A_2 \cup A_3 \cup \cdots) = P(A_1) + P(A_2) + P(A_3) + \cdots.$$    (4.15)

These definitional equations do not tell us what the probabilities of the events are, but only what rules must apply to such a probability function.

An example of rule 3 is provided by the previous example given of double sampling. The probability of reaching a decision on the first sample is the probability of having either zero defectives in the sample of $n$, or else three or more, the former yielding acceptance, the latter rejection. Thus

$$P(\text{decision on first } n)    P(A_a \cup A_r) = P(A_a) + P(A_r)$$

using 3, since acceptance and rejection are mutually exclusive.

An example with an infinity of mutually exclusive events is that of continuing to try an experiment until the first failure occurs. Let $A_1$ be failure F on the first trial, $A_2$ be first failure on the second trial, that is, consecutive outcomes SF, and $A_3$ be SSF, and so on. Then

$$W = A_1 \cup A_2 \cup A_3 \cup \cdots,$$

a countable infinity of mutually exclusive events. Thus by (4.15)

$$P(W) = P(A_1) + P(A_2) + P(A_3) + \cdots.$$

But since by 2, $P(W) = 1$ we have

$$P(A_1) + P(A_2) + P(A_3) + \cdots = 1$$

We also have

$$P(3 \text{ or more trials before first failure})$$
$$= P(A_3 + A_4 + \cdots) = 1 - P(A_1) - P(A_2).$$

The right side may be easier to calculate than the left, since there are only two probabilities.

Many theorems can be proven from the definitional axioms of probability and various properties of sets or events. We shall find it desirable to prove the following theorem:

$$P(A \cup B) = P(A) + P(B) - P(A \cap B). \tag{4.16}$$

First we note from (d) of Fig. 4.1 that since $A \cap \bar{B}$ and $A \cap B$ together make up $A$, but are mutually exclusive, (4.15) gives

$$P(A) = P(A \cap \bar{B}) + P(A \cap B).$$

Likewise

$$P(B) = P(A \cap B) + P(\bar{A} \cap B).$$

Moreover, since $(A \cap B) \cup (A \cap \bar{B}) \cup (\bar{A} \cap B) = A \cup B,$

$$P(A \cup B) = P(A \cap B) + P(A \cap \bar{B}) + P(\bar{A} \cap B).$$

The first two probabilities on the right side give $P(A)$. Thus

$$P(A \cup B) = P(A) + P(\bar{A} \cap B) + P(A \cap B) - P(A \cap B),$$

where we have substituted $P(A)$ for $P(A \cap B) + P(A \cap \bar{B})$ and added and subtracted $P(A \cap B)$. Now the middle two terms on the right side give $P(B)$. Hence

$$P(A \cup B) = P(A) + P(B) - P(A \cap B).$$

This is the general law for the probability of the union of two sets. The particular special case where $A \cap B = \varnothing$, that is, where $P(A \cap B) = 0$, is

$$P(A \cup B) = P(A) + P(B) \qquad \text{if} \quad A \cap B = \varnothing, \tag{4.17}$$

as in (4.15). This and (4.16) are often called additive laws of probability.

Such axioms and laws do not enable us to evaluate probabilities in general. For this we need further information about the particular probability function of sets or events.

## 4.5. PROBABILITIES ON DISCRETE SAMPLE SPACES

We now take up the case of probability functions defined on discrete sample spaces. First suppose that there is a finite number of points in the space. For example, 35 inspectors, or 35 voltmeters or watches. We assign to each person or meter a number 1, 2, ..., 35. Then we draw "at random" one from the space, and thus the outcome can be called $a_1$, or $a_2$, or ..., $a_{35}$, according to which person or meter was chosen. These are individual outcomes or points. We could define the points as events also and use $A_1$, $A_2$, ..., $A_{35}$ if we wish. Each $A_j$ would then be a collection of just one point. Or if there are two meters with "bias" we could define $A$ as the collection of the two biased meters and $B$ the collection of unbiased ones, $B$ thus containing 33 points. It is quite natural to let the probability of drawing each meter be the same, that is, equal probabilities. This is often given as the definition of "drawing at random." Note that such equal probability of drawing does not just happen, but must be actively sought and planned for in practice. Using

$$P(a_1) = P(a_2) = \cdots = P(a_{35})$$

and, since we have mutually exclusive events, $a_j = A_j$

$$P(W) = P(a_1 \cup a_2 \cup \cdots \cup a_{35}) = P(a_1) + P(a_2) + \cdots + P(a_{35}).$$

Then

$$P(W) = 35P(a_j)$$

and since $P(W) = 1$

$$P(a_j) = 1/35, \qquad j = 1, ..., 35.$$

This is an example of "equally likely" drawing from a population of $N$ points, that is, in general

$$P(\text{any one point in } N) = 1/N. \tag{4.18}$$

Then from this we obtain for the above event $A$ of a biased meter being drawn (say, numbers 3 and 12):

$$P(A) = P(a_3) + P(a_{12}) = 2/35.$$

This is an example of the frequently used

$$P(A) = N_A/N, \tag{4.19}$$

where $N$ is the total number of objects or points in the sample space (lot or population) and $N_A$ is the number of the $N$ whose occurrence in drawing would yield event $A$; and "equally likely" drawing is made. To use (4.19) one must somehow be assured that each outcome in $N$ is equally likely. One way is to draw by a table of random numbers such as in Table X in the Appendix.

Thus under such a type of random drawing we have but to count how many outcomes yield event $A$ and how many can occur at all, and divide as in (4.19). This is a common approach in "discrete probability."

### 4.5.1. Countably Infinite Spaces

Some discrete sample spaces are infinite. Thus there may be no limit to the number of defects possible on a ream of paper, or the number of breakdowns of insulation, when 10,000 ft of wire are tested by a suitably high test voltage. Since the number of such defects is for practical purposes unlimited we define the infinite space

$$W = a_0 \cup a_1 \cup \cdots \cup a_j \cup \cdots.$$

Now if we were to try to take equal probabilities for each $a_j$, so as to add up to $1 = P(W)$, we could not let $P(a_j)$ be any positive number, say $e > 0$, for the right side would be infinite. Nor could we let $P(a_j) = 0$, for then the right side would be zero. Thus in the case of a countably infinite sample space, we must use a probability function which gives varying probabilities to the various points $a_j$. But also, since these points are mutually exclusive, we have by (4.14) and (4.15)

$$1 = P(W) = P(a_0 \cup a_1 \cup \cdots) = P(a_0) + P(a_1) + \cdots.$$

Thus

$$\sum_{j=0}^{\infty} P(a_j) = 1 \tag{4.20}$$

is a requirement for whatever law is used. In Chapter 5 we shall see laws such as the Poisson, having the preceding property, (4.20).

### 4.5.2. Events over Discrete Spaces.

In both the preceding cases the probability for an event which includes more than one point is simple to define. Let $A_j$ consist of the distinct outcomes $a_{j1}$, $a_{j2}$, ..., $a_{jk}$, ..., either finite or infinite. Then

$$P(A_j) = P(a_{j1} \cup a_{j2} \cup \cdots) = P(a_{j1}) + P(a_{j2}) + \cdots. \qquad (4.21)$$

The number of points in $A_j$ is usually finite but can easily be infinite. For example, $A_j$ may be the event of three or more collisions of high-energy particles.

## 4.6. INDEPENDENT AND DEPENDENT EVENTS

One cannot go far in applying probability without bringing in the concepts of independent and dependent events. We shall first consider the case of finite discrete spaces to make the ideas clear, and then give general definitions.

Consider 1200 radios which have been manufactured. Suppose there are 60 with at least one poorly soldered connection and 20 with an open circuit. We assume that sampling is with equal likelihood. Let $A$ be the event of drawing a radio with a poorly soldered connection, and $B$ the event of drawing a radio with an open circuit. Then by (4.19)

$$P(A) = 60/1200 = 1/20$$

$$P(B) = 20/1200 = 1/60.$$

Now there are four kinds of radios possible from this viewpoint, those with neither defect, those with defect in soldering but no circuit open, those with no defective soldered connection but an open circuit, and those with both defects. We can call these classes of radios, respectively, $\bar{A} \cap \bar{B}$, $A \cap \bar{B}$, $\bar{A} \cap B$, and $A \cap B$, because, for example, $\bar{A}$ is the collection of radios without a soldering defect, while $B$ is the collection with an open circuit. Those radios with *both* these properties are in the intersection of $\bar{A}$ and $B$. (We choose not to give a separate notation for a set of radios and the event of drawing a radio in such a set.)

Now for the above, what might we expect for the probability of a set with *both* types of defects, that is, $P(A \cap B)$? Since on the average one set out of 20 has a poorly soldered connection and there are just 20 with an open circuit, might we not reasonably expect that one of these 20 has *both* defects, while the other 19 do not have an open circuit. This would imply that but one radio among the 1200 has both defects, and that thus

$$P(A \cap B) = 1/1200.$$

But it could also happen that 0, or 2, 3, ..., 20 radios have *both* types of defects. (Why not more than 20 ?) Hence $P(A \cap B)$ could be 0/1200, 1/1200, 2/1200, ..., 20/1200. Since

$$P(A) P(B) = (1/20)(1/60) = 1/1200$$

there is only one case in which

$$P(A \cap B) = P(A) P(B),$$

namely that in which $A \cap B$ includes just one radio. See Fig. 4.2 a.

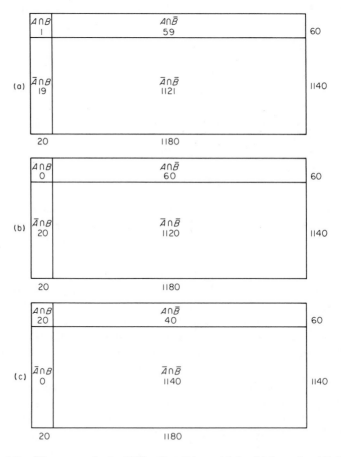

**Fig. 4.2.** Three examples for 1200 radios with two kinds of defects $A$ and $B$. Numbers tell how many radios are in the various sets (combinations of defects). Note that the marginal totals for $A$, $\bar{A}$, $B$, $\bar{B}$ are the same in all three cases.

If $A \cap B$ contains none, then $P(A \cap B) = 0$, Fig. 4.2 b, and

$$P(A \cap B) < P(A) P(B),$$

while if $A \cap B$ contains two or more radios, for example, Fig. 4.2 c, then

$$P(A \cap B) > P(A) P(B).$$

We now define the events $A$ and $B$ to be "independent" when and only when

$$P(A \cap B) = P(A) P(B), \qquad A, B \quad \text{independent.} \qquad (4.22)$$

If (4.22) is not true, then we say that $A$ and $B$ are "dependent":

$$P(A \cap B) \neq P(A) P(B), \qquad A, B \quad \text{dependent.} \qquad (4.23)$$

Let us explain further the idea of "independence" within the context of the preceding example. Suppose that $A$ and $B$ are independent. Then $A \cap B$ contains one radio, $A \cap \bar{B}$ has 59, $\bar{A} \cap B$ has 19, and $\bar{A} \cap \bar{B}$ has 1121, Fig. 4.2a. Suppose we draw a radio and it proves to have a poorly soldered connection; that is, we have observed event $A$. What is the probability that this radio also has an open circuit? Of all the 60 radios with a soldering defect, event $A$, there is only one with an open circuit also. Thus the probability of event $B$ occurring, knowing that $A$ did occur, is 1/60. This probability is called a "conditional probability," $P(B \mid A)$, the probability of "event $B$ given that $A$ did occur."

Now suppose that $A$ did not occur, that is, no soldering defect is in the radio drawn, event $\bar{A}$, then what is the probability of an open circuit being in it? Here there 1140 radios without soldering defects. Of these, 19 have open circuits, hence the conditional probability is

$$P(B \mid \bar{A}) = 19/1140 = 1/60.$$

Note that this is just the same as $P(B \mid A)$ and in fact both are equal to $P(B)$. Thus the occurrence of $B$ is unaffected by whether or not $A$ occurred; $B$ is thus "independent" of $A$.

By similar argument the reader might well verify that in the *independent* case above

$$P(\bar{B} \mid A) = P(\bar{B} \mid \bar{A}) = 59/60 = P(\bar{B})$$

and

$$P(A \mid B) = P(A \mid \bar{B}) = 1/20 = P(A).$$

Now if $A$ and $B$ are not independent but related, then the equalities do not hold. Suppose that there were *no* radios among the 1200 having both types of defects, Fig. 4.2b. Then if $A$ occurs (defective soldering), we know we have one of the 60 such radios. But none have open circuits, as just assumed, hence $P(B \mid A) = 0/60 = 0$. This is not equal to $P(B)$, but is less. Moreover, if $A$ did not occur, then we know we have one of the 1140 such radios. Among them are all 20 with open circuits. Thus $P(B \mid \bar{A}) = 20/1140 = 1/57$, which is greater than $P(B)$.

Let us take the opposite extreme, Fig. 4.2c, where all 20 radios with open circuits *also* had defective soldering. Now we easily find

$$P(B \mid A) = 20/60 = 1/3 \quad \text{and} \quad P(B \mid \bar{A}) = 0/1140 = 0.$$

These are unequal and neither is equal to $P(B)$. Similarly

$$P(\bar{B} \mid A) = 40/60 = 2/3 \quad \text{and} \quad P(\bar{B} \mid \bar{A}) = 1140/1140 = 1,$$

which again are unequal and neither equals $P(\bar{B}) = 59/60$.

### 4.6.1. Conditional Probabilities.

The present section has been concerned with an example in a finite sample space leading up to the definitions given by (4.22) and (4.23) for independence and dependence, which are general. These definitions do not in any way depend on the sample space being finite nor countably infinite. We now make a related definition for the conditional probability of $B$ given that $A$ did occur, namely

$$P(B \mid A) = P(A \cap B)/P(A) \quad \text{if} \quad P(A) > 0. \tag{4.24}$$

Here the outcomes making up event $A$ form the *space* over which *both $A$ and $B$* might occur, and the ratio gives in a sense the proportion of all those times in which $A$ occurs, that $B$ also occurs. Naturally if $P(A) = 0$, then we have not much of a space in which both $A$ and $B$ can occur! Hence the restriction is not confining in applications.

We now have an alternate definition of independence

$$P(B \mid A) = P(B), \quad A, B \quad \text{independent.} \tag{4.25}$$

This could be regarded as a theorem from (4.22) using (4.24), and conversely. Two different forms of (4.24) are convenient

$$P(A \cap B) = P(A)\, P(B \mid A) = P(B)\, P(A \mid B). \tag{4.26}$$

We can call (4.22) and (4.26) two multiplicative laws of probability, just as we can call (4.15) and (4.16) two additive laws.

### 4.6.2. Repeated Trials.

In our first example we illustrated independent events in a sample space of 1200. This served our purposes to introduce the subject. But such cases are not commonly met in practice. We could not even develop a nontrivial example of independence if we had had 1201 radios instead of 1200, because 1201 is a prime number!

The common case is that of repeated trials. Suppose that we make two random drawings from a lot of 300 watches, of which five contain defects, and could therefore be called "defectives." If we let $A$ be the event of a defective watch being drawn on the first draw, then since event $A$ contains five points we have from (4.19)

$$P(A) = 5/300 = 1/60.$$

Suppose that after having drawn the watch and inspected it, we replace it (without repairing any defects). Then the lot again contains five defectives and 295 good watches. Let $B$ be the event of a defective watch on the second draw. Drawing at random we again find

$$P(B) = 5/300 = 1/60.$$

Now from a practical viewpoint it makes sense to assume that these two events are independent; that is, that what happened on the first drawing does not affect the probabilities on the second. Thus that

$$P(B \mid A) = P(B \mid \bar{A}) = P(B)$$

or that

$$P(A \cap B) = P(A)\,P(B \mid A) = P(A)\,P(B) = (1/60)^2 = 1/3600.$$

It seems desirable to describe this in terms of a product space: Let each watch be given a serial number 1 to 300, and let $x$ be the serial number for the first draw, $y$ for the second. Then $(x, y)$ is a point of the "product space" of 90,000 points, 300 by 300. Moreover the independence assumption and equal likelihood on *each* draw leads to equal likelihood on all 90,000, thus a probability of 1/90,000 for each point. Suppose we let serial numbers 1 to 5 be the defective watches. Then all points $(x, y)$ of the product space fulfilling $A \cap B$ lie in the lower left corner, 25 points, and $P(A \cap B) = 25/90,000 = 1/3600$. See Fig. 4.3.

Next suppose we draw "without replacement" from the 300 watches. If we draw number 262 on the first draw, then there are only 299 watches from which to draw a second watch, namely all those 1 to 300, *except* number 262. The sample space now consists of the points in Fig. 4.4,

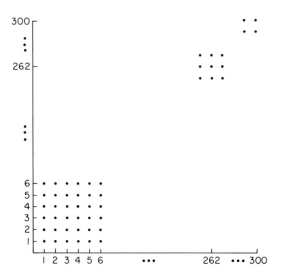

**Fig. 4.3.** Sample space of two consecutive drawings *with replacement* from 300 watches, of which numbers 1 to 5 are defectives, the other all good.

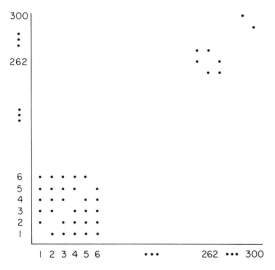

**Fig. 4.4.** Sample space of two consecutive drawings *without replacement* from 300 watches, of which numbers 1 to 5 are defective, the others all good.

which are the same as those in Fig. 4.3 except that the diagonal is missing, since it is impossible to draw the same numbered watch both times because of nonreplacement. (This *was* possible when drawing with replacement, as in Fig. 4.3.) In this case the events $A$, a defective on the first draw, and $B$, a defective on the second draw, are *not* independent. The probabilities on the second draw depend on the watch drawn on the first draw. If the latter was a defective, then the second draw is from a population of 299 watches of which only four are defective, while if the first watch was a good one, then the second draw is from 299 of which five are defective. Thus

$$P(B \mid A) = 4/299 \quad \text{and} \quad P(B \mid \bar{A}) = 5/299.$$

Thus we have, respectively,

$$P(A \cap B) = P(A)\,P(B \mid A) = (5/300)(4/299)$$

$$P(\bar{A} \cap B) = P(\bar{A})\,P(B \mid \bar{A}) = (295/300)(5/299).$$

Now since $A \cap B$ and $\bar{A} \cap B$ are mutually exclusive and together they make up the event $B$ we have

$$P(B) = P(A \cap B + \bar{A} \cap B) = P(A \cap B) + P(\bar{A} \cap B)$$
$$= \frac{5 \cdot 4 + 295 \cdot 5}{300 \cdot 299} = \frac{5 \cdot 299}{300 \cdot 299} = \frac{1}{60}.$$

We note that this is the same as $P(A)$. This is not surprising because it is the probability of a defective watch on the second draw when we are ignorant of what happened on the first draw, and therefore should be the same as on the first draw, even in this nonindependent case.

We finally remark that we may easily extend (4.26) to more than two trials by remembering that $A \cap B$ is an event. Then

$$P[A \cap B \cap C] = P[(A \cap B) \cap C] = P(A \cap B)\,P(C \mid A \cap B)$$
$$= P(A)\,P(B \mid A)\,P(C \mid A \cap B). \qquad (4.27)$$

For example, in the preceding case of 300 watches, of which five are defective, the probability that a random sample of three contains no defective watch is given by

$$P(3 \text{ good}) = P(G)\,P(G \mid G)\,P(G \mid G, G) = \frac{295}{300}\frac{294}{299}\frac{293}{298}.$$

Also the probability of drawing GDG *in this order* is

$$P(GDG) = P(G)\,P(D \mid G)\,P(G \mid G, D) = \frac{295}{300}\frac{5}{299}\frac{294}{298}.$$

These are *dependent* probabilities, or *finite* sampling. If drawing is with replacement so that on each draw the population is always the same, we have *independent* probabilities and *infinite* sampling. The probabilities just found in this opposite case become

$$(295/300)^3 \quad \text{and} \quad (295/300)^2 \, (5/300).$$

The two sets of probabilities are not radically different because the population size (300) was substantial and the sample small (3).

As we see in Chapter 5, the foregoing cases of drawing without replacement and with replacement can be handled, respectively, by the so-called hypergeometric and binomial distributions (Sections 5.5 and 5.3). See also Example 6 of Section 4.7.2.

## 4.7. DISCRETE PROBABILITIES

Many probabilities can be calculated by using the laws and formulas presented in the preceding sections. A basic formula for probabilities is (4.19), $P(A) = N_A/N$. It is used when we can assume equal probabilities for the $N$ points in the sample space $W$, and must count how many points are in $W$, and of these, how many yield event $A$. In applying (4.19) we often need counting techniques, since the number of points in the sample space of an event can easily become gigantic. Thus we find it desirable to present a few such techniques.

### 4.7.1. Permutations and Combinations.

Most readers have met permutations and combinations in previous studies. For the benefit of those who have not, we give a brief presentation.

Both permutations and combinations in simplest form are concerned with objects which are distinguishable, for example, the 26 letters in the English alphabet, 100 people, 500 bicycles with serial numbers, or a deck of 52 playing cards. For permutations we ask how many distinct ordered samples of, say, $r$ may be formed from $n$. Thus for the letters in the alphabet IN and NI are two distinct permutations. Likewise AEG, AGE, EAG, EGA, GEA, and GAE are six distinct permutations. The first case IN, NI are two of the possible permutations of 26 things two at a time, that is, by definition

$$P(26, 2).*$$

* We realize that this may occasionally be confused with a probability. But the permutation notation will always have just two integers in the parentheses. One alternative is to use $P_r(\ )$ for probability, but this is a long-term debt to pay for a very slight improvement in clarity.

Now how many such are possible? There are 26 possible choices for the first letter. Having made the choice of the first letter, there remain 25 letters from which to choose the second letter (duplicates not being permitted). Thus the number is $26 \cdot 25$. Or

$$P(26, 2) = 26 \cdot 25 = 650.$$

Similarly for $P(26, 3)$ we first choose the first two letters in some one of the 650 possible ways. Then from the remaining 24 letters we must choose one for the third letter. Hence, since for each one of the 650 choices there are always 24 choices for the third, we have

$$P(26, 3) = 26 \cdot 25 \cdot 24 = 15,600.$$

In like manner we have in general

$$P(n, r) = n(n - 1)(n - 2) \cdots (n - r + 1), \tag{4.28}$$

there being $r$ factors, so that $n - r + 1$ is the lowest one. (We had already removed $r - 1$ choices from $n$ by the first $r - 1$ in the permutation.) An important special case is

$$P(n, n) = n(n - 1) \cdots 2 \cdot 1 = n!. \tag{4.29}$$

A convenient form of (4.28) using factorials is

$$P(n, r) = n!/(n - r)!, \tag{4.30}$$

found from (4.28) by completing the factorial by multiplying by $(n - r)!$ and then compensating by placing $(n - r)!$ in the denominator. Formula (4.30) can be helpful in connection with a table of $\log(n!)$, such as Table XIV in the Appendix.

Next consider the number of possible "combinations" from a collection of distinguishable objects. The symbol is $C(n, r)$.* A combination is an unordered sample, a collection, a bundle, or a committee perhaps. The order in which they are chosen for a combination makes no difference. Thus I and N can make up just one "combination" of two letters, and A, E, and G just one "combination" of three letters, whereas these sets led, respectively, to two and six distinct "permutations."

Now we may consider the process of forming any one example of $P(n, r)$ as that of (a) choosing a combination or collection of $r$ objects from $n$ (that is, without regard to order), and then (b) having chosen

---

* Another commonly used symbol for $C(n, r)$ is $\binom{n}{r}$, while an older symbol is $C_r^n$. We have chosen to use the one most easily printed.

this collection, to arrange them in an order. Looked at this way we have

$$P(n, r) = C(n, r) P(r, r).$$

We already have the two permutation expressions, but $C(n, r)$ is sought. Solving we have

$$C(n, r) = P(n, r)/P(r, r).$$

Using (4.28)–(4.30) we have

$$C(n, r) = \frac{n(n - 1) \cdots (n - r + 1)}{r!} = \frac{n!}{r!(n - r)!} = \binom{n}{r}. \qquad (4.31)$$

The last formula is especially useful in connection with tables of $\log(n!)$ such as Table XIV. There is also a useful table of $C(n, r)$ [2].

In our two alphabet examples we see that for each case of $C(26, 2)$ combinations there are two distinct permutations of $P(26, 2)$, that is, $P(26, 2) = 2C(26, 2)$. So $C(26, 2) = 325$. Likewise the number of distinct cases of $P(26, 3)$ is six times the number of $C(26, 3)$, since for each of the latter there are six distinct permutations. Here $6 = 3!$, hence $C(26, 3) = 15{,}600/6 = 2600$.

We note that there is a symmetry in (4.29), namely that $C(n, r) = C(n, n - r)$. The latter form is often easier if $r$ is above $n/2$. Thus $C(500, 498) = C(500, 2) = 500 \cdot 499/2$.

We now make explicit a formula which we used in deriving (4.29). It is the multiplicative law

$$N(A \cap B) = N(A) N(B), \qquad (4.32)$$

where $N(A)$ is the number of distinct ways event $A$ can be performed, $N(B)$ is the number of distinct ways event $B$ can occur, for each of $N(A)$, and $N(A \cap B)$ is the number of ways both can occur. Analogously to probability law (4.26) we also have

$$N(A \cap B) = N(A) N(B \mid A), \qquad (4.33)$$

which we used in deriving (4.28), wherein when event $A$ occurred (first letter drawn) the number of ways for $B$ is somewhat restricted by $A$ having occurred.

The generality of permutations and combinations can only be grasped by usage. For example, suppose we have $r$ objects of one kind indistinguishable, and $s$ objects of another kind distinguishable from the first but otherwise indistinguishable. Say, we have $r$ $A$'s and $s$ $B$'s. How many different permutations can be formed? For example, *AABAB*,

*AAABB*, and so on, are two with $r = 3, s = 2$. To solve this *permutation* problem we use a *combination*. Consider a row of $r + s$ boxes or slots. In how many of these may we place the $r$ *A*'s? This is $C(r + s, r)$, which is thus the number of distinct permutations of $r$ objects of one kind and $s$ of another.

Another example is the binomial expansion for

$$(x + y)^n = (x + y)(x + y) \cdots (x + y).$$

For the term containing $x^r y^{n-r}$ we have as coefficient the number of ways we can choose precisely $r$ of the parentheses to contribute $x$'s (the others contributing the $n - r$ $y$'s). Thus the term is

$$C(n, r) \, x^r y^{n-r}.$$

Finally we note that to make formulas such as (4.28) and (4.29) more general we define, as is customary

$$0! = 1 \quad \text{and} \quad m! = 0 \quad \text{if} \quad m \text{ is a negative integer.} \quad (4.34)$$

## 4.7.2. Discrete Probability Examples.

Let us use the machinery we have been building up to solve some problems.

**Example 1.** In choosing two groups of mice one for an experimental treatment $E$ and one for the control treatment $C$, 12 mice are to be used. Presumably the livelier mice will be caught last, as a random choice is made. A coin is flipped and if heads, the first mouse will be used for $E$, or if tails for $C$. The second mouse is then to be given the opposite treatment. Similarly, for the next two mice, and so forth. What is the probability of one of the treatments receiving the first mouse in every pair?

The sample space consists of $2 \cdot 2 \cdot \cdots \cdot 2 = 2^6$ ways in which the coin may fall in its six flips, by (4.32), since the flips are presumed independent. Let us also assume a probability of .5 for a head each time. We may make continued use of (4.22) to find

$$P(\text{event}) = P(HH \cdots H) + P(TT \cdots T) = (1/2)^6 + (1/2)^6 = 1/32.$$

Thus such outcomes, which do not well equalize opportunities, will nevertheless occur one time in 32 on the average.

The reader might have thought of a quite different sample space of C(12, 6) choices of the six mice to be given the experimental treatment. This sample space has $(12 \cdot 11 \cdot 10 \cdot 9 \cdot 8 \cdot 7)/6! = 924$, "points." But if the liveliness theory is valid, these points are not equally likely and we cannot readily use this space.

**Example 2.** In a set of eight thermometers, two are biased the same direction by .1°C, the others being virtually unbiased. An experimenter randomly chooses three thermometers. Upon checking them he finds one reads .1° off from the other two. He eliminates the one. What is the probability that he now has the two biased thermometers?

The sample space consists of C(8, 3) points equally likely (assuming the thermometers are somehow distinguishable). Of these there are only six which contain *both* biased thermometers plus one of the six unbiased ones. Hence

$$P(\text{event}) = 6 \div (8 \cdot 7 \cdot 6/3 \cdot 2 \cdot 1) = 3/28.$$

This probability is rather distressingly large.

**Example 3.** In checking the taste of a manufactured food or drink, one method is first to present the taster a sample of standard product and one of current product, telling him which is which. Then a pair, one of each product, in random order, is given to the taster, and he is to tell which he thinks is the current product; then similarly another pair. (Such results are pooled for many tasters.) If the standard and current products do not differ (which is desired), what is the probability of a taster calling both pairs correctly?

For one taster the sample space is of four equally likely points SS, SF, FS, FF. These have equal probabilities 1/4, that is, also (1/2)(1/2). Hence the probability of both calls being correct is 1/4. For five tasters all making correct calls, the probability is by (4.22)

$$P(\text{all correct}) = (1/4)^5 = 1/1024.$$

If such an event should be observed, there would be clear evidence that the products were distinguishable.

**Example 4.** A count of two-letter words in a standard English dictionary yielded 56. If we draw a two-letter permutation from the English alphabet, duplicates being permitted, what is the probability of hitting an actual word?

The possible number is not the ordinary P(26, 2), which is 650, because duplicates are permitted. Instead the sample space consists of

$26 \cdot 26 = 676$ points, since once the first letter is chosen, there are still available 26 choices for the second. Thus

$$P(\text{hit}) = 56/676.$$

If duplicates are not permitted, only 55 are words and

$$P(\text{hit}) = 55/650.$$

**Example 5.** A witness to an accident caused by a hit-and-run driver, noted the license to be 44A56?? and that the car was a green Lincoln. What is the maximum number of cars the police would need to check upon in the case?

This is a simple permutation of 10 objects, duplicates permitted, namely $10 \cdot 10 = 100$.

**Example 6.** What is the probability that in group of 10 people there will be at least one coincidence in birthdays (day and month)? Also, at what number of people does the probability begin to exceed .5?

Let us first define the sample space for a particular person's birthday. There are 366 possible days including February 29. Suppose that we either eliminate this data or add it to February 28. Then the sample space contains 365 points. We next must set a probability for each date. Undoubtedly these are not quite equal, but slightly seasonal. It would take research to find out. Suppose we assume equal probability, for simplicity, which is likely not far off. Then the sample space consists of 365 equally likely points, and for two independently drawn people $365^2$ equally likely points in the product space, and so forth. (How many points in a sample of 10 people?) The easiest way to figure the probability of at least one coincidence is to figure the probability of the complementary event of no coincidence:

$$P(\text{at least one coincidence}) = 1 - P(\text{no coincidence}).$$

For the last we have

$$P(\text{no coincidence in 2 draws}) = \frac{365}{365} \frac{364}{365},$$

because once the first date is chosen, the second date can fall on any one of 364 dates without yielding a coincidence, a conditional probability. Then for 10 drawings of people (birthdates):

$$P(\text{no coincidence in 10}) \quad \frac{365}{365} \cdot \frac{364}{365} \cdot \ldots \cdot \frac{356}{365} = \frac{365!}{365^{10} 355!}.$$

Using Table XIV and logarithms we find

$$P(\text{no coincidence in } 10) = .883$$

$$P(\text{at least one coincidence}) = .117.$$

For the second part we can solve by trial and error and find that the probability of a coincidence first exceeds .5 at $n = 23$, when it is .507. Most people hearing this probability for the first time are quite surprised that it occurs for $n$ so small. For $n = 50$, which is not as much as a seventh of the year's dates, the probability is already .970. This is a good example of how those untrained in probability might misguess a probability.

**Example 7.** The presence of a microorganism is being looked for. A dilute solution of liquid which might contain the organism is spread in six plates and given time to develop under controlled conditions. If the concentration of the organism is such that the probability of one or more colonies growing on a single plate is .05, independently of the results on the other plates, what is the probability of no colonies on any of the plates?

Here the sample space for a single plate consists of two outcomes: $a$, no colony; $b$, one or more colonies. But these are not equally likely, in fact $P(a) = .95$, $P(b) = .05$. The sample space for the six plates consists of $2^6 = 64$ points which are far from equally likely. We in fact want the probability of precisely one of these 64 points, namely $a, a, ..., a$. Since we have been given independence of outcomes

$$P(A \cap A \cap \cdots \cap A) = P(a) P(a) \cdots P(a) = .95^6 = .7351$$

by (4.22). Thus we have over a 73% chance of missing any indication of the presence of the microorganism.

The probabilities for other outcomes, such as precisely one plate of the six showing one or more colonies, can easily be figured by a probability distribution called the "binomial." It is discussed in the next chapter. There is also another distribution, the Poisson, which could apply in another way in this example. It is a model which would give us the probability of precisely one colony on a plate (not one or more), of precisely two colonies, and so on. Thus we would be splitting up event $b$ above. See Chapter 5 for the Poisson distribution.

The same kinds of approaches can be used on counts from a Geiger counter over time intervals, accidents per month, and defects counted over samples of industrial product.

**Example 8.** What is the probability of a couple having a family with five consecutive girls followed by six consecutive boys? The author has a clipping on such a family.

This is an example of the kind of vague question one often is asked by persons untrained in probability. The desired probability could be the answer to any of several questions, with answers varying widely. Three are as follows: Given that there is a family of 11, what is the probability of five girls and six boys? Given that there is a family of five girls and six boys, what is the probability of the former in succession followed by the latter? Or in this second question, what is the probability of just two "runs" of boys and girls?

Let us answer the second question first. The sample space $W$ consists of all the permutations of 5G's and 6B's. This is, by the paragraph following (4.33), $C(11, 5) = 11 \cdot 10 \cdot 9 \cdot 8 \cdot 7/5! = 462$. These points are assumed equally likely. The event of GGGGGBBBBBB is just one such point. Hence the probability is $1/462$. This has assumed no biological reason for dependence. Since there is another permutation BBBBBBGGGGG which fulfils the third question, this probability is $2/462 = 1/231$.

To answer the first question we need to know something about the probability of a boy baby versus a girl baby. This is not far from $1 : 1$, so we assume $P(B) = .5$, and also assume independence.

Then there being $2^{11}$ equally likely points in $W$, we have

$$P(GGGGGBBBBBB) = [P(G)]^5 [P(B)]^6 = .5^{11} = 1/2048.$$

All three of these probabilities are far bigger than the probability of a young married couple just starting now, having such a family! Why?

**Example 9.** In seeking reliability for equipment, the reliability of a device is the probability of its functioning satisfactorily under specified conditions and for a specified time. Such a reliability or probability can be estimated by suitable experimentation. Suppose that a switch is designed to open a circuit and abort a mission if a certain dangerous condition exists. One type of switch has a reliability of .996 in this application. Using one such switch gives a reliability in the circuit of .996. A second type of switch has a reliability of .995. Suppose we place one switch of each type in series in the circuit. Assuming independence of reaction of the two switches, what is the reliability of the series system now, when a dangerous condition exists?

Suppose that a dangerous condition exists. Then the circuit will open and abort the mission if either switch or both open. Let $A$ be the event of the switch of the first kind opening, $B$ of the second kind opening.

Then three events give an open circuit, that is, success, if an abort condition exists: $A \cap B$, $A \cap \bar{B}$, $\bar{A} \cap B$. Since independence of action is assumed:

$$
\begin{aligned}
\text{reliability} &= P(\text{open circuit}) = P(A \cap B) + P(A \cap \bar{B}) + P(\bar{A} \cap B) \\
&= P(A)\,P(B) + P(A)\,P(\bar{B}) + P(\bar{A})\,P(B) \\
&= .996(.995) + .996(.005) + .004(.995) \\
&= .999980.
\end{aligned}
$$

The reader has probably spotted an easier method already:

$$
\begin{aligned}
\text{reliability} &= 1 - P(\text{neither opens}) = 1 - P(\bar{A} \cap \bar{B}) = 1 - P(\bar{A})\,P(\bar{B}) \\
&= 1 - .004(.005) \\
&= .999980.
\end{aligned}
$$

Note how much more reliable this combined circuit is than a circuit with just one switch would be. This improvement has been achieved by what is called "redundancy." If one switch works, the other was redundant. The improvement came under the assumption of independence of action. What are the limits of the circuit reliability under nonindependence?

**Example 10.** Suppose there are 40 operators in a department, of which eight are men. A sample of five is to be chosen at random without replacement, for a production experiment. What is the probability of the selected sample containing three men?

In the first place, of what does the sample space consist? It is a compound space, a product space of the five drawings. The space for the first drawing consists of the 40 operators. Once the first drawing has been made, the space for the second drawing consists of the 39 people not yet chosen, all of whom are now equally likely to be chosen. Thus the space for the first two drawings has $40 \cdot 39$ points; these points are ordered points and all equally likely. Similarly $W$ consists of $P(40, 5)$ distinct ordered points equally likely. How many of these will provide the event of exactly three men and two women? One such is MMMWW. The number of ways for this is $8 \cdot 7 \cdot 6 \cdot 32 \cdot 31$. But this is only one order in which the event can occur. How many three-man–two-woman orderings are there? By the paragraph following (4.33), we see that there are $C(5, 3) = 10$ orders. Thus using (4.32)

$$
P(3 \text{ men in sample}) = \frac{(8 \cdot 7 \cdot 6 \cdot 32 \cdot 31)10}{40 \cdot 39 \cdot 38 \cdot 37 \cdot 36} = .0422.
$$

The foregoing shows well the sample space and the equally likely character of the points. But an easier and much more often used approach is that of combinations rather than permutations (as in the approach above).

How many combinations of the 40 people taken five at a time are there? $C(40, 5)$. Now of the eight men, in how many ways may we choose three of them? $C(8, 3)$. Finally, in how many ways may we choose two of the 32 women? $C(32, 2)$. The event of three men and two women being chosen in the sample of five (in any order) is the product of the last two combinations. Thus we have

$$P(3 \text{ men} \cap 2 \text{ women}) = \frac{C(8, 3)\, C(32, 2)}{C(40, 5)}$$

$$= \frac{8 \cdot 7 \cdot 6 \cdot 32 \cdot 31 \cdot 5!}{1 \cdot 2 \cdot 3 \cdot 1 \cdot 2 \cdot 40 \cdot 39 \cdot 38 \cdot 37 \cdot 36},$$

which is just the same as before. If stated entirely in terms of factorials, we can use logarithms of factorials from Table XIV.

The solution to problems such as the foregoing example can be handled by the "hypergeometric distribution." See the next chapter.

**Example 11.** In the sampling of a shipment of rivets, 10 are tested to destruction and their 10 strengths determined. Specifications call for $\bar{y}$ for the strengths to exceed 250 lb and also for $s$ to be 12 lb or less. A supplier's process is such that the probability of his meeting the first requirement is .93 and of his meeting the second is .90. What is the probability of a lot of his production being passed?

Let $A$ be the probability of $\bar{y}$ passing, $B$ of $s$ passing. Then we need the event $A \cap B$. If we assume $A$, $B$ independent, we have

$$P(A \cap B) = P(A)\, P(B) = .93(.90) = .837.$$

As we see in Chapter 6, the independence of $\bar{y}$ and $s$ says something about the distribution of rivet strength, namely, that it is "normal."

**Example 12.** Given a draw of four cards without replacement from a standard 52 card deck, what are the probabilities of the following kinds of draws or "hands":

$A$, four cards all of the same rank, for example, four jacks;
$B$, three of one rank, one of another;
$C$, two of one rank, and one from each of two other ranks;
$D$, two of one rank and two of some other rank;
$E$, all four cards of the same suit;

F, four cards in consecutive order, for example, A234, 2345, up to
JQKA;

G, a hand meeting both E and F characteristics.

First we need the number of possible hands. While we could use the
permutation P(52, 4), it is easier to use the combination C(52, 4) since
the order of obtaining the cards is immaterial. By (4.31)

$$N = C(52, 4) = 52 \cdot 51 \cdot 50 \cdot 49/4 \cdot 3 \cdot 2 \cdot 1 = 270,725,$$

which is the number of points in our sample space. It is suggested that
the reader try to work out each probability before looking.

$$P(A) = 13/N,$$

since there are 13 ranks and we must take all four cards of the chosen
rank. Next we have using (4.33)

$$P(B) = \frac{C(13, 1)\ C(4, 3)\ C(48, 1)}{N} = \frac{2496}{270,725}.$$

The first combination gives the choices of rank for the three cards, and
the second gives the ways to choose the three suits in this rank. Then
C(48, 1) is the choice for the odd card. Note that if we used C(49, 1),
we would thereby permit the other card to be of the same rank as that
of the three.

$$P(C) = \frac{C(13, 1)\ C(4, 2)\ C(12, 2)\ C(4, 1)\ C(4, 1)}{N} = \frac{82,368}{270,725}.$$

The combinations give, respectively, the choice of rank for the pair,
choice of suits for the pair, choice of two new ranks for the singles, and
choices of their suits. This uses (4.33). Similarly

$$P(D) = \frac{C(13, 2)\ C(4, 2)\ C(4, 2)}{N} = \frac{2808}{270,725}$$

$$P(E) = \frac{C(4, 1)\ C(13, 4)}{N} = \frac{2860}{270,725}.$$

The first combination gives the choice of suit, and the second gives the
choice of cards in that suit. For event F we first observe that the lowest
ranked card in a run can be A, or 2, or 3, ..., or J, namely 11 possibilities.

Then for each chosen run of four ranks the suit for the lowest card can be any one of four, for the second similarly, and so on. Thus

$$P(F) = \frac{11(4)^4}{N} = \frac{2816}{270, 725} .$$

Finally for $G$ we have 11 choices for the lowest ranked card and four choices of the one suit. Thus

$$P(G) = \frac{11(4)}{N} = \frac{44}{270, 725} .$$

## 4.8. PROBABILITIES ON CONTINUOUS SPACES

We now come to the matter of events and probabilities on continuous sample spaces. The subject matter of Sections 4.3, 4.4, and 4.6 goes through the same way, for example, combinations of events and the defining axioms for a probability function. But now we have a complication: The number of possible outcomes is uncountably infinite. Because of this we must necessarily define the probability for each single point to be zero (unless we have a mixed variables–attributes sample space, which is very rare). Thus if $y$ is any single outcome (point) in $W$, $P(y) = 0$. For example, the probability for a diameter $y$ in inches to be 2.3777... is zero. But the probability for a diameter $y$ to lie in any particular *interval* can be positive. Thus if $A$ is the event of 2.377 in. $< y <$ 2.378 in., we might have $P(A) = .12$, say. In Chapter 6 we shall see how we can define distribution laws so that the probability that the outcome $y$ lies in any given *interval* $(a, b)$ may be found. When the sample space is continuous containing an uncountably infinite number of outcomes, such as measurements, we define events to be combinations of intervals, or in a plane, of rectangles, and so on. Each such "event" then contains an uncountable infinity of points. But if the "event" contains only an integral number of points, its probability is zero.

An example of a simple distribution function over a continuous sample space is that of an angle. The position at which an index point on the rotor of a motor comes to rest upon stopping is a measurable angle between 0 and 360. Thus the sample space $W$ is the set of $Y$'s such that $0° < Y \leqslant 360°$. In the absence of contrary evidence it is reasonable to assume that the probability of the stopping angle $Y$ lying in any interval $(a, b)$ where $0° < a \leqslant b \leqslant 360°$ is proportional to $b - a$.

Since $P(Y \subset W) = 1$, the proportionality constant is $1/360°$. Thus

$$P(a < Y \leqslant b) = (b - a)/360°.$$

The "distribution function" $F(y)$ for any $y$ in $W$ is the probability that the random variable $Y$ takes a value not greater than $y$, that is, $Y \leqslant y$:

$$F(y) = P(Y \leqslant y). \tag{4.35}$$

In our example

$$F(y) = y/360° \qquad \text{for} \quad 0° < y \leqslant 360°.$$

In general it is quite obvious that

$$P(y_1 < Y \leqslant y_2) = F(y_2) - F(y_1) \tag{4.36}$$

since we have for the sets of $Y$'s, $\{y_1 < Y \leqslant y_2\} \cup (Y \leqslant y_1) = \{Y \leqslant y_2\}$, the first two sets being mutually exclusive. Hence the sum of the probabilities of the sets on the left gives that for the right side, or $P(y_1 < Y \leqslant y_2) + F(y_1) = F(y_2)$, which gives (4.36). We note that since $P(Y = y_1)$ and $P(Y = y_2)$ are both zero, we have

$$P(y_1 < Y \leqslant y_2) = P(y_1 \leqslant Y \leqslant y_2)$$
$$= P(y_1 < Y < y_2) = P(y_1 \leqslant Y < y_2).$$

**Example 1.** Suppose that for a certain date in the spring at a weather station the highest recorded maximum is $87°$F. This means that the true maximum is somewhere between 86.5 and $87.5°$. In the present spring, say, the maximum at this station and on the date, being now measured to the nearest $.1°$, is $87.1°$. How valid is the claim that the record was broken? Or what is the probability that it was?

To solve the problem we need to define the distribution somehow. The simplest assumption is that the true maximum temperature $T$ is uniformly distributed between the extremes, just as were the angles, above. This being the case, then the two true maximum temperatures $T_1$ and $T_2$ could yield a point anywhere within the rectangle shown in Fig. 4.5. Assuming independence of $T_1$ and $T_2$, the area of $W$ in the figure carries uniform probability density; that is, for rectangles of equal area and lying wholly within $W$, the probabilities of points $(T_1, T_2)$ within the rectangles are equal (equal area: equal probability). Now the line $T_1 = T_2$ is shown in Fig. 4.5. To the left of this line $T_2 > T_1$. The area to the left within $W$ is $(87.1° - 86.5°)(.1°) = .06$ deg$^2$. Dividing by the area of $W = .1$ deg$^2$ we have $P(T_2 > T_1) = .6$, which is the probability that the record was broken.

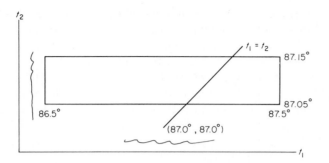

**Fig. 4.5.** Sample space for two maximum temperature readings 87 and 87.1°F. Line shown for $T_1 = T_2$.

In regard to the uniformity of distributions between limits, this assumption may not be too realistic, because when we are dealing with the very highest observation among many, the probability is perhaps greater that the true value lies in the lower rather than the upper half of the interval.

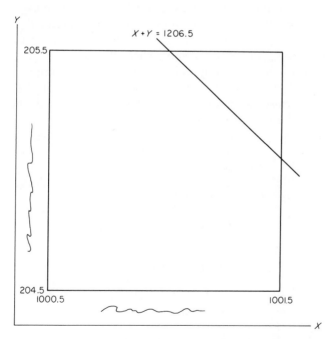

**Fig. 4.6.** Sample space for bottle weight and contents measured as 205 and 1001 g, respectively.

**Example 2.** Two measured weights, each to the nearest gram are as follows: bottle contents, 1001 g, bottle, 205 g. What is the probability that the true combined weight is greater than 1206.5 g?

This problem is quite similar to Example 1. Let $X$ be the true bottle content, $Y$ the true bottle weight. Then the point $(X, Y)$ lies within the rectangle $W$ of Fig. 4.6. Again we assume independence of $X$ and $Y$, and uniformity of distribution between the extremes. This again leads to equal area: equal probability. The area for $W$ is 1 g². The region where $X + Y > 1206.5$ g is the upper right corner of area $.5g(.5 g)/2 = .125$ g². Thus $P(X + Y > 1206.5 \text{ g}) = .125$.

This problem, just as that in Example 1, could have been done by double integration. In fact, but for the simplicity of the uniformity of distribution and independence, we would in general use integration.

### 4.9. APPLIED BAYES' PROBABILITIES—POSTERIOR PROBABILITIES

This approach is named after the Reverend Thomas Bayes, whose famous controversial paper was published, posthumously, in 1763. Mathematically the paper was sound, but he contended that in the absence of any knowledge about the probability of one of two possible events occurring, one might assume that all values for the probability from 0 to 1 are equally likely. This is an *a priori* assumption. Then in the light of observed results, one can modify this equal likelihood into some *a posteriori* distribution of probability. Granted the *a priori* distribution, the *a posteriori* one follows.

For our purpose we assume a set of mutually exclusive conditions or "causes", $A_1, ..., A_k$, for which we know the *a priori* probabilities, $P(A_1), ..., P(A_k)$. Also assume $\sum_{i=1}^{k} P(A_i) = 1$. Then we observe some outcome $B$ which can be associated with the $A_i$'s. Specifically, we know or can find $P(B \mid A_1), ..., P(B \mid A_k)$. Then by (4.26)

$$P(A_i \cap B) = P(B)\,P(A_i \mid B). \tag{4.37}$$

From this

$$P(A_i \mid B) = P(A_i \cap B)/P(B), \tag{4.38}$$

which is the *a posteriori* probability of $A_i$, given that $B$ did occur. But now we have the "event equation"

$$B = B \cap W = B \cap (A_1 \cup \cdots \cup A_k)$$
$$= (B \cap A_1) \cup (B \cap A_2) \cup \cdots \cup (B \cap A_k) \tag{4.39}$$

since the $A_i$'s are mutually exclusive, and together make up the whole space $W$. Thus

$$P(B) = \sum_{i=1}^{k} P(B \cap A_i). \tag{4.40}$$

But using (4.26) again

$$P(B \cap A_i) = P(A_i) P(B \mid A_i), \tag{4.41}$$

hence

$$P(B) = \sum_{i=1}^{k} P(A_i) P(B \mid A_i). \tag{4.42}$$

Substituting (4.42) into the denominator of (4.38), and (4.41) into the numerator we have

$$P(A_i \mid B) = \frac{P(A_i) P(B \mid A_i)}{\sum_{j=1}^{k} P(A_j) P(B \mid A_j)}. \tag{4.43}$$

By repeated use of (4.43) we may find all the *a posteriori* probabilities for the $A_i$ "causes" or conditions, given that we did observe event $B$.

The reader will certainly welcome an example.

Suppose that it is known from past records that a supplier ships lots with percentages of defective pieces .02, .10, and .90. Let these be called events $A_1$, $A_2$, $A_3$, with $P(A_1) = .80$, $P(A_2) = .15$, $P(A_3) = .05$. We take a random sample of just three pieces, and let us suppose that the lot size is large enough that the probabilities on the three draws are virtually constant. Then suppose further that we observe all three pieces to be good ones. Call this event $B$. Then by (4.22) (justified by "large" lot size)

$$P(B \mid A_1) = .98^3, \qquad P(B \mid A_2) = .90^3, \qquad P(B \mid A_3) = .10^3.$$

Now for $A_1$, using (4.43)

$$P(A_1 \mid B) = \frac{(.80)(.98^3)}{(.80)(.98^3) + (.15)(.90^3) + (.05)(.10^3)} = \frac{.75295}{.86236} = .8731.$$

Similarly $P(A_2 \mid B) = .1268$, $P(A_3 \mid B) = .0001$. Thus with just the little sample of three we have increased the probability of $A_1$ to .873, and for practical purposes eliminated $A_3$.

Now suppose the sample contained all three defectives, say event $C$. Then

$$P(C \mid A_1) = .02^3, \qquad P(C \mid A_2) = .10^3, \qquad P(C \mid A_3) = .90^3.$$

Now

$$P(A_1 \mid C) = \frac{(.80)(.02^3)}{(.80)(.02^3) + (.15)(.10^3) + (.05)(.90^3)} = .0002,$$

while $P(A_2 \mid C) = .0041$, $P(A_3 \mid C) = .9957$. Here we have nearly a sure bet that it is a sample from a 90% defective lot.

The main stumbling block in applying Bayes' rule (4.43) is in knowing the *a priori* probabilities $P(A_i)$. But when known one can make applications. In industry, past performance and records may supply these.

## 4.10. INTERPRETATION OF A PROBABILITY

We may well ask at this point as to what is the meaning of some calculated probability? As we see in Chapters 8 and 10 there are two kinds of problems involved. Suppose that there is some calculated or objective probability available such as those found in the examples of Section 4.7.2. This is a population or "true" characteristic which we designate by a Greek letter, here $\phi$. Then suppose

$$\phi = .20,$$

which is the calculated probability of an event $A$, that is,

$$P(A) = \phi_A = .20.$$

Now suppose we try the random experiment which leads either to $A$ or $\bar{A}$. Then in $n$ trials we observe $A$ to occur $n_A$ times. Our observed "success ratio" for $A$ is then, analogous to (4.19):

$$p_A = n_A/n. \tag{4.44}$$

How does $p_A$ compare with $\phi_A$? We would expect that, if $n$ is large enough, $p_A$ would be close to $\phi_A$. Moreover, as $n$ is increased we would anticipate that our success ratio would approach $p_A$ more closely. These anticipations are in fact likely, but not sure to occur. All we *can* say is that if we pick any desired interval around $\phi_A$, for example, .18 to .22, and any desired (large) probability, say, .95, then by taking $n$ sufficiently large, we can have a probability of at least .95 of $p_A = n_A/n$ lying within .18 to .22. The required minimum sample size $n$ depends on (1) $\phi_A$, (2) the width of the interval, and (3) the desired probability or confidence. This follows from the "law of large numbers." But there can be *no absolute guarantee* that $p_A$ does lie in the interval. In fact it is mathematically *possible* for $p_A$ to remain at 0 or 1!

The above is one type of problem, namely when we are given $\phi$, somehow, and wonder what to expect of $p_A$. Conversely, however, is the very common problem in which we do not know $\phi_A$, and only have an observed success ratio of $p_A = n_A/n$. We naturally regard this $p_A$ as some estimate of $\phi_A$. Moreover, the larger $n$ is, the more confidence we have in such an estimate. Such thoughts are made more precise in Chapter 10 by finding from any given $p_A = n_A/n$, a set of limits around $p_A$ such that we can have some specified confidence of $\phi_A$ lying within. This is called "interval estimation."

## 4.11. RANDOM VARIABLES

Statistical reasoning and analysis is basically concerned with random variables. If for every individual outcome in $W$ we have a unique number defined, then such a number is called a random variable over $W$. For example, if $W$ consists of a collection of people, then the height of each is a random variable. Chapters 5 and 6 give probability models of random variables defined over discrete and continuous sample spaces, respectively. Such random variables are denoted by capital letters $Y$ or $X$.

## 4.12. SUMMARY

We have been concerned in this chapter with discrete and continuous sample spaces of outcomes from a random experiment. It is basic in applications to have a clear definition of what sample space we are concerned with. Collections of outcomes from a sample space $W$ are called events, such as $A$, $B$. Then various combinations of events such as $A \cup B$ were given and illustrated with Venn diagrams. The properties of a probability function which will provide the probabilities of various events in a sample space $W$ were next defined as

$$P(A) \geqslant 0, \qquad P(W) = 1,$$

$$P(A_1 \cup A_2 \cup \cdots) = P(A_1) + P(A_2) + \cdots, \qquad A_i\text{'s} \quad \text{mutually exclusive.}$$

These defining axioms apply to both types of sample spaces.

In the case of discrete sample spaces we often use counting techniques, such as permutations and combinations, so that we may use $P(A) = N_A/N$, especially when all points in a finite sample space are equally likely. For continuous sample spaces we used only simple cases which can be handled geometrically, deferring more general treatment till later on.

We have tried to give examples of practical importance, which lead toward statistical applications, rather than emphasizing those in dice, cards, and gambling. Many of these latter are interesting and instructive, but do not seem to the author to be especially useful in approaching statistical problems. Nevertheless, a few of the problems at the end of this chapter are of this type.

REFERENCES

1. W. A. Whitworth, "Choice and Chance." G. E. Stechert, New York, 1934.
2. T. C. Fry, "Probability and Its Engineering Uses." Van Nostrand-Reinhold, Princeton, New Jersey, 1928.

## Problems

4.1.   Evaluate C(100, 3), C(100, 98), P(26, 4), C(11, 11), and give an example of each in real life.

4.2.   A bus trip is to be taken from your home town to Chicago, there being five routes each way. In how many distinct ways can a round trip be made? How many if the return trip is by a different route?

4.3.   In how many distinct ways can we choose three catalysts out of 12 possibilities, to be experimented with, in a chemical reaction experiment? If a random choice is made, what is the probability that a specified two catalysts are among the three chosen?

4.4.   There are 20 different colors decided upon by a manufacturer of automobiles. If only one is chosen for an auto (both body and trim), how many different colors may the car be painted? If two are used, one for body and one for trim, how many ways may the car be painted? (Is this a permutation or a combination?) Suppose all of the foregoing are equally likely (unsound). You order a car and would favor it to be colored from among your four favorite colors. What is the probability of your wish being fulfilled?

4.5.   To make up a concrete mix by a given ratio, there are available three sources of sand, four of gravel, and four brands of cement. How many possible mixes are there? Suppose each possibility is equally likely, what is the probability of any one mix being chosen? Of one made up of a specified sand, one of two specified gravels, and one of three specified brands of cement?

4.6.   There are three disks on a lock box, which must read right for it to open, for example, 309. The owner has forgotten the combination. How many trials are needed to ensure finding the combination? If he knows the first digit is 8, how many? If he knows the first and last digits were the same, how many?

4.7. Not permitting any duplicates, and not permitting an initial zero, how many six-digit license numbers are possible? What is the probability that one of these chosen at random has no zero? No zero nor one?

4.8. How many possible orders are there for the four letters o, p, s, t? Choosing one at random, what is the probability of its being an actual English word? Using three d's and three e's each time, how many possible orders are there? What is the probability that one chosen at random is an English word?

4.9. Prove the identity $C(n, r) + C(n, r + 1) = C(n + 1, r + 1)$, by using factorials on the left side and combining. This is a way of using lower order combinations to calculate recursively the next order combinations, due to Pascal. A suitable table is called Pascal's triangle.

4.10. Show that $\sum_{i=0}^{n} C(n, i) = 2^n$, by using the expansion of $(1 + 1)^n$ in terms of combination symbols.

4.11. For any event $A$ in sample space $W$, what are each of the following: $A \cap A$, $A \cup A$, $A \cap W$, $A \cup W$, $A \cap \varnothing$, and $A \cup \varnothing$?

4.12. Are the parentheses necessary in the following: $(A \cup B) \cup C$, $(A \cap B) \cap C$?

4.13. Show by use of Venn diagrams $(A \cup B) \cap C$ and $(A \cap B) \cup C$. Find an alternate expression for each.

4.14. Analyze whether or not the following events are identical in a series of trials of an experiment (each time experiment is tried, it leads to success or failure, then the experiment is tried again): $A$—third failure occurs before or on the hundredth trial, $B$—three or more failures within first 100 trials.

4.15. A lot consists of 10 circuit breakers of which three are defective. A random sample of two is chosen without replacement and tested. Find the probability of the sample containing two good ones; one good; both defective. What sample space did you use and how many points did it contain? This is a simple example of the hypergeometric distribution, discussed in Section 5.5.

4.16. Do Problem 4.15 when drawing is done with replacement.

This is a simple example of the binomial distribution, discussed in Section 5.3.

**4.17.** A lot of 20 meters contains two defective meters. A random sample of three is drawn with replacement and tested. Find the probability of the sample containing no defectives, one defective, two, and three defectives. What sample space did you use?

**4.18.** Do Problem 4.17 when drawing without replacement.

**4.19.** If $A_i$ is the event of a sample of 100 containing exactly $i$ or fewer defectives, what is $A_3 \cup A_5$? $A_3 \cap A_5$? Suppose you know the probability of each $A_i$, say, $P(A_i)$. How would you find the probability of the sample containing two to four defectives inclusive? The probability of five or more defectives?

**4.20.** Let $A_y$ be the event that a temperature control switch operates at a temperature of $y°$C or less. Suppose that the probability of $A_y$, $P(A_y)$, is called $F(y)$. Find in terms of $F$, the probability that the switch operates at a temperature of over 50°; at a temperature over 40°, but less than or equal to 50°? Find $P(Y \leqslant 40 \cup Y > 50)$.

**4.21.** A committee of three decides to choose a chairman by each flipping a coin. If there is one head and two tails showing, the one flipping the head is chairman, and similarly for one tail and two heads. What is the probability of a decision on the first flipping? If all three coins are the same, a second flipping is made, and so on. What is the probability of a decision not being made on the first or the second flipping?

**4.22.** Similar to Problem 4.21, but there are four on the committee.

**4.23.** A lot contains 150 toasters. Two common types of defects are $A$ open circuit, $B$ timer off. (a) What four kinds of toasters may be in the lot? (b) Give an example in this problem where $A$ and $B$ are independent events; that is, find four cell frequencies giving independence for the events.

**4.24.** In a lot of 14,400 glass bottles there are 20 with "stones" (hard spots) and no "seeds" (air bubbles), 15 with seeds and no stones, and 12 with both defects. Calling these numbers respectively $n(A \cap \bar{B})$, $n(\bar{A} \cap B)$, and $n(A \cap B)$ find for a single random draw of a bottle $P(A)$, $P(B)$, $P(A \mid B)$, $P(B \mid A)$. Are $A$ and $B$ independent?

**4.25.** Suppose that a golfer is of such skill that on a short hole he has a .6 chance of his ball coming to rest within 50 ft of the center of the hole, and that within this area there is an equal probability of the bottom of the ball coming to rest in any square inch. Further suppose that wherever the hole is cut there are 24 in.² where, if the bottom of the ball would come to rest without interference from the hole, the ball would instead have gone into the hole. Find the probability of this player making a hole-in-one, that is, having the ball go into the hole.

4.26.  Two safety devices are placed in series in a circuit such that if a certain dangerous condition exists each is supposed to open the circuit and thereby give an alarm. Thus if either or both operate under this condition, then the circuit is opened. Suppose that the two devices act independently, and have .99 and .97 probabilities of opening under the dangerous condition. Find the reliability of the system (probability of circuit being opened) under the dangerous condition. Was the inclusion of the second device really helpful in improving the reliability?

4.27.  If a circuit is to close and provide an alarm if a dangerous condition $C$ occurs, we can improve reliability by placing two or more switches in parallel. Each should close, and if at least one does, the alarm is given. If two such switches are placed in parallel, and they each have .96 reliability (probability of closing under $C$), what is the reliability for the system (probability of circuit closing)? How many switches would you need to use to have a system reliability of at least .9999? Did you need to assume independence?

4.28.  Suppose that for early records on total monthly rainfall, results were only given to the nearest .1 in. In a certain locality the September 1899 total was 8.1 in. A recent September totaled 8.12 in. Make and state any necessary assumptions and find the probability that the recent total actually broke the former record.

4.29.  The inside diameter of a bearing was measured and found to be 4.01 in. It is supposed to be assembled with a shaft whose outside diameter is also measured to be 4.01 in. Assume both are perfectly cylindrical. Making any necessary assumptions, what is the probability of the shaft having a true diameter less than that of the bearing and thus being capable of assembly?

4.30.  Two resistors are connected in series and their measured resistances are 20.2 and 39.9 ohms. Make and state what assumptions you need, and find the probability of their combined true resistances exceeding 60.15 ohms.

4.31.  Three resistors are connected in series so that their resistances are added. Each is measured to the nearest ohm at 45 ohms. Find the probability of the total of the three resistances exceeding 136 ohms. Assume independence and uniformity of distribution between 44.5 and 45.5 ohms.

4.32.   Three dice are thrown, each with six faces, 1, 2, ..., 6. They fall independently. If we regard the dice as distinguishable, in how many distinct ways may they fall? In how many of these will the total of the three faces showing be 5 or less? 6 or less?

4.33.   For two dice such as in Problem 4.32, a total of 7 or 11 wins immediately for the person rolling the dice. What is the probability that the person rolls a total of 7 or 11? A total of 2, 3, or 12 immediately loses. Find the probability of such an event.

For Problems 4.34–4.44 we assume a well-shuffled deck of cards containing four suits, for each of which there are 13 cards: 2, 3, ..., 9, 10, J, Q, K, A. Drawing is done without replacement. Find the probabilities of the following kinds of hands:

4.34.   Three cards, all of which are 3's.

4.35.   Three cards all of which are spades (which is one of the suits).

4.36.   Three cards all of which are of the same rank, that is, all 2's, all 3's, and so on.

4.37.   Three cards all of the same suit (event may be met by any one of the four suits).

4.38.   Five cards, of which all are of the same suit (unspecified). This is a "flush."

4.39.   Five cards running in any one consecutive order, for example, A 2 3 4 5, 2 3 4 5 6, ..., 9 10 J Q K, 10 J Q K A. Each one of these orders is to be included in the event. This is a "straight."

4.40.   Five cards simultaneously meeting both events in Problems 4.38 and 4.39.

4.41.   Five cards, three of which are of any one rank, and two of which are of any other rank. This is a "full house."

4.42.   Five cards, two of which are of one rank, two more of a second rank, and the fifth card of a third rank.

4.43.   State in terms of combination symbols the probability of a 13-card hand containing precisely seven cards of any one unspecified suit.

4.44.   State in terms of combination symbols the probability of a 13-card hand containing all four of the aces.

**4.45.** A lot of 100 drums is to be tested by opening a random sample of three, and testing them. If any of the three show the presence of an impurity, the lot is rejected. The probability that an impure drum reveals its condition through the test is .95. Moreover the probability that a test on a pure drum erroneously indicates the impurity is .01. If the lot contains four impure drums, what is the probability that it will be passed? State needed assumptions.

**4.46.** Suppose that past records indicate that 99% of the time a supplier submits lots of washers with only 1% of them defective, while 1% of the time 95% are defective. A random sample of two from a lot of 1000 washers is taken and inspected. Both are defective. What are the respective *a posteriori* probabilities that the lot is 1% defective and 95% defective?

**4.47.** A supplier sends lots of screws containing 1% of them defective, 90% of the time, 10% defective 9% of the time, and 100% defective (e.g., wrong size) 1% of the time. A random sample of two from a random lot yields zero defectives. What are the *a posteriori* probabilities of the three kinds of lots? Do the same, if the sample contains two defectives.

**4.48.** Three suppliers send in lots with the following percentages of defective nuts: A 1%, B 3%, C 5%. These distributions being known, A supplies 50% of the lots, B 35%, and C 15%. All lots are of 10,000 nuts. A sample of 10 from a lot of an unknown one of the three is taken at random and inspected. The sample contains no defective nuts. What now are the probabilities that the lot was made by each supplier? Is the sample large enough to distinguish accurately?

**4.49.** Fifty rivets are to be used in making 10 assembled parts. There are three soft rivets in the 50. Only if all three are used in the same part, will it prove to be too weak. What is the probability that one of the parts will prove to be too weak? What assumption have you made? (Find probability of first part being weak, others not; then for second and others not, etc., then add probabilities.)

*Chapter 5*

# SOME DISCRETE PROBABILITY DISTRIBUTIONS

## 5.1. THEORETICAL POPULATIONS

In general when we take a series of observations, either counts or measurements, which vary despite our trying to maintain the conditions fixed, there will sooner or later appear some pattern of variation. The numbers do tend to follow some regular pattern or law. This can easily be seen by looking at Tables 2.2 and 2.10–2.28. There it must be clear that as we get more and more numbers, each frequency distribution would become more regular and smoother. Of course if the conditions under which the numbers arise were to change, then anything could happen. But as long as the conditions are controlled, we could in each of these tables say something about how a future *set* of numbers would run, even if we could not predict the exact value of the *next number*.

Again looking at Tables 2.2 and 2.10–2.28, we see that there are differing types of patterns. Some are quite symmetrical about the greatest frequency, the frequencies falling off at the same rate on each side. Others, however, have varying degrees of asymmetry, the frequencies dropping abruptly on one side but tapering out a long way on the other side. Because of this diversity of types of frequency distributions, we have a number of commonly used theoretical distributions or laws. They are mathematical models for idealizing actual laws of variation. Such theoretical laws, although seldom if ever exactly correct for an actual situation, can be extremely useful as approximations to the actual law to which the data are subject. Our statistical theory is built up for theoretical frequency distributions. To the extent to which they are good approximations, we can use the theoretical properties to apply to the actual problems in the physical world, and draw sound conclusions and inferences.

In this chapter we describe several theoretical probability distributions for discrete data. In Chapter 6 we consider distributions for continuous or measurement data. Then in Chapter 7 we are concerned with the behavior of samples from such theoretical distributions. Subsequent chapters are concerned with the many uses to which these laws and properties may be put, that is, to the application of statistical methods.

A word may well be said here as to notations. In this book we shall follow what is apparently the commonest practice, namely, to use Latin letters for statistical measures for a sample, and insofar as possible the corresponding Greek letter for the population counterpart. Accordingly our notations in (3.1), (3.6), (3.9), and (3.11)–(3.17) were Latin letters, being for sample data. Thus in (3.9), $s$ stands for the standard deviation of the sample, while for the population standard deviation we use the corresponding lower case Greek sigma, $\sigma$. As we have seen, the sample mean for the $y$'s is $\bar{y}$. The population mean will be denoted by $\mu$; or $\mu_Y$, if we want to indicate that it is for $Y$'s. Some authors use $m$ for the sample mean, which leads nicely to the use of $\mu$. But we shall use the commoner $\bar{y}$ for the sample mean. Another system of notations uses any desired letter for the sample value, and then for the corresponding population value, a prime superscript is attached to the sample symbol. While such a system does have flexibility, and has many people currently using it, notably those in quality control and standardization organizations, we shall use the much more common Latin–Greek system.

In Sections 5.3–5.8 we describe several discrete probability distributions. It is very important that the student carefully learn the conditions which give rise to each law studied, so that he uses the applicable one in any given case. In each discrete probability law we shall need to find the probability that the count is 0, that the count is 1, that it is 2, and so on. These probabilities will depend on the *type* of distribution or law appropriate, and also on the one or more population *parameters* which particularize it.

## 5.2. DISCRETE PROBABILITY DISTRIBUTIONS IN GENERAL

In the present chapter we are concerned with probability laws or distributions of a random variable $Y$, which can only take on discrete values, commonly integers. Thus we are dealing with a countable sample space $W$, either finite, as in Sections 5.3, 5.5, and 5.6, or infinite, as in Sections 5.4, 5.7, and 5.8. The capital $Y$ is used for the name of the random variable, and a small $y$ for a realization or observed value of $Y$. Also $y$ is used for limits to $Y$.

Any given theoretical probability law for the discrete random variable $Y$ will provide a probability for each value of $Y$. Thus

$$p(y) = \text{probability function of } Y = P(Y = y). \tag{5.1}$$

Then we require

$$p(y) \geqslant 0 \qquad \text{for all } y \text{ in space } W \tag{5.2}$$

$$\sum_{\text{all } y} p(y) = 1. \tag{5.3}$$

We also define the "cumulative" or "distribution function" using a capital P by

$$P(y) = \sum_{Y \leqslant y} p(Y) = P(Y \leqslant y). \tag{5.4}$$

### 5.2.1. Expected Values for Y and Functions of Y.

It is convenient, in studying the characteristics of a random variable which follows a theoretical probability, to use *expected* or theoretical average values of $Y$ or functions of $Y$. These are very similar to calculations with frequency data from a sample. For a sample mean we have

$$\bar{y} = \frac{\sum_{i=1}^{k} f_i y_i}{n} = \sum_{i=1}^{k} \left( \frac{f_i}{n} \right) y_i, \qquad k \text{ midvalues}. \tag{5.5}$$

Note that here $f_i/n$ is the *observed relative frequency* of $y_i$ in the sample, and that $\sum_{i=1}^{k} f_i/n = 1$. With population calculations we have instead of $f_i/n$, the *theoretical probability* $p(y)$. Then using the Greek $\mu$:

$$\mu_Y = \sum_{\text{all } y} p(y)y = E(Y), \tag{5.6}$$

where the operator $E$ means the *expected* or theoretical average of whatever follows in parentheses.

Also we have for a sample variance from frequency data from (3.8):

$$s^2 = \sum_{i=1}^{k} \left( \frac{f_i}{n-1} \right) (y_i - \bar{y})^2. \tag{5.7}$$

For population variance we use the Greek $\sigma^2$:

$$\sigma_Y^2 = \sum_{\text{all } y} p(y)(y - \mu_Y)^2 = E[(Y - \mu_Y)^2]. \tag{5.8}$$

Then the population standard deviation is defined by

$$\sigma_Y = \sqrt{\sum_{\text{all } y} p(y)(y - \mu_Y)^2} = \sqrt{E[(Y - \mu_Y)^2]}. \tag{5.9}$$

In general

$$E[g(Y)] = \sum_{\text{all } y} p(y) g(y). \tag{5.10}$$

### 5.2.2.* Population Curve-Shape Characteristics.

Analogous to (3.17) we have, replacing $m_k{}'$ by the Greek $\mu_k{}'$ for moments around the origin,

$$\mu_k{}' = \sum_{\text{all } y} p(y) y^k = E(Y^k), \qquad k = 1, 2, 3, \dots . \tag{5.11}$$

Sample central moments were $m_k$, so the population central moments are defined as

$$\mu_k = \sum_{\text{all } y} p(y)(y - \mu_Y)^k = E[(Y - \mu_Y)^k], \qquad k = 2, 3, \dots . \tag{5.12}$$

We note that $\mu_2 = \sigma_Y^2$, and that $\mu_1 = 0$, like $\sum(y - \bar{y}) = 0$. Then the sample curve-shape characteristics $a_3$ and $a_4$ of (3.18) and (3.20) are replaced by the population characteristics $\alpha_k$:

$$\alpha_3 = \mu_3/\sigma_Y^3, \qquad \alpha_4 = \mu_4/\sigma_Y^4. \tag{5.13}$$

### 5.2.3. Algebra of Expectations.

It is often convenient to be able to manipulate the operator E on various functions, without going back to the summation expression. The laws follow those for summations (3.2)–(3.5). Taking $c$ as a constant we have

$$E(c) = c \qquad \text{or} \qquad \mu_c = c \tag{5.14}$$

$$E(cY) = c\,E(Y) \qquad \text{or} \qquad \mu_{cY} = c\mu_Y \tag{5.15}$$

$$E(Y + c) = E(Y) + c \qquad \text{or} \qquad \mu_{Y+c} = \mu_Y + c. \tag{5.16}$$

In proof, for example, of (5.16), we have, using (3.5), (3.4)

$$E(Y + c) = \sum_{\text{all } y} p(y)(y + c) = \sum_{\text{all } y} [p(y)y + p(y)c]$$

$$= \sum_{\text{all } y} p(y)y + c \sum_{\text{all } y} p(y)$$

$$= E(Y) + c \qquad \text{since} \quad \sum_{\text{all } y} p(y) = 1.$$

From (5.15) and (5.16) follows the important

$$E(cY + A) = c\,E(Y) + A \qquad \text{or} \qquad \mu_{cY+A} = c\mu_Y + A. \qquad (5.17)$$

Next for population variances, define the notation

$$\sigma_Y{}^2 = \text{var}(Y),$$

where what is in parentheses is the random variable of which the variance is taken. Then we have the laws

$$\text{var}(Y + c) = \text{var}(Y) \qquad\qquad\qquad (5.18)$$

$$\text{var}(cY) = c^2\,\text{var}(Y) \qquad\qquad\qquad (5.19)$$

$$\text{var}(cY + A) = c^2\,\text{var}(Y). \qquad\qquad\qquad (5.20)$$

Proof of (5.20) follows:

$$\text{var}(cY + A) = E[cY + A - E(cY + A)]^2 = E[cY + A - c\mu_Y - A]^2$$
$$= E[c^2(Y - \mu_Y)^2] = c^2\,E[(Y - \mu_Y)^2] = c^2\,\text{var}(Y).$$

Then (5.18) and (5.19) follow as corollaries. Also

$$\sigma_{cY+A} = c\sigma_Y \qquad \text{if} \quad c > 0. \qquad\qquad (5.21)$$

Also we have the important

$$\sigma_Y{}^2 = E(Y^2) - [E(Y)]^2. \qquad\qquad\qquad (5.22)$$

This follows from the definition of var $Y = \sigma Y^2$, by squaring out the binomial $[y - E(y)]^2$ and using (5.14)–(5.16) on the three terms (see Problem 5.83).

### 5.2.4.* **Further on Population Moments.**

Since it may well be easier to find the $\mu_k'$ than the $\mu_k$ we may wish to use the following which are similar to (5.22):

$$\mu_3 = \mu_3' - 3\mu_2'\mu_1' + 2\mu_1'^3 = E(Y^3) - 3E(Y^2)\,E(Y) + 2[E(Y)]^3 \quad (5.23)$$

$$\mu_4 = \mu_4' - 4\mu_3'\mu_1' + 6\mu_2'\mu_1'^2 - 3\mu_1'^4, \quad (5.24)$$

and then substitute into (5.13).

## 5.3. THE BINOMIAL DISTRIBUTION

The basic situation to which the "binomial" applies is that in which we have a series of $n$ trials, on each of which there can be but two different kinds of outcomes. Examples are success or failure of an experiment, a win or a loss in a game of chance, a good or a defective piece manufactured, or a boy or girl baby born. The first condition for the binomial distribution to be applicable is that the probabilities for each of the two possible outcomes are to remain constant. (Since there are but two possible outcomes on a trial, the sum of the two probabilities is one.) The second condition is that the drawings be independent; that is, the probability on any one drawing does not depend in any manner on the preceding drawing(s). Then our problem is, what is the probability that among the $n$ trials or drawings there occur exactly $y$ of one type of outcome? Thus let us suppose that a process produces fountain pens in such a way that the probability of the next pen being defective is constantly .01, but the defectives occur in random order. Then what is the probability that in 500 pens, there are, say, precisely two defective ones? In order to talk about this problem, let us first set down a few notations:

$$n = \text{number of trials, or the sample size} \quad (5.25)$$

$$
\begin{aligned}
y = &\ \text{number of occurrences of one type, say,} \\
&\ \text{successes} \quad (y = 0, 1, ..., n)
\end{aligned}
\quad (5.26)
$$

$$
\begin{aligned}
p = &\ \text{proportion of occurrences of that type in} \\
&\ n \text{ trials} \\
= &\ y/n.
\end{aligned}
\quad (5.27)
$$

We now come to the choice for a Greek letter for the population proportion, corresponding to $p$ for the sample. Logically it should be the Greek pi ($\pi$), but the author has encountered strong sentiment against its use, because $\pi$ is so well known as 3.14159... . We shall choose to use the lower case phi ($\phi$), instead. Thus

$$\phi = \text{the constant probability of the given type of outcome} \qquad (5.28)$$

$$1 - \phi = \text{the constant probability of the opposite type of outcome.} \qquad (5.29)$$

Also let

$$p(y) = \text{the probability function, that is, the probability of exactly } y \text{ occurrences} \qquad (5.30)$$

$$P(y) = \sum_{i=0}^{y} p(i) = P(y \text{ or fewer occurrences in } n)$$

$$= \text{the distribution function of } Y. \qquad (5.31)$$

Our problem then is to find $p(y)$ in terms of $n$, $\phi$, and $y$. Let us take as an example, $n = 10$, $y = 3$. One way in which there can be three successes is for the first three trials to yield successes and the remainder to all be failures, or symbolically, sssfffffff. By the multiplicative law (4.22) for independent probabilities, we therefore have for the probability of this event (letting $\phi$ be the probability of a success):

$$P(\text{sssfffffff}) = \phi\phi\phi(1 - \phi) \cdots (1 - \phi) = \phi^3(1 - \phi)^7.$$

Now this is obviously not the only way in which there can be exactly three successes and seven failures. For example, we might have ffsfsfffsf. But the probability of this particular ordering or event is also $\phi^3(1 - \phi)^7$, just as it is for all other orderings in which there are exactly three successes and seven failures. Hence to find the probability of exactly three successes (no matter where in the 10 trials) and seven failures, all we need is to find how many distinct such orderings are possible and then to multiply $\phi^3(1 - \phi)^7$ by this number. But this is easy. There are 10 different trial numbers: 1, 2, ..., 10, and we must choose three of these to be the trials yielding the three successes. Thus there are $C(10, 3)$ such orderings of outcomes. Hence the probability of three successes is

$$p(3) = C(10, 3) \phi^3(1 - \phi)^7 = 120\phi^3(1 - \phi)^7.$$

By obvious extension of the foregoing discussion we have in general for any $n$ and $y$

$$p(y) = C(n, y) \phi^y(1 - \phi)^{n-y} \qquad y = 0, 1, ..., n. \qquad (5.32)$$

If $y = 0$, this reduces to $(1 - \phi)^n$; or if $y = n$, to $\phi^n$.

To anyone who has studied the expansion for a general binomial the right-hand expression is obviously a term in the expansion of $[\phi + (1 - \phi)]^n$, hence the name "binomial" for this theoretical distribution. Since the sum of all the terms in this expansion is $1^n = 1$, we have

$$\sum_{y=0}^{n} p(y) = 1,$$

as indeed it should be, since the number of successes is sure to be between 0 and $n$, inclusive.

### 5.3.1. Examples of the Binomial Distribution.

Let us now look at a few examples. As a simple example take

$$\phi = .2 = \text{probability of a defective piece being produced}$$

$$n = 5 = \text{sample size.}$$

By (5.32)

$$p(0) = \text{probability of no defectives in the 5}$$
$$= C(5, 0) .8^5 .2^0 = 5!/(0! \, 5!) \ .8^5 = .32768$$

$$p(1) = \text{probability of 1 defective in the 5}$$
$$= C(5, 1) .8^4 .2^1 = 5 .8^4 .2^1 = .40960$$

$$p(2) = \text{probability of 2 defectives in the 5}$$
$$= C(5, 2) .8^3 .2^2 = 10 .8^3 .2^2 = .20480$$

$$p(3) = 10 .8^2 .2^3 = .05120$$

$$p(4) = 5 .8^1 .2^4 = .00640$$

$$p(5) = .2^5 = .00032.$$

Note that the sum of these probabilities is 1.00000. Also we have, for example,

$$P(2) = p(0) + p(1) + p(2) = .94208$$
$$= \text{probability of 2 or fewer defectives in the 5.}$$

See Fig. 5.1 picturing $p(y)$ and $P(y)$ for this case.
Now consider the probability sought at the beginning of Section 5.3.

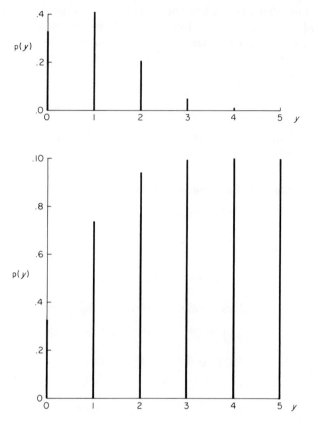

**Fig. 5.1.** Probability function p(y) and distribution function P(y) for the binomial $\phi = .2$, $n = 5$.

There $\phi = .01 =$ probability of a defective pen, $n = 500$, $y = 2$. Then by (5.32) and (4.31)

$$p(2) = \frac{500!}{2!\,498!} .01^2 .99^{498} = \frac{500(499)}{2!} .01^2 .99^{498}.$$

This is a long calculation even after canceling the factorials. To obtain $.99^{498}$, one should use eight-decimal-place logarithms to have five reliable significant figures after raising .99 to such a high power, because of the accumulation of error in log .99. (Thus we usually try to find some tabulation of probabilities for the binomial distribution rather than using direct calculation.) The result comes out .0836 to the best four decimal places.

TABLE 5.1

Examples of the Binomial Distribution. Probabilities p(y)

| y | (a) $n = 500$ $\phi = .01$ | (b) $n = 100$ $\phi = .05$ | (c) $n = 10$ $\phi = .50$ | (d) $n = 10$ $\phi = .10$ | (e) $n = 10$ $\phi = .01$ |
|---|---|---|---|---|---|
| 0 | .0066 | .0059 | .0010 | .3487 | .9044 |
| 1 | .0332 | .0312 | .0098 | .3874 | .0914 |
| 2 | .0836 | .0812 | .0439 | .1937 | .0042 |
| 3 | .1402 | .1396 | .1172 | .0574 | .0001 |
| 4 | .1760 | .1781 | .2051 | .0112 | |
| 5 | .1764 | .1800 | .2461 | .0015 | |
| 6 | .1470 | .1500 | .2051 | .0001 | |
| 7 | .1048 | .1060 | .1172 | | |
| 8 | .0652 | .0649 | .0439 | | |
| 9 | .0360 | .0349 | .0098 | | |
| 10 | .0179 | .0167 | .0010 | | |
| 11 | .0080 | .0072 | | | |
| 12 | .0033 | .0028 | | | |
| 13 | .0012 | .0010 | | | |
| 14 | .0004 | .0003 | | | |
| 15 | .0001 | .0001 | | | |
| $\mu_Y$ | 5 | 5 | 5 | 1 | .1 |
| $\sigma_Y$ | 2.22 | 2.18 | 1.58 | .95 | .31 |
| $\alpha_3$ | .44 | .41 | 0 | .84 | 3.11 |
| $\alpha_4$ | 3.19 | 3.15 | 2.80 | 3.51 | 12.50 |

This second example of the binomial distribution and others are given in Table 5.1. Only case (c) for which $\phi = .5$ has a symmetrical distribution of probabilities. The others are more or less unsymmetrical, especially (d) and (e) where $n$ is small and $\phi$ close to zero. The student would be well advised to make graphs such as Fig. 5.1 for these examples.

## 5.3.2. Population Moments for Binomial Distributions.

In describing a population distribution, certain moments can be most helpful. Probably the most important one is the mean or "expected value." Thus when $\phi = .2$, $n = 5$, it seems perfectly natural to "expect" one defective in a sample of five, since the chance of a defective on a single trial is $1/5$ and we have five trials. To be more precise we may use (5.6).

Then

$$\mu_Y = \mu = E(Y) = \sum_{y=0}^{n} C(n, y) \phi^y (1 - \phi)^{n-y} y = n\phi \qquad \text{(binomial)} \quad (5.33)$$

Thus the expected or population average number of occurrences is the number of trials $n$ times the probability of an occurrence per single trial $\phi$, which makes excellent sense. The proof is not difficult.*

We may also readily prove

$$\sigma_Y^2 = n\phi(1 - \phi) \qquad \text{or} \qquad \sigma_Y = \sqrt{n\phi(1 - \phi)} \qquad \text{(binomial).} \quad (5.34)$$

The results of using (5.33) and (5.34) on the binomial distribution examples of Table 5.1 are shown at the bottom of the table.

We might also prove the curve-shape characteristics (5.13) for the binomial to be

$$\alpha_3 = \frac{1 - 2\phi}{\sqrt{n\phi(1 - \phi)}}, \qquad \alpha_4 = 3 + \frac{1 - 6\phi(1 - \phi)}{n\phi(1 - \phi)}. \quad (5.35)$$

The curve-shape characteristics $\alpha_3$ and $\alpha_4$ are given for the five cases in Table 5.1. In case (c) because of symmetry, $\alpha_3 = 0$. In the other cases $\alpha_3$ shows positive skewness, that is, a greater tendency toward extreme positive skewness, that is, a greater tendency toward extreme positive deviations than negative. This tendency is by far the greatest for case (e).

For $\alpha_4$ the first four examples are not radically different from the "normal" value of about 3. In the last case (e) the tendency toward extreme tailing out in a relative sense is quite pronounced with an $\alpha_4$ of 12.50. A high $\alpha_4$ can occur with the long deviations on only one side as here, or on both sides.

---

* Although there is an easier method making use of more sophisticated background, we have from (5.33)

$$\mu = \sum_{y=0}^{n} \frac{n! \, \phi^y}{y! \, (n - y)!} (1 - \phi)^{n-y} y = \sum_{y=1}^{n} \frac{n! \, \phi^y}{y! \, (n - y)!} (1 - \phi)^{n-y} y,$$

since the $y$ factor is zero for the lower summation limit. Now cancel $y$ ($\geqslant 1$) into $y!$ and factor out from the summation sign the common terms $n$ and $\phi$, giving

$$\mu = n\phi \sum_{y=1}^{n} \frac{(n - 1)! \, \phi^{y-1}}{(y - 1)! \, (n - y)!} (1 - \phi)^{n-y}.$$

But the summation gives exactly the sum of the terms of $[\phi + (1 - \phi)]^{n-1} = 1$, as may be seen by writing out a few terms. Therefore $\mu = n\phi$.

We may note further that for the binomial distribution, the mode, or $y$ value or values having the greatest probability will always be within one unit of $n\phi$, as may be proved.

### 5.3.3. Use of Binomial Tables.

Whenever the probabilities for a binomial distribution are needed the natural thing to do is to see whether someone else has already done the job for us. This is indeed the case for a very large number of combinations of $\phi$ and $n$, with appropriate $y$'s [1–5]. In these tables the probabilities are given in three different forms. The individual terms $p(y)$ may be listed. More often, however, the cumulative sum of terms is given. Thus in the literature [3, 4] the tabulated entries are

$$P(y \text{ or less occurrences}) = P(y).$$

On the other hand, in the literature [1, 2], the entries are

$$P(y \text{ or more occurrences}) = 1 - P(y - 1 \text{ or less occurrences})$$
$$= 1 - P(y - 1).$$

When the entries are individual terms, we can always sum to find any desired $P(y)$, whereas if cumulative sums are given, we can subtract consecutive entries to get any desired $p(y)$. Thus for the example $n = 500, \phi = .01$, we have from Robertson [3]

$$p(2) = P(2 \text{ or less}) - P(1 \text{ or less}) = P(2) - P(1)$$
$$= .12339 - .03975 = .08364.$$

Also similarly

$$P(2 \leqslant Y \leqslant 6) = P(6) - P(1) = .76292 - .03975 = .72317.$$

In all such cases what we need to do is to determine *precisely* which tabulated events will yield the desired one so that subtraction of listed probabilities or complementation (e.g., $A = 3$ or more, $\bar{A} = 2$ or less) will yield the desired probability.

### 5.3.4. Calculation of a Binomial Distribution

If $n$ is small and the probability $\phi$ not too messy, we can use (5.32) to find the desired values of $p(y)$. Or if tables are available, they may be used. But if one must calculate out the terms, calculations can easily become tedious. Use of a table of $C(n, y)$ as presented by Fry [6] can help.

Or we may use logs of $n!$ such as in the literature [6, 7, 10] or Table XIV. High powers of $\phi$ require logarithms, and one must be careful to use enough places in $\log \phi$ and $\log(1 - \phi)$ to have the desired number of significant figures in the $p(y)$'s to be found.

A good way to calculate $p(y)$ terms is to find one of them by direct use of (5.32), and then to find others from this one. If $p(0)$ is not extremely small, then we can begin with it. Otherwise we can begin near the largest $p(y)$'s which occur within one of $y = n\phi$, and go both ways. The recursion formula is*

$$p(y + 1) = p(y) \frac{(n - y)\phi}{(y + 1)(1 - \phi)} \qquad \text{(binomial)}. \qquad (5.36)$$

Suppose we now wish to calculate terms such as for $n = 500$, $\phi = .01$, example (a) of Table 5.1. If we want the best four decimal places for the $p(y)$'s, it would do not to begin with $p(0) = .0066$, since $p(y)$'s calculated from this would only be good to about two significant figures. Suppose for safety we seek five significant figures for $p(0) = (1 - \phi)^n = .99^{500}$. Using eight-place logarithms we have $\log .99 = 9.995,635,19 - 10$. From this $\log .99^{500} = 7.817595 - 10$ and $p(0) = 0.065704$. Even with the eight-place logarithm to start with, the possible error in $\log .99^{500}$ is $2.5 \times 10^{-6}$.

Now using (5.36)

$$y = 0, \qquad p(1) = p(0) \frac{500}{1} \frac{.01}{.99} = .0065704 \frac{(500)}{99} = .033184$$

$$y = 1, \qquad p(2) = p(1) \frac{499}{2} \frac{.01}{.99} = .033184 \frac{(499)}{198} = .083630$$

$$y = 2, \qquad p(3) = .083630 \frac{498}{3} \frac{1}{99} = .14023,$$

and so on. Note that $\phi/(1 - \phi)$ is constant in (5.36). A good check on the calculation is $\sum p(y) = 1$.

---

* This is proved as follows from (5.8):

$$p(y + 1) = \frac{n! \, \phi^{y+1}(1 - \phi)^{n-y-1}}{(y + 1)! \, (n - y - 1)!}$$

$$= \frac{n! \, (n - y)}{y! \, (n - y)! \, (y + 1)} \frac{\phi^y(1 - \phi)^{n-y}\phi}{(1 - \phi)}$$

$$= p(y) \frac{n - y}{y + 1} \frac{\phi}{1 - \phi}.$$

In one big calculation done before electronic computers Carver worked out the terms for $n = 100,000$, $\phi = .008$ starting from p(800), because p(0) would be infinitesimal!

Of course one probably will use an electronic computer to find a series of p(*y*)'s. But in this case (5.36) can again be of much use.

### 5.3.5. Approximations to the Binomial Distribution.

There is a substantial amount of literature on the approximation of a binomial distribution by use of other distributions. The Poisson distribution (see Section 5.4) may be used for individual terms p(*y*) when *n* is "large" and $\phi$ "small." The normal distribution (see Chapter 6) may often be used to evaluate sums of p(*y*)'s, when $\phi$ is not too small relative to *n*. These approximations are discussed later on.

### 5.3.6. Conditions of Applicability of the Binomial.

We here set down the prerequisites to applying the binomial distribution:

(1) two possible outcomes of an experiment, for example, "success" and "failure";
(2) *n independent* trials;
(3) constant probability $\phi$ of one kind of outcome, say success;
(4) seeking the probability p(*y*) of exactly *y* successes in *n* trials.

## 5.4. THE POISSON DISTRIBUTION

This distribution was named after Poisson, a French probabilist of the nineteenth century. It gives the probability of *y* occurrences of some phenomenon over a specified area of opportunity or sample, when the average number $\mu$ of occurrences in such a sample is known. Two conditions are necessary: (1) that the possible number of occurrences is a great deal larger than the average number $\mu$ (actually the possible number is assumed infinite), and (2) that the occurrences are independent, that is, the occurrence of one "defect" does not increase or decrease the chance of another. Typical examples of the use of a Poisson distribution are (1) typographical errors by a good typist on a printed page; (2) defects of all kinds on a piece of equipment or on a subassembly; (3) hospital calls in a plant over a week; (4) collisions of particles or automobiles over a time period; (5) extra petals on a normally five-petaled flower; (6) breakdowns of insulation on 10,000 ft of insulated wire under test conditions; (7) number of lost articles in a day; (8) number of magnetic particles in area of condenser paper; (9) pinholes in painted test area; (10) emission of a type of ray over a time period; (11) arrival of vehicles at a tollgate or a loading area in an hour. The reader would do well to think over the two requirements for a Poisson distribution, in relation to several of these examples.

Let us describe one set of postulates which will justify the Poisson distribution, illustrating with example (6) above. They are as follows:

(1)   The numbers of breakdowns in nonoverlapping lengths of wire are independent.
(2)   The probability of one breakdown in a length $L$ is approximately proportional to $L$ for small $L$, and is independent of the location of $L$ on the 10,000 ft.
(3)   The probability of more than one breakdown in a small length $L$ is negligible in comparison with the probability of one breakdown in $L$.

These postulates may be used to derive the following probability function for the Poisson*:

$$p(y) = \frac{e^{-\mu}\mu^{y}}{y!}, \qquad y = 0, 1, 2, ..., \tag{5.37}$$

where $e = 2.71828...$, and $\mu$ is the average number of occurrences per sample. Note that here we have just one population parameter $\mu$. This parameter is also sometimes designated by $\lambda$, $\gamma$, and $c'$.

As an example, suppose $\mu = 2$. Then

$$p(0) = \frac{e^{-2}2^{0}}{y!} = e^{-2} = .135335$$

by using tables of $e^{-x}$. Continuing,

$$p(1) = \frac{e^{-2}2}{1!} = p(0)\,2 = .270670$$

$$p(2) = \frac{e^{-2}2^{2}}{2!} = p(1)\frac{2}{2} = .270670$$

$$p(3) = \frac{e^{-2}2^{3}}{3!} = p(2)\frac{2}{3} = .180447.$$

These may be rounded to the nearest four decimal places as shown in Table 5.2.

* See Section 5.4.1 for one derivation.

TABLE 5.2

Examples of the Poisson Distribution. Probabilities $p(y)$

| $y$ | (a) $\mu = 5$ | (b) $\mu = 2$ | (c) $\mu = 1$ | (d) $\mu = .2$ | (e) $\mu = .1$ |
|---|---|---|---|---|---|
| 0 | .0067 | .1353 | .3679 | .8187 | .9048 |
| 1 | .0337 | .2707 | .3679 | .1637 | .0905 |
| 2 | .0842 | .2707 | .1839 | .0164 | .0045 |
| 3 | .1404 | .1804 | .0613 | .0011 | .0002 |
| 4 | .1755 | .0902 | .0153 | .0001 | |
| 5 | .1755 | .0361 | .0031 | | |
| 6 | .1462 | .0120 | .0005 | | |
| 7 | .1044 | .0034 | .0001 | | |
| 8 | .0653 | .0009 | | | |
| 9 | .0363 | .0002 | | | |
| 10 | .0181 | | | | |
| 11 | .0082 | | | | |
| 12 | .0034 | | | | |
| 13 | .0013 | | | | |
| 14 | .0005 | | | | |
| 15 | .0001 | | | | |
| $\mu_Y$ | 5 | 2 | 1 | .2 | .1 |
| $\sigma_Y$ | 2.24 | 1.41 | 1.00 | .45 | .32 |
| $\alpha_3$ | .45 | .71 | 1.00 | 2.24 | 3.16 |
| $\alpha_4$ | 3.20 | 3.50 | 4.00 | 8.00 | 13.00 |

Calculation of $p(y)$ values is quite simple when done recursively as above. The general formula for given $\mu$ is

$$p(y + 1) = \frac{p(y)\mu}{y + 1} \qquad \text{(Poisson distribution).} \qquad (5.38)$$

Note that this formula is simpler than that for the binomial (5.36), since here the only varying factor is the $y + 1$.

The moments for Poisson distribution are simple functions of $\mu$ as follows:

$$\mu_Y = E(Y) = \mu \qquad \text{(Poisson distribution)} \qquad (5.39)$$

$$\sigma_Y = \sqrt{\mu} \qquad (5.40)$$

$$\alpha_3 = 1/\sqrt{\mu} \qquad (5.41)$$

$$\alpha = 3 + 1/\mu. \qquad (5.42)$$

### 5.4.1.* A Derivation of the Poisson Probability Function.

From the axioms just given, for breakdowns in lengths of insulated wire, let us derive an expression for the probability of exactly $y$ within a length $x$ of wire. Within any short length $dx$, the probability of one breakdown is $\lambda\, dx$, of more than one is negligible in comparison, and of none is the complementary probability $1 - \lambda\, dx$. Let the probability of *exactly* $y$ in 0 to $x$ be $g_y(x)$, which we seek to find. Then to have had $y + 1$ by length $x + dx$, we can have had

(1)   $y$ by $x$ and 1 breakdown in $x$ to $x + dx$, or
(2)   $y + 1$ by $x$ and 0 in $x$ to $x + dx$.

The compound probabilities by (4.22) are products, and since (1) and (2) are mutually exclusive events, we add. Thus

$$g_{y+1}(x + dx) = g_y(x)\lambda\, dx + g_{y+1}(x)(1 - \lambda\, dx)$$

$$g_{y+1}(x + dx) - g_{y+1}(x) = \lambda\, dx[g_y(x) - g_{y+1}(x)].$$

This is what can be called a "difference–differential equation," being made up of the continuous variable $x$, and *differences* for the discrete variable $y$. To solve we first get rid of $dx$'s by

$$\frac{g_{y+1}(x + dx) - g_{y+1}(x)}{dx} = \lambda[g_y(x) - g_{y+1}(x)].$$

Take the limit as $dx \to 0$, obtaining

$$g'_{y+1}(x) = \lambda[g_y(x) - g_{y+1}(x)]. \tag{5.43}$$

Now, however, we have the boundary condition that $g_y(x) = 0$ if $y = -1$, since we cannot have $-1$ breakdowns in 0 to $x$. Then substituting $y = -1$

$$g_0'(x) = -\lambda g_0(x), \qquad g_0'(x)/g_0(x) = -\lambda.$$

Integrating after multiplication by $dx$ gives

$$\ln g_0(x) = -\lambda x + C.$$

Letting $x = 0$, $g_0(0) = 1$ since we are sure to have no breakdowns at the beginning, thus $C = 0$. Hence

$$\ln g_0(x) = -x\lambda \qquad \text{or} \qquad g_0(x) = e^{-\lambda x}. \tag{5.44}$$

We next may obtain $g_y(x)$ recursively from this start. If $y = 0$, then from (5.43)

$$g_1'(x) = \lambda[g_0(x) - g_1(x)] = -\lambda g_1(x) + \lambda e^{-x}$$

or

$$g_1'(x) + \lambda g_1(x) = \lambda e^{-\lambda x}.$$

Multiplying through by $e^{\lambda x}$ (a trick of differential equations)

$$e^{\lambda x}g_1'(x) + \lambda e^{\lambda x}g_1(x) = \lambda \qquad \text{or} \qquad d[e^{\lambda x}g_1(x)] = \lambda.$$

Hence

$$e^{\lambda x}g_1(x) = \lambda x + C.$$

Since $g_1(x)$ is also a zero probability if $x = 0$, we have $C = 0$. Thus

$$g_1(x) = e^{-\lambda x}(\lambda x). \tag{5.45}$$

Proceeding similarly with $y = 1, 2, ...$, yields

$$g_y(x) = e^{-\lambda x}(\lambda x)^y/(y!). \tag{5.46}$$

If we now let $x \to L$, the entire length, and call $\lambda L = \mu$, then the probability function for $y$ takes the form

$$p(y) = \frac{e^{-\mu}\mu^y}{y!}, \qquad y = 0, 1, ..., \tag{5.37}$$

for the probability of $y$ defects in 0 to $L$.

### 5.4.2. Examples of the Poisson Distribution.

Table 5.2 shows $p(y)$ for five cases of the Poisson with $\mu = 5, 2, 1, .2$, and .1. These are shown graphically in Fig. 5.2. For $\mu = 5$, the distribution is rather symmetrical around the highest $p(y)$'s at $y = 4$ and 5, with somewhat greater tailing out to high $y$'s. As $\mu$ decreases, the distributions become more unsymmetrical with examples (d) and (e) strongly J-shaped. The progression of $\alpha_3$ values as listed in the table goes upward, as the lack of symmetry increases. If $\mu$ increases to, say, 25, the distribution will become quite symmetrical and "normal."

Example (a) might easily be an illustration of the number of Geiger counts for a radioactive compound over a fixed time interval. Then the various probabilities are for there to occur zero counts, just one count, just two, and so on, over the time interval, when the average number is

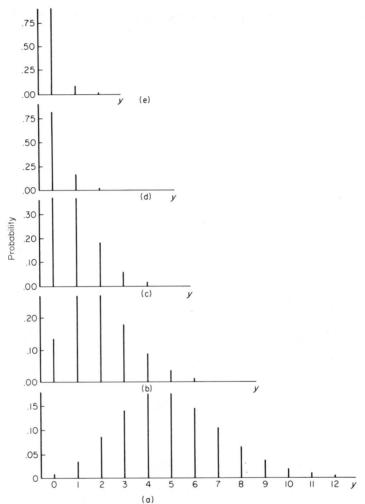

**Fig. 5.2.** Five examples of the Poisson probability function p(*y*), for $\mu_Y$ = (a) 5, (b) 2, (c) 1, (d) .2, (e) .1.

$\mu$ = 5. Example (b) might be the probabilities for various numbers of leaks in auto radiators at initial test after assembly, the average being two. Example (c) might give the probabilities of *y* accidents at a certain intersection over a one-month period, at which the average is one per month. Example (d) could be giving the probability of *y* colonies appearing on an agar plate when treated with a dilute solution of a food. Example (e) might give the probabilities of 0, 1, 2, ..., typographical errors on a page, when the average number is one per 10 pages, $\mu$ = .1.

### 5.4.3. Tables of the Poisson Distribution.

It is much easier to tabulate the Poisson distribution than the binomial. For the latter, to tabulate either p($y$) or P($y$) we need a triple-entry table, that is, we need to specify $n$, $\phi$, and $y$ in order to find p($y$). But for the Poisson we only need $\mu$ and $y$ specified, that is, we have a double-entry table only. Again some tables give values of p($y$) [8, 9]. Others give cumulative sums P($y$ or more) [9]. Our Table XI gives P($y$ or less).

If one cannot find a desired probability p($y$) or P($y$), it is not particularly difficult to calculate using (5.37) and possibly (5.38), especially if $\mu$ is small. Programming a computer to give p($y$) is not at all difficult.

### 5.4.4. Using the Poisson Distribution to Approximate the Binomial.

When $n$ is substantial, say 20 or more, and $\phi$ relatively small, say .05 or less, we may approximate binomial probabilities p($y$) and P($y$) by an appropriate Poisson. The larger $n$ is and the smaller $\phi$ is, the closer the approximation, both in relative and absolute sense.

As an example if $n = 500$, and $\phi = .01$, then the expected number of say "defectives" in 500 is $n\phi = 5$. This is of course case (a) of Table 5.1. What we have just said would imply that case (a) of Table 5.1 may be approximated by case (a) of Table 5.2, since for the latter $\mu = 5$. We see on comparing the two tables that there is indeed a close resemblance between the two sets of p($y$)'s. In fact case (b) of Table 5.1 is also well approximated by the same Poisson. But although case (c) of Table 5.1 has $n\phi = 5$, it is not at all well approximated by the Poisson for $\mu = 5$. This is of course because $\phi$ was large here.

Further, although $n$ is rather small for binomial cases (d) and (e), they are rather well approximated by the corresponding Poisson cases (c) and (e). Later we give a formula to indicate the precision of the approximation, more accurately.

As a further illustration, suppose that we ask for P($1 \leqslant y \leqslant 4 \mid n = 100$, $\phi = .025$). Here $n$ is large, $\phi$ small, so we use the Poisson, with $\mu = n\phi = 2.5$. For this Poisson we look up in appendix Table XI P($y \leqslant 4 \mid \mu = 2.5$) = .8912 and P($y = 0 \mid \mu = 2.5$) = .0821. The difference between these gives the required probability, .8091.

### 5.4.5.* Derivation of the Poisson Distribution as a Limit of the Binomial.

We are to show that for any fixed $y$, p($y$) in (5.32) approaches p($y$) in (5.37), as $n \to \infty$, and $\phi \to 0$ in such a way that $n\phi \to \mu$.

From (5.32), using (4.31) we have

$$p(y) = \frac{n(n-1)\cdots(n-y+1)}{y!} \phi^y (1-\phi)^{n-y}.$$

Multiplying $\phi^y$ by $n^y$ and dividing each term of the numerator of the fraction by $n$ gives

$$p(y) = (1)\left(1 - \frac{1}{n}\right)\cdots\left(1 - \frac{y-1}{n}\right)(n\phi)^y (1-\phi)^{n-y}/y!.$$

Now under the limiting process, each of the $y$ parentheses $1, \ldots, (1 - [y-1]/n)$ approaches one and hence their product does too. Also $(n\phi)^y$ approaches $\mu^y$. Moreover we have

$$(1-\phi)^{n-y} = [(1-\phi)^{-1/\phi}]^{-\phi(n-y)}.$$

Now, as $\phi \to 0$, the quantity within the brackets approaches $e$, since $e = \lim_{z\to 0}(1+z)^{1/z}$ and the limit of $-\phi n + \phi y$ is $-\mu$ since $y$ is fixed. Thus $(1-\phi)^{n-y} \to e^{-\mu}$. Thus we have

$$p(y)_{\text{binom}} \to \frac{e^{-\mu}\mu^y}{y!} = p(y)_{\text{Poisson}} \cdot$$

### 5.4.6. Conditions of Applicability of the Poisson Distribution.

The primary conditions for use of the Poisson probability model are as given in Section 5.4:

(1)   Counting occurrences of events, for example, "defects" which can be of just one kind or of many kinds.
(2)   Considering counts over equal areas of opportunity for "defects."
(3)   The possible number of defects is far greater than the average number $\mu$.
(4)   The defects occur independently.

The Poisson distribution can sometimes be used to approximate the binomial or hypergeometric distribution (see Section 5.5) or others.*

---

* For example, shuffle two decks, and turn over a card from each. If both are the same card, suit and rank, this is a "hit." The number of hits in the 52 opportunities for a "hit" is very well approximated by a Poisson with $\mu = 1$. Thus P(no hits) $= e^{-1} \doteq .3679$.

## 5.5. THE HYPERGEOMETRIC DISTRIBUTION

Whenever one draws randomly and without replacement from finite lots or populations, which consist of just two kinds of objects, the appropriate distribution is the hypergeometric. (Do not get hypertension from this rather fearful name!) This distribution gives the probability p($y$) that a sample of $n$ contains exactly $y$ objects of one of the two types of objects.

We shall need to define a few symbols:

$N$ is the number of objects in the lot or population;
$K$ is the number of objects of one of the two kinds, say, "defective" pieces, as opposed to "good" pieces, in a lot;
$n$ is the number of objects drawn randomly without replacement;
$y$ is the number of "defectives" in the $n$ drawn.

We are now in a position to formulate the desired probability p($y$) that the sample contains exactly $y$ "defectives." This is quite obviously an important probability both for practical application and for theoretical work.

To solve the problem, let us for the present suppose that the $N$ objects in the lot are all numbered so that they can be distinguished from each other. There are $K$ of one kind, say "defectives," and $N - K$ of the second kind, say "good" ones. To obtain the desired probability, we have but to find how many of all the possible samples of $n$ will in fact contain precisely $y$ defectives. Then this proportion will be the probability p($y$) by (4.19) since all samples are equally likely.

The total number of possible samples of $n$ from $N$ without restriction is $C(N, n)$. Now how many ways are there to choose a sample containing $y$ defectives and $n - y$ good? There are $C(K, y)$ ways to choose the former and $C(N - K, n - y)$ ways to choose the latter. We multiply these together to give us the number of possible samples, each with $y$ defectives, by (4.32). Thus we have

$$\mathrm{p}(y) = \frac{C(K, y)\, C(N - K, n - y)}{C(N, n)} \tag{5.47}$$

$$\mathrm{p}(y) = \frac{K!\,(N - K)!\, n!\,(N - n)!}{y!\,(K - y)!\,(n - y)!\,(N - K - n + y)!\, N!}. \tag{5.48}$$

The former is useful if one has small $N$'s or a table of $C(n, r)$ such as that presented by Fry [6]. But the second form is usually used, because if $N$ is sizable, then we must usually resort to tables of log($n!$) such as those in the literature [6, 7] or our Table XIV.

In using either (5.47) or (5.48) it would seem that the limits for $y$ would be 0 and $n$. But there are other limitations, brought out in (5.48) by the fact that the argument for any factorial must be nonnegative. Thus we also have $K - y \geqslant 0$ or $K \geqslant y$. Furthermore $N - K - n + y \geqslant 0$ or $y \geqslant K + n - N$. This could give a minimum greater than zero for $y$. For example, if we take a sample of 10 from a lot of 20 containing 12 defectives, $K + n - N = 12 + 10 - 20 = 2$, and thus the sample must contain at least two defectives since there are but eight good ones in the lot.

As an example let us take the case of a lot of $N = 20$ alarm clocks, of which there are two which will not run. (They make time stand still!) Thus $K = 2$. Now suppose that we take a random sample of $n = 10$, drawn without replacement, so that after each drawing each remaining clock has the same chance of being chosen in the sample. What are the probabilities of the sample containing 0, 1, and 2 defective clocks? (It cannot contain three or more since there are only two defectives in the lot.) Using (5.47) and (4.31) we have

$$p(0) = \frac{C(2, 0)\, C(18, 10)}{C(20, 10)} = \frac{1 \cdot 18 \cdot 17 \cdot \,\cdots\, \cdot 9/10!}{20 \cdot 19 \cdot \,\cdots\, \cdot 11/10!} = \frac{10 \cdot 9}{20 \cdot 19} = .2368$$

$$p(1) = \frac{C(2, 1)\, C(18, 9)}{C(20, 10)} = \frac{2 \cdot 18 \cdot 17 \cdot \,\cdots\, \cdot 10/9!}{20 \cdot 19 \cdot \,\cdots\, \cdot 11/10!} = \frac{2 \cdot 10 \cdot 10}{20 \cdot 19} = .5263$$

$$p(2) = \frac{C(2, 2)\, C(18, 8)}{C(20, 10)} = \frac{18 \cdot 17 \cdot \,\cdots\, \cdot 11/8!}{20 \cdot 19 \cdot \,\cdots\, \cdot 11/10!} = \frac{10 \cdot 9}{20 \cdot 19} = .2368.$$

Note the heavy cancellation in these quotients of factorials.

In Section 4.4, we mentioned the probability of a sample of 10 containing two or less defective blades, when drawn without replacement from a lot of 1000 blades, of which 30 are defective. Thus $N = 1000$, $K = 30$, $n = 10$, and the hypergeometric distribution is applicable. First find $p(0)$:

$$p(0) = \frac{C(970, 10)\, C(30, 0)}{C(1000, 10)} = \frac{970!\, 10!\, 990!}{1000!\, 10!\, 960!} = .7364.$$

For this calculation we have used Table XIV as follows:

$$\log 970! = 2477.7954 \qquad \log 100\ ! = 2567.6046$$

$$\log 990! = 2537.6242 \qquad \log 960! = 2447.9479$$

$$\overline{\log p(0) = 5015.4196} \qquad \overline{\quad -\quad\quad 5015.5525} = 9.8671{-}10.$$

For p(1) we can use p(0) as follows:

$$p(1) = \frac{C(970, 9)\, C(30, 1)}{C(1000, 10)} = \frac{970!\,30!\,10!\,990!}{9!\,961!\,1!\,29!\,1000!}$$

$$= \frac{970!\,30!\,10!\,990!}{10!\,960!\,0!\,30!\,1000!} \cdot \frac{10 \cdot 30}{961 \cdot 1} = p(0)\frac{300}{961} = .2299.$$

Here we first write p(1), then in the next step arbitrarily write p(0) in the first fraction, and see what we must do to it to obtain p(1). Thus the 10 and the 30 in the numerator knock the 10! down to 9! and the 30! to 29! while the 961 and 1 build up the 960! to 961! and the 0! to 1!

Again

$$p(2) = \frac{C(970, 8)\, C(30, 2)}{C(1000, 10)} = \frac{970!\,30!\,10!\,990!}{8!\,962!\,2!\,28!\,1000!}$$

$$= \frac{970!\,30!\,10!\,990!}{9!\,961!\,1!\,29!\,1000!} \cdot \frac{9 \cdot 29}{962 \cdot 2} = p(1)\frac{261}{1924} = .0312.$$

The sum of these three gives the desired probability, .9975. An excellent check is to continue till the probabilities hit .0000. We have as above

$$p(3) = p(2)\frac{8 \cdot 28}{963 \cdot 3} = .0024$$

$$p(4) = p(3)\frac{7 \cdot 27}{964 \cdot 4} = .0001.$$

The sum of the five probabilities is 1.0000.

Although we could give a recursion formula for the hypergeometric distribution similar to (5.36) and (5.38), but more complicated, it is the author's experience that working along numerically as above is easier and safer.

### 5.5.1. Tables for the Hypergeometric Distribution.

After reading the preceding section the reader may be saying, "Help! Bring on the tables." Unfortunately this is more easily said than done. The trouble is that what is required is a quadruple-entry table, that is, one in which one finds an entry corresponding to given values of $N$, $K$, $n$, and $y$. (If there are 10 choices for each of these quantities, we already need 10,000 entries in the table!) There is one substantial table [10]. It has complete coverage up to $N = 50$, then for $N = 60, 70, ..., 100$. Also included are $N = 1000$ with $n = 500$. It gives both $p(y)$ and $P(y)$ to the best six decimal places.

There are certain symmetries which avoid the necessity of tabulating all conceivable cases. For example, if $n > N/2$, one may consider as his sample the pieces *remaining* in the lot, not those chosen. For example, $N = 20$, $n = 15$, $K = 6$, $y = 4$. If our sample of 15 contains four defectives, then the "sample" of five *not* chosen contains two defectives. The most important and interesting symmetry is obtained from (5.48) by regrouping the factorials as follows:

$$p(y) = \frac{C(n, y)\, C(N - n, K - y)}{C(N, K)}. \tag{5.49}$$

Thus $K$ and $n$ play exactly parallel roles, and in tabulation one need only consider cases for which $n \geqslant K$ and not those for which $n < K$ as these are duplicates by (5.49). For example, the probability of one defective in a sample of 10 from a lot of 1000, of which 30 are defective, is just the same as the probability of one defective in a sample of 30 from a lot of 1000 of which 10 are defective, since by (5.47) and (5.49) $K$ and $n$ are interchangeable.

### 5.5.2. Examples of the Hypergeometric Distribution.

Table 5.3 gives the probabilities $p(y)$ for five examples of this distribution, including our first illustration. In examples (a) and (b) we see some effect of lot size, in that the probabilities in (b) are more closely clustered around the maximum, which occurs at $y = 5$ for each of these symmetrical distributions. This is because in (b) the sample of 10 depletes the lot of 20 much more than it does with a lot size of 100. (In the extreme case if $n = 10$ and $N = 10$, the sample will exactly mirror the lot with no variation in $y$ at all.) We may further note that (c) is much more like case (d) of Table 5.1 than is case (d). Each of these three cases, the two hypergeometric and the one binomial, had the same average of 1 and the same $n = 10$, and $\phi = K/N = .1$. It can be shown that with $n$ and $\phi$ fixed, as $N$ increases, the hypergeometric probabilities approach those for the binomial with the given fixed $n$ and $\phi$. This is natural, since in such a case the probability at each draw becomes more and more constant as the population size increases. Thus the situation becomes that of the binomial.

In the light of the foregoing we may now indicate why we require independency for the binomial in (2) of Section 5.3.6, as well as a constant $\phi$ in (3). Without knowing what has happened on the other drawings, the probability of a defective on any one of the $n$ drawings in hypergeometric drawing is constant at $\phi = K/N$. But if we know what has happened prior to the $k$th drawing, the probability on this draw will

**TABLE 5.3**

Examples of the Hypergeometric Distribution. Probabilities $p(y)^a$

| $y$ | (a) | (b) | (c) | (d) | (e) |
|---|---|---|---|---|---|
| | $N = 100$ | $N = 20$ | $N = 100$ | $N = 20$ | $N = 100$ |
| | $n = 10$ | $n = 10$ | $n = 10$ | $n = 10$ | $n = 10$ |
| | $K = 50$ | $K = 10$ | $K = 10$ | $K = 2$ | $K = 1$ |
| | $(\phi = .50)$ | $(\phi = .50)$ | $(\phi = .10)$ | $(\phi = .10)$ | $(\phi = .01)$ |
| 0 | .0006 | .0000 | .3305 | .2368 | .9000 |
| 1 | .0072 | .0005 | .4080 | .5263 | .1000 |
| 2 | .0380 | .0110 | .2015 | .2368 | |
| 3 | .1131 | .0779 | .0518 | | |
| 4 | .2114 | .2387 | .0076 | | |
| 5 | .2593 | .3437 | .0006 | | |
| 6 | .2114 | .2387 | | | |
| 7 | .1131 | .0779 | | | |
| 8 | .0380 | .0110 | | | |
| 9 | .0072 | .0005 | | | |
| 10 | .0006 | .0000 | | | |
| $\mu_Y$ | 5 | 5 | 1 | 1 | .1 |
| $\sigma_Y$ | 1.51 | 1.15 | .90 | .69 | .30 |
| $\alpha_3$ | 0 | 0 | .72 | 0 | 2.67 |
| $\alpha_4$ | 2.84 | 2.91 | 3.22 | 2.11 | 8.11 |

$^a$ $\phi = K/N$ is given for each.

depend on whatever we have already found in the preceding partial sample. The probability may be considerably different from $K/N$, in the extreme case 0 or 1. For example, in $N = 20$, $n = 10$, $K = 2$. If the first drawing of the 10 is a defective, then the conditional probability of a defective on the second is $1/19$, while if we draw a good one on the first then it is $2/19$ on the second draw.

### 5.5.3.* Moments for the Hypergeometric Distribution.

From Kendall [11], we have the following, in which we let $\phi = K/N$;

$$\mu_Y = E(Y) = n\phi = nK/N \quad \text{(hypergeometric distribution)} \tag{5.50}$$

$$\sigma_Y = \sqrt{n\phi(1 - \phi)(N - n)/(N - 1)} \tag{5.51}$$

$$\alpha_3 = \frac{1 - 2\phi}{\sqrt{n\phi(1 - \phi)}} \sqrt{\frac{N - 1}{N - n} \frac{N - 2n}{N - 2}} \tag{5.52}$$

$$\alpha_4 = \frac{(N-1)[N(N+1)-6n(N-n) + 3\phi(1-\phi)\{N^2(n-2)-Nn^2 + 6n(N-n)\}]}{n\phi(1 - \phi)(N - n)(N - 2)(N - 3)} \tag{5.53}$$

It might be mentioned that as $N \to \infty$ with $K/N$ fixed at $\phi$, (5.50)–(5.53) approach the binomial results as given by (5.33)–(5.35).

### 5.5.4.* Binomial Approximations to the Hypergeometric Distribution.

Individual terms $p(y)$ for the hypergeometric may be approximated by a binomial distribution, if the sample size $n$ is small in relation to the lot or population size $N$. The approximation begins to be of some use when $N$ is eight to ten times as large as $n$, for then the drawings to complete the sample do not deplete the lot too much, that is, modify too greatly the probabilities at each draw.

Then to approximate the hypergeometric probability $p(y)$ in (5.47) we may use the binomial probability $p(y)$ in (5.32):

$$p(y; N, K, n) \doteq C(n, y)\,\phi^y(1 - \phi)^{n-y}, \qquad \phi = K/N. \qquad (5.54)$$

For example, for $y = 1$, $N = 100$, $n = 10$, $K = 6$, then by [10], the correct value rounded off is

$$p(1; 100, 6, 10) = .36869,$$

which may be approximated by using $\phi = K/N = 6/100 = .06$

$$C(10, 1)\,.06^1\,.94^9 = .34380$$

from [1], rounded.

Owing to the symmetry between (5.47) and (5.49), we see that we may also use $\phi = n/N$ and approximate as follows:

$$p(y; N, K, n) = C(K, y)\,\phi^y(1 - \phi)^{K-y}, \qquad \phi = n/N. \qquad (5.55)$$

Using (5.55), $\phi = 10/100 = .1$:

$$C(6, 1)\,.1^1\,.9^5 = .35429,$$

which is closer to the true value.

A second degree of approximation is given by [12], which in our present notation is

$$p(y; N, K, n) \doteq C(n, y)\,\phi^y(1 - \phi)^{n-y}\left\{1 + \frac{1}{2K}[y - (y - n\phi)^2]\right\},$$

$$\text{with} \quad \phi = K/N. \qquad (5.56)$$

Or by symmetry

$$p(y; N, K, n) \doteq C(K, y)\phi^y(1 - \phi)^{K-y}\left\{1 + \frac{1}{2n}[y - (y - K\phi)^2]\right\},$$

$$\text{with} \quad \phi = n/N. \tag{5.57}$$

These formulas can be used either to give a closer approximation to $p(y; N, K, n)$ than (5.54) or (5.55), or else to tell approximately how much error to expect in these two approximations. All four approximations (5.54)–(5.57) improve as $N$ increases.*

### 5.5.5.* Poisson Approximations to the Hypergeometric Distribution.

If in addition to having $n$ and $K$ small in relation to $N$, we have $n$ and $K$ 20 or more and $K/N$ or $n/N$ below .05, then instead of the binomial approximations just given, we can use the Poisson

$$p(y; N, K, n) \doteq e^{-\mu}\mu^y/y!, \qquad \mu = nK/N. \tag{5.58}$$

Or by [12], we could use

$$p(y; N, K, n) \doteq \frac{e^{-\mu}\mu^y}{y!}\left\{1 + \left(\frac{1}{2n} + \frac{1}{2K}\right)[y - (y - n\phi)^2]\right\},$$

$$\text{with} \quad \phi = K/N \tag{5.59}$$

as a second approximation.

Table 5.4 compares all of these various approximations for the hypergeometric example.

As a fresh example to show how easy it is to use the Poisson via (5.58), take the case $N = 1200$, $K = 36$, $n = 50$, $y = 2$. Here since $n$ and $K$ are both small relative to $N$ we could use either binomial approximation. But since in addition $K/N = .04$ is less than .05 and $n$ at least 20, we can use the Poisson and (5.59). For this, find $\mu = 50(36)/1200 = 1.5$. Then by straightforward use of $e^{-y}$ tables or Poisson tables

$$p(2) = e^{-1.5}1.5^2/2! = .2510.$$

The exact value is

$$p(2) = \frac{C(36, 2)\, C(1164, 48)}{C(1200, 50)} = .2619$$

---

* As between the first approximations (5.54) and (5.55), the relative errors in $p(y)$'s are the terms added to 1 within the braces. Since for both (5.56) and (5.57) the expression within the square brackets proves to be the same, the amount of relative error is approximately proportional to the multiplier of the square brackets, respectively. Thus (5.54) will in general be a better approximation than (5.55) when $K > n$.

TABLE 5.4

Comparison of Approximations to a Hypergeometric Example[a]

| y | Exact | First-order binomial approximation | | Second-order binomial approximation | | First-order Poisson | Second-order Poisson |
|---|-------|---------------|--------|----------------|--------|---------|---------|
|   | (5.48) | (5.54) | (5.55) | (5.56) | (5.57) | (5.58) | (5.59) |
| 0 | .52230 | .53862 | .53144 | .52246 | .52187 | .54881 | .52247 |
| 1 | .36869 | .34380 | .35429 | .36787 | .36917 | .32929 | .36617 |
| 2 | .09646 | .09875 | .09842 | .09908 | .09862 | .09879 | .09931 |
| 3 | .01183 | .01681 | .01458 | .01294 | .01257 | .01976 | .01248 |
| 4 | .00071 | .00188 | .00122 | .00070 | .00076 | .00296 | [b] |
| 5 | .00002 | .00014 | .00005 | [b] | .00001 | .00036 | [b] |
| 6 | .00000 | .00001 | .00000 | [b] | [b] | .00004 | [b] |
|   |        | $n = 10$ | $n = 6$ | | | | |
|   |        | $\phi = .06$ | $\phi = .1$ | | | | |

[a] $N = 100$, $K = 6$, $n = 10$. Probabilities rounded to the best five decimal places.
[b] Correction, if applied, would yield negative probability.

which is not in tables, and requires use of tables of $\log(n!)$. Most of the latter only go up to 1200 or less; hence if $N > 1200$, we have greater difficulty, unless the work of Owen and Williams [7] or [10] is available. Using (5.59) we would get .2615.

### 5.5.6. Approximations to Sums of Terms of the Hypergeometric Distribution.

If a sum of terms of this distribution is needed, then the methods of the two previous subsections can be used term by term, or by using cumulative sums of terms in a binomial table. But if $n$ is a substantial proportion of $N$, then this is not likely to be accurate. If $\alpha_3$ from (5.52) is about .2 or less in size, we may use the normal distribution for a continuous variable as given in the next chapter. Or if the skewness is greater, the gamma or chi-square distribution may well be used.

### 5.5.7. Conditions of Applicability of the Hypergeometric Distribution.

The hypergeometric distribution is applicable if the following hold:

(1)   The population from which the sample is drawn consists of $N$ objects of two kinds, say, good and defective.

(2) Drawing of objects is at random *without* replacement.*

(3) Interest lies in the probability of a sample of $n$ containing $y$ defectives.

### 5.5.8. Applications of the Hypergeometric Distribution.

The hypergeometric distribution is much used in industry, where we are concerned with acceptance of a series of lots of product. We have some sort of sampling-decision setup; for example, take a sample of $n = 50$ parts, and accept the lot if one or none in the sample is non-conforming. Then we ask what proportion of a series of lots stand to be accepted under this scheme, if each contains 500 parts of which 10 are nonconforming. Other examples are the selection for an experiment of a sample of $n$ quantities of material, or animals, or people, or seeds, of which there are $N$ possibilities consisting of precisely two types.[†] An example in the author's experience is that of sampling a stockpile of material. Sampling of replacement-parts inventory is another common example.

Random sampling is essential, needless to say.

### 5.6. THE UNIFORM DISTRIBUTION

We now discuss three discrete distributions of somewhat more restricted application than the preceding three. The first is the uniform or discrete-rectangular distribution. This is the distribution from which random numbers are drawn. Suppose we have the numbers 0, 1, 2, ..., $M - 1$. All are considered equally likely to be chosen. Thus

$$p(y) = \begin{cases} 1/M, & y = 0, 1, ..., M - 1 \\ 0, & \text{otherwise.} \end{cases} \tag{5.60}$$

One can use formulas for the sums of powers of consecutive integers to show the following, using (5.6), (5.8), (5.12), and (5.13):

$$\mu = E(Y) = (M - 1)/2 \quad \text{(uniform distribution)} \tag{5.61}$$

$$\sigma_Y = \sqrt{(M^2 - 1)/12} \tag{5.62}$$

$$\alpha_3 = 0 \tag{5.63}$$

$$\alpha_4 = .6(3M^2 - 7)/(M^2 - 1). \tag{5.64}$$

* If drawing is *with* replacement, then the binomial distribution is exactly applicable; or if $N$ is very large, the binomial is nearly exact.

† If there are more than two types, we could use the multinomial distribution which we do not include in this book.

A special case occurs when $M = 2$. Then there are but two possible $y$'s with equal probability, and $\alpha_4$ for this simple pure bimodal distribution is 1.0. This is the least possible value which $\alpha_4$ can ever take. An application is when $y$ is the number of heads in a single flip of an unbiased coin.

From the applicational viewpoint, the uniform distribution with $M = 10$ is probably the most important. For it the digits 0, 1, ..., 9 are equally likely. This distribution is the one we hope to generate with a random-digit generating program (at least as closely as possible). For it $\mu = 4.5$, $\sigma = \sqrt{33/4} = 2.87$, and $\alpha_4 = 1.78$. If we take $M = 100$, we have the distribution for a two-digit random number 00, 01, ..., 99.

## 5.7.* THE GEOMETRIC DISTRIBUTION

This distribution is concerned with random independent drawings, with two possible outcomes for each, with constant probabilities for each outcome. So far, this is just the same situation as that for the binomial. But now instead of letting the number of defectives vary and holding $n$ *fixed*, we let $n$ *vary*, calling it $y$, and ask for the probability for each sample size till the first "defective" occurs. Let $\phi$ be the probability of a "defective," and $(1 - \phi)$ be the probability of a "good." Then

$$p(1) = P(Y = 1) = P(d) = \phi$$
$$p(2) = P(Y = 2) = P(gd) = P(g)\,P(d) = (1 - \phi)\phi$$
$$p(3) = P(Y = 3) = P(ggd) = P(g)\,P(g)\,P(d) = (1 - \phi)^2\phi.$$

In general we see that

$$p(y) = (1 - \phi)^{y-1}\phi, \qquad y = 1, 2, \ldots \qquad \text{(geometric distribution).} \quad (5.65)$$

Here, as for the Poisson, we have an infinite sample space, and also only one parameter $\phi$.

We note that no matter how small $\phi$ may be, $1 - \phi < 1$ and so $p(y + 1) < p(y)$, and we have a continually decreasing probability function. Thus the distribution is what we call J-shaped and has its mode at $y = 1$. Also we can expect considerable positive skewness.

The four moments are found to be as follows*:

$$\mu = E(Y) = 1/\phi \qquad \text{(geometric distribution)} \qquad (5.66)$$

$$\sigma_Y = \sqrt{1 - \phi}/\phi \qquad (5.67)$$

$$\alpha_3 = (2 - \phi)/\sqrt{1 - \phi} \qquad (5.68)$$

$$\alpha_4 = (9 - 9\phi + \phi^2)/(1 - \phi). \qquad (5.69)$$

* These are provable by summing infinite series. The direct method for this is one of the techniques of calculus of finite differences.

The mean $1/\phi$ of the geometric distribution makes much intuitive sense, since at each trial there is a probability $\phi$ of a failure or a defective, and it seems natural to expect that we would take $1/\phi$ trials on the average before the first failure. Thus if $\phi = .1$, we average 10 trials before a failure.

For small probabilities of failure $\phi$ we have approximately $1 - \phi = 1$, $\alpha_3 = 2$, and $\alpha_4 = 9$.

The geometric distribution has wide usage in reliability. For example, see the work of Lloyd and Lipow [13]. In this area of application we are likely to be strongly interested in the probability that the first failure occurs at a given trial or operation of some device or part, when we know $\phi$. Suppose $\phi = .001$. Then although p(1) is larger than any other p($y$), the average number of trials before the first failure is 1000.

A rather trivial but interesting example of the geometric distribution is that of there being, say, five different kinds of prizes in a cereal box. What is the average number of boxes before a child will have all of the prizes, assuming random loading of prizes, one per box? The first box is bound to yield a prize. After the first box, $\phi = 4/5$ is the probability of a new prize in the second box. So by (5.66), the average number of additional boxes in order to have two different prizes is $1/(4/5) = 5/4$. Now with two prizes, the value of $\phi$ becomes $3/5$, and the average number of new boxes before a third type of prize is obtained is $5/3$. Thus the total average number of boxes being the sum of these means is

$$1 + 5/4 + 5/3 + 5/2 + 5 = 11.4.$$

A less trivial problem is for two children to pool their prizes till each has at least one of each!

For sums of terms of the geometric distribution, we may use the following approach (in reality merely an application of a geometric progression). We seek, say,

$$P(a \leqslant Y \leqslant b) = p(a) + p(a + 1) + \cdots + p(b).$$

Then let

$$T = \phi(1 - \phi)^{a-1} + \phi(1 - \phi)^a + \cdots + \phi(1 - \phi)^{b-1}.$$

Multiplying by $1 - \phi$ yields

$$(1 - \phi)T = \phi(1 - \phi)^a + \phi(1 - \phi)^{a+1} + \cdots + \phi(1 - \phi)^b.$$

Subtracting the left sides gives $\phi T$. Also we see that the first term of

$(1 - \phi)T$ on the right is identical to the second of $T$, the second for $(1 - \phi)T$ with the third of $T$, and so on; hence the subtraction gives

$$\phi T = \phi(1 - \phi)^{a-1} - \phi(1 - \phi)^{b}.$$

Solving for the desired probability $T$ gives

$$T = \mathrm{P}(a \leqslant Y \leqslant b) = (1 - \phi)^{a-1} - (1 - \phi)^{b}. \qquad (5.70)$$

## 5.8.* THE NEGATIVE BINOMIAL DISTRIBUTION

This distribution is concerned with the same type of drawing as are the binomial and the geometric distributions, that is, independent drawing with a constant probability at each trial, for the two kinds of outcomes or objects.

The problem for the negative binomial becomes that of asking for the probability that the $k$th failure occurs at the $y$th trial. (Note that if $k = 1$, then this specializes to the geometric distribution.)

This problem seems a little difficult at first thought, then one readily notes that for the $k$th failure to occur on the $y$th trial, there must have been $k - 1$ failures scattered along anywhere between the first and the $(y - 1)$th trial, and the $k$th failure *must* occur at the $y$th trial. Thus we have, by (4.22), since there is independence:

$$\mathrm{P}(k\text{th failure at }y\text{th trial}) = \mathrm{P}(k - 1 \text{ in } y - 1 \text{ trials})\phi,$$

where $\phi = \mathrm{P}$ (failure). But the first factor on the right is just an ordinary binomial probability, namely

$$\mathrm{P}(k - 1 \text{ in } y - 1 \text{ trials}) = \mathrm{C}(y - 1, k - 1)\,\phi^{k-1}(1 - \phi)^{y-k}.$$

Hence, combining we have

$$\mathrm{p}(y) = \mathrm{C}(y - 1, k - 1)\,\phi^{k}(1 - \phi)^{y-k}, \qquad y = k, k + 1, \dots . \quad (5.71)$$

The two parameters for the negative binomial are seen to be $\phi$ and $k$.

Note again, as for the geometric distribution, that the sample size is the random variable and the number of defectives is constant, rather than vice versa. We thus have by (5.71) a distribution of sample sizes (till the $k$th failure).

The negative binomial is especially useful in life testing and reliability studies where sample acceptance of product and qualification of designs are two applications [13].

The negative binomial is intimately associated with the geometric distribution. For example, if $k = 2$, then $y$ for the negative binomial distribution is in reality $y_1 + y_2$ where $y_1$ is the number of trials till the first failure, and then $y_2$ is the number of *additional* trials till the second failure. In general $y = y_1 + y_2 + \cdots + y_k$. Using this concept and a little mathematical statistics it is easy to derive the following for the negative binomial, from (5.66)–(5.69):

$$\mu_Y = E(Y) = k/\phi \qquad \text{(negative binomial distribution)} \qquad (5.72)$$

$$\sigma_Y = (1/\phi) \sqrt{k(1 - \phi)} \qquad (5.73)$$

$$\alpha_3 = (2 - \phi)/\sqrt{k(1 - \phi)} \qquad (5.74)$$

$$\alpha_4 = 3 + \frac{6(1 - \phi) + \phi^2}{(1 - \phi)}. \qquad (5.75)$$

Next we may consider the problem of calculating terms of the negative binomial. The easiest way to evaluate a $p(y)$ term (5.71) is to look up in a table the binomial part $P(k - 1, \text{ in } y - 1 \text{ trials})$, and then to multiply by $\phi$.

For sums of terms of the negative binomial, we make use of the "non-trivial" example of two equivalent events given in Section 4.3.1. Suppose that we wish the following probability for the negative binomial

$$\sum_{y=k}^{y=L} p(y \mid k, \phi)_{\text{neg binom}}.$$

This is the probability that the $k$th failure occurs at the $L$th or an earlier trial. But if the $k$th failure occurred between the $k$th and $L$th trial, then surely there were at least $k$ failures up to the $L$th trial and possibly more, up to a maximum of $L$ failures. Thus the event of $k$ or *more* failures in $L$ is equivalent to: the $k$th failure at trial $L$ or *earlier*. But the former of these events is the binomial probability

$$\sum_{y=k}^{L} p(y \mid L, \phi)_{\text{binom}}.$$

Thus we have

$$\sum_{y=k}^{L} p(y \mid k, \phi)_{\text{neg binom}} = \sum_{y=k}^{L} p(y \mid L, \phi)_{\text{binom}}. \qquad (5.76)$$

Hence we may use binomial sums as on the right side in order to calculate sums of negative binomial terms as on the left side.

Finally, we now justify the name "negative binomial distribution." First recall that the binomial expansion

$$(x + y)^p = x^p + px^{p-1}y + \frac{p(p-1)}{2!} x^{p-2}y^2 + \cdots$$

follows, whether or not $p$ is a positive integer. If $p$ is not a positive integer, we shall need convergence of the right side. If $p = -k$ where $k$ is a positive integer, $x = 1$, and $y = -(1 - \phi)$, then the equation above gives

$$[1 - (1 - \phi)]^{-k} = 1 + (-k)[-(1 - \phi)] + \frac{(-k)(-k-1)}{2!}[-(1 - \phi)]^2 + \cdots$$

$$= 1 + k(1 - \phi) + \frac{(k+1)k}{2!}(1 - \phi)^2$$

$$+ \frac{(k+2)(k+1)k}{3!}(1 - \phi)^3$$

$$\cdots + \frac{(X-1)(X-2)\cdots(k+1)k}{(X-k)!}(1 - \phi)^{X-k} + \cdots$$

$$= 1 + C(k, k-1)(1 - \phi) + C(k+1, k-1)(1 - \phi)^2$$

$$+ C(k+2, k-1)(1 - k)^3 + \cdots$$

$$+ C(X-1, k-1)(1 - \phi)^{X-k} + \cdots.$$

But the terms in the last expansion are exactly those of (5.71) except for the factor $\phi^k$. To prove that the sum of the terms in (5.71) is 1, we see from the above that their sum is

$$\phi^k[1 - (1 - \phi)]^{-k} = \phi^k\phi^{-k} = 1.$$

## 5.9. GENERATING SAMPLES FROM DISCRETE DISTRIBUTIONS

There are some occasions when one may wish to sample from some known discrete population. For example, suppose we have a Poisson population with $\mu = 2$, like (b) of Table 5.2. We can figure the probabilities p($y$). Now suppose we take 100 random observations from this population. We would expect there to be 14 zeros, 27 ones, 27 twos, 18 threes and so on. But we would virtually never duplicate the theoretical results so closely. We might, however, like to see how well such a sample of 100 observations mirrors the population. To do this we might use a computer to generate random digits, say four places, 0000 to 9999. Then if the random number is 0000 to 1352 inclusive, we record the observation as a "0," since there are 1353 such random numbers out of 10,000. Next if the random number is 1353 to 4059 inclusive we record a "1," since there are 2707 such numbers, and so on. Since values of $p(y)$ are rounded, it often happens that $\sum p(y) \neq 1$. For this reason it may be best to make each cumulative total $P(y)$ match correctly to within the round-off error, and then find $p(y)$'s by subtraction, $P(y) - P(y - 1)$. This may give us a difference slightly different from a $p(y)$ found directly and rounded off to the agreed upon number of decimals.

## 5.10. SUMMARY

We have here discussed six discrete distributions or populations, applicable to a variety of sampling situations. They are widely useful probability models. They are defined on countable spaces such as 0 to $n$, 0 to $\infty$, or $k$ to $\infty$. They carry one to three parameters, which when specified give particular cases. The student should carefully distinguish the situations to which they are applicable. We have also included information on $\mu$, $\sigma$, $\alpha_3$, and $\alpha_4$ in terms of the parameters, and shown how to approximate the probability functions $p(y)$.

REFERENCES

1. U.S. Dept. of the Army, "Tables of the Binomial Probability Distribution." Nat. Bur. Standards, Appl. Math. Ser. 6. U.S. Govt. Printing Office, Washington, D.C., 1950.
2. Harvard Univ., Comput. Lab., "Tables of the Cumulative Binomial Probability Distribution." Harvard Univ. Press, Cambridge, Massachusetts, 1955.
3. W. H. Robertson, "Tables of the Binomial Distribution Function for Small Values of p." Tech. Services, Dept. of Commerce, Washington, D.C., 1960.
4. H. G. Romig, "Fifty to 100 Binomial Tables." Wiley, New York, 1947.
5. S. Weintraub, "Cumulative Probability Distribution for Small Values of p." Free Press of Glencoe, Collier-Macmillan, London, 1963.
6. T. C. Fry, "Probability and Its Engineering Uses." Van Nostrand-Reinhold, Princeton, New Jersey, 1928.
7. D. B. Owen and C. M. Williams, "Logarithms of Factorials from 1 to 2000," Sandia Corp. Monograph SCR-158. Available from Office of Tech. Services, Dept. of Commerce, Washington, D. C., 1959.
8. T. Kitagawa, "Tables of Poisson Distribution." Baifukan, Tokyo, 1952.
9. E. C. Molina, "Poisson's Exponential Binomial Limit." Van Nostrand-Reinhold, Princeton, New Jersey, 1947.
10. G. J. Lieberman and D. B. Owen, "Tables of the Hypergeometric Probability Distribution." Standard Univ. Press, Stanford, California, 1961.
11. M. G. Kendall, "The Advanced Theory of Statistics," Vol. I. Griffin, London, 1945.

12. I. W. Burr, Some approximate relations between terms of the hypergeometric, bino-
    mial and Poisson distributions. *Communications in Stat.* **1**, 297–301 (1973).
13. D. K. Lloyd and M. Lipow, "Reliability: Management, Methods, and Mathematics."
    Prentice-Hall, Englewood Cliffs, New Jersey, 1964.

ADDITIONAL REFERENCES

J. J. Bartko, A note on the negative binomial distribution. *Technometrics* **4**, 609–610
    (1962).
C. I. Bliss and A. R. G. Owen, Negative binomial distributions with a common k.
    *Biometrika* **45**, 37–58 (1958).
H. D. Brunk, J. E. Holstein and F. Williams, A comparison of binomial approximations
    to the hypergeometric distribution. *Amer. Statist.* **22**, 24–26 (1968).
G. Hadley and T. M. Whitin, Useful properties of the Poisson distribution. *Operations
    Res.* **9**, 408–410 (1961).
P. J. Sandiford, A new binomial approximation for use in sampling from finite
    populations. *Jour. Amer. Statist. Assoc.* **55**, 718-722 (1960).
U.S. Army Ordnance Corps. Tables of cumulative binominal probabilities, Ord-
    nance Corps. Pamphlet ORDP20-1, Office of Tech. Services, Dept. of Commerce,
    Washington, D. C., 1952.

## Problems

In Problems 5.1–5.26 calculate for the given distributions the values
of $p(y)$ to the best three decimal places, carrying along till $p(y) < .0005$,
and check by totaling. It is desirable to work out to four places and
round off. Recurrence relations like (5.36) may be helpful.

5.1.  $n = 4,$      $\phi = .4.$
5.2.  $n = 5,$      $\phi = .3.$
5.3.  $n = 10,$     $\phi = .03.$
5.4.  $n = 10,$     $\phi = .02.$
5.5.  $n = 6,$      $\phi = 1/6$ (applicable to tosses of six dice).
5.6.  $n = 10,$     $\phi = 1/2$ (applicable to tossing 10 coins and to runs
above and below an average).

Poisson problems:

5.7.  $\mu = .3.$      5.10.  $\mu = .02.$
5.8.  $\mu = .5.$      5.11.  $\mu = 1.5.$
5.9.  $\mu = .05.$     5.12.  $\mu = 2.5.$

Hypergeometric problems:

5.13.  $N = 10,$     $K = 2,$      $n = 3.$
5.14.  $N = 12,$     $K = 2,$      $n = 4.$

5.15.  $N = 50$,    $K = 4$,    $n = 8$.
5.16.  $N = 50$,    $K = 3$,    $n = 5$.
5.17.  $N = 150$,   $K = 15$,   $n = 10$.
5.18.  $N = 200$,   $K = 20$,   $n = 10$.

Geometric problems:

5.19.  $\phi = .2$.    5.21.  $\phi = .1$.
5.20.  $\phi = .3$.    5.22.  $\phi = .25$.

Negative binomial problems:

5.23.  $k = 2$,  $\phi = .2$.    5.25.  $k = 3$,  $\phi = .1$.
5.24.  $k = 2$,  $\phi = .1$.    5.26.  $k = 3$,  $\phi = .2$.

5.27.    Use available tables to find the $p(y)$'s for such of Problems 5.1–5.26 as are assigned. Use available tables to find the following probabilities:

Binomial problems:

5.28. $n = 100, \phi = .15$: $p(12)$, $P(Y \leqslant 15)$, $P(Y > 20)$, $P(10 \leqslant Y \leqslant 20)$.
5.29. $n = 40$,  $\phi = .20$: $p(8)$,  $P(Y < 5)$,  $P(Y \geqslant 10)$, $P(5 < Y < 10)$.
5.30. $n = 200, \phi = .04$: $p(8)$,  $P(Y \leqslant 6)$,  $P(Y > 12)$, $P(5 < Y \leqslant 12)$.
5.31. $n = 500, \phi = .03$: $p(20)$, $P(Y \leqslant 15)$, $P(Y > 20)$, $P(12 < Y \leqslant 18)$.
5.32. $n = 50$,  $\phi = .20$: $p(12)$, $P(Y < 5)$,  $P(Y > 7)$,  $P(5 \leqslant Y \leqslant 15)$.
5.33. $n = 100, \phi = .40$: $p(35)$, $P(Y < 30)$, $P(Y > 45)$, $P(30 < Y \leqslant 50)$.

Poisson problems:

5.34.  $\mu = .4$:  $p(1)$, $P(Y > 1)$, $P(Y \leqslant 2)$,  $P(0 < Y \leqslant 3)$.
5.35.  $\mu = 1.8$: $p(3)$, $P(Y \leqslant 1)$, $P(Y > 1)$,  $P(1 \leqslant Y \leqslant 4)$.
5.36.  $\mu = 8$:  $p(6)$, $P(Y \leqslant 8)$, $P(Y > 8)$,  $P(6 \leqslant Y \leqslant 10)$.
5.37.  $\mu = 6$:  $p(7)$, $P(Y \leqslant 5)$, $P(Y \geqslant 10)$, $P(3 \leqslant Y \leqslant 9)$.
5.38.  $\mu = .04$: $p(0)$, $P(Y > 0)$, $U$ for which $P(Y > U) \leqslant .01$.
5.39.  $\mu = .12$: $p(1)$, $P(Y > 0)$, $U$ for which $P(Y \geqslant U) \leqslant .005$.

Hypergeometric problems:

5.40.  $N = 80$,  $K = 10$, $n = 20$: $p(2)$, $P(3)$, $P(Y > 1)$.
5.41.  $N = 60$,  $K = 8$,  $n = 12$: $p(2)$, $P(4)$, $P(Y \geqslant 4)$.
5.42.  $N = 40$,  $K = 4$,  $n = 10$: $p(0)$, $P(2)$, $P(Y > 4)$.
5.43.  $N = 40$,  $K = 3$,  $n = 6$:  $p(1)$, $P(1)$, $P(1 < Y \leqslant 5)$.
5.44.  $N = 100, K = 32$, $n = 10$: $p(4)$, $P(3)$, $P(Y \geqslant 5)$.
5.45.  $N = 100, K = 25$, $n = 10$: $p(5)$, $P(4)$, $P(Y < 6)$.

5.46.    Find $\mu$, $\sigma$, $\alpha_3$ , $\alpha_4$ for such of problems 5.1–5.45 as are assigned.

5.47.    Find $\mu$, $\sigma$, $\alpha_3$ , $\alpha_4$ for the uniform distribution for 0, 1, ..., 9.

5.48.   What is the limit for $\alpha_4$ for a uniform distribution on 0, 1, ..., $M - 1$, as $M \to \infty$? This is the value for the continuous "rectangular" distribution. (See Chapter 6.)

5.49.   What is the limit for $\alpha_3$ and for $\alpha_4$ for the geometric distribution, as $\phi \to 0$? These limits are those for the continuous "exponential" distribution. (See Chapter 6.)

Use binomial tables to find for the following the values of $U$ and $L$ such that $P(Y \leqslant L) \leqslant .005$ and $P(Y \geqslant U) \leqslant .005$:

| | | | | |
|---|---|---|---|---|
| 5.50. | $n = 40$, | $\phi = .20$. | 5.52. | $n = 500$, $\phi = .10$. |
| 5.51. | $n = 100$, | $\phi = .08$. | 5.53. | $n = 50$, $\phi = .08$. |

5.54.   What does (5.76) become when $k = 1$? How might this help on the geometric distribution?

5.55.   Program a computer to generate terms p($y$) of the Poisson distribution and use it for $\mu = 5$, stopping when p($y$) $\leqslant .00005$.

5.56.   Program a computer to generate random $y$'s for the distribution in Problem 5.55 or any of 5.1–5.26 and run off 100 random $y$'s.

5.57.   (a)   Three-digit random numbers are generated by three consecutive random digits from 0 to 9. What is the probability of the next one being 333? What is the sample space and event? What is the probability of the next three-digit random number having all three digits identical?

(b)   What is the probability of the next six random digits from 0 to 9 being 777777? A famous question is as to whether there is at least one such run in the complete numerical representation of $\pi$, 3.14159.... If the digits act like random numbers, is such a sequence probable?

5.58.   Suppose a slot machine has three cylinders, *each* with 15 symbols, including just one lemon. The cylinders are spun at random, and a sizable prize awarded if all three lemons show. What is the probability of such an event?

5.59.   In the first 30 significant figures for $\pi$, there are no zeros. Assuming randomness of digits, what is the probability of no zeros in the 30?

5.60.   In the game of poker a player may only open the betting if his hand contains a pair of jacks, or is even better than this. The probability of such a hand is .206. A player complains that he did not have an opening hand in the first six deals. What is the probability of such an event?

For the following cases, how might you approximate with a simpler distribution:

5.61. Binomial: $n = 175$, $\phi = .04$, $y = 6$.
5.62. Binomial: $n = 60$, $\phi = .03$, $y = 1$.
5.63. Binomial: $n = 2000$, $\phi = .008$, $y = 3$.
5.64. Binomial: $n = 800$, $\phi = .001$, $y = 1$.
5.65. Hypergeometric: $N = 1000$, $k = 500$, $n = 200$, $y = 8$.
5.66. Hypergeometric: $N = 600$, $k = 200$, $n = 150$, $y = 2$.
5.67. Hypergeometric: $N = 3000$, $k = 30$, $n = 15$, $y = 1$.
5.68. Hypergeometric: $N = 2000$, $k = 20$, $n = 40$, $y = 1$.

5.69. The following problem is given by Burr (Ref. [1], Chap. 6): "Having bought a bag of roasted chestnuts, the author walked home in the dark eating then with much gusto. After eating about 20, he arrived home, and, in opening the remaining 10 under the light, he found that 7 contained worms. What is the probability that none of the 20 contained worms? Or to phrase the problem better for statistical analysis: If there were only 7 wormy chestnuts among the original 30, what is the probability of drawing the first 20 all free from worms?"

5.70. Suppose in driving a car you are not afraid to take a chance, where the probability of an accident is "only" one in 100. If you take 50 such chances, what is the approximate probability of your having no accidents in the 50? What is the expected number of accidents in 50?

5.71. Suppose that a certain newspaper averages four typographical errors per page. What is the probability of less than seven on a given page? How well justified is the distribution used likely to be in practice?

5.72. If there is one chance in $10^9$ of a bomb being on a plane, what are the approximate probabilities of there being no bombs on the plane? One bomb? Two bombs?

5.73. In an experiment on two materials $A$ and $B$, being compared under a variety of conditions, an observation of the performance for $A$ is $x_1$ and for $B$ is $y_1$ for the first set of conditions. Likewise for the second set of conditions the performances are $x_2$ and $y_2$, and so forth, for, say, 12 conditions. Variation in performance will occur even for the same material and same conditions, but if the materials are equivalent, $x_i$ is as likely to be above $y_i$ as below. Suppose in your experiment $x_i > y_i$

11 times and $x_i < y_i$ only once. What do you conclude? Indicate your reasoning.

5.74.  An industrial lot of subassemblies will be accepted if upon examining a random sample of 12 of them the total number of defects found is four or less. (There can be more than one defect on a single subassembly, up to a possible large number.) What is the appropriate distribution? Suppose a lot averages only two defects on 12 subassemblies. What is the chance of it being accepted?

5.75.  Suppose in Problem 5.74, a subassembly is called a "defective" if it has one or more defects on it. For acceptance of the lot we now permit four "defectives" to be in the 12 examined. If one-sixth of the subassemblies in the lot contain defects, what is the probability that the lot will be accepted? What is the appropriate distribution?

5.76.  Give one example for each of the distributions: Poisson, binomial, and hypergeometric in your major field of interest.

5.77.  Write out the expression for the probability that a 13-card-hand from a 52-card deck contains exactly seven hearts. Similarly, that it contains exactly seven cards of some one unspecified suit.

5.78.  Write out the expression that a 13-card-hand contains exactly three aces? Can you answer this by a standard probability distribution?

5.79.  Find a simple formula for $P(y)$ for the geometric distribution. Check by $P(\infty) = 1$.

5.80.  For a Poisson distribution with parameter $\mu$, show that $\sum_{y=0}^{\infty} p(y) = 1$, by using the series expansion for $e^x$, where $x = \mu$.

5.81.  For a Poisson distribution with parameter $\mu$, show that $\mu = \sum y\, p(y)$.

5.82.  In attempting to prove that $\sigma_Y^2$ for a Poisson with parameter $\mu$ is also $\mu$, we should like to have $E(Y^2)$ so as to use (5.22). But we unfortunately cannot find $E(Y^2) = \sum Y^2 p(Y)$ directly in the same way $E(Y)$ was found in Problem 5.81. Instead we can find $E(Y^2)$ indirectly by first finding $E[(Y)(Y-1)]$, then from this $E(Y^2)$. Carry through to verify (5.40).

5.83.  Prove (5.22) by taking (5.8) as starting point, squaring out the binomial, then using laws of expectations or summations.

# Chapter 6

# SOME CONTINUOUS PROBABILITY DISTRIBUTIONS

## 6.1. CONTINUOUS PROBABILITY DISTRIBUTIONS

In Chapter 5 we were concerned with mathematical laws of variation for discrete or counted data. In the present chapter we are concerned with laws of variation appropriate to continuous random variables, that is, for application to measured data. By far the most important such model is the so-called normal curve. It is of wide use, has many convenient properties, some of which are unique, and it can be called a first approximation to the law of variation in a great many cases. But there are many situations in which the normal curve is clearly not applicable, and we need to use other frequency or probability curves to obtain a better approximation to the law of variation in question. Thus in this chapter we present the normal curve first, and then a number of other probability distributions. Finally we discuss briefly the central limit theorem, and for completeness an inequality of rather limited *practical* application.

## 6.2. SOME GENERAL PROPERTIES OF CONTINUOUS DISTRIBUTIONS

We first discuss certain general properties of continuous probability distributions, before taking up the normal curve and other distributions. Many of these properties are most naturally expressed through the use of calculus, but we shall try to make the properties understandable to those who have studied little or no calculus.

In general we have what is called a "density function" $f(y)$ for any distribution. It is defined over some range or interval of the $y$ axis, which is the sample space for the random variable. Since the random variable varies continuously, taking all values over this interval, the sample space $W$ is said to be uncountably infinite, a fact which brings in some slight complications or peculiarities. An example of a probability curve is shown in Fig. 6.1.

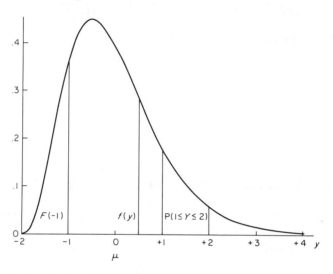

**Fig. 6.1.** A typical probability distribution, showing density function $f(y)$ plotted against the continuous random variable $y$. This distribution is not symmetrical around the mean, $\mu = 0$.

The density function $f(y)$ is the ordinate up to the curve. The sample space $W$ runs from $-2$ to $+\infty$. Using units as shown, the total area is 1, which corresponds to the fact that the probability $P(Y \text{ in } W) = P(-2 \leqslant Y < \infty) = 1$. The complete definition for $f(y)$ is

$$f(y) = \begin{cases} (8/3)(y+2)^3\, e^{-2y-4}, & -2 \leqslant y \\ 0, & y < -2, \end{cases} \tag{6.1}$$

where the second line of (6.1) extends the definition to $(-\infty, \infty)$. Thus $f(y)$ is such that

$$P(-\infty < Y < \infty) = 1 = \int_{-\infty}^{\infty} f(y)\, dy.* \tag{6.2}$$

* This "integral" gives the area under the curve $Z = f(y)$ between the limits $-\infty$ and $\infty$.

As shown in Fig. 6.1 the area bounded by the ordinates at $y = 1$ and 2, the curve, and the $y$ axis is

$$P(1 \leqslant Y \leqslant 2) = (\text{area under curve, } y = 1 \text{ to } 2) = \int_1^2 f(y)\, dy,$$

since the total area is 1. In general

$$P(a \leqslant Y \leqslant b) = (\text{area under curve, } y = a \text{ to } b) = \int_a^b f(y)\, dy. \quad (6.3)$$

One of the peculiarities of probabilities over continuous sample spaces is that the probability of $Y$ taking on any one exact value is zero.* Hence, for example, since $P(Y = a) = 0$, $P(Y = b) = 0$,

$$P(a < Y < b) = P(a \leqslant Y \leqslant b), \quad (6.4)$$

and so on. [For a *discrete* probability distribution, if $a$ and/or $b$ were possible values of $Y$, then (6.4) would not hold.]

A special case of (6.3) occurs when we have a small increment $\Delta y$ added to $y_0$:

$$P(y_0 \leqslant Y \leqslant y_0 + \Delta y) = (\text{area under curve, } y = y_0 \text{ to } y_0 + \Delta y)$$
$$\doteq f(y_0)\, \Delta y \quad (\text{approximately}) \quad (6.5)$$

since the exact probability would be a narrow strip from $y = y_0$ to $y = y_0 + \Delta y$, bounded by the $y$ axis and the curve, whereas the second line of (6.5) would be a rectangular area between the same vertical lines and of height $f(y_0)$. For small $\Delta y$'s the ratio of the areas is very nearly 1.

Finally we define the "distribution function" corresponding to the density function $f(y)$ by $F(y)$ as follows:

$$F(y_0) = P(Y \leqslant y_0) = (\text{area to left of } y = y_0) = \int_{-\infty}^{y_0} f(y)\, dy, \quad (6.6)$$

giving in general

$$\frac{d}{dy} F(y) = F'(y) = f(y). \quad (6.7)$$

Note that

$$F(-\infty) = 0, \qquad F(+\infty) = 1$$

* Unless the probability law is a mixed variables–attributes function, which is very rare in applications.

and that as $y$ increases, $F(y)$ does also, in general. Further, since the two sets of points

$$Y \leqslant a, \qquad a < Y \leqslant b$$

are mutually exclusive and have as their union $Y \leqslant b$, we have by (4.17)

$$P(Y \leqslant a) + P(a < Y \leqslant b) = P(Y \leqslant b).$$

The first and third probabilities are $F(a)$ and $F(b)$, respectively. Hence solving,

$$P(a < Y \leqslant b) = F(b) - F(a). \tag{6.8}$$

This probability equals either member of (6.4) or also $P(a \leqslant Y < b)$. Finally we simply note that

$$f(y) \geqslant 0, \qquad 0 \leqslant F(y) \leqslant 1. \tag{6.9}$$

For the density function of interest in applications there are from one to four "parameters" which particularize the given *type* of $f(y)$ to a special case. In the case of the discrete distributions, the binomial had two, $\phi$ and $n$, while the Poisson had just one, $\mu$. For continuous distributions, one parameter will be used for the mean or average, and another for the variability. Then there may be one or two determining the curve shape (degree of lack of symmetry, etc.).

### 6.2.1. Moments for a Continuous Distribution.

Let us define population moments determined by the population density function $f(y)$. These are simply analogies to (5.6)–(5.8), (5.12), and (5.13), but for a continuous variable $y$ rather than a discrete one. Hence integrals rather than sums are appropriate. However, it may be worthwhile for all readers to see a moment built up as a sum first. Thus suppose that the range where $f(y) > 0$ is from $a$ to $b$, with both finite. We may break up this range into $n$ equal increments of length $\Delta y = (b - a)/n$. Then taking $y_1 = a + \Delta y$, $y_2 = a + 2\Delta y$, ..., $y_i = a + i\Delta y$, ..., $y_n = a + n\Delta y = b$, we can form

$$\frac{\sum_{i=1}^{n} y_i f(y_i) \Delta y}{\sum_{i=1}^{n} f(y_i) \Delta y}.$$

Now $f(y_i) \Delta y$ is by (6.5) the approximate probability of a $y$ value lying between $y_{i-1} = a + (i - 1) \Delta y$ and $y_i = a + i\Delta y$. This is analogous to $p(y_i)$ for a discrete variable. Thus the given ratio is analogous to the last expression of (5.6), namely $\sum_y y\, p(y)$, except that we now have a denominator, because we cannot be sure that $\sum_{i=1}^{n} f(y_i) \Delta y_i = 1$, since each $f(y_i) \Delta y$ is only an approximate probability. Now let $n$

increase while correspondingly $\Delta y$ approaches zero. Then the $\sum f(y_i) \Delta y_i$ approaches a limit which proves to be one; the numerator also approaches a limit. In calculus, by the so-called fundamental theorem of integration, these are $\int_a^b f(y) \, dy = 1$ and $\int_a^b yf(y) \, dy = \mu_Y$, respectively, by definition. Thus letting $a$, $b$ become infinite we have (whenever it exists)

$$\mu = \mu_Y = E(Y) = \int_{-\infty}^{\infty} yf(y) \, dy. \tag{6.10}$$

Analogously to (5.8) we have also (when it exists)

$$\sigma^2 = \sigma_Y^2 = E[(Y - \mu)^2] = \int_{-\infty}^{\infty} (y - \mu)^2 f(y) \, dy. \tag{6.11}$$

These are the weighted average values over space $W$ of $Y$ and $(Y - \mu)^2$, respectively. The alternative form of

$$\sigma_Y^2 = E(Y^2) - [E(Y)]^2 \tag{6.12}$$

follows directly from (6.11) and is often useful. See (5.22) also. Finally we have just as in (5.11), (5.13), (5.23), and (5.24)

$$\alpha_3 = E[(Y - \mu)^3]/\sigma^3 = [E(Y^3) - 3\mu \, E(Y^2) + 2\mu^3]/\sigma^3 \tag{6.13}$$

for skewness, and

$$\alpha_4 = E[(Y - \mu)^4]/\sigma^4 = [E(Y^4) - 4\mu \, E(Y^3) + 6\mu^2 \, E(Y^2) - 3\mu^4]/\sigma^4 \tag{6.14}$$

for kurtosis or contact (if these moments exist).*

## 6.3. THE NORMAL CURVE

As mentioned previously, the normal curve is certainly the best known and most widely used distribution for a continuous random variable. It can often be regarded as a first approximation to a distribution of data. As we shall see it has many convenient properties. Other distributions are often compared with it.

When we use a standardized variable $u$, that is, one having mean 0, and standard deviation 1, the expression for the density function takes the form

$$\phi(u) = e^{-u^2/2}/\sqrt{2\pi} \qquad (-\infty, \infty), \tag{6.15}$$

* Formula (6.12) and the last forms of (6.13) and (6.14) are easily derived by expanding the binomials under the integrals and breaking up into separate integrals.

where $e = 2.71828...$ and $\pi = 3.14159...$ . The lower case Greek phi ($\phi$) is often used in this way, and has no connection with the parameter $\phi$ for the probability in the binomial distribution.

### 6.3.1. Properties of the Normal Distribution.

Some of the properties of the standard normal distribution are now given. See also Fig. 6.2.

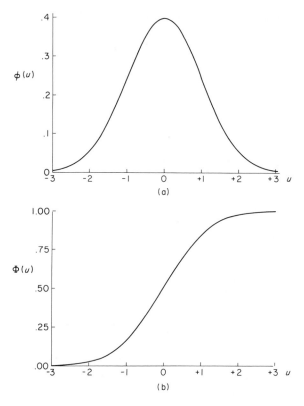

**Fig. 6.2.** The standardized normal distribution. (a) Graph showing the density function $\phi(u)$ plotted against $u$. (b) Graph showing the cumulative probability less than or equal to $u$, that is, the distribution function $\Phi(u)$, plotted against $u$.

1.  $\phi(u)$ reaches a maximum value at $u = 0$, for then the exponent of $e$ is at its highest *algebraic* value, 0.
2.  $\phi(u)$ decreases steadily as $u$ recedes from 0 in either direction, since the exponent of $e$ becomes more strongly negative. The approach toward zero is very rapid.

3.   The curve is symmetrical around the vertical line where $u = 0$, since $u$ occurs to even powers only in (6.15) and thus $\phi(u) = \phi(-u)$ for all $u$'s.
4.   It can be shown that there are two "points of inflection" occurring where $u = -1$ and $u = 1$. These are points such that the curve changes from having its concave side down to having it up, or vice versa.
5.   The total area under the curve from $-\infty$ to $\infty$ is 1, which is the reason for the coefficient $1/\sqrt{2\pi}$.*
6.   Areas under the normal curve, that is, probabilities such as (6.4), are available in tables, all having been worked out by approximate integration methods. The probabilities most often found in tables are the half-central area 0 to $u$, the full central area $-u$ to $u$, and the cumulative area $-\infty$ to $u$, which is also $\Phi(u)$. See Fig. 6.2b. This is given by

$$\Phi(u) = P(-\infty \text{ to } u) = \int_{-\infty}^{u} \phi(v) \, dv. \tag{6.16}$$

$\Phi(u)$ cannot be found in terms of elementary functions.
7.   Some standard areas worth noting are

| $u = -1$ | to | $u = +1$ | area $= .6827$ |
|---|---|---|---|
| $= -2$ | | $= +2,$ | $= .9545$ |
| $= -3$ | | $= +3,$ | $= .9973$ |
| $= -1.960$ | | $= +1.960,$ | $= .9500$ |
| $= -2.576$ | | $= +2.576,$ | $= .9900.$ |

* A nonrigorous sketch of the proof follows. Since the indefinite integral of $e^{-u^2/2}$ cannot be found we proceed as follows: Let $I = \int_{-\infty}^{\infty} e^{-x^2/2} \, dx > 0$, then

$$I^2 = \int_{-\infty}^{\infty} e^{-x^2/2} \, dx \int_{-\infty}^{\infty} e^{-y^2/2} \, dy = \iint e^{-(x^2+y^2)/2} \, dx \, dy,$$

where the double integral is over the whole plane. Letting $x = r \cos \theta$, $y = r \sin \theta$, we have

$$I^2 = \int_{0}^{2\pi} \int_{0}^{\infty} e^{-r^2/2} r \, dr \, d\theta,$$

the limits again covering the plane via polar coordinates. But now the variables are separable and integrable. Since

$$\int_{0}^{2\pi} d\theta = 2\pi, \qquad \int_{0}^{\infty} e^{-r^2/2} r \, dr = -e^{-r^2/2} \Big]_{0}^{\infty} = 1,$$

we have $I^2 = 2\pi$, or $I = \sqrt{2\pi}$.

8. The distribution function $\Phi(u)$ has its cumulative graph starting from 0, at $-\infty$, and steadily rising. It has symmetry around the point $u = 0$, $\Phi(u) = .5$. This means that if the curve for $\Phi(u)$ were rotated 180°, the curve would be unchanged. [This is true of all $F(y)$ distribution functions if the corresponding density function $f(y)$ is symmetrical.]

9. The mean for $\phi(u)$ occurs at $u = 0$, by the symmetry. Likewise the median is also at $u = 0$, since 50% of the area is on each side. Moreover the mode is at $u = 0$ since $\phi(u)$ is a maximum there.

10. The standard deviation can be shown to be 1, $\alpha_3 = 0$, and $\alpha_4 = 3$. [These follow from (6.12)–(6.14), using the footnote to property 5 and integration by parts.] Thus 0 and 3 are "normal" values for the standardized moments $\alpha_3$ and $\alpha_4$.

### 6.3.2. The General Normal Curve.

In practical applications we do not have a random variable with mean 0 and standard deviation 1. Instead we have as a mathematical model a normal distribution with some mean $\mu$ and standard deviation $\sigma$. Then the density function takes the form

$$f(y) = \frac{e^{-(y-\mu)^2/2\sigma^2}}{\sigma\sqrt{2\pi}} \qquad (-\infty, \infty). \qquad (6.17)$$

Again the total area under this curve is 1. The maximum value of $f(y)$ occurs at $y = \mu$, and there is symmetry around the line $y = \mu$, following properties 1 and 3 of the previous section. The connection between $u$ and $y$ is the following:

$$u = \frac{y - \mu}{\sigma} \qquad \text{or} \qquad y = \mu + u\sigma. \qquad (6.18)$$

Thus $u$ is the number of $\sigma$'s that $y$ is away from $\mu$. It is a standard variable corresponding to $y$ via (6.18).

### 6.3.3. Sketching a Normal Curve.

In order to sketch a normal curve for given data having $\mu$ and $\sigma$ known (or at least approximated), we may proceed as follows. Choose any convenient horizontal scale to cover the range and locate the points at $\mu$, $\mu \pm \sigma$, and $\mu \pm 3\sigma$. Then erect an ordinate at any convenient height over $\mu$. At $\mu \pm \sigma$ erect ordinates .6 as high as that at $\mu$. Then connect these three points with a curve resembling that in Fig. 6.2a (concave downward). Continue this curve in each direction so as to practically reach the axis at $\mu \pm 3\sigma$, as in Fig. 6.2a. (Curve is concave upward in these portions.) Note that the points $\mu$, $\mu \pm \sigma$, and $\mu \pm 3\sigma$ on the $y$ scale correspond to 0, $\pm 1$, and $\pm 3$ on the $u$ scale. The .6 figure comes from dividing: $\phi(1)/\phi(0)$, or using (6.15)

$$\frac{e^{-.5}/\sqrt{2\pi}}{e^{0}/\sqrt{2\pi}} = e^{-.5} = .607.$$

Such sketches are often useful in visualizing problems and data, and in writing reports.

For a full graph one can use (6.15) or (6.17) and a table of exponential values of $e$.

### 6.3.4. Approximating Probabilities by a Normal Distribution.

A thousand observations were made on a dimension of an electrical contact. They gave an approximately normal distribution with $\bar{y} = $ .43324 in. and $s = .00865$ in. Suppose that we treat these sample results as population values, that is, we assume $\mu = .43324$ in., $\sigma = $ .00865 in. (In Chapter 7 we shall see that $\bar{y}$ and $s$ as estimates here of $\mu$ and $\sigma$ are subject to errors of .00027 and .00019 in. respectively.) Assuming normality and $\mu$, $\sigma$ as above, let us now answer several typical questions.

1. What percentage of contacts will be .450 in. or below? This kind of question is easily answered using the first form of (6.18). Thus

$$u = \frac{.450 \text{ in.} - .43324 \text{ in.}}{.00865 \text{ in.}} = 1.94.$$

Using Table I in the Appendix, wherein the probabilities $\Phi(u)$ of $u$ or less are given, we find .9738. Thus we can expect 97.38 % of the contacts to have dimensions of .450 in. or less. We could have carried $u$ out to more decimal places, necessitating interpolation in the table. But in practice this is seldom justifiable, since our values of $\mu$ and $\sigma$ are usually more or less fallible estimates rather than the true values, and our assumption of normality is somewhat suspect.

2. What percentage of contacts can be expected to lie between specifications, or requirements, of .420 to .450 in. Proceeding as before, we seek the percentage below .420 in.:

$$u = \frac{.420 \text{ in.} - .43324 \text{ in.}}{.00865 \text{ in.}} = -1.53.$$

Looking up $-1.53$ gives $\Phi(-1.53) = .0630$. Thus we have $97.38\%$ below .450 in. and $6.30\%$ below .420 in. This leaves $91.08\%$ between, using (6.8). Or we have about $9\%$ outside of specifications.

3.  What dimension will have only $1\%$ below it? This is simply a reversal of the first question. We can seek $\Phi(u) = .0100$, looking through the body of Table I, interpolating if desired; $u = -2.327$ or $-2.33$. Or we can use the last line of Table I, in which $u$ values are given for certain simple $\Phi(u)$ probabilities. Thus using (6.18),

$$y = .43324 \text{ in.} - 2.33(.00865 \text{ in.}) = .4131 \text{ in.}$$

4.  Between what two limits can we expect to find the middle 990 of the 1000 contact dimensions? Since we are seeking to enclose the middle 990 or $99\%$, there will be five below the lower limit and five above the upper limit. Thus the probability $\Phi(u)$ will be $5/1000 = .005$ for the lower limit and .995 for the upper. Looking up these in Table I, and interpolating we find $u = \pm 2.575$, or the last line of that table gives the more accurate value of 2.576. Then we use the second form of (6.18), finding

$$X = .43324 \text{ in.} \pm 2.576(.00865 \text{ in.}) = .4110 \text{ in.,} \quad .4555 \text{ in.,}$$

again resisting the temptation to carry too much precision.

Problems such as these occur over and over again in practice.

The normal distribution is also useful in approximating a term or sums of terms of binomial distributions when the probabilities are quite symmetrically grouped around $n\phi$. For example if $n = 400$, $\phi = .20$, then we use $\mu = n\phi = 80$, $\sigma = \sqrt{n\phi(1 - \phi)} = 8$ by (5.33), (5.34). (Also $\alpha_3 = .075$.) Let us now approximate $p(78)$ and $P(78 \leqslant Y \leqslant 83)$. For the former individual probability we need an area under the approximating normal curve, most closely giving the "block" area for $Y = 78$, namely from 77.5 to 78.5.

$$u = \frac{77.5 - 80}{8} = -.312; \qquad u = \frac{78.5 - 80}{8} = -.188.$$

Using Table I for the areas below, we find .3775 and .4255. The difference is .0480. A binomial table gives .0487.

In order to find the area under the normal curve approximating $P(78 \leqslant Y \leqslant 83)$ we note that this should cover the block areas 77.5–78.5, 78.5–79.5,..., 82.5–83.5, that is, from 77.5 to 83.5. Hence

$$u = \frac{77.5 - 80}{8} = -.312; \qquad u = \frac{83.5 - 80}{8} = .438.$$

The respective cumulative areas by Table I are .3775 and .6693, giving a difference of .2918. Again a binomial table gives .2911.

Carefully note the use of the "continuity correction" of .5 in the foregoing. Similar approximations may be made for the Poisson and hypergeometric distributions when the distributions are symmetrical.

### 6.4. THE RECTANGULAR DISTRIBUTION

The "rectangular" or "uniform" distribution has some practical applications. But perhaps its chief interest is as an extreme type of distribution with the probability stopping abruptly, and with uniform $f(y)$ between the limits. The rectangular distribution is the continuous analog of the discrete uniform distribution of Section 5.6. It is defined as follows:

$$f(y) = \begin{cases} 1/(b-a), & a < y < b \\ 0, & \text{elsewhere.} \end{cases} \qquad (6.19)$$

Figure 6.3 shows $f(y)$ and $F(y)$ for this distribution. The constant value of $f(y)$ of course makes the rectangular area 1. In the active portion

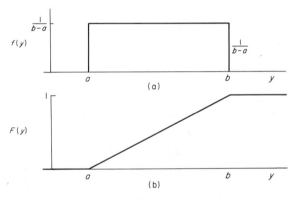

**Fig. 6.3.** (a) Density function $f(y)$, and (b) distribution function $F(y)$, for the general rectangular distribution from $a$ to $b$.

$(a, b)$ of $F(y)$, this function represents the proportion of area under the horizontal line in Fig. 6.3a from $a$ to $y$, that is

$$F(y) = \begin{cases} (y-a)/(b-a), & a < y < b \\ 0, & y < a \\ 1, & b < y. \end{cases} \qquad (6.20)$$

Often it is convenient to take $a = 0$, that is, to start the active range for $y$ at the origin. For calculational purposes it is even easier to run from $-1$ to $+1$ with $f(y) = .5$ between.

One useful application of (6.19) is a random angle from 0 to $\pi$, or 0 to $2\pi$ in radians, or 0 to 180°, or 0 to 360°. Another possible application occurs when we work with a large number of random digits, for example, five or six. In the latter case we have 000000 to 999999. We can take $a = 0$, $b = 1,000,000$, or if we wish to be more accurate in treating such numbers, we can use $a = -.5$, $b = 999,999.5$. (The latter choice of $a$ and $b$ is a "correction for continuity," and often occurs when we use continuous distributions to approximate discrete ones.) In industrial work we can encounter something like a rectangular distribution, if the production process manufactures parts with measurements forming a normal curve which has far too much spread to meet the required limits, say $L$ and $U$. For example, in Fig. 6.2 the limits might be $-.5$ to $.5$. If all parts within these limits are sorted out, the resulting distribution will be much like (6.20). For this reason design engineers sometimes conservatively assume a rectangular distribution between specification limits.

The following are the moment characteristics for (6.20)*

$$E(Y) = \mu_Y = (a + b)/2 \qquad \text{(rectangular distribution)} \qquad (6.21)$$

$$\sigma_Y = (b - a)/2\sqrt{3} \qquad (6.22)$$

$$\alpha_3 = 0 \qquad (6.23)$$

$$\alpha_4 = 1.8. \qquad (6.24)$$

## 6.5. THE EXPONENTIAL DISTRIBUTION

This distribution is very widely used for lengths of life of equipment or parts. In fact it is the standard distribution in this area of reliability. It is the continuous analog of the discrete geometric distribution (5.65). The density function for the exponential distribution is

$$f(y) = \begin{cases} \mu^{-1}e^{-y/\mu}, & 0 \leqslant y \\ 0, & y < 0. \end{cases} \qquad (6.25)$$

---

* The symmetry of (6.20) around $(a + b)/2$ gives (6.21) and (6.23). For (6.22) we can use the $-1$, $+1$ form having $\mu_Y = 0$. Then $\sigma_Y^2 = \int_{-1}^{1} .5y^2 \, dy = .5y^3/3]_{-1}^{1} = 1/3$. Hence $\sigma_Y = 1/\sqrt{3}$, and the range of 2 is thus $2 \div (1/\sqrt{3}) = 2\sqrt{3}$ standard deviations. Thus for an $a$ to $b$ rectangular distribution with range $b - a$, $\sigma_Y = (b - a)/2\sqrt{3}$. For $\alpha_4$ we need $E(Y^4)$, which is similar to $\sigma_Y^2$, and then use (6.14). ($\alpha_4$ is invariant under any *linear* transformation.)

Also we have

$$F(y) = \int_0^y f(z)\, dz = \int_0^y e^{-z/\mu}\, dz/\mu = \left. -e^{-z/\mu} \right]_0^y = 1 - e^{-y/\mu}, \qquad y > 0. \quad (6.26)$$

It starts at $y = 0$, and from there onward $f(y)$ continually decreases just as any negative exponential function. See Fig. 6.4. The mode for the distribution is at $y = 0$, and the mean is at $y = \mu$.* Using (6.12)–(6.14), one has for (6.25)

$$E(Y) = \mu \qquad \text{(exponential distribution)} \qquad (6.27)$$

$$\sigma_Y = \mu \qquad\qquad (6.28)$$

$$\alpha_3 = 2, \qquad \alpha_4 = 9. \qquad\qquad (6.29)$$

Here we see that, curiously enough, the mean and standard deviation are identical. Since the curve starts at $y = 0$, this means that $y$'s can only be as much as one standard deviation below $\mu$, but there is no limit to the positive deviations from $\mu$. The skewness is $\alpha_3 = 2$, which indicates quite extreme lack of symmetry. Also $\alpha_4 = 9$, indicating the long tailing out toward high values of $y$. Note also that neither $\alpha_3$ nor $\alpha_4$ involves $\mu$. Thus whatever the mean $\mu$ is, the basic curve shape is the same. (This is of course also true of the normal and rectangular distributions because any linear transformation of $y$ leaves $\alpha_3$ and $\alpha_4$ unchanged.)

It is possible to start $y$ at some other point than zero, and thus have two parameters, but this is seldom done.

In life testing, the form used is (6.25). The mean life is $\mu$. For applications $\mu$ can be the average amount of time till failure, in minutes, days, years, or other time units. Or we can use this distribution when $y$ is the number of cycles or operations before failure. For this case the correct distribution would be the geometric (5.65), but if $\mu$ is a substantial number of cycles, we will often use the continuous distribution (6.25) instead. Another application is of radioactive material decaying by emission of particles, and we may ask when half of the particles will be gone. This is the "half-life," namely, the solution of $F(y) = \frac{1}{2}$, giving $y = .693\mu$.

---

* Using (6.10),

$$E(Y) = \int_0^\infty y e^{-y/\mu}\, dy/\mu = \mu \int_0^\infty w e^{-w}\, dw,$$

where $w = Y/\mu$. Then using integration by parts one obtains 1 for the integral and $E(Y) = \mu$.

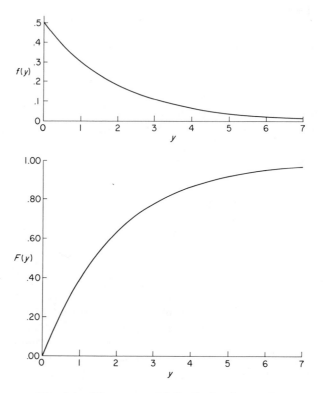

**Fig. 6.4.** The exponential distribution for $\mu = 2$.

## 6.6.* THE GAMMA DISTRIBUTION

A very useful distribution model for measurement data is the so-called gamma distribution or Pearson type III. Whereas the normal curve can be called a first approximation to data, the gamma distribution may be regarded as a second approximation, because with it one can match not only the mean and standard deviation of the observed data but also match an observed skewness $a_3$ (3.18). It involves three parameters.

The simplest form of the density function is as follows:

$$f(y) = \begin{cases} \dfrac{y^{p-1}e^{-y}}{\Gamma(p)}, & y > 0, \quad p > 0 \quad \text{(gamma distribution)} \\ 0, & y < 0. \end{cases} \tag{6.30}$$

Here the gamma function $\Gamma(p)$ is a constant depending on $p$. It is used so as to make the total area under the curve 1 (as it should be). The definition of $\Gamma(p)$ is

$$\Gamma(p) = \int_0^\infty X^{p-1} e^{-X}\, dX, \qquad p > 0. \tag{6.31}$$

Since the value of the integral depends on $p$, it is therefore a function of $p$, called $\Gamma(p)$. This function gives its name to the distribution (6.30). We state the following simple properties of the gamma function*:

$$\Gamma(1) = 1 \tag{6.32}$$

$$\Gamma(k+1) = k\Gamma(k), \qquad k > 0 \tag{6.33}$$

$$\Gamma(n) = (n-1)!, \qquad n \text{ a positive integer} \tag{6.34}$$

$$\Gamma(1/2) = \sqrt{\pi}. \tag{6.35}$$

The gamma distribution has an interesting connection with the exponential distribution. If we let

$$Y = Y_1 + Y_2 + \cdots + Y_p, \tag{6.36}$$

where the $Y_i$'s are all independent and follow the exponential (6.25) with $\mu = 1$, then $Y$ follows (6.30). Thus the gamma distribution bears the same relation to the exponential as the negative binomial (5.71) does to the geometric distribution (5.65). This property is useful in total length of life after replacing each failed unit by a new unit in life testing. The moments for the gamma distribution as given in (6.30) are[†]

$$\mu_Y = p \qquad \text{(gamma distribution)} \tag{6.37}$$

$$\sigma_Y = \sqrt{p} \tag{6.38}$$

$$\alpha_3 = 2/\sqrt{p} \tag{6.39}$$

$$\alpha_4 = 3 + (6/p) = 3 + 1.5\alpha_3^2. \tag{6.40}$$

---

* Formula (6.32) is proved by direct integration from the definition (6.31). Formula (6.33) follows from (6.31) by integration by parts, letting $u = X^k$, $dv = e^{-X}\, dX$. Formula (6.34) follows from continued use of (6.33), then (6.32). Formula (6.35) can be proven by first noting that, by the footnote to property 5 of the normal curve (Section 6.3.1), $\sqrt{2\pi} = \int_{-\infty}^\infty e^{-x^2/2}\, dx$. Then by symmetry of the integrand the latter is $2\int_0^\infty e^{-x^2/2}\, dx$. Now let $x^2/2 = X$, and one has

$$\sqrt{2\pi} = 2\int_0^\infty X^{-.5} e^{-X}\, dX/\sqrt{2} = \sqrt{2}\,\Gamma(1/2),$$

giving the desired result.

[†] These may be derived from (6.27)–(6.29) using the relations for (6.36), for independent and identically distributed variables, that $\mu_Y = p\mu_{Y_i}$, $\sigma_Y = \sqrt{p}\,\sigma_{Y_i}$, $\alpha_{3:Y} = \alpha_{3:Y_i}/\sqrt{p}$, $\alpha_{4:Y} = 3 + [(\alpha_{4:Y_i} - 3)/p]$.

The value of $p$ thus determines the skewness $\alpha_3$ for (6.30). In applications we equate the observed skewness $a_3$ to $2/\sqrt{p}$, and solve for $p = 4/a_3^2$. The larger $p$ is, the less the skewness. In order to match an observed $\bar{y}$ and $s$, we can use a linear transformation. (See below.) Figure 6.1 showed a gamma distribution in standard form, that is, with mean 0 and standard deviation 1, and for which $\alpha_3 = 1$. The equation was

$$f(y) = (8/3)(y + 2)^3 \, e^{-2y-4}.$$

### 6.6.1.* Tables of the Gamma Distribution.

The most useful tables of this distribution are presented by Salvosa [2] and have been reprinted by Carver [3]. These tables give the area below and the ordinate at $u = (y - \mu)/\sigma$ for skewnesses .0 by .1's to 1.1. Values of $u$ are by .01's and areas and ordinates are to the nearest sixth decimal place. These tables appear to have been rather neglected in usage. They can be used for the chi-square distribution, which is introduced in Chapter 7 and has wide applicability in statistics. [The chi-square distribution is a special case of (6.30) wherein $p$ is either an integer or an integer plus .5.] Linear interpolation in these tables is good, and one need not use (6.37)–(6.40) at all, merely entering the table with $u = (y - \bar{y})/s$ and $a_3$ .

### 6.6.2.* Relation to the Normal Distribution.

As $p$ increases, the gamma distribution (6.30) approaches the normal curve. The approximation has already become "good" (except at the extreme tails) when $\alpha_3$ is .2, that is, $p = 100$.

### 6.6.3.* Use of the Gamma Distribution to Approximate Discrete Distributions.

When we do not have a relatively large $n$ and small value of the probability $\phi$, which would permit approximation by the Poisson distribution, we may well use the gamma distribution to approximate sums of terms of the binomial (5.32). An example will illustrate the method.

Suppose we have $\phi = .2$, $n = 36$. Then using (5.33)–(5.35) we have

$$\mu_Y = .2(36) = 7.2, \qquad \sigma_Y = \sqrt{36(.2)(.8)} = 2.4, \qquad \alpha_3 = \frac{.6}{2.4} = .25.$$

Now let us approximate the probability of $Y$ lying between 5 and 12 *inclusive*, say. See Fig. 6.5. There the curve is of the gamma distribution with the three characteristics just described. The blocks picture the

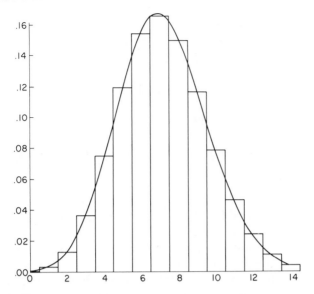

**Fig. 6.5.** Histogram of the binomial distribution for $\phi = .2$, $n = 36$, and an approximating gamma distribution curve having the same first three moments.

exact binomial probabilities to be approximated. Note that the curve goes nearly through the middle of the top of each block except for that at $Y = 7$. Now, the desired area is that of the blocks for $Y = 5, 6, ..., 12$ inclusive. It might seem natural to the student to approximate this block area by obtaining the area under the curve from $Y = 5$ to $Y = 12$. But a closer inspection of Fig. 6.5 reveals that we should find the area under the curve from $Y = 4.5$ to $Y = 12.5$ instead. This is because the block area for $Y = 5$ is well approximated by the nearly trapezoidal area under the curve from $Y = 4.5$ to $Y = 5.5$, not by that from $Y = 5$ to 5.5. The same thing is true at $Y = 12$. Using 4.5 to 12.5 here is called a "correction for continuity," and often comes in when we approximate a discrete distribution by a continuous one.

Proceeding, we use standardized values

$$u = \frac{4.5 - 7.2}{2.4} = -1.125, \qquad u = \frac{12.5 - 7.2}{2.4} = +2.208.$$

For $\alpha_3 = .2$ these give, respectively, the areas below by [2] interpolating,*

$$.127872, \qquad .981978$$

* Using a desk calculator, such interpolation is considerably simplified by taking a weighted average instead of using a difference. Thus to find the interpolated value for a function $g(2.208)$, we use $.2g(2.20) + .8g(2.21)$, giving the larger weight to the value which should be closer to the desired one.

while for $\alpha_3 = .3$, they are

$$.126195, \qquad .979886.$$

Interpolating for $\alpha = .25$ we then find

$$.127033, \qquad .980932.$$

To find the desired area between, we subtract these and round off:

$$.8539.$$

Using the binomial table [2], Chapter 5, we can find .8549, which has thus been well approximated.

Had we used a normal curve table for $u = -1.125$ to $u = +2.208$, we would find the area to be .8561. This is almost as good, but in the opposite direction. The use of the gamma distribution, however, is considerably better if our desired area extends further out on each tail, and/or if $\alpha_3$ is greater than .25. In this particular example we did not need to seek an approximation by the gamma distribution if an appropriate table is available. In fact, binomial tables are likely to be more available than those of Salvosa [2] or Carver [3]. But if $n$ is large, we may well find it desirable to use the foregoing approach. For example, Carver (see Glover and Carver [4]) used it on $n = 100{,}000$, $\phi = .008$, and found in one case the "exact" probability approximated perfectly to the best four decimal places.

Moreover the same approach as just given for approximating the binomial, can also be used for the hypergeometric by using (5.50)–(5.52). This is important for $N > 100$, since the only comprehensive table of the hypergeometric [10], Chapter 5, only goes up to $N = 100$ (except for cases of $N = 1000$ with $n = 500$).

### 6.7.* THE BETA DISTRIBUTION

A distribution which is even more general than the gamma distribution is the so-called beta distribution. With it one can match not only a given $\bar{y}$, $s$, and $a_3$ but also (within certain limits) $a_4$ (3.20). Hence using a beta distribution one can often expect an even better fit, or mathematical model, to observed data, than can be had from the gamma distribution. This distribution was originally called the Pearson type I distribution. It has in general four parameters.

The simplest form of the density function is

$$f(y) = Ky^{q-1}(1 - y)^{r-1}, \qquad 0 < y < 1, \qquad q, r > 0 \qquad \text{(beta distribution)}$$
$$f(y) = 0 \qquad \text{elsewhere,} \tag{6.41}$$

where $K$ is determined to make the total area (0 to 1) equal to 1.*

The four moments are all functions of $q$ and $r$ as follows:

$$\mu_Y = \frac{q}{q + r} \qquad \text{(beta distribution)} \tag{6.42}$$

$$\sigma_Y = \frac{\sqrt{qr/(q + r + 1)}}{q + r} \tag{6.43}$$

$$\alpha_3 = \frac{2(r - q)}{q + r + 2} \sqrt{\frac{q + r + 1}{qr}} \tag{6.44}$$

$$\alpha_4 = \frac{3(q + r + 1)(q + r) qr + 6(q + r + 1)(q^2 - qr + r^2)}{(q + r + 3)(q + r + 2) qr}. \tag{6.45}$$

Theoretically, given an observed pair of curve-shape moments $a_3$ and $a_4$, one equates them to (6.44) and (6.45) and solves for $q$ and $r$. This is obviously a difficult problem, and not very practical. There is another problem involved too, namely, that available tables for areas under the curve for (6.41) are only for $q, r$ positive integers, or positive integers plus .5 [5, 6]. The former has values of (6.42)–(6.45), $(\alpha_3{}^2)$ calculated, so that to approximate $q, r$ for given $a_3, a_4$ one can seek the closest $q$ and $r$ combination. Then by a linear transformation match an observed variable $z$, say, with $\bar{z}, s_z$ by

$$\frac{z - \bar{z}}{s_z} = u = \frac{y - q/(q + r)}{\sqrt{qr/(q + r + 1)}} (q + r).$$

Areas under the curve for the $y$'s .01, .02, ..., .99, are tabulated by Pearson [5]. Harter [6] gives values of $y$ for various values of probability $F(y)$.

The beta distribution is related to some sampling distributions such as those in Chapter 7. Moreover, it is rather often used as an *a priori*

---

* $K$ has a value $1/\beta(q, r)$ where $\beta(q, r)$ is a beta function defined, if $p, q$ are both positive, by $\int_0^1 X^{q-1}(1 - X)^{r-1} dX$. This may be shown to be equal to $\Gamma(q) \Gamma(r)/\Gamma(q + r)$. This function gives its name to the distribution (6.41).

distribution for decision theory approaches to problems of rational decision-making. Previous knowledge or data may suggest an *a priori* distribution like (6.41).

## 6.8.* THE WEIBULL DISTRIBUTION

A widely used distribution, especially in life testing and reliability, is the Weibull distribution. It was named after W. Weibull who first developed and applied it. The *distribution* function is given by

$$F(y) = \begin{cases} P(Y < y) = \int_0^y f(z)\,dz = 1 - e^{-(y/\theta)^k}, & k, \theta > 0, \quad y > 0 \\ 0, & y < 0 \quad \text{(Weibull distribution)}. \end{cases} \quad (6.46)$$

This is the so-called two-parameter form most commonly used, for example, for length of life to failure [7]. One can use a three-parameter form by replacing $y$ by $x - a$, thus having the active part of the distribution begin at $x = a$. But this is less often used.

Many techniques have been developed for estimating $\theta$, the scale parameter, and $k$, the shape parameter. These include use of various Weibull papers for graphical fitting, equating sample moments to population moments (see below), and the so-called method of maximum likelihood (see Chapter 10). We do not try to list this growing literature here.

The density function $f(y)$ is obtained from $F(y)$ in (6.46) by differentiation, giving

$$f(y) = k e^{-(y/\theta)^k} y^{k-1} \theta^{-k}. \quad (6.47)$$

From this we have the following*:

$$\mu_Y = \theta \Gamma\left(\frac{1}{k} + 1\right) \quad \text{(Weibull distribution)} \quad (6.48)$$

$$\sigma_Y = \theta \left[\Gamma\left(\frac{2}{k} + 1\right) - \Gamma^2\left(\frac{1}{k} + 1\right)\right]^{\frac{1}{2}} \quad (6.49)$$

$$\alpha_3 = \frac{\Gamma[(3/k) + 1] - 3\Gamma[(2/k) + 1]\,\Gamma[(1/k) + 1] + 2\Gamma^3[(1/k) + 1]}{\{\Gamma[(2/k) + 1] - \Gamma^2[(1/k) + 1]\}^{\frac{3}{2}}} \quad (6.50)$$

$$\alpha_4 = \frac{\begin{bmatrix} \Gamma[(4/k) + 1] - 4\Gamma[(3/k) + 1]\,\Gamma[(1/k) + 1] \\ + 6\Gamma[(2/k) + 1]\,\Gamma^2[(1/k) + 1] - 3\Gamma^4[(1/k) + 1] \end{bmatrix}}{\{\Gamma[(2/k) + 1] - \Gamma^2[(1/k) + 1]\}^2}. \quad (6.51)$$

* These are proved by use of the transformation $(y/\theta)^k = v$ on (6.10)–(6.14) for (6.47), and using the definition for the $\Gamma$ function (6.31).

The use of Weibull papers is usually sufficient for finding $\theta$ and $k$, since we seldom have large samples of data (100 or more), and thus the error due to sampling is greater than that of the graphical approach.

We note that when fitted ($k$ and $\theta$ approximated) we can easily use (6.46) to find probabilities:

$$P(y_1 < Y < y_2) = F(y_2) - F(y_1) = e^{-(y_1/\theta)^k} - e^{-(y_2/\theta)^k}.$$

Also we note that if $k$ is 1, we have the exponential distribution (6.25).

As $k$ increases from 1 the curve for the density function becomes less and less unsymmetrical.

### 6.9.* THE PEARSON SYSTEM OF DISTRIBUTIONS

In order to cover a broad region of curve-shape characteristics, $\alpha_3$ and $\alpha_4$, K. Pearson developed a whole system of distributions. Using the standardized variable $u = (y - \mu)/\sigma$, (6.18), and setting $z = f(u)$, the density function, the Pearson system curves are all solutions of the differential equation

$$\frac{dz}{du} = \frac{z(a - u)}{b_0 + b_1 u + b_2 u^2}.$$

The slope of the curve for the density function, that is, $dz/du$, is thus zero when $f(u)$ hits zero, and also zero when $u = a$; that is, we will in common cases have a mode when $u = a$. The denominator is a function of $u$, in fact the first part of a series expansion.

If $b_1$ and $b_2$ are zero, then we have the normal curve (6.15), while if $b_2$ only is zero, we obtain the gamma or type III distribution (6.30), a special case of which is the exponential (6.25). The three general cases, types I, IV, and VI, occur according to the type of roots of

$$b_0 + b_1 u + b_2 u^2 = 0.$$

These types cover *areas* in the $\alpha_3$, $\alpha_4$ plane, while transitional types such as III cover only curves in the $\alpha_3$, $\alpha_4$ plane. For example, type III has $\alpha_4 = 3 + 1.5\alpha_3^2$ (6.40).

An excellent exposition of the Pearson system of frequency curves is given by Craig [8].

Two difficulties in using this system are the following: (1) there are about 12 different types, each of which has its own particular functional form for the density function $f(u)$, and (2) for many of the types one cannot find $P(a < u < b) = \int_a^b f(u)\, du$, without resorting to approximate integration methods, or heavy interpolation in tables, such as presented by Pearson [5]. Nevertheless this system is a valuable one, covering the most important region in $\alpha_3$, $\alpha_4$ space, and serving as a standard of comparison for other systems.

## 6.10.* AN EASILY FITTED GENERAL SYSTEM OF FREQUENCY CURVES

The author initiated the development of the following general system of frequency curves [9], [10]. The *distribution* function has the following form in terms of random variable $X$:

$$F(X) = \begin{cases} 1 - (1 + X^c)^{-k}, & c, k > 0, \quad X > 0 \\ 0, & X < 0. \end{cases} \qquad (6.52)$$

One can differentiate to find the respective density function $f(X)$. Note that we can easily find probabilities for an interval $(a, b)$, $P(a < X < b)$, by using (6.8) without resorting to integration.*

Table XIII in the Appendix gives the values of $c$, $k$, $\mu_X$, and $\sigma_X$ for various combinations of $\alpha_3$, $\alpha_4$. Adequate results can usually be obtained by picking the $\alpha_3$, $\alpha_4$ combination nearest to those desired. One can of course interpolate, if desired and/or needed. Then for any given $Y$ we can use $\bar{y}$, $s$ as follows, to find $X$ for substitution into (6.52):

$$\frac{X - \mu_X}{\sigma_X} = u = \frac{y - \bar{y}}{s_y}, \qquad X = \frac{\sigma_X}{s_y}(y - \bar{y}) + \mu_X, \qquad (6.53)$$

where $\mu_X$, $\sigma_X$ are from Table XIII for the chosen $c$, $k$. In this way we are matching the first four moments of the data. In a sense, then, the use of (6.52) provides a third approximation to given data, just as some member of the Pearson system might.

## 6.11. SUMS AND AVERAGES AND A CENTRAL LIMIT THEOREM

The central limit theorem is concerned with an approach to normality of distribution, which occurs when a variable $Y$ is composed of components $Y_1$, $Y_2$, ..., $Y_n$, and $n$ is increased. There are several theorems going by this title. We describe the simplest one.

---

* The values of $E(X^i)$ for use in (6.10)–(6.14) are easily found by using $v = (1 + X^c)^{-1}$, giving beta functions for the moments.

For this we state the following relations which are easily proved (Section 6.14). Given the total

$$T = Y_1 + Y_2 + \cdots + Y_n, \tag{6.54}$$

then

$$\mu_T = \mu_1 + \mu_2 + \cdots + \mu_n, \tag{6.55}$$

where the $\mu$'s are the respective population means of the terms in (6.54). This is true whenever the $\mu$'s all exist, whether or not the $Y_i$'s are independent. If, however, the $Y_i$'s are independent, and all have finite variances $\sigma_i^2$, then we can easily prove

$$\sigma_T^2 = \sigma_1^2 + \sigma_2^2 + \cdots + \sigma_n^2 \quad (Y_i\text{'s independent}). \tag{6.56}$$

From these we thus find for

$$\bar{Y} = (Y_1 + Y_2 + \cdots + Y_n)/n \tag{6.57}$$

that

$$\mu_{\bar{Y}} = (\mu_1 + \mu_2 + \cdots + \mu_n)/n \tag{6.58}$$

$$\sigma_{\bar{Y}}^2 = (\sigma_1^2 + \sigma_2^2 + \cdots + \sigma_n^2)/n^2 \quad (Y_i\text{'s independent}). \tag{6.59}$$

In the above the $Y_i$'s may have differing distributions, each with finite variance. However, an extremely important special case occurs where we take the $Y_i$'s randomly and independently from the *same* population, which has mean $\mu$ and finite variance $\sigma^2$. Then for (6.57) we have from (6.58) and (6.59), respectively, since $\mu_i = \mu_Y$ and $\sigma_i = \sigma_Y$, $i = 1, ..., n$:

$$\mu_{\bar{Y}} = \mu_Y \tag{6.60}$$

$$\sigma_{\bar{Y}}^2 = \sigma_Y^2/n \quad \text{or} \quad \sigma_{\bar{Y}} = \sigma_Y/\sqrt{n}. \tag{6.61}$$

Many people regard these last two formulas as among the most important in all of statistics.

Speaking now of the random variable $\bar{Y}$ having $\mu_{\bar{Y}}$ and $\sigma_{\bar{Y}}$ as given in (6.60) and (6.61) we may standardize as in (6.18):

$$u = \frac{\bar{Y} - \mu_{\bar{Y}}}{\sigma_{\bar{Y}}} = \frac{\bar{Y} - \mu_Y}{\sigma_Y/\sqrt{n}}. \tag{6.62}$$

With the foregoing in mind, we are now in a position to state the theorem:

**A Central Limit Theorem.** Let $Y_1, Y_2, ...,$ be drawn independently from the same population with given mean $\mu_Y$ and finite standard deviation $\sigma_Y$. Then the probability

$$P \left( a < \frac{\overline{Y} - \mu}{\sigma_Y/\sqrt{n}} < b \right) \tag{6.63}$$

approaches the limit $\Phi(b) - \Phi(a) = \int_a^b \phi(u)\, du$, as $n$ becomes infinite. [See (6.15) and (6.16).]

Thus the probability that the standardized variable in (6.62) lies between any limits $a$ and $b$ approaches a fixed quantity given by the standard normal curve.

Several comments on this remarkable theorem are in order:

1.  The theorem does *not* say that the density function for $Y$ approaches a normal density function, but only that the probability (6.63) has a limit. It is true that when $Y_i$ is continuous the density function for $Y$ will become normal. But suppose that we have a *discrete* probability function, such as those in Chapter 5. Their $\overline{Y}$ probability function, being discrete, cannot approach a normal *continuous* density function. But the *probability* approach is still true, as in the theorem, even for discrete variables $Y_i$.

2.  The approach to normality is surprisingly rapid. Samples of as few as four or five will often show very nearly normal distributions for $\overline{Y}$ (see Shewhart [11, p. 182]).

3.  To tell when normality is likely to be good one can use the following formulas, which are extensions of (6.60), (6.61):

$$\alpha_{3:\overline{Y}} = \frac{\alpha_{3:Y}}{\sqrt{n}} \tag{6.64}$$

$$\alpha_{4:\overline{Y}} = 3 + \frac{\alpha_{4:Y} - 3}{n}. \tag{6.65}$$

It is the author's experience that if $|\alpha_{3:\overline{Y}}| < .2$ and $|\alpha_{4:\overline{Y}} - 3| < .3$, the normal-curve probabilities will be quite satisfactory unless one works at the extreme tails. But for approximating a discrete variable, use sums $T$ instead of $\overline{Y}$'s and make the continuity correction as in Section 6.3.4.

4.  Note that we have already mentioned that for $Y_i$'s given by the exponential (6.25), their sum will be given by (6.30), which is more normal; and that in turn (6.30) by the central limit theorem becomes normal. Similarly the geometric distribution for $Y_i$'s leads to the negative binomial which is more "normal" looking,

and probabilities can be increasingly well approximated by the normal curve as $k$ increases. Similarly the binomial becomes more normal looking as $n$ increases while $\phi$ remains fixed.

5.   The $Y_i$'s need not have the same distribution [12, p. 215]. But a restriction is needed to help any one or a few $Y_i$'s from "stealing the show" and determining $\overline{Y}$'s distribution.

### 6.12.* TCHEBYCHEFF'S THEOREM

For the sake of completeness only, we shall include this inequality. But from the viewpoint of applications it seems to be of rather slight importance. There is, however, no question but that it is important in theoretical developments.

To use Tchebycheff's inequality we need to know the mean and standard deviation of the population. Then we have the following:

**Tchebycheff's Theorem.**   Given $k > 1$ and a population with mean $\mu$ and standard deviation $\sigma$, then we have

$$P[\,|Y - \mu| \geqslant k\sigma] \leqslant 1/k^2. \qquad (6.66)$$

Some remarks are in order:

1.   This inequality (6.66) is extremely general: $Y$ can be continuous or discrete; $\sigma$ can even be infinite, but then (6.66) says nothing at all. Why?

2.   The inequality is for the $Y$ values at least $k\sigma$ away from $\mu$. The proportion of such can never exceed $1/k^2$. For example one-fourth or fewer $Y$'s are over $2\sigma$ from $\mu$, and so on. Note that we could have $0 < k < 1$ and (6.66) is still true for it says the probability is less than $1/k^2$ which exceeds one, and this is true of all probabilities!

3.   It seems much more desirable to try to *approximate* a probability than to set a very *extreme upper limit* to the probability. To approach the upper limit of $1/k^2$ we have to have a very impractical distribution never seen in practice, and moreover, a different one for each $k$. Suppose $k = 2$. Define the discrete distribution

$$P(Y = -1) = 1/8 = P(Y = +1)$$
$$P(Y = 0) = 3/4.$$

Because of symmetry $\mu_Y = 0$. Also

$$\sigma_Y^2 = (1/8)[(-1)^2 + 1^2] = 1/4, \qquad \sigma_Y = 1/2.$$

Now

$$P[\,|Y - 0\,| \geqslant 2(1/2) = 1] = 2(1/8) = 1/4.$$

Thus in this case the upper limit of $1/2^2$ is actually attained. But how often will one encounter such a distribution in practice?

4. Somewhat tighter inequalities can be proven if one makes further assumptions on $f(y)$ [12].

### 6.13. SUMMARY

In this chapter we have first presented some general information for the density $f(y)$, and the cumulative or distribution function $F(y)$, for a continuously varying $Y$. Then we have presented the extremely important normal curve, and some of its properties and applications. Other mathematical models were then considered for measurement data cases. We have the information given in the accompanying table.

| Distribution type | Number of available parameters |
|---|---|
| Normal | Two: $\mu$ and $\sigma$ |
| Rectangular | Two: $a, b$ giving $\mu, \sigma$ |
| Exponential | One: $\mu$ (could also translate) |
| Gamma | Three: $\mu, \sigma, p$ (shape) |
| Weibull | Two: $\theta$ (scale), $k$ (shape) (could also translate) |
| Beta | Four: $\mu, \sigma, q, r$ |
| Pearson system | Four: $\mu, \sigma, \alpha_3, \alpha_4$ |
| System (6.52) | Four: $\mu, \sigma, c, k$ |

Several interesting relations between distributions were given, and the use of some distributions (normal and gamma) to approximate the binomial, Poisson, and hypergeometric distributions was indicated.

The important central limit theorem was presented. It shows that the limits of probabilities involving $\overline{Y}$ (standardized) are given by the standard normal distribution function $\Phi(u)$. With this theorem we can often approximate probabilities for $\overline{Y}$, even when the component $Y$'s are not normal.

Moments to the fourth were given for the distributions. Curve shape is characterized by $\alpha_3$ and $\alpha_4$. They also enable one, for example, to tell when the central limit theorem or the gamma distribution can be used to approximate probabilities for $\overline{Y}$.

Tchebycheff's inequality for the upper limit to tail probabilities for $\overline{Y}$ was also briefly covered.

## 6.14.* PROOFS OF SOME RELATIONS IN SECTION 6.11

In order to prove relations such as (6.55) we need to consider the joint density function for random variables $X_1$, $X_2$ $f(X_1, X_2)$. It is nonnegative and has a total volume of 1 beneath $z = f(X_1, X_2)$ and above the $X_1$, $X_2$ plane. Also

$$P(a < X_1 < b \cap c < X_2 < d) = \int_a^b \int_c^d f(X_1, X_2) \, dX_2 \, dX_1.$$

The so-called marginal density functions are defined for $X_1$ and $X_2$ as follows:

$$f_1(X_1) = \int_{-\infty}^{\infty} f(X_1, X_2) \, dX_2, \qquad f_2(X_2) = \int_{\infty}^{\infty} f(X_1, X_2) \, dX_1. \tag{6.67}$$

We shall be concerned mostly with the case where $X_1$ and $X_2$ are independent. By definition this occurs whenever

$$f(X_1, X_2) = f_1(X_1) f_2(X_2) \qquad (X_1, X_2 \text{ independent}) \tag{6.68}$$

identically, for all points $(X_1, X_2)$ in the plane. Generalization to more than two $X_i$'s is straightforward.

Now for $n = 2$ in (6.54), $T = X_1 + X_2$ using $X$'s and

$$E(T) = \int_{-\infty}^{\infty} \int_{-\infty}^{\infty} (X_1 + X_2) f(X_1, X_2) \, dX_1 \, dX_2$$

$$= \int_{-\infty}^{\infty} X_1 \int_{-\infty}^{\infty} f(X_1, X_2) \, dX_2 \, dX_1 + \int_{-\infty}^{\infty} X_2 \int_{-\infty}^{\infty} f(X_1, X_2) \, dX_1 \, dX_2$$

$$= \int_{-\infty}^{\infty} X_1 f_1(X_1) \, dX_1 + \int_{-\infty}^{\infty} X_2 f_2(X_2) \, dX_2,$$

or

$$\mu_T = \mu_1 + \mu_2. \tag{6.69}$$

This has not used independence of $X_1$, $X_2$. Equation (6.55) is a simple generalization. Now by subtraction of this from $T = X_1 + X_2$ we have

$$T - \mu_T = X_1 - \mu_1 + X_2 - \mu_2.$$

For $\sigma_T{}^2 = E[(T - \mu_T)^2]$ we have

$$E[(T - \mu_T)^2] = \int\limits_\infty^\infty \int\limits_\infty^\infty (X_1 - \mu_1 + X_2 - \mu_2)^2 f(X_1, X_2) \, dX_1 \, dX_2$$

$$\int\limits_{-\infty}^\infty \int\limits_{-\infty}^\infty [(X_1 - \mu_1)^2 + 2(X_1 - \mu_1)(X_2 - \mu_2) + (X_2 - \mu_2)^2]$$

$$\cdot f_1(X_1) f_2(X_2) \, dX_1 \, dX_2 \,,$$

where we have squared out the quantity and used (6.68), which assumes independence. Now the first quantity in brackets gives

$$\int\limits_{-\infty}^\infty (X_1 - \mu_1)^2 f_1(X_1) \, dX_1 \int\limits_{-\infty}^\infty f_2(X_2) \, dX_2 = \sigma_1{}^2 \cdot 1 = \sigma_1{}^2.$$

The third is similar. For the second we have

$$2 \int\limits_{-\infty}^\infty (X_1 - \mu_1) f_1(X_1) \, dX_1 \times \int\limits_{-\infty}^\infty (X_2 - \mu_2) f_2(X_2) \, dX_2 \,.$$

Each integral is zero as follows:

$$\int\limits_{-\infty}^\infty (X_1 - \mu_1) f_1(X_1) \, dX = \int\limits_{-\infty}^\infty X_1 f_1(X_1) \, dX_1 - \mu_1 \int\limits_\infty^\infty f_1(X_1) \, dX_1$$

$$= \mu_1 - \mu_1 \cdot 1 = 0. \tag{6.70}$$

Thus we have (6.56) for $n = 2$. The generalization is easy.

Next for any variable $X$, and any $b$ and positive constant $k$ we have for

$$Y = kX + b \tag{6.71}$$

$$\mu_Y = E(Y) = \int\limits_{-\infty}^\infty (kX + b) f(X) \, dX = k \int\limits_{-\infty}^\infty Xf(X) \, dX + b \int\limits_{-\infty}^\infty f(X) \, dX$$

or

$$\mu_{kX+b} = k\mu_X + b. \tag{6.72}$$

Now for $\sigma_Y{}^2$ we have

$$Y - \mu_Y = kX + b - k\mu_X - b = k(X - \mu_X)$$

$$\sigma_Y{}^2 = E[(Y - \mu_Y)^2] = \int\limits_{-\infty}^\infty [k(X - \mu_X)]^2 f(X) \, dX$$

$$= k^2 \, E[(X - \mu_X)^2] = k^2 \sigma_X{}^2,$$

or

$$\sigma_{kX+b} = k\sigma_X .$$ (6.73)

Now consider $T = X_1 + \cdots + X_n$ and $\bar{X} = T/n$. Let us draw $n$ $X_i$'s from the same population having $\mu_X$ and $\sigma_X$. Then for $T = X_1 + \cdots + X_n$ we have by (6.55) and (6.56) for $X$'s

$$\mu_T = n\mu_X$$ (6.74)

$$\sigma_T = \sqrt{n}\,\sigma_X .$$ (6.75)

For the sample mean $\bar{X} = T/n$, we may now use (6.72)–(6.75):

$$\mu_{\bar{X}} = \mu_{(1/n)T} = (1/n)\,\mu_T = (1/n)\,n\mu_X$$

or

$$\mu_{\bar{X}} = \mu_X$$ (6.60)

and

$$\sigma_{\bar{X}} = \sigma_{(1/n)T} = (1/n)\,\sigma_T = (1/n)\,\sqrt{n}\,\sigma_X$$

or

$$\sigma_{\bar{X}} = \sigma_X/\sqrt{n}.$$ (6.61)

The third and fourth moments are merely a bit more complicated. Let us denote the $i$th moment around the mean for any variable $v$ by

$$\mu_{i:v} = E[(v - \mu_v)^i].$$ (6.76)

Then

$$\alpha_{3:v} = \mu_{3:v}/\sigma_v{}^3, \qquad \alpha_{4:v} = \mu_{4:v}/\sigma_v{}^4.$$ (6.77)

As for (6.61), let the $X_i$'s be drawn independently from the same population having $\mu_X$, $\sigma_X$, $\alpha_{3:X}$, $\alpha_{4:X}$. Then consider $T = X_1 + X_2$, say, and find $\mu_{3:T} = E[(T - \mu_T)^3]$:

$$(T - \mu_T)^3 = (X_1 + X_2 - \mu_X - \mu_X)^3 = (X_1 - \mu_X + X_2 - \mu_X)^3$$
$$= (X_1 - \mu_X)^3 + 3(X_1 - \mu_X)^2 (X_2 - \mu_X) + 3(X_1 - \mu_X)(X_2 - \mu_X)^2$$
$$+ (X_2 - \mu_X)^3.$$

The expectation or integral of the first term on the right is $E[(X - \mu_X)^3] = \mu_{3:X}$ for the $X$ population. The double integral of the second term can be broken up into two integrals giving $3 \cdot \sigma_X{}^2 \cdot 0 = 0$, using (6.70). The other terms are exactly similar to these two. Thus

$$\mu_{3:X_1+X_2} = 2\mu_{3:X} .$$

For the standardized moment:

$$\alpha_{3:T} = \frac{\mu_{3:T}}{\sigma_T{}^3} = \frac{2\mu_{3:X}}{(\sqrt{2\sigma_X{}^2})^3} = \frac{\mu_{3:X}}{\sqrt{2}\,\sigma_X{}^3} = \frac{\alpha_{3:X}}{\sqrt{2}}.$$

The only difference in the above when $T = X_1 + \cdots + X_n$ is that the 2 is replaced by $n$. Now it can also be readily shown that for $Y = kX + b$, $\alpha_{3:Y} = \alpha_{3:X}$ and $\alpha_{4:Y} = \alpha_{4:X}$; that is, these measures are independent of any linear transformation with $k > 0$. Thus

$$\alpha_{3:\bar{X}} = \frac{\alpha_{3:X}}{\sqrt{n}}. \tag{6.64}$$

To prove (6.65), we use

$$(T - \mu_T)^4 = (X_1 - \mu_X + \cdots + X_n - \mu_X)^4$$

$$= (X_1 - \mu_X)^4 + \cdots + (X_n - \mu_X)^4 + 6 \sum_{i<j} (X_i - \mu_X)^2 (X_j - \mu_X)^2$$

+ terms having expectation of 0 because they contain at least one like $(X_i - \mu_X)^1$.

Then taking the expectation or integral we have

$$\mu_{4:T} = n\mu_{4:X} + 6C(n, 2)\,\sigma_X{}^2\sigma_X{}^2$$

$$= n\mu_{4:X} + 3n(n - 1)\,\sigma_X{}^4,$$

since there are $C(n, 2)$ different combinations of $i$ and $j$ in $n$. Now divide by $\sigma_T{}^4 = n^2\sigma_X{}^4$ for $\alpha_{4:T}$

$$\alpha_{4:T} = \frac{n\mu_{4:X} + 3n^2\sigma_X{}^4 - 3n\sigma_X{}^4}{n^2\sigma_X{}^4} = 3 + (1/n)(\alpha_{4:X} - 3).$$

But $\alpha_{4:\bar{X}} = \alpha_{4:T}$, so that we have (6.65).

## REFERENCES

1. I. W. Burr, "Engineering Statistics and Quality Control," p. 73. McGraw-Hill, New York, 1953.
2. L. R. Salvosa, Tables of Pearson's Type III function. *Ann. Math. Statist.* 1, 191–198 (1930).
3. H. C. Carver, "Statistical Tables." Edwards, Ann Arbor, Michigan, 1940.
4. J. W. Glover and H. C. Carver, "Introduction to Mathematical Statistics." Edwards, Ann Arbor, Michigan, 1928.
5. K. Pearson, "Tables of the Incomplete Beta-Function," 2nd ed. Cambridge Univ. Press (for the Biometrika Trustees), London and New York, 1968.

6. H. L. Harter, "New Tables of the Incomplete Gamma-Function and of Percentage Points of the Chi-Square and Beta Distributions." US Govt. Printing Office, Washington, D.C., 1964.
7. D. K. Lloyd and M. Lipow, "Reliability, Management, Methods, and Mathematics." Prentice-Hall, Englewood Cliffs, New Jersey, 1964.
8. C. C. Craig, A new exposition and chart for the Pearson system of frequency curves. *Ann. Math. Statist.* **7**, 16–28 (1936).
9. I. W. Burr, Cumulative frequency functions. *Ann. Math. Statist.* **13**, 215–232 (1942).
10. I. W. Burr, Parameters for a general system of distributions to match a grid of $\alpha_3$ and $\alpha_4$ . *Commun. in Stat.* **2**, 1–21 (1973).
11. W. A. Shewhart, "Economic Control of Quality of Manufactured Product." Van Nostrand-Reinhold, Princeton, New Jersey, 1931.
12. H. Cramer, "Mathematical Methods of Statistics." Princeton Univ. Press, Princeton, New Jersey, 1946.

ADDITIONAL REFERENCES

M. J. Beckman and F. Bobkoski, Airline demand: An analysis of some frequency distributions. *Naval Res. Logist. Quart.* **5**, 43–51 (1958).

J. N. Berretoni, Practical applications of the Weibull distribution. *Indust. Quality Control* **21**, 71–79 (1964).

R. E. Clark, Percentage points of the incomplete Beta function. *J. Amer. Statist. Assoc.* **48**, 831–843 (1953).

D. J. Finney, On the distribution of a variate whose log is normally distributed. *J. Roy. Statist. Soc. Suppl.* **7**, 155–161 (1941).

F. A. Haight, Index to the distributions of mathematical statistics. *J. Res. Nat. Bur. Standards Sect. B* **65**, 23–60 (1961).

N. L. Johnson, Systems of frequency curves generated by methods of translation. *Biometrika* **36**, 149–176 (1949).

W. L. Nicholson, On the normal approximation to the hypergeometric distribution. *Ann. Math. Statist.* **27**, 471–483 (1956)

A. Plait, The Weibull distribution, with tables. *Indust. Quality Control* **19**, (No. 5) 17–26 (1962).

D. R. Thoman, L. J. Bain, and C. E. Antle, Inferences on the parameters of the Weibull distribution. *Technometrics* **11**, 445–460 (1969).

M. T. Wasan and L. K. Roy, Tables of inverse Gaussian percentage points. *Technometrics* **11**, 591–604 (1969).

W. Weibull, A statistical distribution function of wide applicability. *J. Appl. Mech.* **18**, 293–297 (1951).

## Problems

6.1.   For 399 tensile strengths, $\bar{y} = 202,000$ psi (pounds per square inch), and $s = 4035$ psi. The distribution was approximately normal. Sketch the curve showing units on the strength axis, and estimate the percentage below the minimum of 195,000 psi. What estimated range will include the middle 90 % of strengths ?

6.2.  The distribution of thicknesses for 1090 cork disks for bottle crowns was approximately normal (with $a_3 = .13$) $\bar{y} = .06547$ in. $s = .00299$ in. Estimate the percentage of disks outside of limits of .058 to .073 in., and show a sketch of the curve with units on horizontal scale. What estimated range will include the middle 90% of the thicknesses?

6.3.  The distribution of weights of metal covers for fiber shipping containers was approximately normal for 675 observations. Measurements were in ounces above $2\frac{3}{4}$ lb, giving $\bar{y} = 6.163$ oz and $s = 1.048$ oz. Sketch the normal curve labeling the weight axis, and estimate the percentage between 3.5 and 8.5 oz.` What range would you expect to include the middle 95% of cases?

6.4.  Lengths of rubber gaskets cut from extruded rubber tubes were measured from the desired nominal, which could be either .150 or .155 in. all data being pooled. The distribution was quite normal, $a_3 = -.15$, $a_4 = 2.85$, for 860 measurements. For results in .001 in. from nominal, $\bar{y} = -.725$, $s = 1.57$. Sketch the normal curve labeling the horizontal scale, and find a range around the nominal to include the middle 99% of lengths. What proportion of lengths will be more than 5(.001 in.) away from nominal?

6.5.  For 2700 eccentricities of needle valves (distance between centers of cone and triangular base) the distribution was skewed positively, since eccentricities can only be positive. Results showed $\bar{y} = .003538$ in., $s = .002073$ in., $a_3 = .966$, $a_4 = 4.428$. Since $a_4$ is close to $3 + 1.5a_3^2$ [see (6.40)], we might expect the gamma distribution to fit well, and this is the case. Estimate the percentage above the upper specification limit of .010 in. by using appropriate tables. Compare the gamma distribution estimate with that assuming normality.

6.6.  For 228 percentages of silicon in blast furnace casts, $\bar{y} = 1.539\%$, $s = .325\%$, $a_3 = .56$. Assume a gamma distribution and estimate the percentage of casts which will be between 1 and 2%. Compare with estimates assuming normality.

6.7.  For fiber strength of Indian cotton, $\bar{y} = 4.088$ oz, $s = 2.352$ oz, $a_3 = .79$, $a_4 = 3.22$, for $n = 1000$. Estimate the percentage below 1.5 oz and above 6.0 oz assuming a gamma distribution and compare with similar normal estimates. [Results are not too good here because $a_4$ is quite a bit below $3 + 1.5a_3^2$; see (6.40). Formula (6.52) could be used.]

6.8.  For 1005 electrical contacts, a dimension had specifications of .020 to .050 in., $\bar{y} = .02815$ in., $s = .00548$ in, $a_3 = .82$, $a_4 = 3.15$.

Since $a_4$ is much below $3 + 1.5a_3^2$ [see (6.40)], the gamma distribution is not ideal. But use it to estimate the percentage of contacts outside of specifications. What is most needed in the way of modification of distribution to increase the percentage within the specifications: change in $\bar{y}$, decrease in $s$, decrease in $a_3$?

6.9. A minimum measurement for the gap in piston rings when compressed by a specified force was being studied. A composite distribution for many rather similar types of rings was obtained ($n = 2100$). Relative to the minimum specification (of 0 as coded) $\bar{y}$ was .0044 in., $s = .0027$ in. The distribution was quite normal. Estimate the percentage below the minimum of zero and above .0100 in. Compare with respective observed data: 57/2100 and 21/2100.

6.10. The distribution of content weight of 1-oz packages of cereal was as follows:

| oz | 1.06 | 1.07 | 1.08 | 1.09 | 1.10 | 1.11 | 1.12 | 1.13 | 1.14 | 1.15 | 1.16 | 1.17 | 1.18 | 1.19 | |
|----|------|------|------|------|------|------|------|------|------|------|------|------|------|------|-----|
| $f$ | 1 | 7 | 16 | 14 | 15 | 12 | 12 | 8 | 8 | 2 | 1 | 1 | 2 | 1 | 100 |

If the minimum weight set by government regulations is 1.00 oz, what recommendations are suggested by statistical analysis?

For the following frequency distributions (a) find $\bar{y}$ and $s$; (b) for each class boundary except the lower boundary for the lowest class and the upper boundary for the upper class, find $u = (y - \bar{y})/s$; (c) for each $u$ find the area (probability) below for the normal curve or an appropriate gamma distribution; (d) by subtraction of results in (c) find the class probabilities (those at the extreme classes include the entire tails); (e) multiply the class probabilities in (d) by $n$ to find calculated class frequencies; and (f) compare fitted frequencies in (e) with the observed frequencies.

6.11. Table 2.10.
6.12. Table 2.13 (6/4/58 data).
6.13. Table 2.17.
6.14. Table 2.21.
6.15. Table 2.22.
6.16. Other tables as assigned.

For the following discrete distributions, approximate the desired probability using the continuity correction and the normal curve or the gamma distribution if $\alpha_3$ is much above .2.

6.17.    Binomial: $\phi = .4$,    $n = 20$,    $P(5 \leqslant Y \leqslant 10)$.
6.18.    Binomial: $\phi = 1/3$,    $n = 18$,    $P(3 < Y < 10)$.
6.19.    Binomial: $\phi = .2$,    $n = 64$,    $P(7 < Y < 15)$.
6.20.    Binomial: $\phi = .5$,    $n = 25$,    $P(6 \leqslant Y \leqslant 12), P(Y > 18)$.
6.21.    Binomial: $\phi = .02$,    $n = 1000$,    $P(Y \leqslant 10), P(Y > 30)$.
6.22.    Binomial: $\phi = .05$,    $n = 475$,    $P(15 \leqslant Y \leqslant 30)$.
6.23.    Binomial: $\phi = .2$,    $n = 25$,    $P(2 \leqslant Y \leqslant 10)$.
6.24.    Binomial: $\phi = .1$,    $n = 25$,    $P(1 \leqslant Y \leqslant 5)$.
6.25.    Poisson:    $\mu = 25$,    $P(15 \leqslant Y \leqslant 35)$.
6.26.    Poisson:    $\mu = 64$,    $P(Y < 50), P(Y > 80)$.
6.27.    Poisson:    $\mu = 100$, $Y$ so that $P(Y$ or greater$) \leqslant .01$, but as close as possible.
6.28.    Poisson:    $\mu = 49$, $Y$ so that $P(Y$ or less$) \leqslant .05$, but as close as possible.

6.29.    Hypergeometric: $N = 500$,    $K = 50$,    $n = 100, P(Y \geqslant 18)$.
6.30.    Hypergeometric: $N = 1000$, $K = 200$, $n = 100, P(Y \leqslant 15)$.

6.31.    Four parts are to be assembled so that their thicknesses add together. Suppose that each part is produced in such a way that the distribution between the specification limits is rectangular, (6.19). The four specifications are (a) $12 \pm .03$ cm, (b) $8 \pm .03$ cm, (c) $8 \pm .03$ cm, and (d) $2 \pm .03$ cm. What is the standard deviation for each part? What are the mean and standard deviation of the total thickness, assuming independence? Can you assume normality? Test by finding $\alpha_3$ and $\alpha_4$ for the combination. Suppose that instead of the rectangular distribution, each distribution is normal with $3\sigma = .03$ cm. What can you now say about the distribution of total thickness? In the two respective cases between what limits would $99\%$ of the cases lie? How do these compare with the simple sums of the limits, that is, $30 \pm .12$ cm?

6.32.    Prove that $a_{3:x}$ and $a_{4:x}$ are invariant under any linear transformation $X = kY + b$, with $k > 0$. [Thus if $X$ data are coded, as for example, $v = (X - A)/c$, the values of $a_3$ and $a_4$ are unchanged.]

6.33.    If we are given $\mu$ and $\sigma$ for an otherwise unspecified distribution, what are the narrowest limits for which there will be not over $4\%$ of the population? $1\%$ of the population? If we know the distribution is normal, what do these limits become?

6.34.    A drive chain is designed to have 100 links, each 1 cm long. Suppose that $\mu = .9980$ cm, $\sigma = .012$ cm (including linkage variability).

What can you say about the total linkage for 100 links, stating any needed assumptions, then using them?

6.35.   For the exponential distribution (6.25), integrate to find $F(y)$. Then find $y$ so that $F(y) = .5$. This is the median. Compare with mode and mean.

# Chapter 7

# SOME SAMPLING DISTRIBUTIONS

## 7.1. DISTRIBUTION OF SAMPLE STATISTICS FROM POPULATIONS

Probably the most basic problem in statistical work, both applied and theoretical, is the relationship between the population and samples from it. We can, in this connection, think of two types of population: (1) a collection of objects, letters, people, animals, or machines, or (2) a collection of numbers. The first case can be merged into the second, because in the first case we can have a sample space $W$ of, say, people and define for each person a random variable, that is, a number, such as the height of the person. To each person there thus corresponds a number, the *random variable*. Then all of these heights for the population of people, form a population of numbers. This is case (2) of the above. Then such a population of numbers may be theoretically described by a distribution such as those in Chapters 5 and 6. We might also simply categorize the people of the population into say male and female, or under 30 years of age, and 30 or over. Here we have a binomial situation.

Then our basic problem is the relationship between population and sample. We shall be especially concerned with populations of numbers such as those in the two preceding chapters. The relationship can well be studied from two viewpoints: (1) What does the population tell us about the behavior of samples from it? (2) What does a sample or series of samples tell us about the population from which the sample came? The second type of question is in general more difficult.

The remainder of this chapter is concerned primarily with problem (1), both for one and two samples. Then the remaining chapters in this book take up various statistical methods and applications stemming from problems (1) and (2).

## 7.2. CHOICE OF SAMPLE

It is of great importance in all statistical applications that samples be properly chosen. In general the basic aim is to avoid sample *bias* insofar as possible. For this the most common technique is to choose our sample *at random* from the population. By definition this means that for a finite collection of people, say, we select the sample in such a way that each person has the same probability of being chosen in the sample. If the selection is made one at a time, then for those people in the population not yet drawn for the sample, each is to have the same chance of being drawn next. This is the way to draw a sample randomly. In order to accomplish such drawing with equal probability, the standard approach is to use a table of random numbers, such as Table X in the Appendix. Each person or object of the population is given a serial number, for example, 00 to 99, or 000 to 999. Then a random spot in the table is chosen and two- or three-digit numbers taken in order in the column(s). Each number in the table in turn will correspond to a person not yet chosen, or the number will be discarded if it corresponds to a person already in the sample or a number beyond those listed. For example, with a population of 80 we can use serial numbers 00 to 79. If the first random number is 16, we take person 16. If the next is 87, we discard it. If the third is 16, we discard it, and so on. In this way we continue till a sample of $n$ is chosen. To the extent that the table of random numbers is really random, all samples of size $n$ of the $N$ people become equally likely. The number $C(N, n)$ of *possible* samples can be fantastically large!

It might be well to point out at this time that it is the manner in which we *draw* a sample that is *random*, not the sample itself. We *might* randomly draw a sample of 10 from a population of 1000 people, and get the 10 tallest people! This sample is just as random as any other *one* sample. The only thing we can control is the manner of drawing, not the outcome. By random sampling we hope to obtain a sample of people whose average height is fairly near that of all 1000 people. In subsequent writing when we loosely speak of a "random sample" we shall mean one chosen in a random manner.

While the foregoing use of a table of random numbers for drawing a sample is probably the safest, other commonly used methods are (1) to tumble or mix thoroughly the pieces, items, or product to be sampled and then draw $n$, not all from one place (just in case mixing was imperfect); (2) to place slips or chips with names or other designations in a box, mix, and draw out $n$ of them; and (3) to flip coins, roll dice, or draw cards from a deck in such a way as to give equal probabilities.

To illustrate what can happen when random sampling is not done we give the following actual case the author came upon [1]. A box of 3000 piston-ring castings was to be sampled for quality. A sample of 100 was drawn and inspected, yielding 25 defective castings. It was thus thought that about 25% were defective. To salvage the product, the 2900 remaining in the box were inspected. One would expect about .25(2900) = 725 to be defective. But in actual fact only four proved to be defective. Was the drawing of the 100 done at random? It is conceivable, but the odds against such an unrepresentative sample, as was observed, being drawn *at random*, are fantastically remote. What random drawing does is to help us properly play the odds against unrepresentative samples. We do not obtain *absolute* protection but *probabilistic* protection, which is good enough, and the only thing practical.

There are still other ways of sampling, in fact entire books are written on the subject, for example the work of Deming [2].

### 7.2.1.* Sampling from a Probability Distribution.

If we wish to draw a random sample from some theoretical probability distribution, such as those in Chapters 5 and 6, we can proceed as follows: Take a table of random numbers of as many digits as our data are to be given. For example, we might sample from a Poisson distribution with mean $\mu = 2$ [see Table 5.2 (b)]. The probability of a 0 observation is .1353. Thus allocate the 1353 four-digit random numbers 0000 to 1352 inclusive to 0. Then for 1 take the next 2707 such random numbers, that is, 1353 to 4059, and so on. Finally 9998 and 9999 will correspond to $y = 9$. We can now use four one-digit columns of a table of random numbers to give us a random sample of observations of $y$ for this Poisson.

If we want to draw a random sample from some continuous population of Chapter 6, we can first decide upon how much precision we wish to have in our sample of $y$'s. Suppose we wish, say, two decimal places. Then setting these down as midvalues, find the corresponding boundaries half-way between, say, $y_{i-\frac{1}{2}}$, $y_{i+\frac{1}{2}}$, and so on. Then whenever the random number $R$ lies between $F(y_{i-\frac{1}{2}})$ and $F(y_{i+\frac{1}{2}})$ we regard $y_i$ as having been observed. This takes some setting up, with the needed integrals.

Whenever the equation $R = F(y)$ may be solved explicitly for $y$ in terms of $R$, then we can convert the random number $R$ directly into $y$. Thus, for example, for (6.26) $F(y) = R = 1 - e^{-y/\mu}$ gives $y = -\mu$ $ln(1 - R)$, which converts $R$ to $y$. We may also similarly solve (6.20) and (6.52). See also Burr [3].

One can also sample from an observed frequency distribution in the same manner as above.

### 7.2.2.* Machine Generation of Random Samples.

Calculators can be programmed in a variety of ways to generate random digits, 0 to 9. These may then be combined into two-digit pairs, three-digit triples, and so on. Then these can be used directly as $R$ values equated to $F(y)$. If, as in the preceding subsection, we may solve for $y$ in terms of $R = F(y)$, then a random sequence of $R$ values yields a random sequence of $y$ values, that is, a sample of as many $y$'s as desired from the population. Thus we are able to sample randomly from a population, by a machine, with great rapidity. In this manner many sampling results and experiments may be run off, when the exact solution is not available. Such an approach is often called a Monte Carlo method. (See also Problems 7.3–7.5.)

## 7.3. SAMPLING DISTRIBUTIONS OF A SAMPLE STATISTIC

This subject is that of the first general problem of Section 7.1; namely, given a probability population of numbers, what can we expect of some sample statistic, for example, $\bar{y}$ or $s$, for random samples of some size $n$? In the case of very small *populations* of five to ten numbers only, it is possible to write down all samples which can occur, then for each, find the statistic in question and tabulate these values of the statistic into a frequency distribution. This distribution is then called the "sampling distribution" of the statistic in question from this population. See Li [4] for a number of simple examples of this type.

But the direct approach just mentioned quickly becomes infeasible owing to the enormous multiplication of the number of possible samples. Another approach is to make a sampling study by collecting a "large enough" number of samples and analyzing the results. An example of this approach was in an unpublished Master's thesis by Niemann [5]. The objective was to study the behavior of means $\bar{y}$, medians, ranges $R$, standard deviations $s$, and mean deviations (3.1), (3.6), (3.7), (3.9), for very small samples from a strongly skewed gamma population. Although the situation for $\bar{y}$'s was known, the others were not known at that time, and still are not! Our information is fragmentary. To perform the work the experimenter drew 4000 samples, each of $n = 4$ observations $y$ from an approximate gamma distribution with skewness $\alpha_3 = 1.15$, using 1000 numbered beads. The work was all done by desk calculators, this being before the flowering of the electronic computer! The sampling distributions thus obtained of 4000 $\bar{y}$'s, 4000 $R$'s, and so on, give useful and reliable information on the sampling distributions of these statistics.

Nowadays such a study as that of Niemann could be done in vastly less time by an electronic computer. Such studies are often called Monte Carlo studies.

The best approach when available is to work out the sampling distributions *exactly* by mathematical development. Most fortunately the exact sampling distributions of $\bar{y}$'s, $s$'s, and $s^2$'s are known exactly for random samples of $y$'s from any normal population. But information from nonnormal populations is still quite incomplete.

In the following sections, we state many results without proof. Derivations of such results are given in the many mathematical statistics books.

## 7.4. DISTRIBUTION OF SAMPLE MEANS

We have already stated some of the information on sample means in Section 6.11. We summarize in the following theorem:

**Theorem 7.1.** For samples of $n$ $y$'s chosen at random and independently from the same normal population having mean $\mu_y$ and standard deviation $\sigma_y$, we have

(1)  $\mu_{\bar{y}} = \mu_y$                                                                        (7.1)

(2)  $\sigma_{\bar{y}} = \sigma_y/\sqrt{n}$                                                          (7.2)

(3)  the distribution of $\bar{y}$'s is normal.

This theorem says that $\bar{y}$'s average the same as $y$'s, but vary much less around that average value. For example, if $n = 25$, they only vary a fifth as much. This theorem is very basic in all inferences and tests concerning means. The three conclusions of the theorem may all be summarized in the abbreviation

$$\bar{y}\text{'s}\quad \text{are}\quad N(\mu_y, \sigma_y{}^2/n),$$

wherein the "N" means a normal distribution, and the mean and variance of the variable $\bar{y}$ are given in consecutive order.

Further, the central limit theorem given in Section 6.11 states that the distribution function $G_n(\bar{y})$ becomes more and more normal as $n$ increases, even though the distribution of $y$'s is nonnormal. And, as was pointed out, the approach to normality is quite rapid. For example, suppose we have 10 independent $y$'s from a rectangular distribution (6.19). For this $\alpha_{3:y} = 0$, $\alpha_{4:y} = 1.8$, (6.23), (6.24). Then using (6.64), (6.65) we have $\alpha_{3:\bar{y}} = 0$ and $\alpha_{4:\bar{y}} = 3 + (1.8 - 3)/10 = 2.88$, which are "very normal" already. Or, consider the strongly J-shaped exponential distribution (6.25), which has $\alpha_{3:y} = 2$, $\alpha_{4:y} = 9$. For $\bar{y}$'s of samples of $n = 25$ we have $\alpha_{3:\bar{y}} = 2/\sqrt{25} = .4$, $\alpha_{4:\bar{y}} = 3 + (9 - 3)/25 = 3.24$. These results are not any too "normal" yet, but are getting fairly close. As has been pointed out, however, the gamma distribution with $\alpha_4 = .4$ would give an exact fit. [This latter example is a case of the statement that $\bar{y}$'s from the exponential and gamma distributions (6.30) are distributed according to another gamma distribution.]

### 7.4.1. Standardized Distribution for Means.

Using the information just given, we can say that when we standardize $\bar{y}$ by (6.62), that is,.

$$u = \frac{\bar{y} - \mu}{\sigma_y/\sqrt{n}}, \qquad (7.3)$$

we have a variable which is nearly normal if not exactly so, and has mean 0 and standard deviation 1. That is, $u$ is approximately $N(0, 1)$. This distribution is of much use for estimations and testing hypotheses, whenever $\sigma_y$ is known, as we see in Chapters 8 and 10.

### 7.4.2. Distribution of Means, when Standard Deviation is Unknown— Student's t.

Probably a more common case than the one in Section 7.4.1, is that in which we do not know $\sigma_y$ . When this is the case we cannot substitute $s$ for $\sigma_y$ in (7.3) and assume the fraction to be normally distributed, even though the $y$'s are, unless $n$ is substantial. But when the population of $y$'s is normal the exact distribution of the following variable is known

$$t = \frac{\bar{y} - \mu}{s/\sqrt{n}} \qquad (n - 1 \text{ degrees of freedom}). \qquad (7.4)$$

This is the widely used $t$ variable.* The $t$ distribution is really a *family*

---

* It is also called "Student's $t$," named for W. S. Gosset who published under the pen name Student.

of symmetrical density functions, with a single parameter to determine the particular member of the family. This parameter is the "degrees of freedom," which in (7.4) is $n - 1 = v$, say. It can be shown that as $v$ increases, the distribution of $t$ becomes more and more normal.

The density function for $t$ is given by

$$f_v(t) = \frac{\Gamma[(v + 1)/2]}{\sqrt{v\pi}\,\Gamma(v/2)} \left(1 + \frac{t^2}{v}\right)^{-(v+1)/2}, \qquad -\infty < t < \infty. \qquad (7.5)$$

For it

$$\mu_t = 0 \qquad (\text{if} \quad v > 1) \tag{7.6}$$

$$\sigma_t = \sqrt{v/(v - 2)} \qquad (\text{if} \quad v > 2) \tag{7.7}$$

$$\alpha_{3:t} = 0 \qquad (\text{if} \quad v > 3) \tag{7.8}$$

$$\alpha_{4:t} = 3(v - 2)/(v - 4) \qquad (\text{if} \quad v > 4). \tag{7.9}$$

The last equation indicates something about the rapidity of approach to $\phi(u)$.*

Figure 7.1 shows a couple of $t$ distributions, that is, for degrees of freedom $v = 3$ and $v = 9$, which would be applicable for single samples

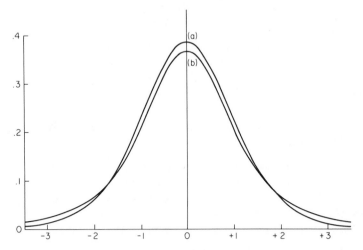

**Fig. 7.1.** Two examples of the $t$ distribution density. (a) Curve for degrees of freedom $v = 3$, and (b) curve for $v = 9$. Note the greater variability in (a) with slower approach to the base line, than for (b). The respective $\sigma_t$'s are 1.73 and 1.13 by (7.7).

* Actually the approach is much more rapid if we were to use instead of (7.4), $t' = t\sqrt{(n - 3)/(n - 1)}$, because $\sigma_{t'} = 1$, whereas $\sigma_t$ is not 1. For $t'$, normal-curve tables of areas can quite well be assumed for $n = 15$, but for $t$ we need at least 30.

of $n = 4$ and $n = 10$. Both curves seem to look quite normal and are symmetrical around $t = 0$. There is much greater variability in the $t$ distribution for $v = 3$ than that for $v = 9$ as shown by the much more gradual approach to the base line. The respective $\sigma_t$'s are 1.73 and 1.13 by (7.7). (The $\alpha_4$'s are, respectively, $\infty$ and 4.2.)

### 7.4.3. Areas for the t Distribution.

Again, as with the normal curve, one finds $t$ tables made up in different ways. One must be very careful to make sure what area is being listed as the column headings. Often the heading is the area from the tabulated $t$ in the body of the table, on out to $+\infty$, that is, a single-tail area. Or it may be the total of two symmetrical tails, $-\infty$ to $-t$ and $+t$ to $\infty$. In our Table III (Appendix), we have the upper tail area $1 - F_v(t)$, where

$$F_v(t) = P(T < t) = \int_{-\infty}^{t} f(T)\,dT. \tag{7.10}$$

Row headings in virtually all tables are degrees of freedom, that is, $v$.

Thus, for example, if $v = 3$, the area or probability of $t$ below 3.182 is .975. This leaves the upper tail area as .025 (from $t = 3.182$ to $\infty$). Meanwhile by symmetry the corresponding lower tail ($t = -\infty$ to $-3.182$) is also .025. Thus the two-tail area is .050. For $v = 9$, the similar $t$ entry is 2.262, which is an indication of the greater concentration of this $t$ distribution about $t = 0$. The normal-curve value for $u$ with a two-tail area is 1.960. This can be found at the bottom line of Table III, where $v = \infty$. The reason for this is that as $v$ becomes larger, the density function $f_v(t)$ approaches the normal $\phi(t)$ (6.15) as a limit.

As another example, find the $t$ value such that the two-tail area is .01, when $v = 30$. Here the single upper tail is $.5(.01) = .005$, and hence $F_v(t) = .995$. Looking down the .005 column to the entry opposite $v = 30$ yields 2.750. The corresponding normal-curve value is 2.576, in the last row.

### 7.4.4.* Interpolation Note.

If one ever finds it necessary to interpolate when $v > 30$, it is a good plan to use reciprocals of $v$ rather than $v$, itself thus:

| $v$ | $1/v$ | $t$ |
|-----|-------|-----|
| 30  | .0333 | 2.750 |
| 33  | .0303 |       |
| 40  | .0250 | 2.704 |

Hence for $\nu = 33$, use $2.750 - (30/83) \cdot (.046) = 2.733$. Interpolation on $\nu$ itself would yield 2.736. One can use the reciprocal approach even for interpolating from $\nu = 1000$ to $\nu = \infty$! Try it.

### 7.4.5. Distribution of Means from Nonnormal Populations.

The material of Section 7.4 has so far been primarily concerned with means from samples chosen randomly and independently from a *normal* population. If the population is not normal, then such random sampling gives a nonnormal distribution of $\bar{y}$'s. But by the central limit theorem (Section 6.11), we are assured of an approach to normality of distribution. Thus $\bar{y}$ is *approximately* $N(\mu, \sigma^2/n)$. Moreover, the approach is in general quite rapid. If $\alpha_3$ and $\alpha_4$ are known for the $y$ distribution, then we can use (6.64) and (6.65) to tell about what $n$ will yield reasonable normality. Hence when $\sigma_y$ is known we will often be able to regard

$$u = \frac{\bar{y} - \mu}{\sigma_y/\sqrt{n}}$$

as $N(0, 1)$.

In like manner, when sampling from a nonnormal population of unknown standard deviation, we cannot assert that

$$t = \frac{\bar{y} - \mu}{s/\sqrt{n}}$$

follows the $t$ density function (7.5). But (7.5) can commonly be regarded as a reasonable first approximation, and can be expected to improve as $n$ increases.

## 7.5. DISTRIBUTION OF SAMPLE VARIANCES

Again let us suppose that we draw a sample of $n$ independent random observations $y$ from some numerical population of $y$'s. Then we may obtain a second such sample of $n$, a third, and so on, until we have a large number of them. For each sample we calculate the sample variance $s^2$. Now what can we expect the distribution of these variances to look like? As is usual with distributions it will have a mean, a standard deviation, and some sort of curve shape.

If the population of $y$'s is normal, then the distribution of $s^2$'s is known exactly. Suppose that the $y$ distribution has standard deviation $\sigma$. Ordinarily instead of talking about the distribution of $s^2$, we standardize to form a random variable which has no physical units, that is, a pure number. Thus we define

$$\chi_\nu^2 = \frac{\nu s^2}{\sigma^2} = \frac{(n-1)s^2}{\sigma^2}. \tag{7.11}$$

This is called "chi-square with $\nu$ degrees of freedom." (The chi-square distribution was originally derived for quite a different application, associated with the goodness with which a theoretical distribution may be made to fit an observed frequency distribution.) The chi-square distribution is in reality a *family* of probability curves, varying according to the degrees of freedom $\nu$. Thus the density function is

$$f_\nu(\chi^2) = \begin{cases} K(\chi^2)^{.5\nu-1} e^{-.5x^2}, & \chi^2 > 0 \\ 0, & \chi^2 < 0 \end{cases} \tag{7.12}$$

where the constant is

$$K = [\Gamma(.5\nu)]^{-1} 2^{-.5\nu}, \tag{7.13}$$

which makes the total area unity. Examination of (7.12) shows that it is a simple transformation of the gamma distribution (6.30).*
  We have the following for the distribution of $s^2$:

$$\mu_{s^2} = \sigma^2 \tag{7.14}$$

$$\sigma_{s^2} = \sigma^2\sqrt{2/\nu} \tag{7.15}$$

$$\alpha_3 = \sqrt{8/\nu} \tag{7.16}$$

$$\alpha_4 = 3 + 12/\nu. \tag{7.17}$$

Thus we see that $s^2$ values average $\sigma^2$. In fact, this is the main reason we divide $\Sigma(y - \bar{y})^2$ by $n - 1 = \nu$ rather than by $n$, to define $s^2$. See Li [4] for a simple example. If we used the denominator $n$, we would have a quantity which would not average $\sigma^2$, but instead $[(n-1)/n]\sigma^2$. Note that as the degrees of freedom increase, $\sigma_{s^2}$ decreases, so that the distribution "pinches" down to the average $\sigma^2$. The curve-shape characteristics approach 0 and 3, respectively, as $\nu$ increases, in line with the fact that the distribution of $s^2$ approaches normality (see Section 6.6.2). But for small $\nu$'s the distribution of $s^2$ is quite nonnormal, being skewed strongly to the right.

  * By Letting $p = .5\nu$, $y = .5\chi^2$, one can transform $f_\nu(\chi^2)\, d(\chi^2)$ to $f(y)\, dy$ in (6.30). Then using (6.37)–(6.40), we obtain for the $\chi^2$ distribution: $E(\chi^2) = \nu$, $\sigma_{\chi^2} = \sqrt{2\nu}$, $\alpha_3 = \sqrt{8/\nu}$, $\alpha_4 = 3 + 12/\nu$.

### 7.5.1. Distribution of Sample Standard Deviation.

In the experiment we conceived of at the beginning of the previous section, we might have found the sample standard deviation $s$ for each sample instead of the variance $s^2$. Then we would have an approximation to the theoretical distribution of $s$. This latter is available exactly for samples from a normal population of $y$'s. In fact it is but a simple transformation from (7.12) by change of variable. We can then show the following

$$\mu_s = c_4\sigma \tag{7.18}$$

$$\sigma_s = \sigma\sqrt{1 - (c_4)^2} = c_5\sigma. \tag{7.19}$$

The values of $c_4$ and $\sqrt{1 - (c_4)^2}$ are given in Table IV, in the Appendix. There it is seen that $c_4$ is always below 1, but rather rapidly approaches 1 as the sample size $n$ increases. Meanwhile $\sigma_s$ decreases as did $\sigma_{s^2}$. The expressions for $\alpha_3$ and $\alpha_4$ for the $s$ distribution are somewhat complicated looking (involving gamma functions), and are not reproduced here. Suffice it to say that the distribution of $s$ has much less skewness and is more "normal" than that for $s^2$ for any given degrees of freedom.

Probably the surprising thing about the distribution of $s$ is that $s$'s do not average $\sigma$, even though $s^2$'s do average $\sigma^2$.* Compare (7.14) and (7.18). This fact has caused much confusion among students and even some textbook writers. Thus we can say that as an estimate of $\sigma$, $s$ tends to underestimate a little bit; that is, it averages slightly low. Such bias of estimation may be removed by dividing our observed $s$ by the corresponding value of $c_4$. Thus the estimates $s/c_4$ average $\sigma$.

### 7.5.2.* Population of y's Nonnormal.

When the population from which our $y$'s come is not normal then the density function (7.12) is no longer exactly correct. In fact the exact distribution of $s^2$ is unknown, except for a few special populations and small $n$'s of rather negligible practical application.

But one can use moment characteristics to some extent. It is easy to show in fact that $\mu_{s^2}$ is *still* $\sigma^2$, for nonnormal populations. The term $\sigma_{s^2}$ and the values of $\alpha_3$ and $\alpha_4$ for $s^2$ depend on those for the $y$'s in the population (see Cramér [6], p. 348). One could thus make allowance for nonnormality.

---

* The reason that this must be so is shown by applying (6.12) where $Y$ is replaced by $s$, giving $\sigma_s^2 = E(s^2) - [E(s)]^2$. Now if $E(s^2)$ were $\sigma^2$, which is true, and simultaneously $E(s)$ were $\sigma$, then the right side of the equation above would be zero, which says that $\sigma_s^2 = 0$, or that $s$ is constant for all samples, which is nonsense.

### 7.5.3. Tables of Chi-Square.

Our table of chi-square values gives for various degrees of freedom $\nu$, and given areas below, the value of $\chi^2$. Thus with nine degrees of freedom, the probability of an observed $\chi^2$ value lying below 16.919 is .95. Hence 5% of the time a sample $\chi^2$ value will exceed 16.919. The value of $\chi^2$ marking the boundary of the lower 5% tail appears in the .05 column and is 3.325 for $\nu = 9$. Note that these two values are not symmetrically placed around the mean $\chi^2$ of 9, due to the asymmetry of the distribution.

### 7.6.* THE JOINT DISTRIBUTION OF $\bar{y}$ AND $s$ FROM A NORMAL POPULATION

When random independent samples each of $n$ $y$'s are drawn from a normal population, we may well consider whether the sample $\bar{y}$ and $s$ are independent or related. As is shown in most mathematical statistics books $\bar{y}$ and $s$ are completely independent, if the population of $y$'s was normal. Moreover there is a companion theorem, more difficult to prove, that if $\bar{y}$ and $s$ are independent, then the distribution of the $y$'s is normal. Hence independence of $\bar{y}$ and $s$ is a *unique* property of the normal distribution. Thus knowledge of what $\bar{y}$ is tells nothing about what $s$ is, and conversely. Also the joint distribution of $\bar{y}$, $s$, say, $f(\bar{y}, s)$, is factorable into $g(\bar{y}) \cdot h(s)$. See Section 6.14.

### 7.7. TWO NORMAL POPULATIONS, INDEPENDENT SAMPLES

Suppose we have two normal populations with $\mu_1$, $\sigma_1$ and $\mu_2$, $\sigma_2$, respectively. We draw at random $n_1$ $y$'s from the first population, and independently, $n_2$ from the second, finding $\bar{y}_1$, $s_1$ ; $\bar{y}_2$, $s_2$ . In subsequent chapters, as we shall see, we are interested in comparing the two populations, and in drawing various inferences about their parameters. For this we need to know something about the distribution of certain combinations of the sample statistics. These are presented in the next three subsections.

### 7.7.1. Sum and Difference of Two Means, Standard Deviations Known.

For these distribuitions we can make use of (6.55) and (6.56), proofs of which were sketched in Section 6.14.

If we replace $y_1$ and $y_2$ in (6.54) by $\bar{y}_1$ and $\bar{y}_2$, we have for the sum

$$\mu_{\bar{y}_1 + \bar{y}_2} = \mu_1 + \mu_2 \tag{7.20}$$

$$\sigma_{\bar{y}_1 + \bar{y}_2} = \sqrt{\sigma_1{}^2/n_1 + \sigma_2{}^2/n_2} \tag{7.21}$$

since by (6.61) $\sigma_{\bar{y}_1} = \sigma_1/\sqrt{n_1}$, and so on. Similarly for $\bar{y}_1 - \bar{y}_2$

$$\mu_{\bar{y}_1 - \bar{y}_2} = \mu_1 - \mu_2 \tag{7.22}$$

$$\sigma_{\bar{y}_1 - \bar{y}_2} = \sqrt{\sigma_1{}^2/n_1 + \sigma_2{}^2/n_2}. \tag{7.23}$$

Note that the right sides of (7.21) and (7.23) are identical. A beginner might think there would be a minus under the radical in (7.23).

Now what about the curve shape of $\bar{y}_1 \pm \bar{y}_2$? If the $Y$ distributions are normal then, so are both $\bar{y}_1 \pm \bar{y}_2$. This fact is one of the comparatively easily proven theorems of mathematical statistics. Thus we know all there is to know about the distributions.

Next let us standardize these distributions analogously to (6.18):

$$u = \frac{\bar{y}_1 \pm \bar{y}_2 - (\mu_1 \pm \mu_2)}{\sqrt{\sigma_1{}^2/n_1 + \sigma_2{}^2/n_2}} \tag{7.24}$$

where $u$ is $N(0, 1)$. Formula (7.24) for differences is often used to compare two people's performances, for example, in a time study, or two formulas, two materials, two methods, two machines, and so forth. For sums it is often used for assemblies, mixtures, combining of industrial parts, and so on.

### 7.7.2. Sums and Differences of Two Means, Standard Deviations Unknown, but Equal.

In this situation we do not have $\sigma_1$ and $\sigma_2$ with which to work. Thus we use what is accessible, namely $s_1$ and $s_2$. Then we can show that the following rather formidable looking formula is applicable. Setting $\nu_1 = n_1 - 1$, $\nu_2 = n_2 - 1$:

$$t_{\nu_1 + \nu_2} = \frac{\bar{y}_1 \pm \bar{y}_2 - (\mu_1 \pm \mu_2)}{\sqrt{[(\nu_1 s_1{}^2 + \nu_2 s_2{}^2)/(\nu_1 + \nu_2)][1/(\nu_1 + 1) + 1/(\nu_2 + 1)]}}. \tag{7.25}$$

Under our assumptions of independence, normality of $y$'s and $\sigma_1 = \sigma_2 = \sigma$, say, this expression follows the $t$ distribution with $\nu_1 + \nu_2$ degrees of freedom. The denominator makes good sense. The first fraction is a weighted average of $s_1{}^2$ and $s_2{}^2$, in fact an unbiased estimate of the common $\sigma^2$. The parenthesis is $1/n_1 + 1/n_2$. So the expression is (est $\sigma^2$) $(1/n_1 + 1/n_2)$, much like that in (7.24).

If $n_1 = n_2 = n$, also, then (7.25) simplifies to

$$t_{2n-2} = \frac{\bar{y}_1 \pm \bar{y}_2 - (\mu_1 \pm \mu_2)}{\sqrt{(s_1{}^2 + s_2{}^2)/n}} \qquad (2n - 2 \quad \text{df}).* \qquad (7.26)$$

### 7.7.3. Two Variances, F Distribution.

We now come to the last of our present crop of standard random variables. Suppose we have $y_1 : N(\mu_1, \sigma_1{}^2)$ and $y_2 : N(\mu_2, \sigma_2{}^2)$. Then we might take a large series of pairs of independent samples of $n_1$ and $n_2$ $y$'s, respectively. For each pair we could find $s_1$ and $s_2$. Now to compare any two numbers we could consider their difference or their ratio (for example, two salaries). For $\bar{y}$'s we commonly use the former, $\bar{y}_1 - \bar{y}_2$, while for standard deviations we use the latter, in fact, $s_1{}^2/s_2{}^2$.

Once again the theory for normal populations proves tractable, and it is possible to find explicitly the *exact* distribution of the following variable:

$$F_{\nu_1, \nu_2} = \frac{s_1{}^2/\sigma_1{}^2}{s_2{}^2/\sigma_2{}^2}. \qquad (7.27)$$

Here $\sigma_1$ and $\sigma_2$ can be equal or even drastically unequal. In any case division by $\sigma_i{}^2$ standardizes the respective $s_i{}^2$, so that both numerator and denominator will each average 1. The $F$ distribution is completely determined by the pair of degrees of freedom $\nu_1$, $\nu_2$. In general when $\nu_2$ is at all small, $F$'s distribution is extremely skewed to the right. The precise density function for $F$ takes the following form:

$$f(F; \nu_1, \nu_2) = \begin{cases} KF^{.5\nu_1-1}(\nu_1 F + \nu_2)^{-(\nu_1+\nu_2)/2}, & F > 0 \\ 0, & F < 0, \end{cases} \qquad (7.28)$$

with $K$ making the total area $(0-\infty)$ equal to 1. There is a simple transformation which will carry this distribution into a beta distribution (Section 6.7 and Cramér [6, p. 243]).

For $F$ we have

$$\mu_F = \frac{\nu_2}{\nu_2 - 2}, \qquad \nu_2 > 2 \qquad \text{or} \qquad n_2 > 3 \qquad (7.29)$$

$$\sigma_F{}^2 = \frac{2\nu_2{}^2(\nu_1 + \nu_2 - 2)}{\nu_1(\nu_2 - 2)^2(\nu_2 - 4)}, \qquad \nu_2 > 4 \qquad \text{or} \qquad n_2 > 5. \qquad (7.30)$$

* The abbreviation df will be used for degrees of freedom throughout this book.

The expressions for $\alpha_3$ and $\alpha_4$ of $F$ are considerably more complicated, and require for finiteness $n_2 > 7$ and $n_2 > 9$, respectively. Note that interestingly the mean of $F$ depends *only* on $\nu_2$, not $\nu_1$ at all. All criteria for existence are also in terms of $\nu_2$ because it is the denominator $s_2^2$, which by approaching zero can cause the trouble, not $s_1^2$. See (7.27).

Two examples of the $F$ distribution are shown in Fig. 7.2, $\nu_1 = 4$,

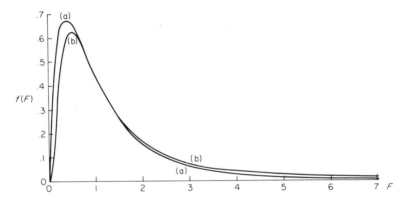

**Fig. 7.2.** Two density functions for the $F$ distribution. (a) Curve for $\nu_1 = 4$, $\nu_2 = 8$ degrees of freedom, and (b) curve for $\nu_1 = 8$, $\nu_2 = 4$. The points for which 1% of the probability lies above are 7.01 and 14.8, respectively. Also the points where 1% lies below are $1/14.80 = .068$ and $1/7.01 = .143$, respectively.

$\nu_2 = 8$, and conversely, $\nu_1 = 8$, $\nu_2 = 4$. As may be obtained from (7.29) and (7.30)

(a)   $\nu_1 = 4$, $\nu_2 = 8$, $\mu_F = 1.33$, $\sigma_F = 1.49$, finite skewness;
(b)   $\nu_1 = 8$, $\nu_2 = 4$, $\mu_F = 2.00$, $\sigma_F = \infty$, infinite skewness.

Also using Table V, we have the 99% point for (a) at 7.006 and for (b) it is much further out, at 14.80. But for the 1% points we use the reciprocals of table entries, *reversing* degrees of freedom. Thus for (a) we have $1/14.8 = .068$, while for (b) we have $1/7.006 = .143$. This is because $F_{4,8} = 1/F_{8,4}$. See (7.27). One has to be careful in using $F$ tables for the lower points to reverse the degrees of freedom before finding the entry to be divided into 1.

### 7.7.4.* Two Variances, Large Samples.

Whenever the sample sizes $n_1$, $n_2$ are sufficiently large, $s_1$ and $s_2$ will be quite normal, and hence also $s_1 - s_2$. Although the expected value of a sample standard deviation $s$ is $c_4\sigma$ with $c_4 < 1$, by the time $n = 25$,

$c_4$ is close to 1. Moreover, $\sigma_s = \sqrt{1 - c_4^2}\,\sigma$, but this can be shown to be quite well approximated by $\sigma_s = \sigma/\sqrt{2n}$. [Note the similarity between this and (6.61).] Thus the formally correct standardized variable for two independent samples from $N(\mu, \sigma^2)$

$$u = \frac{s_1 - s_2 - (c_{4,n_1} - c_{4,n_2})\sigma}{\sigma\sqrt{2 - c_{4,n_1}^2 - c_{4,n_2}}} \tag{7.31}$$

becomes the following:

$$u = \frac{s_1 - s_2}{\sigma\sqrt{1/2n_1 + 1/2n_2}}. \tag{7.32}$$

Formula (7.32) can be assumed nearly $N(0, 1)$ for equal $n$'s of 25 or more, while (7.31) can be assumed $N(0, 1)$ for equal $n$'s of 10 or more.

## 7.8. SAMPLING ASPECTS OF THE BINOMIAL AND POISSON DISTRIBUTIONS

We have two simple and related sampling properties of the binomial and Poisson distributions, which we now describe.

Suppose that we take a random sample under binomial conditions (constant probability $\phi$ of one of the two possible kinds of outcomes). Let the sample size be $n_1$ and the number of the one kind of outcome be $y_1$. Now take a second independent sample of $n_2$ from this population, yielding $y_2$. Now what can we say about the random variable $y_1 + y_2 = y$, say? It can be easily shown by a little theory that $y$ follows the binomial with $\phi$ and with $n = n_1 + n_2$. In fact this is so reasonable as to be nearly self-evident.

For the Poisson distribution we have a somewhat similar situation. Suppose we have two Poisson distributions with parameters $\mu_1$ and $\mu_2$. Then let the respective number of "defects" be $y_1$ and $y_2$. For example, suppose that the area of opportunity for $y_1$ is 1000 glass jars and that $\mu_1$ is 2, that is, the average number of defects over 1000 jars is two, and $y$ is distributed according to (5.37). Meanwhile an area of opportunity for defects on covers for the jars is again 1000 covers, with parameter $\mu_2 = 3$, say. Now in assembling the 1000 jars and covers we have 1000 combinations (jar and cover). The average number of defects in such combinations is $2 + 3$ or 5, because $y_1 + y_2 = y$, say. Now does $y$ follow the Poisson (5.37) with $\mu = 5$? The answer is yes. Hence we not only can say that $\mu = 5$ but also that $\sigma_y = \sqrt{\mu} = \sqrt{5}$. Thus the sum of two independent Poisson variables is also a Poisson variable with parameter the sum of the two separate parameters.

## 7.9.* THE SUM OF TWO INDEPENDENT CHI-SQUARE VARIABLES

It is easy to show that if two random variables $\chi_1^2$ and $\chi_2^2$ follow the chi-square distribution with $\nu_1$ and $\nu_2$ degrees of freedom and they are independent, then their sum $\chi_1^2 + \chi_2^2$ follows the chi-square distribution with $\nu_1 + \nu_2$ degrees of freedom.

For example, if we have two independent random samples of $n_1$ and $n_2$ $y$'s from $N(\mu_1, \sigma^2)$ and $N(\mu_2, \sigma^2)$, then we may combine our sample data for the purpose of estimating $\sigma^2$ from all the data. In fact we have

$$\chi_1^2 = \nu_1 s_1^2/\sigma^2, \qquad \chi_2^2 = \nu_2 s_2^2/\sigma^2.$$

Then

$$\chi^2 = \nu_1 s_1^2/\sigma^2 + \nu_2 s_2^2/\sigma^2 = (\nu_1 s_1^2 + \nu_2 s_2^2)/\sigma^2.$$

Since this $\chi^2$ has $\nu_1 + \nu_2$ degrees of freedom,

$$E(\chi^2) = \nu_1 + \nu_2.$$

Hence

$$E[(\nu_1 s_1^2 + \nu_2 s_2^2)/\sigma^2] = \nu_1 + \nu_2$$

or

$$E(\nu_1 s_1^2 + \nu_2 s_2^2) = (\nu_1 + \nu_2)\,\sigma^2,$$

and finally

$$E\left(\frac{\nu_1 s_1^2 + \nu_2 s_2^2}{\nu_1 + \nu_2}\right) = \sigma^2. \tag{7.33}$$

Thus the quantity in parentheses is an unbiased estimate of the common $\sigma^2$. This was used in (7.25).

## 7.10.* NONCENTRAL DISTRIBUTIONS

Sometimes in applications of the foregoing distributions to significance testing problems a modified distribution is needed. For example, if in the $t$ variable of (7.4), we use $(\bar{y} - \mu')/(s/\sqrt{n})$, where $\mu'$ is not the expected value of $\bar{y}$, that is, is not $E(\bar{y})$, then the distribution of this variable is no longer the ordinary $t$ distribution (7.5). Instead it follows the so-called noncentral $t$, which depends not only on the degrees of freedom but also the amount of noncentrality $(\mu' - \mu)\sqrt{n}/\sigma$. The distribution function of the noncentral $t$ is considerably more complicated than (7.5).

There are also noncentral chi-square and $F$ distributions modifying the assumptions behind (7.12) and (7.28) and for which the latter are special cases.

## 7.11. SUMMARY

In this chapter we have been concerned with the manner in which various sample statistics such as $\bar{y}$ and $s$ are distributed. This is vital in all of the subsequent chapters in this book, in which we work out statistical techniques for application. In fact the rest of the book is primarily how to apply what we have already learned.

Both $\bar{y}$'s and $s$'s need to be standardized into some dimensionless variable such as $u$, $t$, $\chi^2$, and $F$, which can be tabulated. This gives us a pure number with which to work and gives as few parameters as possible. These can then be tabulated in standard tables of wide applicability. Even when normality is not present, the distributions of sample statistics such as $u$, $t$, $\chi^2$ and, $F$ are used. Precise criteria are difficult to give unless one works out the curve-shape characteristics, but this can be very complicated. Moreover, if we do not know $\mu$ and/or $\sigma$, it is unlikely we will know $\alpha_3$ and $\alpha_4$ for the population. We can say, however, that the larger the $n$'s are, the more confidence we can place in using distributions assuming a normal population of $y$'s. Or conversely, the more nonnormal the population, the larger the $n$'s must be to treat by normal-curve techniques.

In summary, $u$ is used when we have exact normality, for example, problems of one or two $\bar{y}$'s, when the $\sigma$ or $\sigma$'s are *known*. Or also when there is approximate normality. The variable $t$ comes in when we are concerned with one or two $\bar{y}$'s, and $\sigma$'s are *unknown*. We use $\chi^2$ for a single variance $s^2$, and $F$ for two $s^2$'s. Fortunately we often know the exact distribution of sums when the population is binomial or Poisson.

REFERENCES

1. *Acceptance Sampling Symp.*, p. 59. Amer. Statist. Assoc., Washington, D.C., 1950.
2. W. E. Deming, "Some Theory of Sampling." Wiley, New York, 1950.
3. I. W. Burr, A useful approximation to the normal distribution function with application to simulation. *Technometrics* **9**, 647–651 (1967).
4. J. C. R. Li, "Statistical Inference I." Edwards, Ann Arbor, Michigan, 1968.
5. L. J. H. Niemann, An experiment in sampling from a Pearson Type III distribution. Master of Science Thesis, Purdue Univ., 1949, unpublished.
6. H. Cramér, "Mathematical Methods of Statistics." Princeton Univ. Press, Princeton, New Jersey, 1946.

## Problems

7.1. Class experiment. The following approximately normal population is to be used for repeated sampling with replacement:

| $y$ | −5 | −4 | −3 | −2 | −1 | 0 | +1 | +2 | +3 | +4 | +5 | |
|---|---|---|---|---|---|---|---|---|---|---|---|---|
| $f$ | 1 | 3 | 10 | 23 | 39 | 48 | 39 | 23 | 10 | 3 | 1 | 200 |

(For this $\mu = 0$, $\sigma = 1.715$, $\alpha_3 = 0$, $\alpha_4 = 3.02$.) Two hundred chips or beads may be numbered according to the frequencies given. Fiber chips work well, thick ones of small diameter mix best; beads mix even better.

Each member of the class draws a sample of five with replacement, then three more such samples of five. He is then to take great care in calculating the four $\bar{y}$'s, four $s^2$'s, and four $s$'s.

Someone is to tabulate the three distributions, namely, for $\bar{y}$, $s^2$, and $s$. Then either that person or the entire class can figure out the average and standard deviation of the three distributions. Comparison should then be made with what the theory gives, using $\mu = 0$, $\sigma = 1.715$ in the right-hand sides of (7.1), (7.2), (7.14), (7.15), (7.18), and (7.19).

7.2. As an extension of the preceding problem, each student finds the following for each of his four samples: $t = (\bar{y} - 0)\sqrt{n}/s$ and $\chi^2 = 4s^2/1.715^2$. Also for the first two samples he finds $F = s_1^2/s_2^2$ [$\sigma_i^2$ of (7.27) canceling], and next $F = s_3^2/s_4^2$. The student must be careful to divide the two $s^2$'s in the manner shown, not, for example, the larger of a pair by the smaller. Comparison can then be made with the right sides of (7.6), (7.7), $\mu_{\chi^2} = 4$, $\sigma_{\chi^2} = \sqrt{2 \cdot 4}$, and the right side of (7.29). Note that $\sigma_F$ here is $\infty$, by (7.30).

7.3. Program a computer to use some "random" number generating program and the distribution function $F(y)$ for the population frequencies in Problem 7.1, so as to generate samples of $n = 5$ $y$'s. Then as in Problem 7.1, the computer is to find the mean and the standard deviation of $\bar{y}$'s, $s^2$'s, and $s$'s.

7.4. Continuation of 7.3 as in Problem 7.2. The computer is to find $t$ and $\chi^2$ for each sample and $F$ for each pair of samples, finding also the mean and the standard deviation for all samples for $t$ and $\chi^2$, and the mean $F$. Care must be taken to eliminate $s$'s of 0, when for example, the five $y$'s are all +1's or all 0's, and so on.

7.5. A sampling problem as in Problem 7.3, given the population whose cumulative or distribution function $F(y) = 1 - (1 + y^5)^{-6}$.

For this distribution $\mu = .65513$, $\sigma = .16103$ (and also $\alpha_3 = -.013$, $\alpha_4 = 3.010$) which is quite highly "normal." For each random number $M$ between 0 and 1 generated by the computer, we can solve $M = 1 - (1 + y^5)^{-6}$, finding $y = [(1 - M)^{-\frac{1}{6}} - 1]^{\frac{1}{5}}$. Then generate a sample of five such $y$'s, programming to find $\bar{y}$, $s^2$, $s$, $t = (\bar{y} - \mu)\sqrt{n}/s$, $\chi^2 = 4s^2/.16103^2$, and $F = s_1^2/s_2^2$ for two such samples. Find the respective means and variances. There should be no trouble with any $s = 0$ if enough random digits are taken in each $M$.

7.6. Use the table of random numbers in the Appendix, and start each class member at a random spot in the table. Then each class member takes the next five random numbers 0 to 9 as a sample of $n = 5$ $y$'s and finds $\bar{y}$, $s^2$, and $s$. He repeats for three more such samples of $n = 5$. Then someone tabulates the results as in Problem 7.1. Finally the means and standard deviations are to be found. We have $\mu_y$ and $\sigma_y$ from (5.61), (5.62), and can find $\mu_{\bar{y}}$, $\sigma_{\bar{y}}$ from these using $\mu_{\bar{y}} = \mu_y$, $\sigma_{\bar{y}} = \sigma_y/\sqrt{n}$. Also the $\bar{y}$'s will be quite normal. Much less is known about $s^2$ and $s$ for this uniform population. But we do know $\mu_{s^2} = \sigma_y^2$. Some light is shed by Cramér [6].

7.7. Suppose that we try a test on some foodstuff of such quality that one test in 10 will show a "positive," that is, the presence of some bacterium. We make 20 tests. What distribution law will give us the probability of none, of one, of two and so on, positives? Now suppose we make 30 more tests and combine the numbers of positives in the two series. What precise distribution law gives the probability of each possible total of "positives"?

7.8. Suppose that defects on uniform areas of some given size follow the Poisson law with $\mu = 2$. Now let us treble the size of area studied. What law now governs the probability for each possible number of defects on the larger areas?

7.9. Suppose we have 1000 sheets with a Poisson distribution with $\mu = 1$ (insofar as a finite number of sheets can follow the law). Also we have 1000 sheets with $\mu = 3$. Combining all 2000 sheets, use Table XI to find the proportion of the 2000 with 0, 1, 2,... defects. Is this distribution the same as we obtain from 2000 sheets following a Poisson with $\mu = 2$?

7.10. Suppose we have 10,000 screws of which 50 have no slot in the head. We take a random sample of 100 without replacement. What is the exact law governing the probability for each possible number of slotless screws in the 100? Could this be well approximated by a binomial distribution? Now let us mix 10,000 more screws, having 150 without slots. Now take a random sample of 200 from the 10,000. How would you find the probability for each number of slotless screws in 200?

7.11.   Using (7.12), find $E(\chi^2)$ and $\sigma_{\chi^2}$ from (6.10)–(6.12) and (6.31), (6.33).

*Chapter 8*

# STATISTICAL TESTS OF HYPOTHESES—
# GENERAL AND ONE SAMPLE

## 8.1. INTRODUCTION

In political and social science, biological science, physical science, and industry, and even in studies in the humanities, we find wide application for two complementary types of problems. One of these is to estimate what the true distribution is like, from the characteristics of a sample of data. The second is to determine whether our observed sample could within reason have come from some hypothesized population or condition. Is it compatible with the population, or does it throw doubt on that population? The second of these two problems is the main subject of this chapter. These two problems are both concerned with decision making in the face of variation. If there is no variation present, then the two are not problems at all, but reduce to the very simplest of arithmetic! But with variation present we need the statistical machinery we have been building up in the preceding chapters.

Statistical hypotheses are of many kinds. The most common type would make a statement about a parameter of the population, for example, $\sigma = .002\%$, for some analytical technique. Such a hypothesis may have been suggested by the standard deviation error on analyses using a similar analytical technique. Then we take repeated analyses on homogeneous material using the new technique, and determine whether a sample such as we *did* obtain could readily occur by chance, if indeed $\sigma = .002\%$, or whether on the contrary the probability of obtaining a sample such as we did observe is "small." If the former is the case, we "accept" the hypothesis; but if the latter is the case, we "reject" the hypothesis. In this way we can make a decision on the analytical technique. We shall explain in subsequent sections the meaning of the words in quotations.

## 8.2. AN EXAMPLE

Let us consider an actual case of decision-making on an industrial product. This is a case in which $\sigma$ is known and $\mu$ is unknown.* The product was rubber tubes about 30 in. long, to be sliced into gaskets .10 in. length which are then assembled into caps to enclose food products in glass jars. If the rubber is too thin, air may leak into the vacuum-packed jar; whereas if the rubber is too thick, the cap may "pump off" the jar, again causing spoilage. The former is the more troublesome because it is less easily detected.

Originally every tube from a shipment of several hundred would have its doubled thickness measured at about six places, and then the tube would be placed in one of several piles, according to a mental average of the measurements. This method was costly and not effective because of the great overlapping of distributions of gaskets from the several piles of tubes. (See Table 2.28.) It was decided to have the rubber supplier run at one level without adjusting the average thickness at all, then to use a sample of tubes to decide for which specification range to use the lot. This brought two immediate benefits[†]: (a) a much smaller $\sigma$ which proved to be about .002 in., and (b) normality of distribution of doubled-thickness measurements of tubes.

Three specification ranges had middle or nominal values as follows: $A$, .098 in., $B$, .103 in., and $C$, .108 in. Thus, if $\mu$ is close to .098 in., the lot should be used in caps assuming the $A$ specification, or if near .103 in. the $B$, and so on. But if $\mu$ is .100 or .101 in., it would not be too bad to use the rubber for either range, $A$ or $B$.

We might set up a sample size of $n = 4$, say, and seek to test the hypothesis that $\mu = .103$ in. Such a hypothesis is often designated as $H_0$ and called a "null hypothesis." Then the two alternate specification ranges $A$ and $C$ provide two "alternate hypotheses": $\mu = .098$ in. and $\mu = .108$ in.

* Cases of this type may well be more common in industrial production than in other fields. Past records of production can often give evidence on $\sigma$, but the mean $\mu$ may be harder to control, and hence unknown.

† These benefits both came from the fact that the supplier previously would run at some level, then adjust the level several times, during production of one shipment. This caused nonnormality and excessive variability.

### 8.2.1. Approach 1 Given *n*, Set Significance Level α.

We now need a definite rule for acceptance or rejection of the hypothesis $\mu = .103$ in. This involves choosing a statistic to use for the decision. When the population is normal, the most "efficient" statistic concerned with $\mu$ is the sample mean $\bar{y}$; that is, it is subject to a smaller standard error than is *any other* estimate of $\mu$, for example, the median. A second job is to choose limits for $\bar{y}$, so that if $\bar{y}$ lies outside the limits, we will reject the hypothesis, and if $\bar{y}$ lies inside, we will accept the hypothesis. The region outside such limits is called the "critical region" or region of rejection for $\bar{y}$. The region between the limits is the acceptance region. Now surely the higher $\bar{y}$ lies above .103 in., the more we *tend* to reject $H_0 : \mu = .103$ in., and to take action appropriate if $\mu = .108$ in.* Thus we should set the upper limit to the acceptance region somewhere between .103 and .108 in., and similarly the lower limit between .103 and .098 in. Where shall we place them? In the present approach we shall set these limits so that, if in fact $\mu = .103$ in., we shall erroneously "reject" this hypothesis a "small" proportion of the time, $\alpha$. We may choose $\alpha$ at any small value we wish, say .01 here. We may thus define

$$\alpha = P(\text{rejecting } H_0 \mid H_0 \text{ is true}), \tag{8.1}$$

the vertical bar meaning "given" the condition $H_0$ is true.

We now have a completely specified problem. We can use (7.3) to find the limits. Taking symmetrical limits, since the alternate values .098 and .108 in. are equally spaced around .103 in. $= \mu_0$ , we use

$$u = \frac{\bar{y} - \mu_0}{\sigma/\sqrt{n}},$$

where $u$ is to have two values so as to cut off equal tails of probability $\alpha/2 = .005$. (See Fig. 8.1.) Using Table I we find $u = \pm 2.576$. Then

$$\pm 2.576 = \frac{\bar{y} - .103 \text{ in.}}{.002 \text{ in.}/\sqrt{4}}.$$

Or the limits for $\bar{y}$ are

$$\bar{y} \text{ limits} = .103 \text{ in.} \pm 2.576(.001 \text{ in.}) = .10042 \text{ in.}, \quad .10558 \text{ in.}$$

---

* In the actual case, we can be virtually certain that $\mu$ will never be so far above .108 in. that we cannot use the rubber even for specification $C$.

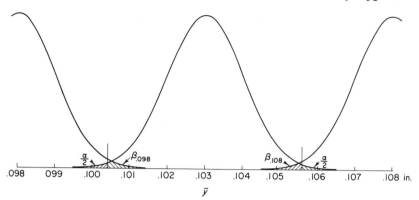

**Fig. 8.1.** This figure shows three distribution curves for *means* $\bar{y}$'s ($\sigma_{\bar{y}} = .001$ in.), when $\mu = .103$ in. $= \mu_0$, that is, $H_0$ is true, and when $\mu = .098$ or .108 in. The acceptance region is .10042 to .10558 in. Whenever $\mu = \mu_0$, we wish to accept, but will take the small risk $\alpha = .01$ of erroneous *rejection*. The graph shows that $\alpha$ is split into two small parts, each $\alpha/2 = .005$. Then if $\mu = .098$ or .108 in. we have the other two curves shown. For such $\mu$'s it is an error to accept hypothesis $H_0 : \mu = .103$ in. As shown on the two graphs, each such error of decision occurs rarely, and shows up as a *single*-tail area between the limits .10042 and .10558 in. These risks are each .0078 as in the text.

Thus the decision rule we arrive at is the following:
Take a random sample of four tubes and find the typical doubled thickness $y$ for each tube. Find the average thickness $\bar{y}$ for the four tubes.

> If $\bar{y} < .10042$ in., use the lot for specification $A$.
> If $\bar{y} > .10558$ in., use the lot for specification $C$.
> If $\bar{y}$ is between,  use the lot for specification $B$.

This decision rule has been designed to give acceptance for specification $B$ 99% of the time whenever $\mu$ is *exactly* .103 in.

Now let us analyze this decision rule further. What are the probabilities of erroneously *accepting* hypothesis $H_0 : \mu = .103$ in., when in fact $\mu = .098$ or .108 in.? Such risks are denoted by the letter $\beta$. Thus

$$\beta_\mu = P(\text{accepting } H_0 \mid H_0 \text{ false: mean at } \mu, \text{ not } \mu_0). \tag{8.2}$$

Note that $\beta_\mu$ depends on how far off the parameter $\mu$ actually is from that specified in $H_0$, hence the subscript on $\beta$. Here $\mu$ is .108 or .098 in.

Suppose $\mu = .098$ in. Then putting both limits into (7.3), using $\mu = .098$ in. we have

$$u = \frac{.10042 \text{ in.} - .098 \text{ in.}}{.002 \text{ in.}/\sqrt{4}} = 2.42, \qquad u = \frac{.10558 \text{ in.} - .098 \text{ in.}}{.001 \text{ in.}} = 7.58.$$

The latter $u$ value is completely "off the map," so that the probability of erroneous acceptance $\beta_\mu$ is simply the tail area above $u = 2.42$, that is, .0078. This is quite a small risk of a wrong decision if $\mu = .098$ in. By symmetry it is also the risk of erroneous acceptance of $H_0 : u = .103$ in., if in fact $\mu = .108$ in.

Now surely in practice we probably will *never* be offered a lot with $\mu$ exactly .098, .103, or .108 in. What then? To study a spectrum of values of $\mu$ we can easily draw an "operating characteristic" or *OC* curve, which shows the probability of acceptance of $H_0$ when in fact $\mu$ takes on *different* values. Table 8.1 shows the easy calculations needed to draw the operating characteristic curve. Figure 8.2 shows the OC curve for this case. The latter somewhat resembles a normal curve but is more "square-shouldered" appearing. The calculations in Table 8.1

**TABLE 8.1**

Calculations for OC Curve for Sampling Decision Plan for Rubber Thicknesses[a]

| (1) | (2) | (3) $u_L =$ (2)/.001 in. | (4) $\Phi(u_L)$ | (5) | (6) $u_U =$ (5)/.001 in. | (7) $\Phi(u_U)$ | (8) $P_a =$ (7)–(4) |
|-----|-----|-----|-----|-----|-----|-----|-----|
| $\mu$ (in.) | $L - \mu$ (in.) | | | $U - \mu$ (in.) | | | |
| .098 | +.00242 | +2.42 | .9922 | +.00758 | +7.58 | 1.0000 | .0078 |
| .099 | +.00142 | +1.42 | .9222 | +.00658 | +6.58 | 1.0000 | .0778 |
| .100 | +.00042 | +.42 | .6628 | +.00558 | +5.58 | 1.0000 | .3372 |
| .101 | −.00058 | −.58 | .2810 | +.00458 | +4.58 | 1.0000 | .7190 |
| .102 | −.00158 | −1.58 | .0571 | +.00358 | +3.58 | .9998 | .9427 |
| .103 | −.00258 | −2.58 | .0049 | +.00258 | +2.58 | .9951 | .9902 |
| .104 | −.00358 | −3.58 | .0002 | +.00158 | +1.58 | .9429 | .9427 |
| .105 | −.00458 | −4.58 | .0000 | +.00058 | +.58 | .7190 | .7190 |

[a] $\sigma_y = .002$ in., so that with $n = 4$, $\sigma_{\bar{y}} = .001$ in. Acceptance region for $\bar{y}: L = .10042$ in. to $U = .10558$ in. Column (8) gives $P_a$.

are straightforward. Column (1) lists $\mu$ values in the range of interest. Then columns (2), (3), (5), and (6) find the standard score values for the two boundaries, $L$ and $U$, of the acceptance region for $\bar{y}$'s. Note that to standardize we divide $L - \mu$ by $\sigma_{\bar{y}}$ not $\sigma_y$. Next we find from Table I the areas below the two $u$ values. Then subtraction (7)–(4) gives the area between, or probability of acceptance $P_a$ of the hypothesis $H_0 : \mu = .103$ in. This is what is plotted against $\mu$ in Fig. 8.2.

When $\mu = .103$ in., we want a high $P_a$, and also a high $P_a$ if $\mu$ is near .103 in. But the farther away $\mu$ is from $\mu_0 = .103$ in., the smaller we want $P_a$ to be, which is exactly the way the OC curve behaves. Defining

$$\beta = P_a \tag{8.3}$$

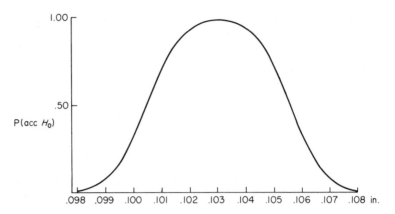

**Fig. 8.2.** Operating characteristic (OC) curve for sampling plan on rubber thicknesses. Sample size $n = 4$, $\sigma_y = .002$ in., $\sigma_{\bar{y}} = .001$ in. Acceptance region for $\bar{y}$'s for $H_0 : \mu = .103$ in., is .10042 to .10558 in. See Table 8.1 for method of calculating the probability of acceptance $P_a$ for each $\mu$.

we can read $\beta$ directly out of Table 8.1 for any of the $\mu$'s unless $\mu = .103$ in. Why not then? It only makes sense to talk about $\beta$ when $\mu$ is actually a practically important distance away from $\mu_0$.

Please read this section again!

### 8.2.2. Approach 2   Set Two Risks: $\alpha$ and a $\beta_\mu$ for some $\mu$ and Find $n$.

Suppose in the example in Section 8.2.1, we set $\alpha = .001$ and ask that $\beta_{.098\,\text{in.}}$ be .001 also. Now $n = 4$ is not big enough to give such sharp discrimination. How do we proceed? We first determine the lower limit $L$ for the acceptance region by using (7.3) twice. Thus

$$u = \frac{L - .098 \text{ in.}}{.002 \text{ in.}/\sqrt{n}} = +3.090, \qquad u = \frac{L - .103 \text{ in.}}{.002 \text{ in.}/\sqrt{n}} = -3.291.$$

Here $u = +3.090$ cuts off from the $\bar{y}$ distribution having $\mu = .098$ in., a *single upper* tail area of $\beta = .001$. On the other hand, $u = -3.291$ cuts off, from the $\bar{y}$ distribution for which $\mu = .103$ in., a *lower* tail area of $\alpha/2 = .0005$, so that the *two-tail area* is $\alpha = .001$. See Fig. 8.1 for the general picture, but now the $\alpha$, $\beta$ risks are even smaller. We now must solve the two equations above for $L$ and $n$. Rewriting, we have

$$(L - .098 \text{ in.}) \sqrt{n} = +3.090(.002 \text{ in.})$$

$$(L - .103 \text{ in.}) \sqrt{n} = -3.291(.002 \text{ in.}).$$

Either $L$ or $n$ could be eliminated and the other solved for. But since $n$ must be an integer, we can best solve for it first and round *up* to the next higher integer, if it does not come out a whole number. Thus we eliminate $L$ by subtracting the second equation from the first, obtaining

$$.005 \text{ in.} \sqrt{n} = 6.381(.002 \text{ in.}) \quad \text{or} \quad \sqrt{n} = 2.552.$$

Thus $n = 6.51$, and to obtain at least the prescribed power of discrimination we must take $n = 7$.

Having settled upon $n = 7$, we can find $L$ from either equation. If we substitute into the first equation, we shall be preserving the $\beta$ risk of .001 at $\mu = .098$ in. and will then be cutting $\alpha$ slightly. Or, if we substitute into the second equation to find $L$, we shall be preserving the $\alpha$ risk at .001, and will slightly cut the $\beta$ risk, at .098 in. Let us do the former:

$$L = .098 \text{ in.} + 3.090(.002 \text{ in.})/\sqrt{7} = .10034 \text{ in.}$$

We might ask what $\alpha$ has now become, using $n = 7$:

$$\frac{.10034 \text{ in.} - .103 \text{ in.}}{.002 \text{ in.}/\sqrt{7}} = -3.52.$$

The tail area below $u = 3.52$ is .0002, so $\alpha$, being the *two-tail* area, is .0004, rather than .001 as originally asked for.

Using the symmetry around $\mu_0 = .103$ in. we find $U = .10566$ in. Thus the plan is as follows:

Take a random sample of seven tubes, and find the typical doubled thickness for each tube. Find the average thickness $\bar{y}$ for the seven tubes.

If $\bar{y} < .10034$ in., use the lot for specification $A$.
If $\bar{y} > .10566$ in., use the lot for specification $C$.
If $\bar{y}$ is between,   use the lot for specification $B$.

The complete OC curve may be calculated just as in Table 8.1, and plotted as in Fig. 8.2. It would be flatter topped than that in Fig. 8.1 and would fall more steeply when the mean $\mu$ is near critical limits $L$ and $U$. Thus sharper discrimination is obtained with the present case in which $n = 7$.

Compare carefully these two approaches.

Owen [1] gives the required sample sizes for tests with $\mu_1$ at various distances from $\mu_0$, in terms of $\sigma$.

## 8.3. SUMMARY OF THE ELEMENTS OF TESTS OF HYPOTHESES ON ONE PARAMETER

In the applied problem discussed in Section 8.2, many of the elements of hypothesis testing were introduced. It was an example of a decision concerned with one parameter, with the population otherwise completely specified. Let us now summarize these elements.

1. *Assumptions.* We usually assume that the population is of a certain general type or form. We may also assume that we know the values of parameters other than the one under test. (In the example we assumed the population normal, and that $\sigma = .002$ in.)

2. *Null hypothesis—the basic hypothesis to be tested.* This usually takes the form of a statement that the parameter has a particular value. Thus we may say, $\mu = \mu_0$ is to be tested. Here $\mu$ is the true parameter value whatever it may be, while $\mu_0$ is a particular value, a number appropriate to the problem at hand. (In the example $\mu$ was the actual population mean, whatever it might happen to be, while $\mu_0$ was .103 in.)

3. *Alternate hypotheses.* Such hypotheses are that the parameter has some value other than the null value as in 2. Such alternative values may be

   (a)   possible values,
   (b)   ones we may be guarding against,
   (c)   ones we might like to have be true,
   (d)   ones which have economic importance or practical significance.

4. *Kinds of hypotheses.* A hypothesis that a parameter has just one particular value is often called a "simple hypothesis," while one which specifies an interval for the parameter is a "composite hypothesis." Thus in our example we could have had the null hypothesis .102 in. $< \mu < .104$ in., and alternative hypotheses $\mu \leqslant .102$ in. and $\mu \geqslant .104$ in. It is, however, the author's experience that most practical problems can be set up, at least initially, using simple null hypotheses, and then the OC curve studied to see whether the decision rule makes practical sense.

5. *Errors of first and second kind.* If it should happen that the null hypothesis $H_0$ actually is true and yet through bad luck we reject it, then we have made an error of the "first kind."[*] On the other hand, if it should happen that an alternative hypothesis $H$ is actually true, and we

---

[*] Historically it was studied earliest, and its probability would be specified. It was not till later that other kinds of errors were considered.

erroneously accept $H_0$, then we have made an error of the "second kind." Thus

error first kind: reject $H_0 \mid H_0$ true

error second kind: accept $H_0 \mid H_0$ false.

6. *Risks of wrong decisions and significance level.* We define two types of risks, or probabilities, of wrong decisions by (respectively, errors of first and second kind)

$$\alpha = \mathrm{P}(\mathrm{rej}\ H_0 \mid H_0\ \mathrm{true}) \qquad (8.4)$$

$$\beta_\theta = \mathrm{P}(\mathrm{acc}\ H_0 \mid H_0\ \mathrm{false, with\ parameter}\ \theta). \qquad (8.5)$$

Note that in the second case the probability of acceptance of $H_0$ depends on what value the parameter $\theta$ really has, hence $\beta$ carries the subscript $\theta$. In the example (see Table 8.1), if $\mu = .099$ in., $\beta_{.099\mathrm{in.}} = .0778$, and so on.

7. *Statistic for test.* We choose a sample statistic which bears upon the parameter under test. Usually we try to pick the one which is most "efficient," that is, has the least variation around the parameter, from sample to sample. Thus for $\mu$ we use $\bar{y}$ rather than the median, or for $\sigma^2$ we use $s^2$ rather than some estimate from the square of the range, say.

8. *Distribution of statistic.* The assumptions in 1, the null parameter value in 2, and the chosen statistic in 7, now determine the distribution of the statistic (whether or not we can find the distribution exactly). It will also depend on the sample size $n$.

9. *Acceptance region and critical region.* The acceptance region is an interval of values of the chosen statistic such that if the statistic falls within the interval, we "accept" the null hypothesis. The critical region or rejection region for the parameter is the interval or intervals such that if the statistic falls within, we "reject" the null hypothesis. There will be only one interval if the alternative values of the parameter are only on one side of the null value, but if there are alternative values on both sides, the critical region will be split into two intervals, both above and below the acceptance region.

10. *Approaches.* We are now in a position to adopt one or another approach and set up the test (before any data are obtained):

(a) Set sample size $n$ and significance level $\alpha$.
   Find limits to the critical region.
(b) Set significance level $\alpha$, and for an alternative hypothesis set $\beta$. Then find $n$ and the limits to critical region.

11. *The OC curve.* It is always a good idea to at least sketch the operating characteristic curve. It will show us the probability of accepting the null hypothesis for any given value the parameter may have. Does the curve make practical sense? Does it have adequate discriminating power? Or is it more discriminating than necessary, so that we can perhaps decrease the sample size $n$?

12. *Summary of decisions and errors*

|                   | Actual situation | |
| Decision reached  | $H_0$ true | $H_0$ not true |
|-------------------|------------|----------------|
| Accept $H_0$      | Correct decision | Error of second kind<br>Probability: $\beta_\theta$ |
| Reject $H_0$      | Error of first kind<br>Probability: $\alpha$ | Correct decision |

## 8.4. SUMMARY OF SIGNIFICANCE TESTING FOR ONE MEAN WITH σ KNOWN

At some risk of too much repetition we now summarize tests of the mean with $\sigma$ known.

1.  Assumption random independent sample of $n$ $y$'s from a normal population with given $\sigma$.
2.  Test is of the hypothesis $\mu = \mu_0$.
3.  Cases of alternative hypotheses:
    (a)  $\mu > \mu_0$, specifically $\mu = \mu_2 > \mu_0$.
    (b)  $\mu < \mu_0$, specifically $\mu = \mu_1 < \mu_0$.
    (c)  $\mu \neq \mu_0$, specifically $\mu = \mu_1 < \mu_0$
         or $\mu = \mu_2 > \mu_0$, $(\mu_1 + \mu_2)/2 = \mu_0$.
4.  Risk $\alpha$ of error of first kind is set.
5.  Set either $n$ or else a risk $\beta$ at some $\mu_1$ or $\mu_2$.
6.  Test statistic: $u = (\bar{y} - \mu_0)/(\sigma/\sqrt{n})$.
7.  If $H_0 : \mu = \mu_0$ is true, then $u$ is N(0, 1). But if an alternative hypothesis in 3 is true, $u$ is not N(0, 1).
8.  Critical regions corresponding to alternate hypotheses 3:
    (a)  $u > u_{\alpha\,\text{above}}$ or $\bar{y} > \mu_0 + (\sigma/\sqrt{n})u_{\alpha\,\text{above}}$.
    (b)  $u < -u_{\alpha\,\text{above}}$ or $\bar{y} < \mu_0 - (\sigma/\sqrt{n})u_{\alpha\,\text{above}}$.
    (c)  $u$ outside $(-u_{\alpha/2\,\text{above}}, u_{\alpha/2\,\text{above}})$ or
         outside $\mu_0 - (\sigma/\sqrt{n})u_{\alpha/2\,\text{above}}, \mu_0 + (\sigma/\sqrt{n})u_{\alpha/2\,\text{above}}$.

9. The approaches in 8, go through if $n$ is known. But if $n$ is unknown but $\beta$ is given, say, for $\mu_2 > \mu_0$, then solve equations

$$\frac{U - \mu_0}{\sigma/\sqrt{n}} = u_{\alpha \text{ above}}, \qquad \frac{U - \mu_2}{\sigma/\sqrt{n}} = -u_{\beta \text{ above}}$$

for $n$ and critical value $U$ for $\bar{y}$'s.

10. It is desirable to sketch an OC curve as was done from Table 8.1.

11. Gather data and interpret decision as in the next section.

## 8.5. INTERPRETATION OF DECISIONS IN HYPOTHESIS TESTING

It is of great importance that we know precisely what it means to "accept" a null hypothesis, and what it means to "reject" it.

Let us consider the former first. We are virtually never faced with a situation in which the null hypothesis is *exactly* true. For example, in the rubber thickness example the plant probably never has received a lot of rubber with $\mu = .103$ in. *exactly*. Thus when we "accept" the hypothesis we by no means have proved that the parameter value actually is equal to the null value hypothesized. There is *no way* in which we can possibly *prove* this. What then can we say when the statistic lies in the acceptance region and we "accept" the null hypothesis? The following are several ways of interpreting the decision to "accept" the null hypothesis $\theta = \theta_0$:

1. The sample result is compatible with the null hypothesis value $\theta_0$. The null hypothesis is tenable.

2. The discrepancy between the sample statistic and $\theta_0$ is *readily* explainable by chance.

3. The null hypothesis perfectly well could be true and yield a sample such as we did observe.

4. We take practical and/or scientific action appropriate, if the null hypothesis $\theta = \theta_0$ were true.

5. $\theta$ is probably "quite close" to $\theta_0$. This is discussed under "interval estimation" in Chapter 10.

In our practical example, Section 8.2, we would use interpretation 4 when accepting the hypothesis $\mu = .103$ in., and thus use the rubber for specification $B$.

Suppose, on the contrary, we "reject" the hypothesis $\theta = \theta_0$ (or in the example reject $\mu = .103$ in.). Now we have not proven $\theta$ is not $\theta_0$ in any absolute mathematical sense, but only have thrown sufficient doubt upon it that from a practical viewpoint we conclude that $\theta$ is not $\theta_0$, and instead conclude that an alternative hypothesis is true. More specifically when we reject the hypothesis $\theta = \theta_0$ by the statistic (e.g., $\bar{y}$) lying in the critical region we can interpret in the following ways:

1. The sample result is incompatible with the null hypothesis value $\theta_0$. The null hypothesis is untenable.
2. The discrepancy between the sample statistic and $\theta_0$ is *not readily* explainable by chance, within our chosen significance level $\alpha$.
3. The null hypothesis, if true, could not readily yield a sample such as we did observe.
4. We take practical and/or scientific action appropriate, if some alternative hypothesis is true, examples being $\theta = \theta_1$ or $\theta < \theta_0$.
5. $\theta$ is probably not "very close" to $\theta_0$.

In our practical example, if $\bar{y} < .10042$ in., we reject the hypothesis $\mu = .103$ in., and use the rubber for specification $A$. Or, if $\bar{y} > .10558$ in., we use it for specification $C$.

## 8.6. NONNORMAL POPULATIONS OF y's

When we are testing hypotheses on mean $\mu$, with $\sigma$ *known*, we can use the approach in Section 8.2.1 or 8.2.2. with much confidence. This is because the distribution of $\bar{y}$ is quite normal for any sizable $n$, as we have pointed out in Section 7.4.5. If $\alpha_{3:y}$ is known approximately, $\alpha_{3:\bar{y}}$ can be checked by (6.64) for the desired $n$. If $|\alpha_{3:\bar{y}}| < .2$, normality is a good approximation in general.

## 8.7. SIGNIFICANCE TESTING FOR MEAN $\mu$, WITH $\sigma$ UNKNOWN

This case is undoubtedly more common than that given in Section 8.2, since if we do not know $\mu$, it is somewhat unlikely that we will know $\sigma$. One approach is, however, to assume an upper limit to $\sigma_y$, from some practical considerations, and use this $\sigma_y$ in Section 8.2.1 or 8.2.2. This then will be a conservative approach, but probably calls for a larger $n$ than necessary.

The ordinary approach used, when we can assume that the population is normal, is to use the $t$ distribution described in Section 7.4.2. We then simply modify the approach summarized in Section 8.4 by substituting $t_{\alpha\,\text{above}}$ for $u_{\alpha\,\text{above}}$. Then everything is much the same.

Let us review the $t$ distribution here. By definition we have (7.3) and (7.4) using $\mu_0$ :

$$u = \frac{\bar{y} - \mu_0}{\sigma_y/\sqrt{n}} \quad \text{and} \quad t_{n-1} = \frac{\bar{y} - \mu_0}{s/\sqrt{n}}.$$

When $\sigma_y$ is known we use the former of course, but when $\sigma_y$ is unknown we are forced into using the latter, which carries degrees of freedom $n - 1$, and is tabulated in Table III. Now while we could use the following form of test, let us say for the case $H_0 : \mu = \mu_0$ versus the one alternate $\mu = \mu_2 (>\mu_0)$

$$\text{accept} \quad H_0 \quad \text{if} \quad \bar{y} < \mu_0 + st_\alpha/\sqrt{n}$$

$$\text{reject} \quad H_0 \quad \text{if} \quad \bar{y} > \mu_0 + st_\alpha/\sqrt{n}.$$

We usually do not make the test in this form. This is because the boundary of the critical region, $\mu_0 + st_\alpha/\sqrt{n}$, on the right, is not the same from sample to sample but varies since sample $s$ varies. Thus, instead of acceptance criterion

$$\bar{y} < \mu_0 + st_\alpha/\sqrt{n},$$

we use the exactly equivalent

$$\frac{\bar{y} - \mu_0}{s/\sqrt{n}} < t_\alpha.$$

And for rejection we use

$$\frac{\bar{y} - \mu_0}{s/\sqrt{n}} > t_\alpha.$$

Hence for the three cases of alternate hypotheses we have the criteria shown in the accompanying table.

| Alternate hypotheses | Rejection criterion, $t$ with $n - 1$ degrees of freedom |
|---|---|
| (a) $\mu = \mu_2 > \mu_0$ | $\dfrac{\bar{y} - \mu_0}{s/\sqrt{n}} > t_\alpha$ |
| (b) $\mu = \mu_1 < \mu_0$ | $\dfrac{\bar{y} - \mu_0}{s/\sqrt{n}} < -t_\alpha$ |
| (c) $\mu = \mu_2 > \mu_0$ | $\dfrac{\bar{y} - \mu_0}{s/\sqrt{n}} > t_{\alpha/2}$ |
| or | or |
| $\mu = \mu_1 < \mu_0$ | $\dfrac{\bar{y} - \mu_0}{s/\sqrt{n}} < -t_{\alpha/2}$ |

Note carefully that in (c) we have $\alpha/2$ in each tail area, while in (a) and (b) $\alpha$ is all in the one single tail. To find the needed critical value of $t$ we must enter Table III, with the degrees of freedom $\nu = n - 1$ which gives the row, then the subscript on $t$, for example, $\alpha$ gives the column. Now, $\alpha$ is the upper tail area, so that $t_\alpha$ is positive if $\alpha < .5$. To find the $t$ value with a *lower* tail of $\alpha$, say, we simply find $t_\alpha$ and by symmetry give it a negative sign: $-t_\alpha$. For example, with $\nu = 10$ ($n = 11$), $t_{.025} = 2.228$, so that $t$ with .025 on the lower tail is $-2.228$. Thus if $\alpha = .05$, we would use these two values for case (c) of alternate hypotheses.

### 8.7.1. Example.

Five current determinations of percent sulfur in the ladle in a steel-making process were

$$.0307, \qquad .0324, \qquad .0314, \qquad .0311, \quad \text{and} \quad .0307.$$

Do these results give evidence that the process average under present conditions exceeds .0300%? Let us set this into the statistical framework, as in Section 8.3. We first assume that the population drawn from is normal, and that $\sigma$ is unknown. It is natural that we take $\mu = .0300\% = \mu_0$. Since our main concern is as to whether $\mu > .0300\%$, this is our alternate hypothesis. (High sulfur gives corrosion problems, for example.) We shall set $\alpha = .005$ because of the expense involved in a process change. Naturally $\bar{y}$ is the statistic to use. If the null hypothesis is true, its distribution will be $N(.0300\%, \sigma^2/5)$. Since $\sigma$ is unknown we must use $s$. The critical region is set up for $t$, with $n - 1 = 4$ degrees of freedom. Using Table III to find the critical value for $t$, we seek the entry with a single tail of $\alpha = .005$. This is 4.604. If the observed $t_4$ from (7.4) exceeds 4.604, we shall reject the hypothesis $\mu = .0300\%$ and conclude that present conditions give a higher percent of sulfur. But if $t_4$ is below 4.604, then we shall say that there is no *reliable* evidence that $\mu$ is actually above .0300%. All of this setup can be done, and in fact should be done, before the data are collected, so that the data will not tend to dictate the significance level, and so on.

We easily find the following for the data by using the coding methods of Chapter 3:

$$\bar{y} = .03126\%, \qquad s = .00702\%, \qquad n = 5.$$

Then (7.4) gives

$$t_4 = \frac{.03126\% - .0300\%}{.000702\%/\sqrt{5}} = 4.01.$$

Since this lies below 4.604, we cannot reject the null hypothesis. Note that if we had specified a larger $\alpha$, for example, .025, we would have rejected the null hypothesis. But with such a small $\alpha$ as .005, we need stronger evidence than $\bar{y} = .03126\%$ provided.

A second point is that if we had known $\sigma$ to be .000702%, then $u = (.03126\% - .0300\%)/(.000702\%/\sqrt{5}) = 4.01$ would have exceeded the normal-curve critical value of 2.576 (see Table I or the bottom line of the $t$ Table III), and we would have rejected the null hypothesis.

Finally we might observe that if $\bar{y}$ were to turn out to be below .0300%, then we could *never* reject the null hypothesis $\mu = .300\%$ in favor of the alternative $\mu > .0300\%$. If such were the case, we would not even need to make the test.

### 8.7.2. Operating Characteristic Curve.

Now another peculiarity between this case, $\sigma$ unknown, and that of Section 8.2 where $\sigma$ is known, lies in the OC curve. We cannot proceed as in Table 8.1 because $\sigma$ is unknown. As we have seen in Section 8.2 and Table 8.1, if $\sigma$ is *known* it is possible to find the probability of acceptance of the null hypothesis $H_0$, for whatever value $\mu$ may have, just by using a normal curve table. Hence we can plot $P_a$ versus $\mu$ for the OC curve. But when $\sigma$ is *unknown* this is no longer possible. Instead all we can do is to say what the probability of acceptance $P_a = \beta$ is, if $\mu = \mu_0 + \lambda\sigma$, where $\lambda$ might be 1 or 1.5, for example. Or as in Table VI in the Appendix, we use a horizontal scale of $\delta$, where (see Section 7.10)

$$\delta = \frac{(\mu - \mu_0)\sqrt{n}}{\sigma}, \quad \text{or} \quad \mu = \mu_0 + \delta\frac{\sigma}{\sqrt{n}}. \tag{8.6}$$

The four parts of this table give $P_a = \beta$ for a good spread of degrees of freedom $f = \nu$, for tests $\mu = \mu_0$ versus $\mu > \mu_0$, and for $\alpha = .005$, .01, .025, and .05. For example, if $n = 4$ or $f = 3$, and $\alpha = .05$, then the probability of accepting the null hypothesis $\mu = \mu_0$, when in fact $\mu = \mu_0 + 2\sigma/\sqrt{4} = \mu_0 + \sigma$, is found by the right-hand vertical scale, with $\delta = 2, f = 3$, to be .54. Thus if the sample size is only 4, we will in this test reject the null hypothesis only about half the time if $\mu = \mu_0 + \sigma$. On the other hand, if we again use $\alpha = .05$, but $n = 25$ and $\delta = 5$, then if $\mu = \mu_0 + 5\sigma\sqrt{25} = \mu_0 + \sigma$, we find the probability of accepting the null hypothesis to be only .0007. Thus the $\beta$ risk has fallen from .54 to .0007. Similarly if we were to be using an $\alpha$ risk of .005, then the first chart in Table VI gives for $n = 4, f = 3, \delta = 2, P_a = \beta = .92$. This example shows how when we decrease the $\alpha$ risk we increase the $\beta$ risk for the same $n$. Notice also that with $\alpha = .005$, but $n = 25$, $f = 24$, $\delta = 5$, so that $\mu = \mu_0 + \sigma$, we find $P_a = \beta = .02$, which is a small value, perhaps a good balance between $\alpha$ and $\beta$ with this degree of discrimination $\mu_0$ to $\mu_0 + \sigma$.

## 8.8. SIGNIFICANCE TESTS FOR VARIABILITY

Tests on the population mean $\mu$ occur in practice more often than those on the population standard deviation $\sigma$. But the latter occur frequently and are often of much importance. The general idea is to test whether observed results, that is, a sample of $n$ $y$'s, gives a standard deviation $s$ which is compatible with some hypothetical value of $\sigma$, say $\sigma_0$. A common example is a study of a new or proposed analytical technique or measuring instrument. One takes homogeneous material and measures it several times, then finds the standard deviation $s$ of the results. If one hopes that the new technique or instrument is better, he will probably use the null hypothesis $\sigma = \sigma_0$ which is the typical standard deviation measurement error, for the old technique, and as an alternative hypothesis $\sigma < \sigma_0$. But, if the new approach is less expensive, and the hope is that it will not be worse than the old method or instrument, then the alternate hypothesis will be taken as $\sigma > \sigma_0$. If $\sigma = \sigma_0$ is not disproved, he can justifiably switch to the less expensive technique if his sample size is large enough. In either case a one-tail test is appropriate.

Another application is to some process which is continuing to run, for example, manufacturing some kind of industrial part. We want to know whether there is any reliable evidence of a change in the variability in the process. If there is an improvement, we wish to know why so we can do it again; or if an increase in variability, we wish to know why so we can eliminate the cause of excessive variability. Here we use the alternate hypothesis $\sigma \neq \sigma_0$, and a two-tail test.

The basis of the test is from Section 7.5, where we recall that if the distribution of $y$'s is normal ($\mu$, $\sigma^2$) and we draw a random sample of $n$ $y$'s, then

$$(n - 1) s^2/\sigma^2 = \chi_\nu^2,$$

where the degrees of freedom $\nu = n - 1$. Thus the critical values of $\chi^2$ come from one or two tails of the chi-square distribution, Table II.

Let us outline the general setup as in Section 8.3 :

1.  Assume a random independent sample $y_1$, $y_2$, ..., $y_n$ from $N(\mu, \sigma^2)$.

2. Null hypothesis $H_0 : \sigma = \sigma_0$.
3. Alternate hypothesis, one of
   (a) $\sigma < \sigma_0$,     (b) $\sigma > \sigma_0$,     (c) $\sigma \neq \sigma_0$.
4. Two approaches:
   (a) set risk $\alpha = P(\text{rej } H_0 \mid H_0 \text{ true})$, and $n$.
   (b) set risk $\alpha = P(\text{rej } \sigma = \sigma_0 \mid \sigma = \sigma_0)$ and
   $\beta = P(\text{acc } \sigma = \sigma_0 \mid \sigma = \sigma_1)$ for some $\sigma_1$.

5. Statistic used $s^2$.*

6. The critical regions for the three alternate hypotheses in 3 are
   (a) $(n - 1)s^2/\sigma_0^2 < \chi^2_{\nu,\alpha \text{below}}$,
   (b) $(n - 1)s^2/\sigma_0^2 > \chi^2_{\nu,1-\alpha \text{below}}$,
   (c) $(n - 1)s^2/\sigma_0^2 < \chi^2_{\nu,.5\alpha \text{below}}$, and
       $(n - 1)s^2/\sigma_0^2 > \chi^2_{\nu,1-.5 \text{below}}$.

### 8.8.1. An Example of First Approach.

The stall speed for transmissions is a measure of the capability of the transmission. It is to be controlled between two limits, 37 and 77 rpm. Thus low variability is desirable, and high variability is to be guarded against. The current manufacturing process was running at $\sigma = 5.4$ rpm. It is desirable to know whenever there is either an increase or a decrease in $\sigma$ so that the cause can be sought. If $\sigma$ increases, the cause should be eliminated; or if $\sigma$ decreases, the cause can possibly be incorporated so that the process becomes more uniform.

Following the outline just given, we use the following:

1. Assume random sample from a normal population.
2. Null hypothesis $\sigma = 5.4 = \sigma_0$.
3. Alternate hypothesis $\sigma \neq 5.4$.
4. Take sample size $n = 4$, let $\alpha = .01$, to avoid too frequent search for causes when in fact $\sigma$ has not changed.
5. Use statistic $s^2 = \Sigma(y - \bar{y})^2/(n - 1)$.
6. Critical region for $\chi_3^2 = 3s^2/\sigma_0^2 = \Sigma(y - \bar{y})^2/5.4^2$ are from the chi-square table, putting $\alpha/2 = .005$ in each tail:

$$\text{below} \quad \chi^2_{3 \text{ df}, .005 \text{below}} = .072$$

$$\text{above} \quad \chi^2_{3 \text{ df}, .995 \text{below}} = 12.838.$$

---

* If it somehow happens that we do know $\mu$, then we can replace $s^2$ by $S^2 = \Sigma_{i=1}^n (y_i - \mu)^2/n$, and use $nS^2/\sigma^2 = \chi_n^2$, having $n$ degrees of freedom. This gives a little added power of discrimination.

Everything is set up. Now all we need is data. An actual sample was 64, 55, 58, 37. To find $\sum(y - \bar{y})^2 = \sum y^2 - (\sum y)^2/n$ we can either find the sums without coding or we can code from, say, 57, finding $v = +7$, $-2, +1, -20$. Then $\sum v = -14$, $\sum v^2 = 454$. Now since no dividing for $v$ was done,

$$\sum (y - \bar{y})^2 = \sum (v - \bar{v})^2 = 454 - (-14)^2/4 = 405.$$

Thus our observed value of $\chi^2 = 405/5.4^2 = 13.9$. Since this observed value lies above the upper critical value 12.84, we reject the hypothesis that $\sigma = 5.4$ and conclude that, at the time this sample was taken, $\sigma$ exceeded 5.4. We should try to see what was different in the process at this time than at other times.*

### 8.8.2. The Second Approach of 4, in Section 8.8.

A useful approach in which we specify a desired discriminating power and find the required sample size and critical region is easily designed for a one-way test. Suppose our null hypothesis is $H_0 : \sigma = \sigma_0$ and we wish to test against the simple alternate hypothesis $\sigma = \sigma_1$, with $\sigma_1 > \sigma_0$. Let the following risks be defined

$$\alpha = P(\text{rej } H_0 \mid \sigma = \sigma_0), \qquad \beta = P(\text{acc } H_0 \mid \sigma = \sigma_1). \qquad (8.7)$$

We can now use Table VIII, provided we can agree to let $\alpha = \beta$ and settle upon one of the values .10, .05, .02, or .01. (Other values, or unequal ones, could be used, but the same operating characteristic curves could be obtained by slight adjustment of $\sigma_0$ and/or $\sigma_1$ and $\alpha$ and/or $\beta$ to make them equal and match the values just mentioned.)

Now the procedure is the following simple one, similar to that of Burr [3]:

1. Find $\sigma_1/\sigma_0$.
2. In Table VIII, the desired $\alpha = \beta$ tells which of columns (2)–(5) to use. In the chosen column find the first entry which is less than or equal to $\sigma_1/\sigma_0$. Column (1) for this row then gives the required sample size $n$.
3. For this $n$ and that column (6)–(9) corresponding to $\alpha$, find the multiplier of $\sigma_0^2$ to give $K$.

---

* One might be curious as to whether the 37 reading in the sample was compatible with the other three. A simple test for this is given by Dixon and Massey [2]. The departure of the 37 from the others is not even significant at the 10% level of significance.

4.    Then the test is as follows:
    (a)    take $n$ observations $y_i$ , and find $s^2$;
    (b)    if $s^2 \leqslant K$, accept $\sigma = \sigma_0$ ; if $s^2 > K$, reject $\sigma = \sigma_0$ , conclude $\sigma > \sigma_0$ .

For example, suppose that a standard analytical technique gives a standard error of analysis of $.02\% = \sigma_0$ . A less time-consuming one is being considered, and an alternative hypothesis value of $\sigma_1$ is $.03\%$. We would like to use $\alpha = \beta = .02$, $\sigma_1/\sigma_0 = 1.5$. Looking for 1.5 in column (4) we find 1.52 for $n = 50$ and 1.46 for $n = 60$. Interpolating gives $n = 54$, rounding up to obtain the desired discrimination. Then for the coefficient of $\sigma_0^2$ to give $K$ we interpolate between 1.46 and 1.41, and we have 1.44. Then

$$K = 1.44(.02\%)^2 = .000576(\%)^2$$

or

accept    if  $s \leqslant .024\%$,    reject    if  $s > .024\%$.

The derivation of Table VIII is quite simple [3].*
    Finally for this approach, if we are interested in $H_0 : \sigma = \sigma_0$ versus some alternate hypothesis $\sigma < \sigma_0$ , we can still use Table VIII by merely renaming the null hypothesis value $\sigma_1$ and the lower alternative value $\sigma_0$ . Then proceed as above but

$$s^2 \leqslant K \qquad \text{reject} \quad H: \sigma = \sigma_1$$

$$s^2 > K \qquad \text{accept} \quad H: \sigma = \sigma_1 .$$

---

* We have $\alpha = P(s^2 > K \mid \sigma = \sigma_0)$. Multiplying both sides of the inequality by $(n-1)/\sigma_0^2$ gives $\alpha = P((n-1)s^2/\sigma_0^2 > K(n-1)/\sigma_0^2)$. But $(n-1)s^2/\sigma_0^2 = \chi^2_{n-1}$ . Hence $\alpha = P(\chi^2_{n-1} > K(n-1)/\sigma_0^2)$. But this says that $K(n-1)/\sigma_0^2 = \chi^2_{n-1 \text{ df, } 1-\alpha \text{ below}}$ . Similarly $1 - \beta = P(s^2 > K \mid \sigma = \sigma_1)$ yields $K(n-1)/\sigma_1^2 = \chi^2_{n-1 \text{ df, } \beta \text{ below}}$ . To solve, we divide and obtain

$$\frac{\sigma_1^2}{\sigma_0^2} = \frac{\chi^2_{n-1 \text{ df, } 1-\alpha \text{ below}}}{\chi^2_{n-1 \text{ df, } \beta \text{ below}}} .$$

Thus the $\sigma_1/\sigma_0$ may be found for each $\alpha = \beta$ and $n$ by taking the square root. Thus we find columns (2)–(5) of Table VIII. For columns (6)–(9), use

$$K = \frac{\chi^2_{n-1 \text{ df, } 1-\alpha \text{ below}}}{n-1} \sigma_0^2,$$

from which the coefficient of $\sigma_0^2$ gives the appropriate entries.

### 8.8.3.* Operating Characteristic Curves for Variability Tests.

Such curves give the probability of acceptance of the null hypothesis for any desired $\sigma$. For the approach in Section 8.8.2, we already have two points on the OC curve, which, with the obvious probabilities $P_a = 1$ when $\sigma = 0$, and $P_a = 0$ when $\sigma$ is relatively large, are usually sufficient for a sketch.

If further points are desired, we use the acceptance–rejection criterion $K$, say in the form

$$s^2 \leqslant K \quad \text{accept}, \qquad s^2 > K \quad \text{reject}$$

$$P_a = P(s^2 \leqslant K \mid \sigma)$$

$$= P[(n-1)\,s^2/\sigma^2 \leqslant (n-1)\,K/\sigma^2]$$

$$= P(\chi^2_{n-1} \leqslant (n-1)\,K/\sigma^2).$$

The way one might expect to draw an OC curve is by substituting values of $\sigma$ in and finding $P_a = P(\chi^2 \leqslant (n-1)K/\sigma^2)$. But this involves interpolation in Table II for the probability below, which is messy and not accurate. Instead, choose for $P_a$ any of the column headings in Table II, and substitute the corresponding $\chi^2$ value for this P and $\nu$ and solve for $\sigma^2$, then $\sigma$. For example, if we want $P_a = .90$, $n = 20$, and $K = 3.80$, then $\nu = 19$. The entry with .90 below is 27.204. Thus $(n-1)K/\sigma^2 = 19(3.80)/\sigma^2 = 27.204$, or $\sigma = \sqrt{2.654} = 1.63$. We then plot $P_a = .90$ versus $\sigma = 1.63$, and so on.

### 8.8.4.* Large Samples.

For samples with large $n$'s we may use the fact that $s$ has an approximately normal distribution with mean approximately $\sigma$ and standard deviation approximately $\sigma/\sqrt{2n}$. We may therefore regard

$$u = \frac{s - \sigma_0}{\sigma_0/\sqrt{2n}} \tag{8.8}$$

as approximately normal $(0, 1)$ if $\sigma = \sigma_0$. This is already quite good if $n$ is 30, since then $\alpha = .13$. An even closer approximation to $N(0, 1)$ may be had by using Table IV in the Appendix. Since $E(s) = c_4\sigma$, $E(s/c_4) = \sigma$, and also $\sigma_s = c_5\sigma$. Thus we may use

$$u = \frac{s/c_4 - \sigma_0}{c_5\sigma_0/c_4} \tag{8.9}$$

as $N(0, 1)$, and make tests similar to those for $\bar{y}$ versus $\mu$.

## 8.9. SIGNIFICANCE TESTING FOR ATTRIBUTES

The problem of significance testing for counted data is very similar to that for variable data, but is associated with the appropriate distributional model: binomial, Poisson (or hypergeometric possibly). We still make the same sort of null hypothesis, for example, $\phi = \phi_0$ is the hypothetical proportion "defective" in a binomial distribution, or $\mu = \mu_0$ is the hypothetical average number of defects in a Poisson distribution. The alternative hypotheses might be $\phi > \phi_0$, $\phi < \phi_0$, or $\phi \neq \phi_0$, and so on, leading to one- or two-tail tests.

The simplest approach to such hypothesis testing is to use tables of the assumed distribution, such as those listed in Chapter 5, or our Tables XI and XII. These are used to find the probability of as great as or a greater discrepancy between the observed result and the null hypothesis value, in the one observed direction, for example, $y = 17$, $np_0 = 10$, find P(17 or more). If this probability is less than $\alpha/2$ if a two-tail test, or $\alpha$ if a one-tail test, we will reject the null hypothesis. Otherwise we "accept" it. This is the "exact" approach.

Another approach is to assume normality, if the parameter value is sufficiently large to justify it, then use normal-curve tables. We shall illustrate both approaches.

### 8.9.1. The Binomial Tests.

As an example let us suppose that we are concerned with the probability $\phi$ of a sample of material yielding the presence of a certain trace element upon analysis. Suppose that $\phi$ might reasonably be supposed to run at .10, so we test the hypothesis $\phi = .10 = \phi_0$. If we may assume independence of tests and constant probability, we can use the binomial distribution. If we are interested in both higher and lower values of $\phi$, we will use the alternate hypothesis $\phi \neq .10$. Suppose we next set $\alpha = .01$, not wishing to refute $\phi = .10$ without substantial evidence. Let us now take $n = 50$ tests. Using binomial tables [4], we find

$$P(y = 0 \mid \phi = .10, n = 50) = .00515$$

$$P(y \geqslant 12 \mid \phi = .10, n = 50) = .00322.$$

Technically, we cannot reject $\phi = .10$ in favor of $\phi < .10$, because .00515 $> \alpha/2$. In practice, however, since this is so close to $\alpha/2 = .005$, the experimenter might regard the observance of zero occurrences of the trace element in 50 tests as sufficient evidence of $\phi < .10$, but this should be fully explained in any report. The occurrence of 12 or more is conclusive of course that $\phi > .10$. [The table gives $P(y \geqslant 11) = .00935$, so the upper critical region starts with 12.] This example illustrates the fact that in significance tests with discrete random variables, we cannot in general hit the desired $\alpha$ exactly.

When tables are available and contain the desired $n$ and $\phi_0$, this approach is very easy. One simply finds boundaries $y_L$ and $y_U$ to critical regions as close to $n\phi_0$ as possible, without exceeding $\alpha$ or $\alpha/2$, as given in the accompanying tabulation.

| Alternate hypothesis | Boundaries to critical regions |
|---|---|
| $\phi \neq \phi_0$ | $P(y \leqslant y_L \mid n, \phi_0) \leqslant \alpha/2, P(y \geqslant y_U \mid n, \phi_0) \leqslant \alpha/2$ |
| $\phi > \phi_0$ | $P(y \geqslant y_U \mid n, \phi_0) \leqslant \alpha$ |
| $\phi < \phi_0$ | $P(y \leqslant y_L \mid n, \phi_0) \leqslant \alpha$ |

Note especially the equality–inequality in the $y$ relation. Thus one rejects $\phi = \phi_0$ if $y = y_L$ exactly.

When $n$ and/or $\phi$, as needed, are not in the table, then one may well use interpolation, and this is likely to be quite sufficient since we cannot expect to hit $\alpha$ or $\alpha/2$ exactly, and some small error in our interpolated value of the probability will probably not change the boundaries of our critical region, that is, $y_L$ and/or $y_U$.

Now suppose that binomial tables are not available or are inadequate. Then if the expected number of "defectives" under the null hypothesis $\phi = \phi_0 \leqslant .5$ is at least 10, we can make use of normal-curve tables as follows:

1.  Find normal-curve critical values appropriate. For example, if the alternate hypothesis is $\phi \neq \phi_0$, and $\alpha = .01$, the two boundaries of the critical region are $\pm 2.576$. Or if $\phi < \phi_0$, then the one critical value is $-2.327$, and so on.

2.  The objective here is to determine whether the probability of as great as or a greater discrepancy from $n\phi_0$ is less than or equal to $\alpha/2$ or $\alpha$ as the case may be. We proceed as in Section 6.3.4, using the correction for continuity described there. Thus for $P(y$ or less) we use a cutoff point of $y + .5$, so as to include the $y$ block. Or if we seek $P(y$ or more), we use a cutoff point of $y - .5$, again to include the $y$ block. Thus

$$y < n\phi_0, \quad \text{use} \quad u = \frac{y + .5 - n\phi_0}{\sqrt{n\phi_0(1 - \phi_0)}}$$

$$y > n\phi_0, \quad \text{use} \quad u = \frac{y - .5 - n\phi_0}{\sqrt{n\phi_0(1 - \phi_0)}}.$$

3. Compare the observed $u$ value just given with the critical values in 1, to see whether $u$ is in the critical region or acceptance region. Make decision.

**Example.** From heterozygous parents, one-fourth of the offspring should have the recessive trait in both genes and thus show the recessive characteristic, if the offspring are all equally likely to survive, that is, equally viable. Suppose that under these circumstances, only 17 in 100 show the recessive characteristic. We will be testing the hypothesis $\phi = .25 = \phi_0$. If prior to sampling we have no sound reason to expect a deviation in either direction, we shall take the alternative hypothesis $\phi \neq .25$. Also we take $\alpha = .01$. Then the boundaries of the two-tail critical region for the normal curve are $\pm 2.576$. Now since $17 < 100 \times (.25) = 25$, from

$$u = \frac{17 + .5 - 25}{\sqrt{100(.25)(.75)}} = -1.73.$$

Since $-1.73$ lies between $\pm 2.576$ we regard $\phi = .25$ as tenable and have no reliable evidence of unequal viability.

## 8.9.2. The Poisson Tests.

We must first assure ourselves that it is reasonable to assume a Poisson distribution. See Section 5.4.6. As for the binomial distribution we again have exact and approximate methods. For the exact method we resort to tables, especially Molina's [5], or that in the Appendix.

To illustrate the exact approach, data on errors on trucks were compiled, showing an average of 1.43 errors per truck at final inspection. On 2-hours production of 35 we would thus expect just about 50 errors of all kinds. Assuming these are independent, and noting that the *possible* number of errors is huge, we may assume the Poisson distribution. To test for a departure from the expected, we may test the hypothesis $\mu = 50$ versus the alternative hypothesis $\mu \neq 50$, for total errors on 35 trucks. Thus we are seeking sound evidence of a shift either up or down in $\mu$. Let us use the rather conservative $\alpha = .01$. Now suppose we only find 32 errors on the next 2 hours of production of 35 trucks. We may use Molina's work [5] to find the probability

$$P(X \leqslant 32 \mid \mu = 50) = 1 - .995607 = .004393.$$

Since this is less than $\alpha/2 = .005$ we conclude that there is evidence of a real improvement in manufacturing conditions (or conceivably poor inspection).

The approximate method proceeds very similarly to the outline in Section 8.9.1, in which we first pick out the appropriate boundaries for the critical region for the normal curve and use $\alpha$ or $\alpha/2$ according to the alternate hypothesis. Then since for the Poisson, $\sigma_y = \sqrt{\mu}$:

$$y < \mu_0, \quad \text{use} \quad u = \frac{y + .5 - \mu_0}{\sqrt{\mu_0}}$$

$$y > \mu_0, \quad \text{use} \quad u = \frac{y - .5 - \mu_0}{\sqrt{\mu_0}}.$$

For the example just given, the critical values would be $\pm 2.576$. Since $y < 50$ form

$$u = \frac{32 + .5 - 50}{\sqrt{50}} = -2.475.$$

Here, contrary to the exact method, we would accept the hypothesis $\mu = 50$. But if we were to use the better approximation than the normal curve, namely, the gamma distribution (Section 6.6.3), with skewness $\alpha_3 = 1/\sqrt{50} = .143$, we would find for $u = -2.475$ a probability of .0045 below. Thus using the closer approximation, the gamma distribution, we reach the same conclusion as the "exact" method.

Another example illustrates a further simple principle. In 1,750,000 rubber gaskets for caps, there were 139 scrap gaskets or gasket material. Is this clear evidence of an improvement over the typical 120 per million ? Using the latter we expect to find $\mu_0 = .000120(1,750,000) = 210$. Then we test $\mu = 210$ versus $\mu < 210$, with $\alpha = .001$, say. Now with a one-way normal value of $-3.090$ we compare

$$u = \frac{139 + .5 - 210}{\sqrt{210}} = -4.87.$$

Hence we have very strong evidence of a real process improvement.

### 8.9.3.* Other Attribute Distributions.

When some attribute distribution other than the binomial and Poisson is assumed, the test is not at all different in principle than those just given. If the distribution is hypergeometric, then perhaps the exact probabilities are available in the work of Lieberman and Owen [6]. Or if not, then a normal-curve approximation may suffice if $\alpha_3$ in (5.13) is small enough. A continuity correction should be used. The negative binomial may be tested through binomial tables, using (5.76). For such tests some care and ingenuity must be used.

### 8.9.4. Operating Characteristic Curves.

An operating characteristic curve for an attribute significance test is quite easy to construct, if available tables are complete enough. For example, consider the test of an industrial lot in which a random sample of $n = 100$ temperature control switches is not to contain over four nonconforming switches, that is, switches operating at a temperature outside the desired range. This might be considered a test of the hypothesis: $H_0 : \phi = .02 = \phi_0$ with $\alpha = .05$, since if $\phi = .02$ we find in the literature [4] that P(4 or less | 100, .02) = .94917 = .95. Now for

$$P(\text{acc } H_0 \mid \phi) = P(4 \text{ or less} \mid 100, \phi) = 1 - P(5 \text{ or more} \mid 100, \phi),$$

the last being of use in the literature [4]. Thus we find Table 8.2, the results of which are shown in Fig. 8.3. As $\phi$ increases we have a continuously decreasing P(acc $H_0$). Note also that if we were testing $H_0$ : $\phi = .02$ versus $\phi = .09$, we would have had $\alpha = .051$ and $\beta = .047$ for our rule: accept on four or less nonconforming switches, reject on five or more nonconforming switches.

**TABLE 8.2**

Calculation for an OC Curve for Binomial Significance Test of Hypothesis $H_0 : \phi = .02$[a]

| $\phi$ | $P(y \geqslant 5 \mid 100, \phi)$ | $P(\text{acc } H_0) = 1 - P(y \geqslant 5 \mid 100, \phi)$ |
|---|---|---|
| .01 | .003 | .997 |
| .02 | .051 | .949 |
| .03 | .182 | .818 |
| .04 | .371 | .629 |
| .05 | .564 | .436 |
| .06 | .723 | .277 |
| .07 | .837 | .163 |
| .08 | .910 | .090 |
| .09 | .953 | .047 |
| .10 | .976 | .024 |

[a] With $n = 100$. Accept $H_0$ if $y \leqslant 4$. Use tables from the literature [4].

If we have a two-sided alternative as in the first example of Section 8.9.1, we would have an operating characteristic curve starting low for low $\phi$'s, then rising to a maximum of about .992 when $\phi = .10$, then falling away toward zero above $\phi = .10$. The calculation requires us to find

$$P(\text{acc } H_0: \phi = .10) = P(1 \leqslant y \leqslant 11) = P(1 \leqslant y) - P(12 \leqslant y).$$

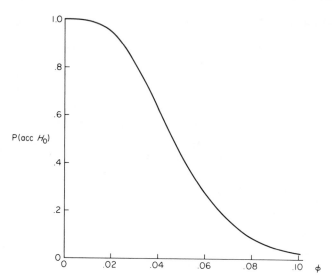

**Fig. 8.3.** Operating characteristic (OC) curve for attribute sampling test of hypothesis $H_0 : \phi = .02$, versus $\phi > .02$ at about $\alpha = .05$, yielding the rule: accept $H_0$ if $y \leqslant 4$, reject $H_0$ if $y \geqslant 5$.

Thus, for example, from the tables [4] if $\phi = .07$ and $.13$, $n = 50$, we find

$$P(\text{acc } H_0 : \phi = .07) = .97344 - .00014 = .97330$$
$$P(\text{acc } H_0 : \phi = .13) = .99905 - .02419 = .97486.$$

These are nearly equal, being symmetrically placed relative to $\phi_0 = .10$.

## 8.10. RELATION OF SIGNIFICANCE TESTING TO DECISION THEORY. IMPORTANCE OF OC CURVE

A modern development in statistics is that of "decision theory." This approach aims at making as rational as possible each process of decision-making from statistical data. For this it is necessary to form some definite model to describe the practical consequences of making a wrong decision. This might be stated in monetary terms or on some relative scale.

For example, the decision might be as to whether some kind of rivet averages 500 kg in tensile strength. If the lot average is indeed 500 kg = $\mu_0$, and we test and reject the lot, we have made an error of the first kind, and there is a loss. But if $\mu < 500$ kg and yet we accept the lot, we have made an error of the second kind, and the consequences could be serious or even disastrous. These two errors are not at all of the same kind. But it certainly is true that the farther $\mu$ lies below $\mu_0$, the more dangerous it is to accept erroneously. And conversely, the farther $\mu$ is above $\mu_0$, the better the consumer likes them. But greater average strength is likely to cost more money. A mathematical model of the consequences when the actual $\mu$ differs from $\mu_0$ is needed. One which is often used is that the *loss* $L = k(\mu - \mu_0)^2$, whenever we accept the hypothesis $\mu = \mu_0$.* The loss $L$ in accepting $\mu = \mu_0$ is thus small when $\mu$ is near $\mu_0$ but rapidly grows the further $\mu$ is from $\mu_0$. Note that this loss function is symmetrical around $\mu_0$, which might not be desirable in the above.

A second requirement for a fully rational decision is knowledge of the *a priori* distribution of the parameter or parameters. Thus in the rivet example, it might be known from past records that the producer makes 98 % of his lots with $\mu \geqslant 500$ kg, and only 2 % of them with $\mu$ being 490 to 500 kg. If such is the case, it might be possible to use a very small sample size $n$, or to test only some percentage of lots, not all.

A third requirement which needs consideration is the cost of testing. In the case of rivets we have the cost of obtaining a random sample, the cost of testing, the value of the rivets tested since the test is destructive, and the cost of any calculations, or setting up of the decision-making process.

When one considers the job of setting mathematical functions for all of these requirements, even if obtainable, and then determining the best test to use as an individual case comes along, the use of the decision theory approach seems limited in usefulness. There is no doubt that in some large-scale, top-management decisions such an approach may well prove fruitful. But for the many day-to-day decisions in industry, science, and engineering, decison theory approaches seem impractical and unnecessary.

It is the author's firm conviction that by subjectively taking account of the three elements: (a) consequences of the two kinds of errors, (b) *a priori* distribution of parameters in question, and (c) costs of testing, he can develop a statistical test with an OC curve taking these into account. Again in our rivet example, the consumer will likely use a one-way test, the critical region being of course in the lower tail, since he does not care if $\mu > 500$ kg, in fact is glad. He is really using $H_0$: $\mu \geqslant 500$ kg versus the alternate $H$: $\mu < 500$ kg. He might set $\alpha = .02$ if $\mu = 500$ kg, and depending on the "factor of safety" in the use of the rivet and the variability $\sigma$ likely to be present, he might set $P(\text{acc } H_0 \mid \mu =$

---

* The number $k$ is a constant of proportionality which depends on economic factors.

475 kg) $= \beta = .01$. He is then in a position to use Section 8.2.2, if he knows $\sigma$ approximately. If not, he may set $\sigma$ at an upper limit. This test may seem to the consumer to involve too large an $n$ and thus excessive cost of testing. The way to cut down on this is to increase $\alpha$ and/or $\beta$, and/or widen the gap between the two critical levels 475 and 500 kg. Now as time goes on evidence may well accumulate on the *a priori* distribution of $\mu$, and on the size of $\sigma$. We may be able to use a typical value of $\sigma$, rather than a maximum. If it develops that the producer has almost no lots rejected or none at all, we may well increase $\beta_{475}$ to .05, .10, or even .50, with accompanying smaller $n$'s and lower sampling costs.

Looked at from the producer's viewpoint, however, he may well be using a two-way test for $\bar{y}$ to control his process. He does not want $\mu$ to be much below 500 kg because he will be in danger of a rejected lot. But on the other hand, he may not wish to produce rivets with $\mu$ considerably above 500 kg because it might be more costly due to expensive additives, and so on. He might perhaps use $\mu_0 = 500$ kg with alternate hypotheses $\mu = 475$ and 525 kg. One still has said *nothing* with these three $\mu$ values till one sets the risks. If he is tempted to use different $\beta$'s at the two rejectable levels, he can accomplish the same thing with equal $\beta$'s by adjusting one or the other alternate value of $\mu$, and also adjusting $\mu_0$. Or he can just work out with the two unequal $\beta$'s at 475 and 525 kg and forget $\mu_0$. There will still be a maximum point on the OC curve.

The main point is to set up levels and risks, develop the test, sketch the OC curve and see whether it seems to give a good balance between (a) small risks and large sample size, and (b) large risks and small sample size.

## 8.11. SUMMARY

In this chapter on significance testing we have been concerned with drawing inferences about populations from observed samples. The question is whether the sample could reasonably have come from the hypothetical population. If this is the case, then we cannot disprove this population and we regard it as tenable. But if the probability of a sample such as we did obtain is small enough, then we regard the hypothetical population as untenable and reject it, in favor of some alternative population or class of populations.

Significance tests discussed were all concerned with decisions on parameters rather than decisions on the *type* of population (for example normal versus gamma). For the normal population the parameters were $\mu$ and $\sigma$, while for the binomial and Poisson, they were for $\phi$ and $\mu$, respectively. See Table 8.3.

Summary of Tests

**TABLE 8.3**

| Hypothesis | Conditions | Test statistic | Distribution of test statistic under Null hypothesis | Alternate hypothesis | Critical region |
|---|---|---|---|---|---|
| 1. $\mu = \mu_0$ | $y$'s normal, $\sigma$ known | $u = (\bar{y} - \mu_0)\sqrt{n}/\sigma$ | $N(0, 1)$ | $\mu \neq \mu_0$ | $u < -u_{\alpha/2}$ and $u > u_{\alpha/2}$ |
| | | | | $\mu < \mu_0$ | $u < -u_\alpha$ |
| | | | | $\mu > \mu_0$ | $u > u_\alpha$ |
| 2. $\mu = \mu_0$ | $y$'s normal, $\sigma$ unknown | $t_{n-1} = (\bar{y} - \mu_0)\sqrt{n}/s$ | $t$ distribution, $\nu = n - 1$ | $\mu \neq \mu_0$ | $t < -t_{\alpha/2}$ and $t > t_{\alpha/2}$ |
| | | | | $\mu < \mu_0$ | $t < -t_\alpha$ |
| | | | | $\mu > \mu_0$ | $t > t_\alpha$ |
| 3. $\sigma = \sigma_0$ | $y$'s normal, $\mu$ unknown | $\chi^2_{n-1} = (n-1)s^2/\sigma_0^2$ | Chi-square, $\nu = n - 1$ | $\sigma \neq \sigma_0$ | $\chi^2 < \chi^2_{\alpha/2}$ and $\chi^2 > \chi^2_{1-\alpha/2}$ |
| | | | | $\sigma < \sigma_0$ | $\chi^2 < \chi^2_\alpha$ |
| | | | | $\sigma > \sigma_0$ | $\chi^2 > \chi^2_{1-\alpha}$ |
| 4. $\phi = \phi_0$ | Binomial, Section 5.3.6 | $y$ observed | $C(n, y)\,\phi_0^y(1 - \phi_0)^{n-y}$ | $\phi \neq \phi_0$ | $y \leqslant y_{L(\alpha/2)}$ and $y \geqslant y_{U(\alpha/2)}$ |
| | | | | $\phi < \phi_0$ | $y \leqslant y_{L(\alpha)}$ |
| | | | | $\phi > \phi_0$ | $y \geqslant y_{U(\alpha)}$ |

*Continued on following page*

**TABLE 8.3** (*continued*)

Summary of Tests

| | | | | |
|---|---|---|---|---|
| 5. $\phi = \phi_0$ | Binomial, Section 5.3.6 $n\phi_0 \geq 10$ | $u = \dfrac{y + .5 - n\phi_0}{\sqrt{n\phi_0(1 - \phi_0)}}$ | N(0, 1) | $\phi \neq \phi_0$ | $u < -u_{\alpha/2}$ and |
| | | $u = \dfrac{y - .5 - n\phi_0}{\sqrt{n\phi_0(1 - \phi_0)}}$ | N(0, 1) | $\phi \neq \phi_0$ | $u > u_{\alpha/2}$ |
| | | $u = \dfrac{y + .5 - n\phi_0}{\sqrt{n\phi_0(1 - \phi_0)}}$ | N(0, 1) | $\phi < \phi_0$ | $u < -u_{\alpha}$ |
| | | $u = \dfrac{y - .5 - n\phi_0}{\sqrt{n\phi_0(1 - \phi_0)}}$ | N(0, 1) | $\phi > \phi_0$ | $u > u_{\alpha}$ |
| 6. $\mu = \mu_0$ | Poisson, Section 5.4.6 | $y$ observed | $e^{-\mu_0}\mu_0^{y}/y!$ | $\mu \neq \mu_0$ | $y \leqslant y_{L(\alpha/2)}$ and $y \geqslant y_{U(\alpha/2)}$ |
| | | | | $\mu < \mu_0$ | $y \leqslant y_{L(\alpha)}$ |
| | | | | $\mu > \mu_0$ | $y \geqslant y_{U(\alpha)}$ |
| 7. $\mu = \mu_0$ | Poisson, Section 5.4.6 $\mu_0 \geq 10$ | $u = \dfrac{y + .5 - \mu_0}{\sqrt{\mu_0}}$ | N(0, 1) | $\mu \neq \mu_0$ | $u < -u_{\alpha/2}$ and |
| | | $u = \dfrac{y - .5 - \mu_0}{\sqrt{\mu_0}}$ | N(0, 1) | $\mu \neq \mu_0$ | $u > u_{\alpha/2}$ |
| | | $u = \dfrac{y + .5 - \mu_0}{\sqrt{\mu_0}}$ | N(0, 1) | $\mu < \mu_0$ | $u < -u_{\alpha}$ |
| | | $u = \dfrac{y - .5 - \mu_0}{\sqrt{\mu_0}}$ | N(0, 1) | $\mu > \mu_0$ | $u > u_{\alpha}$ |

Two approaches may be used. We may set the null hypothesis value of the parameter, the sample size, and the $\alpha$ risk. Then we may sketch the OC curve and see whether it gives a good balance between sampling costs and $n$ versus protection against wrong decisions. If not, we must adjust $n$ and/or $\alpha$. The second approach is to set two levels of the parameter in question and corresponding $\alpha$ and $\beta$ risks, then find the needed sample size and critical value for the statistic to be used. Thus in this case we have already specified two points on the OC curve. Does the desired protection against wrong decisions cost us too much sampling, or is there a good balance? If not, then some adjustment is needed. The latter approach can readily be used on attribute distributions using tables although it was not discussed.

### REFERENCES

1. D. B. Owen, "Handbook of Statistical Tables." Addison-Wesley, Reading, Massachusetts, 1962.
2. W. J. Dixon and F. J. Massey, "Introduction to Statistical Analysis." McGraw-Hill, New York, 1957.
3. I. W. Burr, "Engineering Statistics and Quality Control." McGraw-Hill, New York, 1953.
4. Harvard Univ., Comput. Lab., "Tables of the Cumulative Binomial Probability Distribution." Harvard Univ. Press, Cambridge, Massachusetts, 1955.
5. E. C. Molina, "Poisson's Exponential Binomial Limit." Van Nostrand-Reinhold, Princeton, New Jersey, 1947.
6. G. J. Lieberman and D. B. Owen, "Tables of the Hypergeometric Probability Distribution." Stanford Univ. Press, Stanford, California, 1961.
7. L. R. Salvosa, Tables of Pearson's Type III function. *Ann. Math. Statist.* 1, 191–225 (1930).
8. H. C. Carver, "Statistical Tables." Edwards, Ann Arbor, Michigan, 1940.

### ADDITIONAL REFERENCES

T. A. Bancroft, Probability values for the common tests of hypotheses. *J. Amer. Statist. Assoc.* 45, 211–217 (1950).

D. D. Bien, Tables of component reliability for binomial redundancy applications. NASA Tech. Note TND-5549. NASA, Washington, D.C., 1969.

J. M. Cameron, Tables for constructing and for computing the operating characteristics of single-sampling plans. *Indust. Quality Control* 9, (no. 1), 37–39 (1952).

T. Colton, A test procedure with a sample from a normal population when an upper bound to the standard deviation is known. *J. Amer. Statist. Assoc.* 55, 94–104 (1960).

O. L. Davies, "The Design and Analysis of Industrial Experiments." Oliver & Boyd, Edinburgh, 1954.

C. D. Ferris, F. E. Grubbs, and C. L. Weaver, Operating characteristics for the common statistical tests of significance. *Ann. Math. Statist.* 17, 178–197 (1946).

W. C. Guenther and P. O. Thomas, Some graphs useful for statistical inference. *J. Amer. Statist. Assoc.* **60**, 334–343 (1965).

H. L. Harter, A new table of percentage points of the Pearson Type III distribution. *Technometrics* **11**, 177–187 (1969).

M. G. Natrella, "Experimental statistics." *Nat. Bur. Stand. (U.S.) Handb.* **91**, 1963.

B. Ostle, "Statistics in Research." Iowa State Univ. Press, Ames, 1963.

E. S. Pearson, Comments on the assumption of normality involved in the use of some simple statistical techniques. *Rev. Belge Statist. Recherche Operationelle* **9** (no. 4), 2–18 (1969).

H. L. Rietz, On the distribution of the "student" ratio for small samples from certain non-normal populations *Ann. Math. Statist.* **10**, 265–274 (1939).

H. Vahle and G. Tews, Probabilities of a $\chi^2$-distribution. *Biometrische Z.* **11**, 175–202 (1969).

## Problems

8.1.   Sketch the OC curve for the example given in Section 8.2.2, finding a few points $(P_a, \mu)$ as needed.

8.2.   The diameter of a "low-speed plug" was to meet limits of .190–.194 in. Thus the process of manufacturing should be running at .192 in. $= \mu_0$ to minimize the number of plugs out of limits. For 160 plugs measured in one day $\bar{y} = .192{,}272$ in. and $s = .000{,}692$ in. Test at the $\alpha = .005$ level $\mu = \mu_0$ versus $\mu > \mu_0$, and interpret. Estimate the percentage of plugs above .194 in.

8.3.   The steel-making range for untreated steel wheels was .70–77% carbon at the open hearth ladle. For 500 observations $\bar{y} = .7380\%$ and $s = .01883\%$. To meet the desired range "nearly all the time" it would be desirable to have $\mu$ at .735%, and $6s = .07\%$ as a maximum. Test at the $\alpha = .01$ level $H_0 : \sigma = .0117\%$ versus $\sigma > .0117\%$. Also test at the same level $H_0 : \mu = .735\%$ versus $\mu \neq .735\%$. Interpret results.

8.4.   The percentage of silicon in a blast furnace making pig iron was supposed to be between .70 and 1.20. For 450 analyses (five per day for 90 days), $\bar{y} = 1.031\%$ and $s = .150\%$. Test $\mu = .95\%$ versus $\mu \neq .95\%$ at $\alpha = .01$. Also since the range of .50% should be at least $6\sigma$, use $\sigma_0 = .50\%/6$ and test against $\sigma > \sigma_0$ at $\alpha = .01$. What conclusions do you draw?

8.5.   A hydraulic stoplight switch for an automobile is supposed to operate at between 45 and 100 psi. Thus the desired mean is 72.5 psi. A sample of $n = 5$ switches is tested, yielding $\bar{y} = 80.5$ psi and $s = 5.05$ psi. Test the hypothesis that $\mu = 72.5$ psi versus $\mu \neq 72.5$ psi. at $\alpha = .02$. Suppose we know that the long-run standard deviation is $\sigma = 5.38$ psi; then again make the test, and interpret. Since $100 - 45$

is over $10\sigma$, do we need to maintain $\mu$ close to 72.5 psi? How far might you let it be off from 72.5 psi?

8.6. As a cutting tool wears, an outside dimension tends to increase for a series of machined parts. In .0001 in. units, $\sigma = 1.30$. In the same unit the upper specified limit is 1250. If $\mu$ is as low as 1246, there will be very few pieces above 1250; but if $\mu = 1247.3$, there will begin to be a few pieces above 1250. Set up a test for $\mu = 1246$ versus $\mu = 1247.3$ with $\alpha = .10$. If either $\alpha$ or $\beta$ must be preserved and the other slightly decreased, due to making $n$ an integer, preserve $\alpha$. Sketch the OC curve.

8.7. The diameter of an exhaust valve is coded in .0001 in. from 1.1540 in. In the coded units the specified limits of 1.1540 in. and 1.1560 in. become 0, 20. A process average of $\mu = 17.5$ is considered satisfactory because of the low variability in the process. As the cutting tool wears the diameters tend to increase. A sample of five yields diameters 19, 20, 19, 18, 19. Test the hypothesis $\mu = 17.5$ versus $\mu > 17.5$ at $\alpha = .10$. Is there evidence to justify resetting $\mu$ downward? Does the precision of measurement raise any question about the validity of your $s$?

8.8 The desired average dimension for a "turbine bucket" is .3975 in. A sample of pieces gives .382, .380, .386, .390, and .382 in. Using $\alpha = .01$, test $\mu = .3975$ in. versus $\mu < .3975$ in., interpreting results. Does this sample suggest that some pieces are below .380 in.?

8.9 The desired average tin-coating weight for cans was 12.0 in appropriate units. A sample of analyses on one strip gave 16.0, 14.3, and 14.4. Test $\mu = 12.0$ versus $\mu > 12.0$ at the $\alpha = .01$ level, stating assumptions and interpreting conclusions.

8.10 Weight of filling of a container is important. Overfill is costly; underfill is illegal or at least undesirable. An insecticide dispenser had specifications of $454 \pm 27$ g weight. A sample of four weights was 476, 478, 473, and 459 g. Find $\bar{y}$ and $s$ and test at the 1% level whether $\mu$ exceeds 454 g. If we use the standard deviation of 200 weights (over four days) of $s = 10.6$ g, calling it $\sigma$, test the same hypotheses. Using $\sigma = 10.6$ g, what is the lowest $\mu$ safely meeting the minimum weight of 427 g?

8.11. In weight of fill of an insecticide dispenser the nominal weight is 454 g. Consider that $\sigma$ is known to be 10 g. It is desired to reject lots averaging 444 g 99% of the time, and also reject the hypothesis $\mu = 454$ g 95% of the time when $\mu = 464$ g. Set up two equations incorporating these conditions and using an acceptance region for $\bar{y}$ from $k_1$ to $k_2$ ($k_1 < k_2$), with $n$ unknown. A third condition is needed, but setting $\alpha$ at $\mu = 454$ g is difficult. Set instead $n = 9$, and find $k_1$ and $k_2$. Now find $\alpha$. Do you think it is too large?

8.12.    A sample of 100 bearing diameters yielded $\bar{y} = .984,032,3$ in. and $s = .000,017,0$ in. The specifications are .98390 to .98410 in., so the desired average is .98400 in. Test at the $\alpha = .02$ level $\mu = .98400$ in. versus $\mu \neq .98400$ in. If the latter is adopted, would there be evidence that the bearings are not lying between the specified limits?

8.13.    In connection with the example given in Section 8.2.1, wherein $\sigma = .002$ in. and the desired mean for specification $B$ was .103 in., suppose we decide that we can afford $n = 16$ tubes to be measured. Also we decide that we will use the points half-way between the three desired means, as the critical limits, namely $(.098 \text{ in.} + .103 \text{ in.})/2 = .1005$ and .1055 in. If $\bar{y}$ lies between these, accept for $B$, and so on. Draw the OC curve for this plan and comment. (This is in a sense a third approach.)

8.14.    Make a general formulation of Section 8.2.2, for testing the hypothesis $\mu = \mu_0$ versus the alternate $\mu = \mu_1$ with $\mu_1 > \mu_0$, respective risks $\alpha$ and $\beta$, and $\sigma$ known. It is convenient to use $K_p$ defined by $\int_{K_p}^{\infty} \phi(u) \, du = p$; that is, $K_p$ cuts off a tail area of $p$. Assuming normality derive a formula for $n$ and one for the acceptance–rejection boundary $K$.

8.15.    The dimension in question for a rheostat knob was from the back of the knob to the far side of a pinhole. The desired average was (in .001 in.) 140, with specified limits $140 \pm 3$. For a sample of $n = 5$ knobs, $\bar{y} = 138.8$ and $s = 2.40$. Test $\mu = 140$ versus $\mu \neq 140$ at $\alpha = .01$ and interpret your result. If we assume that the range from 137 to 143 must be $6\sigma$ as a minimum to be met, then we need $\sigma = 1$. Test at the 1% level $\sigma = 1$ versus $\sigma > 1$ for the sample data. (Actually the long run average for $s$ was 3.54.)

8.16.    A new process for fiber orientation in a fiberglass process results in about the same average "z-factor" (.160), but it is hoped that there will be less variability. The usual standard deviation is $\sigma = .0026$. Twenty checks on the new process give $s = .0014$. Using $\alpha = .01$, what do you conclude? State hypotheses used.

8.17.    For an inexpensive thermometer, a standard deviation for the temperature readings is $\sigma_0 = .25°C$, while $\sigma_1 = .50°C$ is undesirable. Set up appropriate tests with respective risks $\alpha = \beta = .10$ and $\alpha = \beta = .02$.

8.18.    An analytical technique is being considered by using it several times on carefully homogenized material. Test whether it gives an improvement over the old technique $\sigma_1 = .015\%$, $\sigma_0 = .010\%$ being considered a worthwhile improvement. Use $\alpha = \beta = .10$ and set up the test.

8.19. The diameter of a pin has specified limits of .1235 to .1240 in. We need to have $\sigma \leqslant .0005$ in./6 $= .000,083$ in. Suppose that we take $\sigma_0 = .000,075$ in. or .75 in .0001 in. units. Using an $\alpha$ risk of .05, and $n = 50$, set up the critical value for $s^2$ to test $\sigma = .75$ versus $\sigma > .75$. For this test find what $\sigma$ has only a 10% risk of acceptance. How do $s$'s of .846, .965, and .694 (.0001 in.) on consecutive shipments fare? (Of course the lot mean must be at the proper level.)

8.20. A difficult vacuum tube runs at a percentage defective of $\phi_0 = .15$. Currently a sample of 200 shows 44 defectives. Test whether this supplies evidence of a deterioration of the manufacturing process, using $\alpha = .02$, by approximating the probability of 44 or more. Interpret your finding.

8.21. At one time 7% of tungsten disks for brazing to a steel rivet showed cracks. Process improvements had brought the proportion down to $1\% = \phi_0$, say. A recent sample of 1200 showed but three with cracks. Is this clear evidence of still further improvement ($\phi < .01$) at $\alpha = .02$?

8.22. In a letter to a sportswriter: "The Catholic League has now trimmed the Public School League 17 out of a possible 22 championship basketball contests. I don't think one could be considered rash in concluding that the Catholic League is a far stronger league. Why, then, doesn't the Catholic League get a chance to represent Chicago in the annual state tourney?" Test this with $\phi_0 = .5$ versus $\phi \neq .5$ at $\alpha = .01$ and interpret results, giving assumptions.

8.23. In a thesis on highway research the following question was posed: Who should have the responsibility for railroad crossing signals at highways, the railroad or the highway system? Of 259 answering, 116 thought the railroad, 143 the highway system. Is this significantly different from $\phi = .50$? Use $\phi \neq .5$ and $\alpha = .05$, and test.

8.24. With one's two thumbs rapidly spin a Jefferson nickel on a smooth hard surface 20 times and count the number of heads in the 20. Test at the $\alpha = .02$ level the hypothesis P(head) $= .5$ versus P(head) $\neq .5$. Interpret your result.

8.25. A length of insulated wire is tested at an excessive test voltage, and the number of breakdowns of insulation counted. A typical average is $3 = \mu_0$, say. Assuming a Poisson distribution and desiring to test

$\mu = \mu_0$ versus $\mu \neq \mu_0$, we find that with $\alpha = .02$ we cannot have a lower critical region, so we put all of $\alpha$ in the upper one. Determine the critical region. What is the probability of acceptance if actually $\mu = 5$? $\mu = 8$?

8.26.   In Problem 8.25, suppose that to obtain a lower critical region as well, we count the total breakdowns on three lengths of wire. (It may be shown that such counts still follow a Poisson distribution.) Set up the critical region for this test with $\alpha = .02$.

8.27.   If 10 collisions of high-energy particles over standard time and conditions is typical, how many collisions observed would make us conclude there is an influential change yielding $\mu \neq 10$, at $\alpha = .05$. What is the probability of such conclusion if $\mu$ is 6, or if 13?

8.28.   The typical number of "pinholes" occurring in a test area of painted surface is 5. Suppose we call this $\mu_0$. It is desired to test the hypothesis $\mu = 5$ versus $\mu \neq 5$, assuming a Poisson distribution, and using $\alpha = .02$. Determine the critical region for the test.

8.29.   In the text we have been rejecting the null hypothesis when the probability of as great as or a greater discrepancy between the observed sample result and that expected under the null hypothesis is small, say less than or equal to $\alpha$. Why add "or a greater discrepancy"; that is, why not just say when "the probability of obtaining our sample under the null hypothesis is small"?

*Chapter 9*

# SIGNIFICANCE TESTS—TWO SAMPLES

## 9.1. THE GENERAL PROBLEM

In Chapter 8 we were concerned with significance tests using a single sample (or *n* observations), and comparing the observed sample statistic with the parameter value given in the null hypothesis. Such tests require that we be able to place a null value of the parameter at some reasonable level and then decide upon alternative values of interest for the parameter. Sometimes these are both above and below the null value, and sometimes just in one direction from it. Such cases occur often in industry where standards or requirements are dictated, as we have seen in the problems to that chapter. However, in research it may be less easy to obtain a reasonable hypothetical value of a parameter for the null hypothesis. A situation considerably more common may well be that in which we wish to compare two potentially different conditions. For example, the two can be two distinct materials, or two analysts, or test methods, catalysts, formulas, treatments, fertilizers, medicines, or processing temperatures or times. Often one of these is a "control," the other the "experimental" or proposed alternative. Our problem then is to take two samples and to determine whether the observed discrepancy between the two sample results is large enough to be significant, indicating a real difference. Or instead, could the observed difference be readily explained by chance, when we take account of the natural variability? This is our present problem.

Let us consider the statistics involved. As in Chapter 8 we shall of course be concerned with the *type* of population, for example, normal or binomial, a particularizing parameter, and a sample statistic related to that parameter. (See Fig. 9.1.) But now we have *two* populations. If they are normal, we have parameters $\mu_1$, $\mu_2$ estimated by $\bar{y}_1$, $\bar{y}_2$, or $\sigma_1$, $\sigma_2$ estimated by $s_1$, $s_2$. Or, if a binomial population, we have $\phi_1$, $\phi_2$, estimated by $p_1 = y_1/n_1$ and $p_2 = y_2/n_2$. Or, if a Poisson, we have parameters $\mu_1$, $\mu_2$ which are estimated by counts $y_1$ and $y_2$.

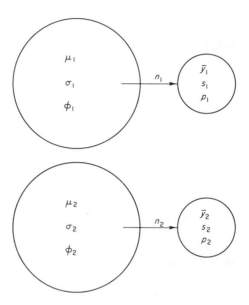

**Fig. 9.1.** Two populations are shown which might be normal, binomial, Poisson, and so on. Samples of $n_1$ and $n_2$ observations are drawn at random and the respective estimates shown. Question: Do the sample results provide reliable evidence that the parameters are different, or could the sample statistics readily have come from identical populations?

Now, even if we drew our two samples from the *same* population, for example, a normal population, we should not expect identical values of $\bar{y}_1$ and $\bar{y}_2$ or of $s_1$ and $s_2$. We have in fact from (7.23)

$$\sigma_{\bar{y}_1-\bar{y}_2} = \sigma\sqrt{1/n_1 + 1/n_2}$$

so we should "expect" to observe this large a difference and might fairly readily observe twice as large a difference. Moreover since $\sigma_s$ is approximately $\sigma/\sqrt{2(n-1)}$, we have for $n_1 = n_2 = n$

$$\sigma_{s_1-s_2} \doteq \sigma/\sqrt{n-1},$$

and thus our two $s$'s can readily differ this much purely by chance, even if both samples came from the same population.

Hence we need to make sure that any observed discrepancy between our two sample statistics cannot be *readily* explained purely by chance. *Only then* can we regard the sample results as clearly indicative of a real difference between parameters.

## 9.2 TESTS ON TWO VARIANCES—THE *F* TEST

Let us now consider tests when the underlying distribution is normal. Since, when testing $\bar{y}_1$ versus $\bar{y}_2$, it makes a difference whether $\sigma_1 = \sigma_2$ or $\sigma_1 \neq \sigma_2$, but the tests on $s_1$ versus $s_2$ are completely independent of $\mu_1$ and $\mu_2$, we take up tests of variability first.

Now to compare any two numbers, we may either find their difference or their ratio. It proves feasible and desirable to compare $s_1^2$ and $s_2^2$ by their quotient, which is $F$. See Section 7.7.3. Under the null hypothesis that $\sigma_1 = \sigma_2$

$$F_{\nu_1, \nu_2} = \frac{s_1^2/\sigma_1^2}{s_2^2/\sigma_2^2} \tag{7.27}$$

becomes

$$F_{\nu_1, \nu_2} = \frac{s_1^2}{s_2^2}. \tag{9.1}$$

The general approach is as follows:

1. We make the assumption that we have random, *independent* samples from two normal populations. "Independent" is important because we cannot, for example, have two matched-pair samples such as those we will see in Section 9.6. No assumption is made about $\mu_1$ and $\mu_2$. They may be equal or even radically different.

2. The null hypothesis is that $\sigma_1 = \sigma_2$. The alternative hypothesis is that

   (a) $\sigma_1 \neq \sigma_2$, two-tail critical region for $F$;
   (b) $\sigma_1 > \sigma_2$, upper tail critical region for $F$;
   (c) $\sigma_1 < \sigma_2$, lower tail critical region for $F$.

3. Set significance level $\alpha$. If 2a is used, put an $\alpha/2$ probability into each tail, but if 2b or 2c is used, put all of the $\alpha$ probability into the one tail.

4. Use the test statistic

   $$F_{\nu_1, \nu_2} = \frac{s_1^2}{s_2^2} \tag{9.1}$$

   where $\nu_1 = n_1 - 1$, $\nu_2 = n_2 - 1$.

5. Enter Table V with $\nu_1$, $\nu_2$ and the desired probability P below. Thus $P = 1 - \alpha$ or $1 - .5\alpha$.
6. If the observed value of $F$ in 4 lies in the critical region, reject $\sigma_1 = \sigma_2$ and conclude $\sigma_1 > \sigma_2$ or $\sigma_1 < \sigma_2$, as the case may be. But if $F$ is not in the critical region, then $\sigma_1 = \sigma_2$ is tenable, and we have no reliable evidence (at the $\alpha$ level) that they differ. We take action appropriate if $\sigma_1 = \sigma_2$. Note that an observed $F$ can easily be significant if $\alpha = .05$, and not significant if $\alpha = .01$ (as is always true in significance testing).
7. In finding critical values from the $F$ table, we note that only $F$ values with cumulative probability .75 to .995 are given, that is, upper tail boundaries for $F$. Hence such tabular entries give the *lower* boundary to the *upper tail* critical region for $F = s_1^2/s_2^2$, which is used when the alternate hypothesis is $\sigma_1 > \sigma_2$. Note that if $s_1$ should happen to be less than $s_2$, then in this case $F < 1$, and we do not even need to look up a critical value of $F$ since we automatically "accept" $\sigma_1 = \sigma_2$. If, however, the alternate hypothesis is $\sigma_1 < \sigma_2$, we could use $F = s_1^2/s_2^2$ and compare this with a *lower tail* critical $F$, which is less than 1. Since such $F$'s are not tabulated,* the usual thing is to use the test statistic $F = s_2^2/s_1^2$ and enter the $F$ table with area $1 - \alpha$ below, being careful to use $\nu_2$ as degrees of freedom for the numerator, and $\nu_1$ for the denominator. Now if the alternate hypothesis is $\sigma_1 \neq \sigma_2$, divide the larger $s_i^2$ by the smaller and compare to the upper tail $F$ for $1 - (\alpha/2)$ below and appropriate degrees of freedom for the $s^2$'s actually in numerator and denominator. Note particularly the use of $\alpha/2$.

### 9.2.1. An Example.

It has often been thought that the same lathe working on similar material and at the same rate of stock removal will have greater variability in turning large diameters than it does in small diameters. In one study on this the following two sets of data occurred, listed in .0001 in. units from the base shown:

| .6600 in. | 7 | 7 | 7 | 6 | 6 | 8 | 7 | 6 | 6 | 7 | $s^2 = .456$ |
| 2.4541 in. | 1 | −1 | 0 | 0 | 0 | 2 | 0 | 1 | 2 | 2 | $s^2 = 1.122$ |

* This is because they are merely reciprocals of the upper tail $F$'s with the degrees of freedom reversed.

Now we are testing $\sigma_1 = \sigma_2$. Since the preconceived notion is that the latter $\sigma$ should be larger we will use a one-tail test, say, at $\alpha = .05$. Now to avoid the trouble in use of tables pointed out in step 7 of Section 9.2, we name the second variance $s_1^2$ and the first $s_2^2$. Then the critical value is $F_{9,9,.05\,\text{above}} = 3.179$. But our observed $F = 1.122/.456 = 2.46$. Since this lies below 3.179 we have no reliable evidence that the two true $\sigma$'s differ. If indeed they do, we shall need more data to establish the fact. ($F = 2.46$ would be significant if $n_1 = n_2 = 16$.)

### 9.2.2. Large Sample Tests.

The exact $F$ test (assuming normality) is available whenever sufficiently complete tables are at hand, but a large sample test is often convenient. It is *necessary* for cases not covered by an $F$ table, and even for cases in an available $F$ table it may still be *useful* because a standardized variable like the normal $u$ is quite easily interpreted.

Our test statistic for comparing $s_1$ with $s_2$ is the following*

$$u = \frac{s_1 - s_2}{\sqrt{[s_1^2/2(n_2 - 1)] + [s_2^2/2(n_1 - 1)]}} \tag{9.2}$$

Note that in (9.2) the $n_1$ term is beneath $s_2^2$ and the $n_2$ term is below $s_1^2$; that is, the denominators are crossed. Now using this change in the test statistic the first three steps in the outline in Section 9.2. are the same, and we make our test using the standard normal distribution.

### 9.2.3. A Large Sample Example.

Data quoted by Freeman [1], coming from a British Cotton Industry Report are given in the accompanying table.

---

* The formula, which is seldom seen in precisely this form, is derived as follows: Recall that $\sigma_s^2 \doteq \sigma^2/2(n - 1)$ for samples from a normal population. Then by (6.56)

$$\sigma_{s_1 - s_2} = \sqrt{\sigma_{s_1}^2 + \sigma_{s_2}^2} \doteq \sqrt{\sigma^2[1/2(n_1 - 1) + 1/2(n_2 - 1)]}.$$

Now since we assume $\sigma_1 = \sigma_2 = \sigma$ in the null hypothesis, we estimate $\sigma^2$ unbiasedly by (7.33), that is, by

$$\frac{(n_1 - 1) s_1^2 + (n_2 - 1) s_2^2}{n_1 + n_2 - 2}.$$

Substituting this for $\sigma^2$ yields

$$\frac{(n_1 - 1) s_1^2 + (n_2 - 1) s_2^2}{n_1 + n_2 - 2} \cdot \frac{n_1 + n_2 - 2}{2(n_1 - 1)(n_2 - 1)} = \frac{s_1^2}{2(n_2 - 1)} + \frac{s_2^2}{2(n_1 - 1)}.$$

Thus (9.2) is essentially $(s_1 - s_2)/\sigma_{s_1 - s_2}$.

| Type of yarn | Mean breaking load (in oz) | Standard deviation (in oz) | Sample size |
|:---:|:---:|:---:|:---:|
| 1 | 6.83 | 1.23 | 1782 |
| 2 | 7.48 | 1.33 | 1914 |

Let us suppose that prior to testing we did not have any reason to suppose one yarn more variable than the other. Then we would test $H_0 : \sigma_1 = \sigma_2$ versus alternate $H$: $\sigma_1 \neq \sigma_2$. Let us adopt the significance level $\alpha = .01$. The difference between $s_1$ and $s_2$ seems small, but let us see. We do have large samples.

The appropriate critical values for $u$ here are $\pm 2.576$. Now use (9.2):

$$u = \frac{1.23 - 1.33}{\sqrt{1.23^2/2(1913) + 1.33^2/2(1781)}}$$

$$= \frac{-.10}{.0299} = -3.34.$$

Hence we have clear evidence at the $\alpha = .01$ level of a real difference in $\sigma_1$ versus $\sigma_2$; the latter is clearly greater.* We might well interpret the $-.10$ as the "observed difference" and the .0299 denominator as the "only to be expected difference."

## 9.3. DIFFERENCES BETWEEN MEANS

The general problem of testing differences between two means is rather more complicated than that of comparing two variances. We still use the observed difference in means $\bar{y}_1 - \bar{y}_2$. But to test whether this difference is significant we must compare it with some standard error $\sigma_{\bar{y}_1 - \bar{y}_2}$ or $s_{\bar{y}_1 - \bar{y}_2}$. The former is available if we know $\sigma_1$ and $\sigma_2$, but the latter must be used if we only have $s_1$ and $s_2$ available. Moreover there are two cases of the latter: (a) where we expect that $\sigma_1 = \sigma_2$, and (b) where we have concluded that $\sigma_1 \neq \sigma_2$.

### 9.3.1. Standard Deviations Known.

This is a rather uncommon case. But when we do know $\sigma_1$ and $\sigma_2$ we are in a splendid position to make our tests, because we can assume

---

* It is interesting to note that if we compare "coefficients of relative variation," $V = s/\bar{y}$, we do *not* have a significant difference, using methods of Kendall and Stuart [2]. That is, the stronger yarn simply was more variable but only proportionally so.

normality for the test statistic given below, even for quite nonnormal populations, because of the central limit theorem (Section 6.11). (Only strong population skewness and very small samples will nullify this favorable situation.) Also since (7.2) is true in general, not just for normal populations, we can use (7.24) for our test:

$$u = \frac{\bar{y}_1 - \bar{y}_2 - (\mu_1 - \mu_2)}{\sqrt{\sigma_1^2/n_1 + \sigma_2^2/n_2}} . \qquad (9.3)$$

Now if, as our null hypothesis states, $\mu_1 = \mu_2$, (9.3) becomes

$$u = \frac{\bar{y}_1 - \bar{y}_2}{\sqrt{\sigma_1^2/n_1 + \sigma_2^2/n_2}} . \qquad (9.4)$$

We have the following general approach to difference of means testing:

1. We make the assumption that we have random *independent* samples from two normal populations.
2. The null hypothesis is that $\mu_1 = \mu_2$. The alternative hypothesis is that

    (a) $\mu_1 \neq \mu_2$, two-tail critical region;
    (b) $\mu_1 > \mu_2$, upper tail critical region;
    (c) $\mu_1 < \mu_2$, lower tail critical region.

3. Set significance level $\alpha$. If 2a is used, put $\alpha/2$ probability into each tail; but if 2b or 2c is used, put $\alpha$ probability into the one tail.
4. Cases of variability:

    (a) $\sigma_1$, $\sigma_2$ known whether equal or not, use (9.4) with a normal-curve table.
    (b) If $\sigma_1$, $\sigma_2$ are unknown, test $s_1$ versus $s_2$ as in Section 9.2. If not significantly different by the $F$ test, assume $\sigma_1 = \sigma_2$, and proceed to use a $t$ test as in Section 9.3.2. If significantly different by the $F$ test, assume $\sigma_1 \neq \sigma_2$ and proceed as in Section 9.3.3.

5. Use the appropriate distribution in 4 and compare the observed test statistic versus the appropriate one- or two-tailed critical region, and make the decision.

**Example.** The dimension in question was the distance from the back of a rheostat knob to the far side of a pinhole. Using .001 in. as a unit, considerable past experience yielded the process standard deviation as 3.54 (.001 in. units). One day the first two consecutive hourly samples, each of $n = 5$ knobs, gave $\bar{y}_1 = 137.8$ and $\bar{y}_2 = 143.0$. Is there evidence of a real change in the process level? Following the outline above we may well assume 1. We shall test $\mu_1 = \mu_2$ versus $\mu_1 < \mu_2$, at the .02 level. The $\sigma_1$, $\sigma_2$ are considered known to be 3.54, so we may use (9.4):

$$u = \frac{137.8 - 143.0}{\sqrt{3.54^2(2/5)}} = \frac{-5.2}{2.239} = -2.32.$$

Comparing this with $u._{02\,\text{below}} = -2.054$, we may regard the process as having had a real change in level $\mu$.

### 9.3.2. Standard Deviations Equal but Unknown.

This is the most common case of significance testing on means. To make the appropriate $t$ test, we have but to use the general formula (7.25) on $\bar{y}_1 - \bar{y}_2$ and, letting $\mu_1 - \mu_2 = 0$ from the null hypothesis, $\mu_1 = \mu_2$. Then

$$t_{n_1+n_2-2} = \frac{\bar{y}_1 - \bar{y}_2}{\sqrt{\dfrac{(n_1 - 1)\,s_1^2 + (n_2 - 1)\,s_2^2}{n_1 + n_2 - 2}\left(\dfrac{1}{n_1} + \dfrac{1}{n_2}\right)}}. \qquad (9.5)$$

Or if $n_1 = n_2 = n$, say, this specializes to the simple form [see (7.26)]:

$$t_{2(n-1)} = \frac{\bar{y}_1 - \bar{y}_2}{\sqrt{(s_1^2 + s_2^2)/n}}. \qquad (9.6)$$

A third case is also useful where the samples are large and/or the $t$ table entries are not available. Letting $n_1$ and $n_2$ both become "large" (perhaps 50 or more), then (9.5) is closely approximated by

$$u = \frac{\bar{y}_1 - \bar{y}_2}{\sqrt{s_1^2/n_2 + s_2^2/n_1}}, \qquad (9.7)$$

which we refer to the normal-curve table. Note the crossing of variances and sample sizes, just as in (9.2).

**Example 1.** An experiment was run on an open hearth furnace to determine whether a proposed change in practice would increase the yield (usable steel divided by metallics charged in the furnace, expressed in percent). A "heat" of approximately 200 tons would be made by the standard practice and the yield found. Then a heat under the proposed practice was run on the same furnace and yield found. All other conditions were held as alike as feasible. Then the experiment was continued, alternating the two practices. (An even better design, instead of pure alternation, is to decide at random for each consecutive pair of heats which of the practices to use first.) The following are the percent yields for the first 10 heats under each practice:

(1)   standard: 78.1, 72.4, 76.2, 74.3, 77.4, 78.4, 76.0, 75.5, 76.7, 77.3;
(2)   proposed: 79.1, 81.0, 77.3, 79.1, 80.0, 79.1, 79.1, 77.3, 80.2, 82.1.

For these data we have

$$n_1 = 10, \qquad \bar{y}_1 = 76.23, \qquad s_1^2 = 3.325$$
$$n_2 = 10, \qquad \bar{y}_2 = 79.43, \qquad s_2^2 = 2.225.$$

We may first test the variances. It was not known which practice would be likely to give the greater variance. Hence we test $\sigma_1 = \sigma_2$ versus $\sigma_1 \neq \sigma_2$, at, say, the .01 level:

$$F_{9,9} = 3.325/2.225 = 1.49.$$

Compare this with $F_{9,9,.005\,\text{above}} = 6.54$. Thus there is no reliable evidence of any real difference in $\sigma$'s. So we use the simple formula (9.6). Now we are only interested in adopting the new practice if there is a real gain in yield. So we test $\mu_1 = \mu_2$ versus $\mu_1 < \mu_2$, and will be conservative about making a change, by using $\alpha = .005$. Then

$$t_{18} = \frac{76.23 - 79.43}{\sqrt{(3.325 + 2.225)/10}} = \frac{-3.20}{.7450} = -4.30.$$

Since $t_{18,.005\,\text{below}} = -2.878$, we have extremely good evidence in favor of the proposed practice. Now how much of a gain can we expect? This is the subject of Chapter 10. But suffice it to say that with the given data we can be 90% sure that the true gain will lie between

$$-3.20 \pm 1.734 \times .7450 = -4.49, \quad -1.91.$$

The actual history of this example was that the experimenters, being statistically untrained, did not use tests like the above, using only a graphical approach. They continued on with this furnace for 37 more heats by each practice. The overall gain was a bit over 2% per heat. So what did they lose by not making a decision on only 10 heats of each? They lost $37(.02)200 = 148$ tons in this *one* furnace by not adopting the proposed practice. Meanwhile other furnaces in the corporation (some 20) were making about 74 heats each, and could have been using the new practice but were not. This adds up to over 6000 tons of steel lost by not adopting the new practice on the basis of the very conclusive evidence above.

**Example 2.** A study was made of the effect of two drugs on the oxygen consumption of kidney slices. The two different drugs (hormones) gave the following:

$$(1) \quad n_1 = 9, \qquad \bar{y}_1 = 27.92, \qquad s_1^2 = 8.673$$

$$(2) \quad n_2 = 6, \qquad \bar{y}_2 = 25.11, \qquad s_2^2 = 1.843.$$

Testing $\sigma_1 = \sigma_2$ versus $\sigma_1 \neq \sigma_2$ first, at $\alpha = .05$ we find

$$F_{8,5} = \frac{8.673}{1.843} = 4.71,$$

with critical values $1/4.817 = .21*$ and $6.757$, so we can accept $\sigma_1 = \sigma_2$, and thus use (9.5). We will test $\mu_1 = \mu_2$ versus $\mu_1 \neq \mu_2$ at $\alpha = .05$, as the experimenter did:

$$t_{13} = \frac{27.92 - 25.11}{\sqrt{\{[8(8.673) + 5(1.843)]/13\}(1/9 + 1/6)}} = 2.168.$$

Since $t_{13,.025 \text{ above}} = 2.160$, the observed difference in means is barely significant at this level.

### 9.3.3.* Standard Deviations Unknown and Possibly Unequal.

Following an $F$ test on our variances from which we conclude $\sigma_1 \neq \sigma_2$, we have a more complicated situation. No longer can we assume that (9.5) or (9.6) follows the $t$ distribution exactly. We must use approximate methods. A number of solutions and approximations have been developed and used. These go by the general name of the "Behrens–Fisher problem." A serviceable test was developed by Aspin and Welch [3–5], and is known as the Aspin–Welch test. It makes use of $t$ tables, but the degrees of freedom to be used in the test depend on (a) the degrees of freedom $\nu_1$ and $\nu_2$ for the separate samples, and (b) the degree of inequality of $s_1^2/n_1$ and $s_2^2/n_2$. More specifically, if we let

$$k = \frac{s_1^2/n_1}{s_1^2/n_1 + s_2^2/n_2} = \frac{s_{\bar{y}_1}^2}{s_{\bar{y}_1}^2 + s_{\bar{y}_2}^2},$$

then the degrees of freedom to use for $t$ in (9.9) are given by

$$\nu = \frac{1}{(k^2/\nu_1) + [(1 - k)^2/\nu_2]}. \tag{9.8}$$

* The value 4.817 is $F_{8,5,.025 \text{ below}}$, its reciprocal giving $F_{8,5,.025 \text{ below}}$.

Then we assume that

$$t = \frac{\bar{y}_1 - \bar{y}_2}{\sqrt{s_1^2/n_1 + s_2^2/n_2}} \tag{9.9}$$

is distributed approximately as $t$, with degrees of freedom $\nu$. If sample sizes are large enough, we can assume (9.9) to be a standardized normal variable under the null hypothesis. Compare (9.7) and (9.9) carefully.*

**Example.** Blood sugar determinations for two similar groups of rats gave the following results, for normal treatment and using a drug:

normals: $n_1 = 12,$ $\bar{y}_1 = 109.17,$ $s_1^2 = 97.43;$

pitocin: $n_2 = 8,$ $\bar{y}_2 = 106.88,$ $s_2^2 = 7.268.$

It seems quite obvious that there is no significance difference in means but that there is in variabilities. But let us make the tests to illustrate the method. First we test $\sigma_1 = \sigma_2$ versus $\sigma_1 \neq \sigma_2$ because *prior to testing* it would likely not be known in which direction the $\sigma$'s would differ, if they did. Also we set $\alpha = .01$. Then

$$F_{11,7} = s_1^2/s_2^2 = 97.43/7.268 = 13.41.$$

Since this lies above $F_{11,7,.995\,\text{below}} = 8.27$, we have reliable evidence of $\sigma_1 > \sigma_2$. (It is up to the experimenter to determine why.) Now let us test $\mu_1 = \mu_2$ versus $\mu_1 \neq \mu_2$, assuming the direction of inequality, if present, would not be known in advance. Again let $\alpha = .01$. Now since $s_1^2/n_1 = 8.119$, $s_2^2/n_2 = .909$, (9.8) gives

$$k = 8.119/(8.119 + .909) = .899, \qquad 1 - k = .101,$$

hence

$$\nu^{-1} = .899^2/11 + .101^2/7 = .0749, \qquad \nu = 13.35.$$

Using (9.9) gives

$$t_{13.35} = \frac{109.17 - 106.88}{\sqrt{8.119 + .909}} = \frac{2.29}{3.00} = .76.$$

This is so small that no critical value is needed. But if it were, we would use 3.00, which is found by interpolating in the $t$ table for $\nu = 13.35$, with .005 tail area.

---

* Theoretical considerations indicate that the approximation is especially good if $n_1 = n_2$.

## 9.4. SIGNIFICANCE OF DIFFERENCES—BINOMIAL DATA

The general approach for testing binomial populations is similar to that for means. Thus we have two binomial populations with proportions $\phi_1$, $\phi_2$. We test the hypothesis $\phi_1 = \phi_2$ versus alternate hypotheses $\phi_1 \neq \phi_2$, $\phi_1 < \phi_2$, or $\phi_1 > \phi_2$, leading to one- or two-tail tests. We take samples of $n_1$ and $n_2$ observations, yielding, respectively, $y_1$ and $y_2$ occurrences of the type whose probability is $\phi$. Then since we have $p_1 = y_1/n_1$ and $p_2 = y_2/n_2$ as estimates of $\phi_1$ and $\phi_2$, respectively, we might say that our problem is to determine whether or not $p_1$ and $p_2$ differ significantly; that is, whether or not their difference can readily be attributed to chance. This is indeed a feasible approach. The observed difference is $p_1 - p_2$. The "only-to-be-expected" difference is $\sigma_{p_1-p_2} = \sqrt{\phi(1-\phi)(1/n_1 + 1/n_2)}$, where $\phi$ is the common parameter under the null hypothesis. Since $\phi$ is unknown, we use instead the weighted average

$$\bar{p} = \frac{n_1 p_1 + n_2 p_2}{n_1 + n_2} = \frac{y_1 + y_2}{n_1 + n_2}. \tag{9.10}$$

Using this for $\phi$ we have a standardized variable

$$u = \frac{p_1 - p_2}{\sqrt{\bar{p}(1-\bar{p})(1/n_1 + 1/n_2)}}, \tag{9.11}$$

where the mean value of $p_1 - p_2$, namely, $\phi_1 - \phi_2$, is missing from the numerator, since it is 0 under the null hypothesis. We can then assume that $u$ in (9.11) is approximately normally distributed $(0, 1)$, if the $n$'s are "large enough." Thus (9.11) is a large sample test. If both $n_1\bar{p}$ and $n_2\bar{p}$ are at least 10, we can proceed as above.

But there is a somewhat better way to proceed through using a "continuity correction," which is desirable since $y_1$ and $y_2$ are integers. Let us first suppose that $n_1 = n_2 = n$ and that $y_1 > y_2$ (renaming the samples if necessary). Then our problem is basically that of determining whether or not the probability of *as great as* or a greater discrepancy between $y_1$ and $y_2$ is "small." Since $y_1 - y_2$ is an integer we use $y_1 - y_2 - .5$, so as to include the entire $y_1 - y_2$ block in our normal distribution approximation to the distribution of $y_1 - y_2$. (See Section 6.3.4.) Now using (5.33), (5.34), $E(y_1) = n_1\phi$, and

$$\sigma_{y_1} = \sqrt{n_1\phi(1-\phi)}, \quad \text{so } E(y_1 - y_2) = 0, \quad \sigma_{y_1-y_2} = \sqrt{\phi(1-\phi)2n}.$$

Then substituting $\bar{p}$ from (9.10) we obtain

$$u = \frac{y_1 - y_2 - .5}{\sqrt{\bar{p}(1 - \bar{p})\, 2n}}, \qquad y_1 > y_2, \qquad n_1 = n_2 = n. \qquad (9.12)$$

This formula gives substantially better results than (9.11), especially for relatively small $n$'s.

It does not seem to be recognized that we can use an approach similar to (9.12) even when the $n$'s are not equal. For this we note that using the binomial (5.33) and (5.34) again, we find $E(y_1 - y_2) = n_1\phi - n_2\phi = (n_1 - n_2)\phi$ and $\sigma_{y_1-y_2} = \sqrt{\phi(1 - \phi)(n_1 + n_2)}$. Hence naming $y_1$ and $y_2$ so that $y_1/n_1 > y_2/n_2$, we have

$$u = \frac{y_1 - y_2 - .5 - \phi(n_1 - n_2)}{\sqrt{\phi(1 - \phi)(n_1 + n_2)}}.$$

Again substituting $\bar{p}$ by (9.10) for $\phi$ we thus have under the null hypothesis

$$u = \frac{y_1 - y_2 - .5 - \bar{p}(n_1 - n_2)}{\sqrt{\bar{p}(1 - \bar{p})(n_1 + n_2)}}, \qquad y_1/n_1 > y_2/n_2. \qquad (9.13)$$

This we assume to be N(0, 1) under the null hypothesis, and can proceed to find from a normal-curve table the probability of as great as or a greater discrepancy, or else compare it to a critical value from the table.

This approach seems to the author to be fully as accurate as the "2 × 2 contingency tables with a continuity correction" (See Chapter 14), and considerably more natural. The author suggests that it can be used all the way down to $y$'s of 5 or so (when the $p_i$'s are below .5).

**Example.** On one type of practice, a check at 280 random times was made, revealing 110 instances of nonconformance. After a quality control moving picture was shown, 300 checks were similarly made, and 92 nonconformances noted. Test $\phi_1 = \phi_2$ against $\phi_1 > \phi_2$, at the $\alpha = .01$ level.

Since $y_1/n_1 = 110/280 = .393$ is greater than $y_2/n_2 = 92/300 = .307$, we use (9.13) directly. First $\bar{p} = (110 + 92)/(280 + 300) = .348$. Then

$$u = \frac{110 - 92 - .5 - .348(280 - 300)}{\sqrt{.348(.652)\, 580}} = \frac{24.46}{11.47} = 2.13.$$

This is compared to $u_{.01\,\text{above}} = 2.327$ and found to be not significant. There is thus no reliable evidence (at this level) of any real effect of the moving picture.

For another operation the $p$'s were 13/280 and 29/280. Thus in this instance $p_1 < p_2$, and we do not do any calculating at all, since we had an alternative hypothesis $\phi_1 > \phi_2$, and with $p_1 < p_2$ we could never reject $\phi_1 = \phi_2$ in favor of $\phi_1 > \phi_2$. We might want to "reverse our field" and test whether the picture may have had an undesirable effect. But this is letting the data dictate the test, which is bad practice.

In completing this section we may say that there is a so-called exact method of making this test a "2 × 2 contingency table," which calls for often rather extensive calculations of the hypergeometric type. They can be readily programmed on a digital computer, however. See Fisher's work [6], in most any edition.

## 9.5. SIGNIFICANCE OF DIFFERENCES—POISSON DATA

Once again we have the same general approach, now assuming random samples from two Poisson populations with parameters $\mu_1$, $\mu_2$. The null hypothesis is that $\mu_1 = \mu_2$ versus $\mu_1 \neq \mu_2$, or others. Now there is, however, one thing we must be careful about. This is that there be equal areas of opportunity for Poisson counts or occurrences. For example, we must not compare directly the number of bacterial colonies from 1 g of food product with that from .5 g, or the number of pits on 20 m² with the number on .5 m² of metal plate, though all of these distributions may well be Poisson. See Section 9.5.1 for tests on unequal areas of opportunity.

Just as for binomial data, we form a standardized variable, and then use the normal curve as an approximation. Recalling that for a Poisson variable with mean $\mu$, $\sigma_y = \sqrt{\mu}$, we have under the null hypothesis $\mu_1 = \mu_2 = \mu$

$$u = \frac{y_1 - y_2}{\sigma_{y_1 - y_2}} = \frac{y_1 - y_2}{\sqrt{\sigma_{y_1}^2 + \sigma_{y_2}^2}} = \frac{y_1 - y_2}{\sqrt{\mu + \mu}} = \frac{y_1 - y_2}{\sqrt{2\mu}}.$$

But we do not know $\mu$. However, under the null hypothesis both counts of occurrences $y_1$ and $y_2$ are unbiased estimates of $\mu$, and thus we use their unweighted average $(y_1 + y_2)/2 = \bar{y}$ to replace $\mu$ in the above. This gives

$$u = \frac{y_1 - y_2}{\sqrt{y_1 + y_2}}. \tag{9.14}$$

Once again, however, we may obtain improved results in seeking the probability of as great as or a greater discrepancy between $y_1$ and $y_2$, if we use the continuity correction of .5. Thus

$$u = \frac{y_1 - y_2 - .5}{\sqrt{y_1 + y_2}} \tag{9.15}$$

to test the hypothesis $\mu_1 = \mu_2$ versus $\mu_1 > \mu_2$ or $\mu_1 \neq \mu_2$. If the alternate is $\mu_1 < \mu_2$, simply rename $y_1$ to $y_2$, and so on.*

**Example.** In a series of large aircraft assemblies a record was made of the number of missing rivets found by inspectors. On number 1 there were $y_1 = 8$ missing rivets, while on number 54 there were only $y_2 = 3$. Let us test the hypothesis $\mu_1 = \mu_2$ against $\mu_1 > \mu_2$; that is, we are anticipating an improvement as the workmen gain in experience. Also set $\alpha = .05$, since we can be rather readily convinced. Using (9.15)

$$u = \frac{8 - 3 - .5}{\sqrt{8 + 3}} = 1.36.$$

The appropriate critical normal-curve value is 1.645, and so the evidence from just these two assemblies is not conclusive.

Suppose, however, we had used the first two, giving 8 and 16, and the last two, giving 4 and 3. Then since the two samples, each of two assemblies, still provide equal areas of opportunity, we can use $y_1 = 24$, $y_2 = 7$, and have

$$u = \frac{24 - 7 - .5}{\sqrt{24 + 7}} = 2.96.$$

This is significant even at the $\alpha = .002$ level! So this meager amount of data has supplied very clear evidence of an improvement in manufacturing conditions relative to missing rivets. One must be careful in such a comparison not to pick out purposely a time when there were few and then select another time for comparison when there were many.

### 9.5.1. Unequal Areas of Opportunity.

An assumption used in the preceding section, was that the two Poisson counts of occurrences $y_1$ and $y_2$ were made over equal areas of opportunity. Thus we could not use (9.14) or (9.15) if $y_1$ were the number of accidents in a section of a city over six months, and $y_2$ were the

---

* It may be shown that if $y_1$ and $y_2$ are drawn from the same Poisson distribution of parameter $\mu$, that $\alpha_{3:y_1-y_2} = 0$ and $\alpha_{4:y_1-y_2} = 3 + 1/2\mu$, which is only 3.1 if $\mu$ is as much as 5.

number for the next single month. We would expect $y_1$ to be much greater than $y_2$, in fact about six times as large. Thus we must somehow make the variables comparable.

This is quite easily done by converting to occurrences or "defects" per unit area of opportunity. In the example just given we could divide $y_1$ and $y_2$ by 6 and 1, respectively, or else by the respective number of days in each time period. We will designate such numbers by $k_1$ and $k_2$. Then we define the average occurrences per unit by $U$:

$$U = y/k. \tag{9.16}$$

Other examples are as follows: (1) the number of breakdowns of insulation under a test voltage in two lengths, 1000 and 1500 m, in which if the unit is 1000 m, $k_1 = 1$, $k_2 = 1.5$, or if 500 m, $k_1 = 2$, $k_2 = 3$; (2) the number of "pinholes" in 200 and 500 ft$^2$ of painted test areas, where $k_1 = 2$, $k_2 = 5$ if the unit area is 100 ft$^2$ or 1 and 2.5 if 200 ft$^2$; (3) bacterial colonies from .1 and .3 g of a food where $k_1 = 1$ and $k_2 = 3$.

The appropriate formula for the comparison uses (9.16) for the data to give $U_1$ and $U_2$; then

$$u = \frac{U_1 - U_2}{\sqrt{U_1/k_2 + U_2/k_1}}. \tag{9.17}$$

Under the null hypothesis we may assume this to be $N(0, 1)$ if $y_1$, $y_2$ both exceed 5, say, and use normal tables to supply critical values to test $\mu_1 = \mu_2$ versus $\mu_1 \neq \mu_2$, or a one-way alternative. Here the $\mu_i$'s are the true average *defects per unit*.*

**Example.**   In the rolling of hot steel ingots into slabs or plates, if such a slab becomes jammed in the rolling process, it is called a "cobble." This event creates considerable trouble and delay. On one mill with one man in charge of the rolling, and using two different crews of men, the following results were noted over six months:

Crew 1:   $k_1 = 38$ heats of a grade of steel,   $y_1 = 63$ cobbles

Crew 2:   $k_2 = 31$ heats of same grade of steel,   $y_2 = 30$ cobbles.

Is this a significant discrepancy?

---

* Formula (9.17) is quite easily derived. Under the null hypothesis $E(U_i) = \mu$, $i = 1, 2$, so $E(U_1 - U_2) = 0$. Then if $k_1$ is a whole number of units $y_1 = c_1 + c_2 + \cdots + c_{k_1}$, where $c_i$ is the count of occurrences over the $i$th unit. Then $E(c_i) = \mu$, and $\sigma_{c_i}^2 = \mu$ also. From this $\sigma_{y_1}^2 = k_1 \cdot \sigma_{c_i}^2 = k_1 \mu$. Then $\sigma_{U_1}^2 = \sigma_{(y_1/k_1)}^2 = \sigma_{y_1}^2/k_1^2 = \mu/k_1$. From this $\sigma_{U_1-U_2}^2 = \mu(1/k_1 + 1/k_2)$. But now $\mu$ is unknown, so we estimate it by $(k_1 U_1 + k_2 U_2)/(k_1 + k_2)$. Substituting this for $\mu$ we find $\sigma_{U_1-U_2}^2 \doteq U_1/k_2 + U_2/k_1$.

Hypothesize $\mu_1 = \mu_2$ versus $\mu_1 \ne \mu_2$, per heat. Let $\alpha = .02$. Then $U_1 = 63/38 = 1.658$, $U_2 = 30/31 = .968$. We have from (9.17)

$$\frac{1.658 - .968}{\sqrt{1.658/31 + .968/38}} = \frac{.690}{.281} = 2.46.$$

We compare this with $\pm 2.327$. Since 2.46 does not lie between, we reject the null hypothesis and say that somehow the number of "cobbles" is related to the crew.

As was true in the use of other formulas .690 is the observed difference and .281 the only-to-be-expected difference explainable by chance.

## 9.6. MATCHED-PAIR DATA. IMPORTANCE OF EXPERIMENTAL DESIGN

We now come to a most interesting experimental design, which is often used. Table 9.1 gives some typical data of this type. The test was

**TABLE 9.1**

Corrosion Measured in Weight Loss for Pairs of Steel Pipe of Two Kinds, Buried about Eight Years in Various Soils[a]

| | Corrosion—weight loss (oz/ft² yr) | | Difference |
| Soil type | Lead-coated steel (1) | Bare steel (2) | $d = (1)–(2)$ |
|---|---|---|---|
| Very fine sandy loam | .18 | 1.70 | −1.52 |
| Gravelly sandy loam | .08 | .21 | −.13 |
| Muck | .61 | 1.21 | −.60 |
| Clay | .44 | .89 | −.45 |
| Tidal marsh | .77 | .86 | −.09 |
| Alkali | 1.27 | 2.64 | −1.37 |
| | $\Sigma d_i = -4.16$, | $\Sigma d_i^2 = 4.7748$ | |

[a] Matched-pair data. See Logan and Ewing [7].

designed to compare corrosion of steel pipe (a) lead coated versus (b) bare steel. Two pipes $2 \times 17$ in., one of each, were buried in similar positions, at each of several soil-type locations. The pairs were then dug up about eight years later. Table 9.1 gives the data, where the measured variable is weight loss over the burial period. Within each soil type each type of pipe had the same conditions and thus the same opportunity to yield to corrosion. Both pipes might show considerable weight loss from corrosion as, for example, in the alkali soil, or both might have little loss as in gravelly sandy loam. But if one type is better over such a collection of different conditions as we have here, it should show a difference basically in the same direction in pair after pair. We see that this is indeed the case here, since in all six cases the weight loss for the bare pipe exceeded that for the lead-coated pipe; that is, all the differences in the last column were negative.

What we are basically doing is to take the *two* samples of six observations each, and through using differences, making it into a *single* sample of six differences. Then the question of the comparison of the two types of pipe becomes one of testing whether such differences $d$ as we have here observed could readily have come from a population of mean 0. Thus we are testing the null hypothesis $\mu_d = 0$. As an alternative hypothesis we could have $\mu_d \neq 0$, $\mu_d > 0$, or $\mu_d < 0$. (What would you use here?) It seems reasonable to suppose that bare steel pipe would likely show greater corrosion than lead coated, else why coat it? Hence we will use $\mu_d < 0$. Now we have thrown our problem involving two samples into one involving only one sample (of $d$'s). We may thus proceed as in Section 8.7. Let us use $\alpha = .025$. Then, letting $\mu_d = 0$ by the null hypothesis,

$$t_{n-1} = \frac{\bar{d}}{s_d/\sqrt{n}} \qquad \text{(matched-pair differences).} \qquad (9.18)$$

For our data we then have

$$t_5 = \frac{-4.16/6}{\sqrt{[6(4.7748) - (-4.16)^2]/6^2(5)}} = -2.76.$$

This is to be compared with $t_{5,.025 \text{ below}} = -2.571$, and so the difference is significant at this level.

It is instructive to look at Fig. 9.2 in which the data of Table 9.1 are plotted. The corrosion weight loss points for bare steel pipe lie above the corresponding points for lead-coated pipe, but at varying distances. The distributions of the two sets, each of six points, would show much overlap. But *site by site* the differences are all in one direction. Part (b) of the figure shows the differences by vertical lines (in this case all being negative). Our test is as to whether the mean difference $\mu_d$ could within reason be 0 and still give such results.

The technique just described is for *nonindependent* samples, that is, matched pairs. At any given site or soil type, if one type of pipe shows

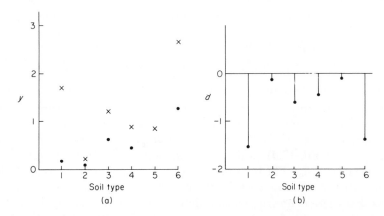

**Fig. 9.2.** Data of Table 9.1 plotted two ways. In part (a) are given the original corrosion weight losses for the six soil types. The crosses are for bare steel pipe and the dots for the lead-coated steel pipe. The former points all lie above the *corresponding* latter points, but the differences apart vary considerably. These differences are shown in part (b) with vertical lines extending down from the 0 line, all being negative. These vertical lines picture the distances between the points in part (a).

considerable corrosion, the other is likely to also. Note the alkali soil, sixth site. Conversely, note the second site, where neither pipe showed much corrosion. In other words the amount of corrosion tends to be "correlated" in the two kinds of pipe. If we had treated the data as though they were two independent samples, *which they were not*, then we would have used (9.6) and found $t_{10} = -1.80$. We would not have had significance even at $\alpha = .05$. When there is closer correlation the drop in the size of $t$ from using (9.18) to using (9.6) is often much greater. The only possible advantage to (9.6) is that the degrees of freedom are twice as great. We study correlation in Chapter 11.

### 9.6.1.* The Matched-Pair Model.

The details of using the matched-pair technique should now be clear. But since interpretation is so important, and there is more than meets the eye and mind at first acquaintance with this method, let us describe the model involved.

Let $y_{ij}$ be the observation for the $i$th kind of "material" and $j$th site or experimental condition. Then $i = 1, 2; j = 1, 2,..., n$; that is, we have $n$ pairs. Letting the true errors be $\varepsilon$'s, and the true site means be $\mu_j$, $\mu_j + \varDelta_j$, $j = 1, ..., n$, for the two "materials," we have the following table:

| Site | Material 1 | Material 2 | Difference $d$ |
|------|-----------|-----------|----------------|
| 1 | $y_{11} = \mu_1 + \varepsilon_{11}$ | $y_{21} = \mu_1 + \Delta_1 + \varepsilon_{21}$ | $d_1 = \Delta_1 + \varepsilon_{11} + \varepsilon_{21}$ |
| 2 | $y_{12} = \mu_2 + \varepsilon_{12}$ | $y_{22} = \mu_2 + \Delta_2 + \varepsilon_{22}$ | $d_2 = \Delta_2 + \varepsilon_{12} + \varepsilon_{22}$ |
| $j$ | $y_{ij} = \mu_j + \varepsilon_{1j}$ | $y_{2j} = \mu_j + \Delta_j + \varepsilon_{2j}$ | $d_j = \Delta_j + \varepsilon_{1j} + \varepsilon_{2j}$ |
| $n$ | $y_{1n} = \mu_n + \varepsilon_{1n}$ | $y_{2n} = \mu_n + \Delta_n + \varepsilon_{2n}$ | $d_n = \Delta_n + \varepsilon_{1n} + \varepsilon_{2n}$ |

Now we make our test on the $d_j$'s by (9.18) which comes from the one-sample $t$ test (7.4). What are the hypotheses tested there, and assumptions? For (7.4) as described in Section 8.7, we are testing the hypothesis $\mu = \mu_0$. For the present case we are assuming $\mu_0$ is zero, that is, that all the $\Delta_j$'s are zero. This means we are testing the null hypothesis that there is no true mean difference at any of the sites. In Section 8.7 the alternative hypothesis is $\mu \neq \mu_0$. Here this takes the form $\Delta_j = \Delta \neq 0$ for all $j$'s, or that the true mean difference between pairs is the same for all sites. This is a somewhat unfortunate restriction. There is a way around this restriction, by taking *two* pipes of *each* type at each site, as we shall see in Chapter 12 using analysis of variance. The other assumption of Section 8.7 is that the distribution of errors is normal, $N(0, \sigma^2)$. Here at site $j$ the error is $\varepsilon_{1j} + \varepsilon_{2j}$. These errors combine to give $\sigma_j = \sqrt{\sigma_{1j}^2 + \sigma_{2j}^2}$. These standard deviations $\sigma_j$ are assumed to be the same at all sites. This assumption will be satisfied if $\varepsilon_{11}$, $\varepsilon_{12}$, ..., $\varepsilon_{1n}$ all have the same variance $\sigma_1^2$ and also if $\varepsilon_{21}$, $\varepsilon_{22}$, ..., $\varepsilon_{2n}$, have the same variance $\sigma_2^2$.

This is summarized by saying that we are testing no true difference versus a constant nonzero difference at all sites, under constant error variance site to site.

## 9.7. SAMPLE SIZES NEEDED FOR TESTS OF TWO MEANS

There are many possible cases for this type of problem, equal or unequal $\sigma$'s, equal or unequal $n$'s, and known or unknown $\sigma$'s. Let us consider the simplest case, namely, where we have equal and known $\sigma_1 = \sigma_2 = \sigma$, say, and for the desired test $n_1 = n_2 = n$, say. Then suppose that we wish to distinguish between the two hypotheses: $\mu_1 - \mu_2 = 0$, $\mu_1 - \mu_2 = \sigma\Delta$, with risks $\alpha$ and $\beta$ as follows:

$$\alpha = P(\text{rej } \mu_1 - \mu_2 = 0 \mid \mu_1 - \mu_2 = 0) \tag{9.19}$$

$$\beta = P(\text{acc } \mu_1 - \mu_2 = 0 \mid \mu_1 - \mu_2 = \Delta\sigma). \tag{9.20}$$

It is natural to use a one-sided test on the null hypothesis, with the following type of rule:

$$\text{accept} \quad \mu_1 - \mu_2 = 0 \quad \text{if} \quad \bar{y}_1 - \bar{y}_2 \leqslant k, \quad \text{otherwise reject.} \quad (9.21)$$

Now letting $K_\alpha$ be defined by $\alpha = \int_{K_\alpha}^\infty \phi(u) \, du$, and recalling that under the null hypothesis $\bar{y}_1 - \bar{y}_2$ is $N(0, 2\sigma^2/n)$, we have for the boundary of the critical region for the variable $\bar{y}_1 - \bar{y}_2$, say $k$:

$$k = 0 + K_\alpha \cdot \sigma\sqrt{2/n}. \quad (9.22)$$

Under the alternate hypothesis $\bar{y}_1 - \bar{y}_2$ is $N(\Delta\sigma, 2\sigma^2/n)$ so that we now have

$$u = \frac{k - \Delta\sigma}{\sigma\sqrt{2/n}} = -K_\beta$$

or that

$$k = \Delta\sigma - K_\beta \cdot \sigma\sqrt{2/n}. \quad (9.23)$$

Now we equate (9.22) and (9.23):

$$K_\alpha \cdot \sigma\sqrt{2/n} = \Delta\sigma - K_\beta \cdot \sigma\sqrt{2/n}. \quad (9.24)$$

Solving for $n$ yields

$$n = 2 \left( \frac{K_\alpha + K_\beta}{\Delta} \right)^2. \quad (9.25)$$

Now if it should happen that one wishes to use a *two-sided* test of the null hypothesis and alternate hypothesis $\mu_1 - \mu_2 = \pm\Delta\sigma$, the only change is that we use $K_{\alpha/2}$, but still use $K_\beta$.

A number of combinations of risks for the case above, and a spectrum of $\Delta$'s from .1 to 4 is given by Owen [1], Chap. 8. If the common $\sigma$ is unknown (but still both are equal) and $n_1$ is to equal $n_2$, then the problem becomes more complicated, making use of the noncentral $t$ distribution. Again solutions are presented by Owen [1] Chap. 8. But if one can set a reasonable or practical upper bound on the two $\sigma$'s assuming they are reasonably equal, he can use (9.25). This will be likely to give larger $n$'s than needed.

**TABLE 9.2**

Summary of Tests[a]

| Hypothesis | Conditions | Test statistic | Distribution of test statistic under Null hypothesis | Alternate hypothesis | Critical region |
|---|---|---|---|---|---|
| 1. $\mu_1 = \mu_2$ | $y$'s normal and independent, $\sigma$'s known | $u = \dfrac{\bar{y}_1 - \bar{y}_2}{\sqrt{\sigma_1^2/n_1 + \sigma_2^2/n_2}}$ | $N(0,1)$ | $\mu_1 \neq \mu_2$<br><br>$\mu_1 > \mu_2$ | $u > u_{\alpha/2}$ and<br>$u < -u_{\alpha/2}$<br>$u > u_\alpha$ |
| 2. $\mu_1 = \mu_2$ | $y$'s normal and independent, $\sigma$'s unknown and equal | $u = \dfrac{\bar{y}_1 - \bar{y}_2}{\sqrt{\dfrac{\nu_1 s_1^2 + \nu_2 s_2^2}{\nu_1 + \nu_2}\left(\dfrac{1}{n_1} + \dfrac{1}{n_2}\right)}}$<br>or $u = \dfrac{\bar{y}_1 - \bar{y}_2}{\sqrt{(s_1^2 + s_2^2)/n}}$ | $t_{\mathrm{df}\ \nu_1 + \nu_2}$<br><br>$t_{\mathrm{df}\ 2\nu}$ | $\mu_1 \neq \mu_2$<br><br>$\mu_1 > \mu_2$ | $t > t_{\alpha/2}$ and<br>$t < -t_{\alpha/2}$<br>$t > t_\alpha$ |
| 3. $\mu_1 = \mu_2$ | $y$'s normal and independent, $\sigma$'s unknown and unequal | $u = \dfrac{\bar{y}_1 - \bar{y}_2}{\sqrt{s_1^2/n_1 + s_2^2/n_2}}$ | Approx. $t$, df by (9.8) | $\mu_1 \neq \mu_2$<br><br>$\mu_1 > \mu_2$ | $t > t_{\alpha/2}$ and<br>$t < -t_{\alpha/2}$<br>$t > t_\alpha$ |
| 4. $\sigma_1 = \sigma_2$ | $y$'s normal and independent | $F = s_1^2/s_2^2$ | $F_{\nu_1, \nu_2}$ | $\sigma_1 \neq \sigma_2$<br><br>$\sigma_1 > \sigma_2$ | $F > F_{1-\alpha/2}$ below and<br>$F < \dfrac{1}{F_{\nu_2,\nu_1,1-\alpha/2}\ \text{below}}$<br>$F > F_{1-\alpha}$ below |

| Null hypothesis | Conditions | Test statistic | Distribution | Alternate hypothesis | Rejection region |
|---|---|---|---|---|---|
| 5. $\sigma_1 = \sigma_2$ | $y$'s normal and independent $n$'s large | $u = \dfrac{s_1 - s_2}{\sqrt{[s_1{}^2/2(n_2-1)] + [s_2{}^2/2(n_1-1)]}}$ | N(0, 1) | $\sigma_1 \neq \sigma_2$ | $u > u_{\alpha/2}$ and $u < -u_{\alpha/2}$ |
| | | | | $\sigma_1 > \sigma_2$ | $u > u_\alpha$ |
| 6. $\phi_1 = \phi_2$ | $y$'s binomial, $p_i = y_i/n_i$, $\bar{p}$ weighted average | $u = \dfrac{p_1 - p_2}{\sqrt{\bar{p}(1 - \bar{p})(1/n_1 + 1/n_2)}}$ or better (9.12), (9.13) | N(0, 1) | $\phi_1 \neq \phi_2$ | $u > u_{\alpha/2}$ and $u < -u_{\alpha/2}$ |
| | | | | $\phi_1 > \phi_2$ | $u > u_\alpha$ |
| 7. $\mu_1 = \mu_2$ | $y$'s Poisson. Equal areas of opportunity | $u = \dfrac{y_1 - y_2}{\sqrt{y_1 + y_2}}$ or better (9.15) | N(0, 1) | $\mu_1 \neq \mu_2$ | $u > u_{\alpha/2}$ and $u < -u_{\alpha/2}$ |
| | Unequal "areas" | See (9.17) | | $\mu_1 > \mu_2$ | $u > u_\alpha$ |
| 8. $\mu_{1j} - \mu_{2j} = \mu_{dj} = 0$ | Matched pairs of $y$'s. Equal variability per site | $t = \dfrac{d}{s_d/\sqrt{n}}$ | $t_{n-1}$ | $\mu_{dj} \equiv \mu_d \neq 0$ | $t > t_{\alpha/2}$ and $t < -t_{\alpha/2}$ |
| | | | | $\mu_{dj} \equiv \mu_d > 0$ | $t > t_\alpha$ |

[a] If alternate hypothesis is that parameter $\theta_1 < \theta_2$, rename samples and populations 1, 2 to 2, 1.

## 9.8. SUMMARY

This chapter has been concerned with inferences from two samples of data. The samples have been chosen from two populations. Our problem is to determine whether or not the population parameters in question differ. (See Table 9.2.) The evidence at hand is the two sample statistics, which estimate the respective parameters. Even if the two populations were absolutely identical, we could not expect the sample statistics from two samples to be identical. Some difference is only to be expected, purely by the workings of chance. We can thus compare observed difference with that explainable by chance, to see whether or not the observed difference is attributable to chance. If it is, then we cannot justifiably declare that the parameters differ; but if not attributable to chance, then we can claim a real difference in parameters.

We use the $t$ test for differences in means when $\sigma$'s are unknown: exact, if $\sigma_1 = \sigma_2$ is assumed, otherwise an approximate $t$. For comparing variances we use the $F$ ratio $s_1^2/s_2^2$. In all the other cases we use the normal population as an exact or approximate test. We can also use it in place of the $t$ and $F$ tests when large sample sizes are being used.

Mostly we do not need to be concerned with slight to moderate nonnormality of the original populations of $y$'s.

REFERENCES

1. H. A. Freeman, "Industrial Statistics." Wiley, New York, 1942.
2. M. G. Kendall and A. Stuart, "The Advanced Theory of Statistics," Vol. I. Hafner, New York, 1969.
3. B. L. Welch, The generalization of "Student's" problem when several different population variances are involved. *Biometrika* **34**, 28–35 (1945).
4. A. A. Aspin, An examination and further development of a formula arising in the problem of comparing two mean values. *Biometrika* **35**, 88–96 (1948).
5. A. A. Aspin, Tables for use in comparisons whose accuracy involves two variances, separately estimated. *Biometrika* **36**, 290–296 (1949).
6. R. A. Fisher, "Statistical Methods for Research Workers." Oliver & Boyd, Edinburgh, 1941.
7. K. H. Logan and S. P. Ewing, Soil corrosion studies, 1934. Field tests of non-bituminous coatings for underground use. No. RP982. *J. Res. Nat. Bur. Standards* **18**, 361–388 (1937).
8. K. A. Brownlee, "Industrial Experimentation." Chem. Publ. Co., New York, 1947.

ADDITIONAL REFERENCES

D. J. Bartholomew, A comparison of some Bayesian and frequentist inferences. *Biometrika* **52**, 19–35 (1965).

W. G. Cochran, Approximate significance levels of the Behrens-Fisher test. *Biometrics* **20**, 191–195 (1964).

J. H. Darwin, Note on a three-decision test for comparing two binomial populations. *Biometrika* **46**, 106–113 (1959).

K. Detre and C. White, The comparison of two Poisson-distributed observations. *Biometrics* **26**, 851–854 (1970).

J. H. Fairfield, A rapid method of comparing two percentages. *Indust. Quality Control* **16** (no. 5), 20–21 (1959).

J. H. A. Ferguson, A nomogram for the power of the F test modified after Keuls. *Statistica Neerlandica* **16**, 177–180 (1962).

M. Fox, Charts of the power of the F-test. *Ann. Math. Statist.* **27**, 484–497 (1956).

J. Gurland and R. S. McCullough, Testing equality of means after a preliminary test of equality of variances. *Biometrika* **49**, 403–417 (1962).

N. L. Johnson, On an extension of the connection between Poisson and $\chi^2$ distributions. *Biometrika* **46**, 352–363 (1959).

R. F. Mayhood, Applying statistics to instrumentation problems. *Internat. Statist. Assoc. J.* **3**, 76–83 (1956).

R. H. Noel and M. A. Brumbaugh, Applications of statistics to drug manufacture. *Indust. Quality Control* **7** (no. 2), 7–14 (1950).

F. E. Satterthwaite, Comparison of two fractions defective. *Indust. Quality Control* **13** (no. 5), 17–18 (1956).

H. Scheffé, Practical solutions of the Behrens-Fisher problem. *J. Amer. Statist. Assoc.* **65**, 1501–1508 (1970).

A. B. L. Srivastava, Effect of non-normality on the power of the t-test. *Biometrika* **45**, 421–429 (1958).

B. V. Sukhatme, A two-sample distribution free test for comparing variances. *Biometrika* **45**, 544–548 (1958).

J. W. Tukey, A quick, compact, two-sample test to Duckworth's specifications. *Technometrics* **1**, 31–48 (1959).

L. Van Valen, Combining the probabilities from significance tests. *Nature (London)* **201**, 642 (1964).

J. B. de V. Weir, Significance of the difference between two means when the population variances may be unequal. *Nature (London)* **187**, 438 (1960).

B. L. Welch, The significance of the difference between two means when the population variances are unequal. *Biometrika* **29**, 350–362 (1938).

## Problems

9.1. In testing the strength of chains under heat treatments $A$ and $B$ the following two sets of strengths were found: $A$: 3610, 3510, 3670; and $B$: 3560, 3400, 3400. Use two-tail tests and $\alpha = .05$ to check for significance.

9.2. A study of fatigue properties of concrete was being made, specifically on "air entrainment." After 34 days, breaking stress in pounds per square inch was measured for two types of concrete.

air entrained:        $n_1 = 8,$    $\bar{y}_1 = 4492,$    $s_2{}^2 = 16,610$
non-air entrained:  $n_2 = 8,$    $\bar{y}_2 = 4150,$    $s_2{}^2 = 29,030.$

Decide on your alternate hypotheses and $\alpha$ risks and interpret results.

9.3.   A biologist took his dog's temperature five times during thunderstorms in May and June, with the following results in degrees Fahrenheit: 102.4, 102.4, 102.6, 102.5, 102.2. He also took the same dog's temperature in the same months in ordinary weather finding 101.3, 101.2, 101.2, 101.3, 101.1. Test $\bar{y}$'s and $s$'s at $\alpha = .01$, using two-tail tests.

9.4.   In a flour mill, the "percent of flour" is that proportion of "carbon particles which will pass through a 200 mesh screen." The desired percentage is 45%, in this problem. Same mill, two methods: (1) without dusting, (2) with dusting. Otherwise the methods were identical. Samples of 24 determinations each were made giving $\bar{y}_1 = 46.58$, $s_1 = 3.77$, $\bar{y}_2 = 45.60$, $s_2 = 4.43$. Does dusting appear to have a significant effect? Test at the $\alpha = .05$ level.

9.5.   Records were kept of different input variables for heats of rimmed steel with .12% carbon or under. For $n_1 = 17$ heats the following results were obtained for "good" heats (zero tons rejected) on percent sulfur: $\bar{y}_1 = .03047\%$, $s_1 = .00304\%$. But for 20 heats with an average of 26.2 tons rejected (range 16.4–43.3 tons): $\bar{y}_2 = .03420\%$, $s_2 = .00552\%$. Make two-way tests with $\alpha = .05$, and interpret.

9.6.   Records are made on the thickness of tin coating on steel sheets. The scale of measurement is pounds per base box (equivalent of both sides of 112 sheets 14 × 20 in.). Estimates are from the average of three measurements taken across the strip. On stack $A$, $n_1 = 31$, $\bar{y}_1 = 155.8$, $s_1 = 4.09$, while on stack $B$, $n_2 = 41$, $\bar{y}_2 = 142.7$, $s_2 = 2.99$. Compare these results with $\alpha = .01$ using two-tail tests.

9.7.   The breaking strength on bonded mats of fiberglass was studied. A change in the binder formula was made. It is not known which direction $\mu$ or $\sigma$ might go, so use two-way tests, and set $\alpha = .01$. Interpret results. Before: $\bar{y}_1 = 17.3$, $s_1 = 2.3$, $n_1 = 220$; after: $\bar{y}_2 = 18.0$, $s_2 = 2.4$, $n_2 = 300$.

9.8.   Thicknesses of cork disks for soft drink crowns were being studied. On two consecutive days the following results were obtained (make two-tail tests using $\alpha = .05$), with measurements in .001 in.

units: $n_1 = 540$, $\bar{y}_1 = 65.441$, $s_1 = 2.96$; $n_2 = 550$, $\bar{y}_2 = 65.429$, $s_2 = 3.07$. (Distributions were very normal.)

9.9. Speeds of cars were measured at a number of different stations during the day and also at night. At one station the following results were obtained: day, $n_1 = 88$, $\bar{y}_1 = 46.23$, $s_1 = 5.60$; night, $n_2 = 142$, $\bar{y}_2 = 45.51$, $s_2 = 5.63$ mph. Test for significance at $\alpha = .05$ using two-tail tests, and interpret.

9.10. For the size of heats, averaging 200 tons of steel, variability is of much importance. For 127 "scrap" heats, $s_1 = 10.80$ tons, while for 116 "flush" heats $s_2 = 10.45$ tons. Test $\sigma_1 = \sigma_2$ versus $\sigma_1 \neq \sigma_2$, with $\alpha = .02$.

9.11. Data were given by Brownlee [8] for analyses by two different methods. Uniform material was analyzed separately by each method. Data were in percents. $A$: $n_1 = 6$, $s_1^2 = .324$; $B$: $n_2 = 7$, $s_2^2 = 5.14$. Test $\sigma_1 = \sigma_2$ versus $\sigma_1 \neq \sigma_2$ at $\alpha = .01$.

9.12. In studying stopwatch errors in time studies by experienced industrial engineers under audible stimuli, two techniques were compared: (1) "continuous" and (2) "snap-back." For 1850 observations each, $s_1 = .00617$ min, $s_2 = .00514$ min. Using an alternate hypotheis $\sigma_1 \neq \sigma_2$, because the relative repeatabilities were unknown in advance, and $\alpha = .01$, make the test and interpret. Also test $s_3 = .00813$ min for continuous technique and visual stimulus, $n_3 = 650$, versus $s_1$. Why might you not wish to test $s_2$ versus $s_3$?

9.13. Compare the matched-pair and independent sample techniques for significance of difference between means for the following data using two-way tests and $\alpha = .01$. Which technique is correct? Why the discrepancy? The same machine was used in the test and most conditions were the same, but two different operators reading microamperes obtained:

| Lamp | 1 | 2 | 3 | 4 | 5 |
|---|---|---|---|---|---|
| Operator 1 | 162 | 164 | 146 | 155 | 166 |
| Operator 2 | 164 | 167 | 148 | 159 | 169 |

9.14. Corrosion of 1 1/2 × 13 in. steel pipe, buried approximately two years, was studied by measured rate of loss (oz/ft² yr) of weight. At each site one lead-coated pipe and one bare steel pipe were buried in similar positions. Results were as shown [7] in the accompanying

|                    | Weight Loss |       |
| Soil type          | Lead cover  | Bare  |
| ------------------ | ----------- | ----- |
| Acadia clay        | 2.72        | 3.75  |
| Cecil clay loam    | .40         | 1.37  |
| Hagerstown loam    | .25         | 1.28  |
| Lake Charles clay  | .53         | 2.03  |
| Merced city adobe  | .24         | 2.33  |
| Muck               | .87         | 1.85  |
| Peat               | 1.59        | 3.24  |
| Sharkey clay       | .38         | .86   |
| Susquehanna clay   | .41         | 2.12  |
| Tidal marsh        | .01         | 1.88  |
| Salinas loamy sand | .19         | 6.62  |
| Alkali soil        | .20         | 3.87  |
| Mojave sandy loam  | .41         | 4.02  |
| Cinders            | 16.66       | 10.65 |

Make an appropriate test at $\alpha = .01$. Do you think the results at the cinder site are compatible with the others?

9.15. In a thesis on forces in a blanking operation at varying strokes per minute, the hardness (Rockwell C) was measured for punches as received, and after the last run. Each pair of observations was on the same punch. Punches were supplied by three manufacturers and were made slightly differently. The results were as follows:

| Before | 63.5 | 63.0 | 63.0 | 63.0 | 62.0 | 62.0 | 63.5 | 65.0 | 64.0 | 64.0 | 64.0 |
| ------ | ---- | ---- | ---- | ---- | ---- | ---- | ---- | ---- | ---- | ---- | ---- |
| After  | 63.0 | 63.5 | 63.5 | 63.5 | 64.0 | 63.0 | 63.5 | 65.0 | 65.0 | 65.0 | 64.5 |

| Before | 62.0 | 63.0 | 64.0 | 60.0 | 60.0 | 62.0 | 60.0 |
| ------ | ---- | ---- | ---- | ---- | ---- | ---- | ---- |
| After  | 64.0 | 64.0 | 65.0 | 62.0 | 61.0 | 62.0 | 61.0 |

Test appropriately at $\alpha = .01$ and interpret your results.

9.16. Cork disks for bottle crowns were measured at the center once each by two gages. Results in .001 in. units follow.

| Disk   | 1  | 2  | 3  | 4  | 5  | 6  | 7  | 8  |
| ------ | -- | -- | -- | -- | -- | -- | -- | -- |
| Gage 1 | 67 | 68 | 64 | 66 | 60 | 63 | 66 | 66 |
| Gage 2 | 64 | 67 | 61 | 61 | 57 | 62 | 63 | 61 |

Analyze for significance of difference between means (1) by matched-pair technique, (2) as though independent samples, using $\alpha = .01$ and two-way tests. Which is the correct approach? (Gage 1 had top metal disk 3/4 in., bottom 1/2 in.; gage 2 had 1/4 in. disks and stronger spring.)

9.17. Compare two days' samples of 15 poorly plated sheets out of $n_1 = 280$ versus 33 out of $n_2 = 220$. Use $\phi_1 = \phi_2$ versus $\phi_1 \neq \phi_2$ at $\alpha = .01$ level, and interpret results.

9.18. Compare the following for significant differences using $\phi_1 = \phi_2$ versus $\phi_1 > \phi_2$ at $\alpha = .01$: Ingots rejected for poor surface, cracks, and so on, and "bloomed" in order to salvage: May, 440 out of 7868; June, 272 out of 8843. (Improvement by increased rolling temperatures.)

9.19. In an audit of caps for food jars, functionally defective caps were being looked for. In February, 40 were found in 153,220, while in March, 54 were found in 137,818. Test $\phi_1 = \phi_2$ versus $\phi_1 \neq \phi_2$, at $\alpha = .05$.

9.20. To test whether an advertising envelope with the name of the magazine in the corner is more effective than one with the publisher's name in the corner, all conditions were held fixed insofar as possible, except this one. Then 8409 of each advertisement were mailed. There were 99 of the former and 84 of the latter returned with an order. The difference of 15 was convincing to the company. Are you convinced?

9.21. For 25 years prior to the author coming to Purdue University, this university's basketball teams had either won or shared the "Big Ten" championship 11 times. During the next 25 years the teams won no championships. Is this difference significant at $\alpha = .05$? Interpret.

9.22. A lot of 1573 pieces was sampled supposedly at random, $n = 410$ giving 54 defective pieces. Upon 100% inspection of the remainder of the lot, 289 more defectives were found. What do you think of the randomness of the sampling and equivalence of inspection?

9.23. A record was kept of missing rivets in large aircraft assemblies. A series of 54 was studied. Eight missing rivets were found in the first and only three in the last one. Is this reliable evidence of an improvement at $\alpha = .01$? For the *first five* a total of 68 missing rivets was found, while for the *last five* only 20 were found. How good is this evidence of improvement?

9.24. The number of "defects" of all types found in large aircraft subassemblies by a crew of inspectors was studied. For the first one in a series, 157 defects were found, while for the last one 103 were found. Test $\mu_1 = \mu_2$ versus $\mu_1 > \mu_2$, that is, anticipating an improvement in the 54 subassemblies manufactured. Use $\alpha = .01$. Also test the total of 854 defects for the first five versus the total of 593 for the last five, the same way. What do we assume to justify the Poisson distribution?

9.25.   The presence of conducting particles on capacitor paper ruins the dielectric strength of the paper and can cause failures. Tests were made by counting the number of such particles in a square foot of paper. A company wanted to compare their paper with a vendor's. Results were, respectively, 820 and 480 in 400 such square-foot areas each. Test $\mu_1 = \mu_2$ versus $\mu_1 \neq \mu_2$ at $\alpha = .001$. If there were 142 on 70 ft$^2$ and 68 on 50 ft$^2$ respectively, make the same test. (These were the first parts of the respective samples.)

*Chapter 10*

# ESTIMATION OF POPULATION
# CHARACTERISTICS

## 10.1. POINT ESTIMATES—GENERAL IDEA

In this chapter we are looking at the relationship of sample to population in a different way than in Chapters 8 and 9. There we were looking from the hypothesized population or populations toward the sample or samples, to see whether they could *reasonably* have come from such populations. If so, then we "accepted" the hypothesis; but if not, then we "rejected" the hypothesis. Now we are concerned with taking the observed sample and using it to make estimates of the characteristics of the population: looking from the sample toward the population.

A population of numbers, continuous or discrete, has in general an average, variability, and curve shape (that is, *type* of distribution). Some idea of the curve shape can be obtained from a substantial sized sample taken under homogeneous conditions. See, for example, Tables 2.10–2.28 at the end of Chapter 2. In the present chapter we are mostly concerned with estimation of the population *parameters* from sample statistics, in particular averages and standard deviations.*

"Point estimates" are single values such as $\bar{y}$, $s$ for measurement data,

---

* It is also true that we can estimate the curve-shape parameters $\alpha_3$ and $\alpha_4$ from the sample statistics $a_3$ and $a_4$. [See (3.18), (3.20), and (5.13).] But we need large samples, at the *very* least 100, to justify such estimation.

or $p = y/n$ for binomial data, or $y$ or $\bar{y}$ for Poisson data. They are calculated from the data at hand. The statistics just mentioned would be single-number estimates of $\mu$, $\sigma$, $\phi$, and $\mu$ (Poisson), respectively. An "estimate" is a single number derived from a sample, whereas an "estimator" is a method (formula if you please) of deriving such an estimate.

## 10.2. WHICH ESTIMATOR TO USE—CHARACTERISTICS OF ESTIMATION

There are in general many ways of estimating, say, $\mu$ or $\sigma$. For example, $\mu$ may be estimated by the sample mean $\bar{y}$, median, or mode, while $\sigma$ may be estimated by $s$, by $\sqrt{\sum(y - \bar{y})^2/n}$, or even by using the range. Which estimator is "best" in a given case? This is an important question which has led to much important and interesting research.

One characteristic of an estimator is its *average value*. Suppose that we have a population and take a large series of samples, each of $n$ observations. Then for each sample we find $\bar{y}$ and $s$. From these we may find $\bar{\bar{y}}$ and $\bar{s}$. These should be close to the theoretical average values $E(\bar{Y})$ and $E(S)$, which are determined by $n$ and the population in question. Now we might well hope that $E(\bar{Y})$ is equal to $\mu$, and that $E(S)$ is equal to $\sigma$. The first of these hopes is true, namely by (6.60)

$$E(\bar{Y}) = \mu_Y. \tag{10.1}$$

Whenever the theoretical average value of a statistic is exactly equal to the parameter, then the statistic is called an "unbiased estimator" of the parameter. Thus

$$T \quad \text{an unbiased estimator of } \tau, \quad \text{whenever} \quad E(T) = \tau. \tag{10.2}$$

Hence $Y$, $\bar{Y}$, and $\bar{\bar{Y}}$ are all unbiased estimators of $\mu_Y$, and this is moreover true for all populations with finite $\mu_Y$'s.

Likewise whenever the population *variance* $\sigma_Y^2$ is finite we may prove (see 7.14)

$$E(S^2) = \sigma^2. \tag{10.3}$$

Thus the sample variance $S^2$ (which uses $n - 1$ in its denominator) is an *unbiased* estimator of $\sigma^2$ (for virtually all populations).

But now what about $S$ as an estimator of $\sigma$? Unfortunately we cannot say that $S$ is an unbiased estimator of $\sigma$, even though $S^2$ is an unbiased estimator of $\sigma^2$. Thus

$$E(S) \neq \sigma. \tag{10.4}$$

Moreover the extent of the relative bias varies according to the population type and the sample size, in general becoming less as $n$ increases. The distribution of $S$'s from a normal population is described in Section 7.5.1. There we have

$$E(S) = c_4\sigma \qquad (10.5)$$

$$\sigma_S = c_5\sigma, \qquad (10.6)$$

where $c_4$ and $c_5$ are tabulated in Table IV in the Appendix. In this table note that $c_4$ rapidly approaches 1 as $n$ increases.*

We may, however, easily remove the bias in $S$ as an estimator of $\sigma$, by merely dividing $S$ by $c_4$ (if the population is normal). Thus

$$E(S/c_4) = \sigma, \qquad S/c_4 \quad \text{unbiased estimator of } \sigma. \qquad (10.7)$$

Moreover for a series of samples of $n$

$$E(\overline{S^2}) = \sigma^2 \qquad (10.8)$$

$$E(\bar{S}/c_4) = \sigma. \qquad (10.9)$$

The immediately preceding discussion is concerned with whether the statistic shoots in the "right direction" or else is biased. Bias is fairly easily removed in most cases, hence a more important matter is as to whether the estimator is subject to much variability or little from sample to sample. Does it shoot *close* to the mark (parameter), or is there a large amount of scatter? As between two different estimators of a parameter (both of which are unbiased, let us say) we certainly prefer the one which is subject to less scatter around the parameter. In fact we would like to use that unbiased estimator having the smallest variability, if one can be found. Such an estimator is called an "efficient" estimator. Fortunately the following are both *unbiased and efficient estimators*:

$\overline{Y}$ as an estimator of $\mu$ (10.10)

(normal population)

$S^2$ as an estimator of $\sigma^2$ (10.11)

$p = Y/n$ as an estimator of $\phi$ (binomial population) (10.12)

$Y$ or $\overline{Y}$ as an estimator of $\mu$ (Poisson population). (10.13)

* If we were to define the sample standard deviation by $\sqrt{\sum(Y - \overline{Y})^2/n}$, we would still have a biased estimator of $\sigma$, in fact more biased than $s$ is. A very nearly unbiased estimate of $\sigma$ is provided by $\sqrt{\sum(Y - \overline{Y})^2/(n - 1.5)}$, as given by Brugger [1].

Our problems come when we do not have one of these population types. For measurement populations we usually use the estimators in (10.10) and (10.11) anyway. To estimate $\sigma$ for a normal population we use $s/c_4$ unless $n$ is substantial (when $c_4$ is near 1). This estimator though unbiased is not perfectly efficient, but a more efficient one is difficult to find, if it exists.

### 10.2.1.* Consistency and Sufficiency.

There are two other desirable characteristics of estimators. An estimator is said to be "consistent" if, when we take any narrow $\pm$ range around the parameter, we can make the probability of an estimate lying inside this range become as close to 1 as we wish by taking $n$ sufficiently large. An estimator is called "sufficient" if it extracts from the sample every bit of available information in the sample, relative to the parameter. Estimators (10.10)–(10.13) are consistent and sufficient. Moreover for normal populations, $s$ as an estimator for $\sigma$ is consistent and sufficient. The median is a consistent estimate of the population median, in general.

## 10.3.* HOW TO FIND A DESIRABLE ESTIMATOR

It is beyond the scope of this book to discuss this interesting topic. Suffice it to say that one good method is to take as our estimator, or estimate of a parameter, that value which, for the population in question, gives the maximum relative probability of obtaining the sample we actually *did* observe. This is called the method of maximum likelihood, and we speak of a "maximum likelihood estimator." All of those (10.10)–(10.13) are of this type. See books on mathematical statistics.

## 10.4. POINT ESTIMATES—COMMON CASES

The common cases of point estimates with which we shall be concerned have already been given in (10.10)–(10.13). To these we may add (10.7)–(10.9), two of which are estimators of $\sigma$. In practical situations these are our "best" single guesses as to the population parameter in question.

There are two practical precautions we should keep in mind, however. One is that our sample should be drawn at random from the population, otherwise we are not estimating this parameter, and may be wildly biased. The other is especially important if we have a series of samples as in (10.8) or (10.9), or have a "large" sample. We must somehow be assured that the conditions have remained constant, so that there is in fact but one parameter value to be estimated, instead of many!

## 10.5. INTERVAL ESTIMATION IN GENERAL

As we have been pointing out, the scattering or variation of the estimates around the parameter value is of much importance. There are two ways to cut down on this variation: (1) to use the most efficient estimate of the parameter for a given sample size $n$, and (2) to increase the sample size.

Now how do we make use of the variability of an estimator? The common way is to determine from the sample, "confidence limits" for the parameter so that we can have a specified degree of confidence that the parameter lies between them. Such a usage of the variability of the estimator is called "interval estimation" as distinct from "point estimation."

What we do is to set a desired degree of confidence, say 90%. Then from the data, we find two limits between which the parameter is supposed to lie. This is done in such a way that, in the long run, limits determined in this manner will actually contain the true parameters (in the respective cases) 90% of the time. We cannot tell, in a particular case, whether or not the parameter lies between the limits, but we can in the above sense have 90% confidence that it does.

Consider the example in Section 8.2 on doubled thickness of rubber tubes. The *known* standard deviation was $\sigma = .002$ in. If for a sample of 10 tubes $\bar{y} = .1044$ in., set limits between which we can be 90% confident that the true lot mean $\mu$ lies.

Let us give an algebraic approach. No matter what $\mu$ is, if we have a normal population with $\sigma = .002$ in., we may state that, prior to sampling,

$$P[\mu - 1.645(.002 \text{ in.})/\sqrt{10} \leqslant \bar{y} \leqslant \mu + 1.645(.002 \text{ in.})/\sqrt{10}] = .90,$$

since $u = -1.645$ to $+1.645$ contains .90 probability, and $\sigma_{\bar{y}} = \sigma_Y/\sqrt{n}$. Or

$$P(\mu - .00104 \text{ in.} \leqslant \bar{y} \leqslant \mu + .00104 \text{ in.}) = .90.$$

Now whenever $\mu - .00104 \text{ in} \leqslant \bar{y}$, then also $\mu \leqslant \bar{y} + .00104 \text{ in.}$ Likewise whenever $\bar{y} \leqslant \mu + .00104 \text{ in.}$, then also $\bar{y} - .00104 \text{ in.} \leqslant \mu$. Thus, if and only if the first and third of these inequalities are true, as in the probability statement, then the second and fourth inequalities are true. Thus we have $\bar{y} - .00104$ in. $\leqslant \mu \leqslant \bar{y} + .00104$ in. Since this double inequality represents the same identical event as that in the probability statement, we have, prior to sampling,

$$P(\bar{y} - .00104 \text{ in.} \leqslant \mu \leqslant \bar{y} + .00104 \text{ in.}) = .90.$$

So that 90% of the time, limits formed this way will contain $\mu$.

Now why did we use "prior to sampling"? This was done in the interest of accuracy. Because, once the sample is drawn, it yields a $\bar{y}$ and limits which either contain $\mu$ or do not contain $\mu$. So in this sense the probability *after sampling* is either 0 or 1. But since $\mu$ is not known (and probably never will be) we cannot tell whether the probability is 0 or 1. (In fact, 90% of the time it will be 1!) It has always seemed to the author that, from the practical viewpoint, this is a rather pedantic point. We assuredly are setting up the kind of limits which will contain $\mu$ 90% of the time and hence have "90% confidence" that in our particular case $\mu$ does lie between them.

### 10.5.1.* Geometrical Argument for Confidence Intervals.

In order to give further depth to the interpretation of confidence limits we shall give the classical geometrical picturization. Suppose that as in Fig. 10.1, for each possible $\mu$ we plot two points, symmetrically around $\mu$, so that if $\mu$ were the mean, 90% of the $\bar{y}$'s would lie between the points. These are at $\pm 1.645(.002 \text{ in.})/\sqrt{10} = \pm .00104$ in. from $\mu$. See the figure, where two normal curves for $\bar{y}$ are shown with two tails each of .05 probability per curve. All such pairs of points lie along two straight lines parallel to the line $\mu = \bar{y}$, and .00104 in. to the left and right. That is, the equations are $\bar{y} = \mu \pm .00104$ in. Now suppose we run a long series of experiments taking random samples of $n = 10$ from $N(\mu, .002^2 \text{ in.}^2)$. We can fix $\mu$ at just one value, at five or six values, or if desired change $\mu$ after each sample of 10. It does not matter. But in the long run 90% of the points $(\bar{y}, \mu)$ will lie between the two outer lines. Thus 90% of the points will be like $A$ in the figure and 10% like $B$. Now, not knowing $\mu$, but only knowing $\bar{y}$, we always choose as confidence interval for $\mu$, the interval from the lower line to the upper. Whenever the $(\bar{y}, \mu)$ point lies *in* the band the confidence interval *contains* $\mu$, as at $A$. But whenever the $(\bar{y}, \mu)$ point lies *outside* the band, the confidence interval *does not contain* $\mu$, as at $B$. Thus we are using for our interval the lines whose equations are $\mu = \bar{y} \pm .00104$ in. If for each $\bar{y}$ as it comes along, we state "$\mu$ lies between $\bar{y} \pm .00104$ in.," we stand to be correct 90% of the time.

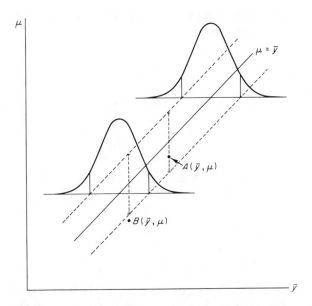

**Fig. 10.1.** Confidence interval boundaries for $\mu$, given a population $N(\mu, .002^2 \text{ in.}^2)$, $n = 10$, and 90% confidence. For any $\mu$, the probability of a $\bar{y}$ between the outer lines is .90. Thus 90% of all $(\bar{y}, \mu)$ points stand to lie between the lines. For each $\bar{y}$, not knowing $\mu$, we use the interval between the lines as the confidence interval. Ninety percent of the time $\mu$ will lie in such intervals, that is, be like the situation at point $A$ rather than at point $B$.

## 10.6. CONFIDENCE INTERVALS FOR $\mu$

Let us now consider the cases of confidence limits for $\mu$. Assuming that the population of $y$'s is normal, we have just two cases, $\sigma$ known and $\sigma$ unknown. We have been using a particular case of the former as an illustration in Sections 10.5 and 10.5.1. We summarize as follows for $1 - \alpha$ confidence limits on $\mu$:

$$P[\bar{y} - (u_{\alpha/2})\,\sigma/\sqrt{n} \leqslant \mu \leqslant \bar{y} + (u_{\alpha/2})\,\sigma/\sqrt{n}] = 1 - \alpha. \qquad (10.14)$$

Thus the $1 - \alpha$ confidence limits for $\mu$ are

$$\bar{y} \pm (u_{\alpha/2})\,\sigma/\sqrt{n}. \qquad (10.15)$$

These limits are quite accurate, even if the $y$'s are not very normally distributed, because by the central limit theorem (Section 6.11), the $\bar{y}$'s are still quite normal.

Now if $\sigma$ is unknown as is the more common case, then we are forced into using the $t$ distribution. Since by (7.4)

$$t = \frac{\bar{y} - \mu}{s/\sqrt{n}}$$

follows the $t$ distribution with $n - 1$ degrees of freedom, we have

$$\mathrm{P}\left(-t_{\alpha/2} \leqslant \frac{\bar{y} - \mu}{s/\sqrt{n}} \leqslant t_{\alpha/2}\right) = 1 - \alpha.$$

Now solve the two inequalities so as to isolate $\mu$. Then we find

$$\mathrm{P}[\bar{y} - t_{\alpha/2}s/\sqrt{n} \leqslant \mu \leqslant \bar{y} + t_{\alpha/2}s/\sqrt{n}] = 1 - \alpha, * \qquad (10.16)$$

so that $1 - \alpha$ confidence limits for $\mu$ are

$$\bar{y} \pm t_{\alpha/2}s/\sqrt{n}, \qquad t \quad \text{with} \quad n - 1 \quad \text{df.} \qquad (10.17)$$

As an example of the use of (10.16) and (10.17), let us use the data of Section 8.7.1, and set 99% confidence limits on $\mu$, the process average percent sulfur. There we found $\bar{y} = .03126\%$, $s = .000702\%$. Then with four degrees of freedom, $\alpha = .01$, $t_{.005, \nu=4} = 4.604$, (10.17) gives 99% confidence limits of

$$.03126\% \pm 4.604(.000702\%)/\sqrt{5} = .02981\%, \quad .03271\%.$$

We are 99% confident of the true mean percent sulfur lying between these limits. We might also have used the upper value of .03271% as a 99.5% upper confidence limit on $\mu$; that is, we would be 99.5% confident of $\mu \leqslant .03271\%$. Similarly we could have used the lower limit as a 99.5% confidence limit on $\mu$. As a tie-in with the significance test of Section 8.7.1, we could compare the 99.5% lower confidence limit on $\mu$ of .02981% with $\mu_0 = .300\%$ sulfur. Since $\mu_0$ lies *above* the lower limit, we can say that $\mu$ could readily (within an $\alpha$ risk of .005) be as low as .0300%. Thus the data are compatible with such a $\mu_0$. This illustrates the intimate relation between significance testing and interval estimation.

A point worth noting is that the length of the $1 - \alpha$ confidence interval in (10.17) is

$$2t_{\alpha/2}s/\sqrt{n}.$$

This is a constant multiple of $s$, but will vary as $s$ varies from sample to

---

* We note that (10.14) and (10.16) apply *strictly*, only as probabilities prior to sampling. Such limits will, however, contain the $\mu$ in question $1 - \alpha$ of the time in the long run.

sample. Thus we do not have a constant width of band as in Fig. 10.1. There the interval is constantly of length

$$2(u_{2\,\text{tails}=\alpha})\,\sigma/\sqrt{n}.$$

In the literature [2] there are shown the results of 100 samples each of $n = 4$ observations. In that experiment $\mu$ was constant as was $\sigma$. But both $\bar{y}$ and $s$ varied. Vertical intervals were drawn for each sample; 90% confidence was specified. The lengths of the intervals varied directly with $s$. In some cases $\mu$ lay in the interval, in others it did not. It happened by good luck that in precisely 90 out of the 100 samples the former was true. Thus the .90 probability was hit exactly.

If the population of $y$'s is nonnormal but not badly so, the $t$ distribution is still useful, but we may wish to be more conservative in interpreting our probability, at least saying, for example, "approximate 90% confidence limits are..."

## 10.7. CONFIDENCE LIMITS FOR σ

For a normal population it is easy to set confidence limits on $\sigma$ from knowledge of $s$ and $n$. We have but to recall from (7.11) that ($n - 1 = \nu$)

$$\frac{\nu s^2}{\sigma^2} = \chi_\nu{}^2,$$

and hence in seeking $1 - \alpha$ confidence limits

$$P[\chi^2_{\alpha/2\,\text{below}} \leqslant \nu s^2/\sigma^2 \leqslant \chi^2_{1-\alpha/2\,\text{below}}] = 1 - \alpha. \qquad (10.18)$$

But we want $\sigma$ in the middle between two terms depending on $s$. Now

$$\chi^2_{\alpha/2\,\text{below}} \leqslant \nu s^2/\sigma^2$$

is exactly equivalent to

$$\sigma^2 \leqslant \nu s^2/\chi^2_{\alpha/2\,\text{below}}$$

since we may multiply or divide both sides of an inequality by the same *positive* quantity. Further

$$\sigma \leqslant \sqrt{\nu s^2/\chi^2_{\alpha/2\,\text{below}}}$$

since we are only taking positive roots. Likewise the second inequality yields

$$\sqrt{\nu s^2/\chi^2_{1-\alpha/2\,\text{below}}} \leqslant \sigma.$$

Putting these two inequalities together we have for an event identical to that in (10.18):

$$P[\sqrt{vs^2/\chi^2_{1-\alpha/2\,below}} \leqslant \sigma \leqslant \sqrt{vs^2/\chi^2_{\alpha/2\,below}}] = 1 - \alpha. \qquad (10.19)$$

Hence $1 - \alpha$ confidence limits for $\sigma$ are

$$s\sqrt{v/\chi^2_{1-\alpha/2\,below}}, \qquad s\sqrt{v/\chi^2_{\alpha/2\,below}} \qquad (10.20)$$

where $\chi^2_v$ is for $n - 1 = v$ degrees of freedom.

For an example, consider the first six cases of Table 2.7. Then the sample size was $n = 6$, $\bar{y} = 31.668$, and $s = .0264$ in fluid ounces. (Consistency of manufacture of bottle capacity is important.) Let us set $90\%$ confidence limits on $\sigma$ for the data. Using (10.20) we have $v = 5$, $\chi^2_{.95} = 11.071$, $\chi^2_{.05} = 1.145$:

$$.0264\sqrt{5/11.071} = .0177 \quad \text{(fl oz)}$$

$$.0264\sqrt{5/1.145} = .0552 \quad \text{(fl oz)}.$$

So we can have $90\%$ confidence that $\sigma$ lies between .0177 and .0552 fl oz, on the basis of this small sample.

### 10.7.1.* Large Sample Confidence Limits for σ.

Recall that, as in (8.8)

$$u = \frac{s - \sigma}{\sigma/\sqrt{2n}}$$

is very nearly $N(0, 1)$ when $n$ is "large," say 100 or more. (Table IV in the Appendix gives $\alpha_3 = .072$ at $n = 100$.) Below that $n$ we may use the foregoing approach. Thus

$$P\left[-u_{2\,tails=\alpha} \leqslant \frac{s - \sigma}{\sigma/\sqrt{2n}} \leqslant u_{2\,tails=\alpha}\right] \doteq 1 - \alpha. \qquad (10.21)$$

All we need now do is solve each separate inequality so as to isolate $\sigma$. (This is left as an exercise for the reader.) Then we obtain

$$P[s/(1 + u/\sqrt{2n}) \leqslant \sigma \leqslant s/(1 - u/\sqrt{2n})] \doteq 1 - \alpha \qquad (10.22)$$

and thus approximate $1 - \alpha$ confidence limits for $\sigma$ are

$$\frac{s}{1 + (u/\sqrt{2n})}, \qquad \frac{s}{1 - (u/\sqrt{2n})}, \qquad (10.23)$$

where $u$ has a two-tail probability of $\alpha$.

In Table 3.1, there were given data for 125 percentages of manganese. In Section 3.41 we calculated $s = .163\%$. Set 90% confidence limits on $\sigma$. Using (10.23) with $u = 1.645$

$$\frac{.163\%}{1 + 1.645/\sqrt{250}} = .148\%, \qquad \frac{.163\%}{1 - 1.645/\sqrt{250}} = .182\%.$$

Notice that here these two approximate 90% confidence limits are relatively much closer to $s = .163\%$ than was the case for the small sample. The interpretation is the same, however, namely that we are about 90% confident of the true $\sigma$ lying between. (Had we used more extensive $\chi^2$ tables we would have found the more exact limits to be identical to the above to the best three significant figures!)

## 10.8. HOW TO HAVE NARROWER CONFIDENCE INTERVALS

The experimenter may often wish that he had a narrower confidence interval for a parameter, $\mu$ or $\sigma$, say. Careful inspection of the length of the interval in (10.15), (10.17), (10.20), and (10.23) will reveal that there are three ingredients determining the lengths of these intervals. First there is the natural variability in the population, that is, $\sigma$. This appears directly in (10.15) and indirectly in the others, through governing $s$. If it is possible to decrease $\sigma$ by more carefully controlling conditions and/or improving technique, the length of the interval can be cut. A quite obvious way to cut the length is by increasing the sample size. This helps materially, but only in general as $\sqrt{n}$. Finally we may narrow the length of the interval by being willing to accept a larger $\alpha$ risk in our limits. That is, 90% confidence limits will naturally be narrower than 99% confidence limits. In any given case these are basically the available avenues. There is one other avenue for $\sigma^2$, namely, that owing to the asymmetry of the $\chi^2$ distribution, it is possible to obtain slightly narrower limits than are given by (10.20). See the work of Tate and Klett [3] where the necessary multipliers for $\Sigma(y_i - \bar{y})^2$ are given so as to provide shortest intervals for $\sigma^2$ up to $n = 30$.

## 10.9 CONFIDENCE INTERVALS FOR FUNCTIONS OF TWO PARAMETERS—TWO SAMPLES

The present section is concerned with interval estimation from results from two samples, that is, analogous to the significance tests in Chapter 9. The common problem is that of setting confidence limits on $\mu_1 - \mu_2$.

This gives us limits to the amount of gain and/or loss in means in using population 1 instead of population 2. We find that there are several cases of this, depending on whether or not the standard deviations are known. (Less commonly we might wish confidence limits on $\mu_1 + \mu_2$ or other functions.) For $\sigma_1$ versus $\sigma_2$ we are likely to be more interested in setting limits upon $\sigma_1/\sigma_2$ than upon $\sigma_1 - \sigma_2$. The former is easy, by using the $F$ distribution.

### 10.9.1. Confidence Limits on the Difference of Means.

Our problem is to set an interval such that we may have $1 - \alpha$ confidence that $\mu_1 - \mu_2$ lies within the interval. There are the same three cases as in Sections 9.3.1–9.3.3. If the true $\sigma$'s are known, then using (7.24) or (9.3) we have in seeking $1 - \alpha$ confidence limits:

$$\mathrm{P}\left[-u_{\alpha/2} \leqslant \frac{\bar{y}_1 - \bar{y}_2 - (\mu_1 - \mu_2)}{\sqrt{\sigma_1^2/n_1 + \sigma_2^2/n_2}} \leqslant u_{\alpha/2}\right] = 1 - \alpha. \qquad (10.24)$$

Solving the inequalities so as to isolate $\mu_1 - \mu_2$ we find an identical event is for $\mu_1 - \mu_2$ to lie between the $1 - \alpha$ confidence limits

$$\bar{y}_1 - \bar{y}_2 \pm u_{\alpha/2}\sqrt{\sigma_1^2/n_1 + \sigma_2^2/n_2}. \qquad (10.25)$$

On the other hand, if the $\sigma$'s are unknown, then we cannot use the approach above unless the sample sizes are both large, in which case we may substitute $s_1$ for $\sigma_1$ and $s_2$ for $\sigma_2$. But the usual way is to use the $t$ distribution. Using (7.25) or (7.26), which assumes $\sigma_1 = \sigma_2$ (even though both are unknown), we find the $1 - \alpha$ confidence limits to be

$$\bar{y}_1 - \bar{y}_2 \pm t_{\alpha/2}\sqrt{\frac{(n_1 - 1)s_1^2 + (n_2 - 1)s_2^2}{n_1 + n_2 - 2}\left(\frac{1}{n_1} + \frac{1}{n_2}\right)}, \qquad (10.26)$$

where $t$ has $n_1 + n_2 - 2$ degrees of freedom, and

$$\bar{y}_1 - \bar{y}_2 \pm t_{\alpha/2}\sqrt{(s_1^2 + s_2^2)/n}, \qquad (10.27)$$

$t$ having $2n - 2$ degrees of freedom. [If confidence limits are desired for $\mu_1 + \mu_2$, we need but use $\bar{y}_1 + \bar{y}_2$ in (10.25)–(10.27).]

In the last case we analogously use (9.9) with $\mu_1 - \mu_2$ included in the numerator, and find $1 - \alpha$ confidence limits for $\mu_1 - \mu_2$ to be

$$\bar{y}_1 - \bar{y}_2 \pm t_{\alpha/2}\sqrt{s_1^2/n_1 + s_2^2/n_2} \qquad (\sigma_1 \neq \sigma_2), \qquad (10.28)$$

where $t$ carries degrees of freedom given by (9.8).

**Example 1.** Let us use the example on yield of steel in an open hearth furnace by the standard method versus a proposed method, as given in Section 9.3.2. There the executive might well say, "It is all very well to say that the proposed practice will give a greater yield beyond reasonable doubt. But I want to know how big the gain in yield will be." So we will set confidence limits in answer to his implied question. Since neither $\sigma$ was known but the $\sigma$'s were compatible as judged by $F$, we will use (10.27). Let us take $\alpha = .01$ for 99% confidence limits on $\mu_1 - \mu_2$. We have $\bar{y}_1 = 79.43\%$ for the yield for the proposed practice and $\bar{y}_2 = 76.23\%$ for the standard. The respective variances were 2.225 and 3.325$(\%)^2$. Then the limits are ($n = 10$, $t_{.005} = 2.878$ for 18 degrees of freedom):

$$79.43 - 76.23 \pm 2.878\sqrt{(2.225 + 3.325)/10}$$

$$= 3.20 \pm 2.14 = 1.06, \quad 5.34 \quad \text{(in \%)}.$$

Hence the executive may now be given 99% assurance of a gain in yield of at least 1.06% but not more than 5.34%. Or if the executive had only asked for a 99% *lower* confidence limit, we would have given him

$$79.43 - 76.23 - 2.552\sqrt{(2.225 + 3.325)/10} = 1.30 \quad \text{(in \%)}.$$

(Or if he had wanted a 99.5% *lower* confidence limit, it would have been the 1.06%.)

Note finally in passing that the two-way 99% confidence limits did not include 0, which says that if we had tested the hypothesis $\mu_1 = \mu_2$ versus $\mu_1 \neq \mu_2$, at an $\alpha = .01$ level we would have rejected $\mu_1 = \mu_2$, because $\mu_1$ could not reasonably equal $\mu_2$.

**Example 2.** Consider the second example also of Section 9.3.2. We use (10.26) for this, and will let $1 - \alpha = .95$ for 95% confidence limits:

$$27.92 - 25.11 \pm 2.160\sqrt{\frac{8(8.673) + 5(1.843)}{13}\left(\frac{1}{9} + \frac{1}{6}\right)}$$

$$= 2.81 \pm 2.80 = +.01, \quad +5.61.$$

So we are 95% confident of the true difference in oxygen consumption lying between these two limits (under these conditions). Since both limits are positive, we could say that the hypothesis $\mu_1 = \mu_2$ has been refuted at the $\alpha = .05$ level, but just barely, as we saw in Section 9.3.2.

### 10.9.2.* Confidence Limits on the Ratio of σ's.

As noted before we are usually more likely to be interested in the ratio $\sigma_1/\sigma_2$ than the difference $\sigma_1 - \sigma_2$. Letting the respective degrees of freedom be $\nu_1$, $\nu_2$ as usual, we use the $F$ distribution for (7.27) as follows for $1 - \alpha$ confidence:

$$P[F_{\nu_2,\nu_1,\alpha/2\,\text{below}} \leqslant s_2{}^2\sigma_1{}^2/s_1{}^2\sigma_2{}^2 \leqslant F_{\nu_2,\nu_1,1-\alpha/2\,\text{below}}]$$
$$= 1 - \alpha. \tag{10.29}$$

Now the only trick is to isolate $\sigma_1/\sigma_2$ in the middle. Taking the left inequality with all quantities being positive, it yields

$$(F_{\nu_2,\nu_1,\alpha/2\,\text{below}})\, s_1{}^2/s_2{}^2 \leqslant \sigma_1{}^2/\sigma_2{}^2,$$

while the right one gives

$$\sigma_1{}^2/\sigma_2{}^2 \leqslant (F_{\nu_2,\nu_1,1-\alpha/2\,\text{below}})\, s_1{}^2/s_2{}^2.$$

Putting these together and taking the positive square root of each yields the event equivalent to that in (10.29) as

$$(s_1/s_2)\sqrt{F_{\nu_2,\nu_1,\alpha/2\,\text{below}}} \leqslant \sigma_1/\sigma_2 \leqslant (s_1/s_2)\sqrt{F_{\nu_2,\nu_1,1-\alpha/2\,\text{below}}}\,.$$

In order to use the table of $F$ we use the reciprocal of the left $F$ and have as $1 - \alpha$ confidence limits for $\sigma_1/\sigma_2$:

$$(s_1/s_2)/\sqrt{F_{\nu_1,\nu_2,1-\alpha/2\,\text{below}}}\,, \qquad (s_1/s_2)\sqrt{F_{\nu_2,\nu_1,1-\alpha/2\,\text{below}}}\,. \tag{10.30}$$

*Be very careful* in your use of $\nu_1$, $\nu_2$ in (10.30).

**Example.** Let us use the technique on the example of Section 9.2.1, and set 95% confidence limits on $\sigma_1/\sigma_2$. (Degrees of freedom are equal—9, 9—so that it is harder to make an error!) Use the $F$ table for $F_{9.9,.975\,\text{below}} = 4.03$. Then since $s_1 = \sqrt{.456} = .675$ and $s_2 = \sqrt{1.122} = 1.059$ we have as limits on $\sigma_1/\sigma_2$:

$$(.675/1.059)/\sqrt{4.03} = .32$$

$$(.675/1.059)\sqrt{4.03} = 1.28.$$

These do not narrow down the ratio very tightly; in fact the larger is four times as great as the smaller.

Also note that the limits include the ratio $\sigma_1/\sigma_2 = 1$, and hence if $\alpha = .05$, we would accept $\sigma_1 = \sigma_2$ (versus $\sigma_1 \neq \sigma_2$) at this level.

### 10.9.3. Paired Differences.

Finally let us consider the confidence interval case analogous to the matched-pairs significance test, as described in Section 9.6. The idea is to set confidence limits on the true consistent difference, pair by pair, if there seems to be one. In Section 9.6, we took the pairs of data and subtracted $y$'s for each separate condition to obtain the difference $d$. We then had one sample of $d$'s instead of two samples of $y$'s (See Table 9.1.) The object then is to set $1 - \alpha$ confidence limits on $\mu_d$, the consistent difference in means assumed to exist in each pair. We thus have thrown this two-sample problem into a one-sample one, namely, that of Section 10.6. Hence $1 - \alpha$ confidence limits on $\mu_d$ are

$$\bar{d} \pm (t_{\alpha/2, \mathrm{df}n-1}) \, s_d/\sqrt{n}. \tag{10.31}$$

Let us use 95% confidence limits for $\mu_d$. Using (10.31) we have for Table 9.1,

$$s_d/\sqrt{n} = \sqrt{[6(4.7748) - (-4.16)^2]/6^25} = .251, \qquad t_{.025,5} = 2.571,$$

$$\bar{d} = 4.16/6 = .693, \qquad \text{bare} - \text{lead coated,}$$

so that the limits are

$$.693 \pm 2.571(.251) = .693 \pm .645 = +.048, \quad +1.338.$$

We thus have 95% confidence in the consistent difference, pair to pair, lying between $+.048$ and $+1.338$.

Note that since both limits are positive we can be confident that the bare steel pipe will show greater corrosion than the lead covered under the studied conditions. Thus we would reject $\mu_d = 0$ in favor of $\mu_d \neq 0$, specifically $\mu_d > 0$.

It is also worth mentioning that, had we asked for a 97.5% lower limit on $\mu_d$, we would have found the one limit $+.048$. If, on the other hand, we had sought a 95% confidence lower limit, it would have been

$$.693 - 2.015(.251) = .693 - .506 = .187.$$

## 10.10. CONFIDENCE LIMITS FOR ATTRIBUTE DATA

The problem of setting confidence limits from counted data is much the same as that for measurement data, especially those for the mean.

This is because when we are assuming the binomial or Poisson distributions, the parameters are, respectively, $\phi$ and $\mu$. These are the population averages of $p$ and $y$, respectively. Given a sample value of $p = y/n$ for the binomial, or a count $y$ for the Poisson, we desire to set confidence limits on $\phi$ or $\mu$. One thing tending to make a little trouble is that the observed data are discrete. But this does not prevent our setting an *exact* pair of limits.

### 10.10.1. Exact Method for Binomial Population.

Suppose we observe $y$ occurrences of one type in a sample of $n$ independent trials, and hence $n - y$ of the other type. Of course, $p = y/n$ is our point estimate of $\phi$, that is, our one guess at $\phi$. But we now want to set limits on each side.

The exact approach in setting $1 - \alpha$ confidence limits on $\phi$ is as follows. We first ask how low can $\phi$ be and still have an $\alpha/2$ probability of yielding a sample such as we did observe, namely, $y$ in $n$. By a sample "such as we did observe," we mean $y$ or more in $n$, not just $y$. Thus we ask $\phi_L$ to satisfy

$$\alpha/2 = P(y \text{ or more in } n \mid \phi_L) = \sum_{i=y}^{n} C(n, i) \phi_L{}^i (1 - \phi_L)^{n-i}. \quad (10.32)$$

Likewise we also ask how high can $\phi$ be and still have an $\alpha/2$ probability of yielding a sample such as we did observe. And so we seek $\phi_U$ to satisfy:

$$\alpha/2 = P(y \text{ or less in } n \mid \phi_U) = \sum_{i=0}^{y} C(n, i) \phi_U{}^i (1 - \phi_U)^{n-i}. \quad (10.33)$$

*Observe carefully* that the observed $y$ is included in *both* summations.

Except for very small $n$'s the job of a direct solution of (10.32) and (10.33) for the $\phi$ limits is prohibitive because they lead to polynomial equations of the $n$th degree. Therefore we normally resort to tables of the binomial, such as those in the literature [1–5], Chap. 5, to help us obtain a quick approximation, or else we use tables providing the desired confidence limits directly. See Section 10.10.5.

Let us illustrate the direct exact method on a simple problem. A top executive in walking through a plant picks up five parts from a pan containing many of them. He looks them over and finds one defective. He immediately makes trouble for some one! Just what is his *point estimate* of the fraction defective in the pan? It is obviously $1/5 = .20$.

But now how much variation is there around this .20? Let us set 95% confidence limits on $\phi$. Then (10.32) and (10.33) become

$$.025 = P(1 \text{ or more in } 5 \mid \phi_L) = \sum_{i=1}^{5} C(5, i)\phi_L{}^i(1 - \phi_L)^{5-i}$$

$$= 1 - (1 - \phi_L)^5$$

and

$$.025 = P(1 \text{ or less in } 5 \mid \phi_U) = \sum_{i=0}^{1} C(5, i)\phi_U{}^i(1 - \phi_U)^{5-i}.$$

The former of these is readily solved by taking logarithms of both sides of $(1 - \phi_L)^5 = .975$. We find $\phi_L = .0051$ easily enough. But the second is

$$(1 - \phi_U)^5 + 5(1 - \phi_U)^4\phi_U = .025.$$

This requires the use of successive approximations for solution. Or we can seek in a binomial table .025 for P(0 or 1 in 5 | $\phi$), or .975 for P(2 to 5 | $\phi$). The solution proves to be .7160.

So, on the basis of his sample of $n = 5$, yielding one defective, the executive can only be 95% sure of the fraction defective $\phi$ for the lot, lying between .5 and 71.6%, not a very tremendous acquisition of knowledge! It certainly cannot form the basis for much action. (He *has* learned there were at least four good parts and at least one defective in the lot!)

### 10.10.2. Normal Approximation for Confidence Limits for the Binomial.

There are several ways to use the normal curve for an approximation to the binomial confidence limits. This is quite accurate if $y$ is at least 10 ($y/n \leqslant .5$). Recalling from (5.33) and (5.34) that $E(y) = n\phi$ and $\sigma_y = \sqrt{n\phi(1 - \phi)}$, and using a continuity correction of .5 on $y$, we have from (10.32) the standardized variable

$$\frac{y - .5 - n\phi_L}{\sqrt{n\phi_L(1 - \phi_L)}} = u_{\alpha/2}.$$

This is designed to make the probability $\alpha/2$ for *exactly y or more*, given $\phi_L$. Likewise from (10.33) we have

$$\frac{y + .5 - n\phi_U}{\sqrt{n\phi_U(1 - \phi_U)}} = -u_{\alpha/2}$$

to make the probability $\alpha/2$ for exactly $y$ *or less*, given $\phi_U$. Squaring, and then solving for the two $\phi$'s, tedious algebra yields*

$$\phi_L = \frac{2y - 1 + u^2 - u\sqrt{[(2y-1)(2n-2y+1)/n] + u^2}}{2(n + u^2)} \quad (10.34)$$

$$\phi_U = \frac{2y + 1 + u^2 + u\sqrt{[(2y+1)(2n-2y-1)/n] + u^2}}{2(n + u^2)}, \quad (10.35)$$

where $u = u_{\alpha/2}$.

As an example, suppose we observe 10 left-clawed lobsters out of a weekly catch of 50. Set 95% confidence limits on the true proportion left clawed. Here $n = 50$, $y = 10$, $u = 1.96$. The formulas yield .105 and .341 for the limits. These are quite close to the precise values of .100 and .337 from a table of confidence limits in the work of Mainland *et al.*, [4]. (What practical assumptions would you make here and how would you interpret the results?)

### 10.10.3. Exact Method for Poisson Population.

The Poisson parameter is $\mu$. Confidence limits for $\mu$ can be found in precisely the same way as for the binomial by using appropriate tables such as Molina's [9], Chap. 5, or Kitagawa's [8], Chap. 5. We simply ask how low $\mu_L$ could be and still reasonably give a sample such as we did observe, namely, $y$ *or more* occurrences. "Reasonably" means to have a probability of $\alpha/2$ of $y$ or more, given $\mu_L$; $\mu_U$ is similar. Thus

$$\alpha/2 = P(y \text{ or more} \mid \mu_L) = \sum_{i=y}^{\infty} e^{-\mu_L}\mu_L{}^i/i! \quad (10.36)$$

$$\alpha/2 = P(y \text{ or less} \mid \mu_U) = \sum_{i=0}^{y} e^{-\mu_U}\mu_U{}^i/i! \quad (10.37)$$

$$1 - \alpha/2 = P(y + 1 \text{ or more} \mid \mu_U). \quad (10.38)$$

One usually has to interpolate between two tabled entries for such limits. For example, if eight occurrences are observed in a sample, and we seek 95% confidence limits, then $\alpha/2 = .025$. We find from Molina [9], Chap. 5,

$$P(8 \text{ or more} \mid \mu = 3.4) = .0231$$

$$P(8 \text{ or more} \mid \mu = 3.5) = .0267.$$

---

* In using the quadratic equation formula, we must use the negative of the square root in the quadratic formula for $\phi_L$ and the positive for $\phi_U$.

So $\mu_L = 3.45$. Using (10.38), we seek .975 for P(9 or more):

$$P(9 \text{ or more} \mid \mu = 15.0) = .963$$
$$P(9 \text{ or more} \mid \mu = 16.0) = .978.$$

Hence $\mu_U = 15.8$.

### 10.10.4 Normal Approximation for Confidence Limits for the Poisson.

Proceeding similarly to Section 10.10.2, we assume the observed count $y$ to be large enough for normality to begin to be satisfactory, perhaps about 12 (it depends on the $\alpha$ chosen and how close we wish our results). We have to solve the equations for the two standardized variables, using the continuity correction so as to include the entire $y$ block in each of the two tails:

$$\frac{y - .5 - \mu_L}{\sqrt{\mu_L}} = u_{\alpha/2}, \qquad \frac{y + .5 - \mu_U}{\sqrt{\mu_U}} = -u_{\alpha/2}.$$

The solutions are for $1 - \alpha$ confidence limits on $\mu$:

$$\mu_L = (2y - 1 + u^2 - u\sqrt{4y - 2 + u^2})/2 \qquad (10.39)$$

$$\mu_U = (2y + 1 + u^2 + u\sqrt{4y + 2 + u^2})/2, \qquad (10.40)$$

where $u = u_{\alpha/2}$ for a normal curve.

As an example, let us find 95% confidence limits for $\mu$, having observed a count of 12 occurrences. From (10.39) and (10.40) we obtain

$$\mu_L = 6.50 \quad \text{and} \quad \mu_U = 21.61.$$

These may be compared with the correct results of 6.69 and 20.34, as found in the work of Crow and Gardner [5].

### 10.10.5. Tables of Confidence Limits.

A number of useful tables of confidence limits for discrete distributions have been published. For the Poisson distribution there is the work of Crow and Gardner [5] which gives 80, 90, 95, 99, and 99.9% two-way confidence limits, for $y$, for all integers 0–300 inclusive. The limits may be interpreted also as one-way limits with confidence coefficients 90, 95, 97.5, 99.5, and 99.95%.

We often obtain data in the form of an average number of occurrences per item, and may want limits on $\mu$ per item. Thus suppose $\bar{y} = 1.50$

defects or errors per truck after the final assembly, when the total production was 200 trucks. The observed total number of errors was therefore 1.50(200) = 300. Set confidence limits for $\mu$ over a field of opportunity of 200 trucks. From Crow and Gardner [5], the 95% limits prove to be 266.71 and 334.62. Dividing these by 200 we have 1.334 and 1.673 for the population average $\mu$ where the field of opporunity for errors is a *single* truck. These limits are quite close together and fairly symmetrical around the point estimate of 1.5.

For the binomial, we have probably the most useful table in Mainland *et al.* [4], but there are other tables in the literature [6, 7]. Such tables must be triple entry: $y$, $n$, and $1 - \alpha$, whereas for the Poisson only a double-entry table is needed for $y$ and $1 - \alpha$. Available two-way confidence coefficients in the work of Mainland *et al.* [4] are 80, 95, and 99%. These also provide confidence coefficients of 90, 97.5, and 99.5% if we interpret any limit as a single limit.

As an example, consider a common problem in reliability. We test a large number of pieces for some rocket engine or space vehicle. We find one defective in 1000. The point estimate for $\phi$ is $1/1000 = .001$. But reliability engineers are often more interested in an upper confidence limit on $\phi$. Let the confidence coefficient be 97.5% and use the work of Mainland *et al.* [4]. We find $\phi_U = .0056$. Since "reliability" is the probability of successful functioning, it is $1 - \phi$. Hence we can be 97.5% confident of a reliability of at least $1 - .0056 = .9944$.* We may mention that, because of the small count of defectives, only one, the *upper limit* $\phi_U$, was over five times as great as the *point estimate* .001.

The reader may have noticed that in the reliability problem above, $n$ was large and $\phi$ small, which suggests a Poisson approximation to the binomial. The illustration just given utilized the largest $n$ in the table. If $n$ is larger, we may use two-way 95% confidence limits [5] as follows. If a single defective is observed in 5000 tested, then $y = 1$ and the 95% Poisson limits to $\mu$ are .051 and 5.323. The latter is $\mu_U$. Dividing by 5000 yields an approximate $\phi_U = .0011$, or a minimum reliability of .9989.

### 10.10.6.* Confidence Limits for Two Samples of Attribute Data.

It may sometimes occur that we wish to set confidence limits on $\phi_1 - \phi_2$ for two binomial populations, or on $\mu_1 - \mu_2$ for two Poisson populations, from random independent samples from the two populations. This may be especially desirable when a significance of difference test as in Sections 9.4 and 9.5 refutes the null hypothesis. Since exact

---

* Sounds like a certain soap, which for years boasted of 99.44% purity.

methods are not available, to the author's knowledge, we rely on "large sample" or normal-curve approximation methods. How large is "large" is not too well known.

To proceed, having observed $y_1$ of one type of outcome out of $n_1$ trials from a population with $\mu_1$, and $y_2$ in $n_2$ from $\mu_2$, we use $p_1 = y_1/n_1$ and $p_2 = y_2/n_2$. Then we merely solve

$$\frac{p_1 - p_2 - (\mu_1 - \mu_2)}{\sqrt{[p_1(1 - p_1)/n_1] + [p_2(1 - p_2)/n_2]}} = \pm u_{\alpha/2} \qquad (10.41)$$

for $1 - \alpha$ confidence limits on $\mu_1 - \mu_2$. Note that we really should have $\mu_i$'s rather than $p_i$'s in the denominator, but they are not available to us. Also we made no continuity correction.

For the Poisson, the buildup is similar. Let there be $k_1$ and $k_2$ equal areas or units of opportunity in the two samples. We observe $y_1$ and $y_2$ defects or occurrences in the two samples. Then $U_1 = y_1/k_1$ and $U_2 = y_2/k_2$ are the respective defects per unit observed. Let $\mu_1$ and $\mu_2$ be the two population average defects per unit. Then we use

$$\frac{U_1 - U_2 - (\mu_1 - \mu_2)}{\sqrt{U_1/k_1 + U_2/k_2}} = \pm_{\alpha/2} \qquad (10.42)$$

and solve for $1 - \alpha$ confidence limits on $\mu_1 - \mu_2$. This differs from (9.17) because here we are not assuming $\mu_1 = \mu_2$ as in the null hypothesis, and thus make separate estimates of $\sigma_{U_1}^2$ and $\sigma_{U_2}^2$, as in the denominator of (10.42).

## 10.11. RELATION BETWEEN INTERVAL ESTIMATION AND SIGNIFICANCE TESTING

As we have been pointing out in this chapter, significance testing on one or two parameters and confidence interval estimation are intimately related and complementary techniques. We commonly test a null hypothesis at some $\alpha$ level which has been decided upon by practical considerations. Let us say we are making a two-way test, for example, $\mu_1 = \mu_2$ versus $\mu_1 \neq \mu_2$. Then we make the test arriving at a categorical (yes–no) decision to either "accept" the null hypothesis or "reject" it. We can interpret the result as in Section 8.5. However, such an interpretation does not in general make much of a distinction between a $t$ of 2.68, which is, say, barely significant, and one of 15.2, which is enormously so. A more positive conclusion and one of a type often more appealing to the experimenter is to set $1 - \alpha$ confidence limits on

**TABLE 10.1**

Interval Estimation—Confidence Level: $1 - \alpha$

| Assumptions | Parameter estimated | Two-way limits | Equation number |
|---|---|---|---|
| I. Normal population | | | |
| A. $\sigma$ known | $\mu$ | $\bar{y} \pm (u_{\alpha/2})\sigma/\sqrt{n}$ | (10.15) |
| B. $\sigma$ unknown | $\mu$ | $\bar{y} \pm (t_{\alpha/2,\, n-1 \text{ df}})s/\sqrt{n}$ | (10.17) |
| C. degrees of freedom within $\chi^2$ table | $\sigma$ | $s\sqrt{\nu/\chi^2_{1-\alpha/2,\, \nu \text{ df}}},\ s\sqrt{\nu/\chi^2_{\alpha/2,\, \nu \text{ df}}}$ | (10.20) |
| D. "Large" $n$ | $\sigma$ | $s/[1 \pm (u_{\alpha/2}/\sqrt{2n})]$ | (10.23) |
| E. $\sigma$'s known | $\mu_1 - \mu_2$ | $\bar{y}_1 - \bar{y}_2 \pm u_{\alpha/2}\sqrt{\sigma_1^2/n_1 + \sigma_2^2/n_2}$ | (10.25) |
| F. $\sigma_1 = \sigma_2$ but unknown | $\mu_1 - \mu_2$ | $\bar{y}_1 - \bar{y}_2 \pm t_{\alpha/2}\sqrt{\dfrac{\nu_1 s_1^2 + \nu_2 s_2^2}{\nu_1 + \nu_2}\left(\dfrac{1}{n_1} + \dfrac{1}{n_2}\right)}$ | (10.26) |
| G. $\sigma_1 \neq \sigma_2$ unknown | $\mu_1 - \mu_2$ | $\bar{y}_1 - \bar{y}_2 \pm t_{\alpha/2}\sqrt{(s_1^2/n_1) + (s_2^2/n_2)}$, df by (9.8) | (10.27) |
| H. $\sigma_{1i},\ \sigma_{2i}$ constant; constant $\varDelta$ | $\varDelta = \mu_{1i} - \mu_{2i}$ | $d \pm (t_{\alpha/2,\, n-1 \text{ df}})s/\sqrt{n}$ | (10.31) |
| I. $\mu_1 = \mu_2$ unnecessary | $\sigma_1/\sigma_2$ | $(s_1/s_2)/\sqrt{F_{\nu_1,\nu_2,1-\alpha/2}}$ below$\quad (s_1/s_2)\sqrt{F_{\nu_2,\nu_1,1-\alpha/2}}$ below | (10.30) |

II. Binomial population

A. Tables of confidence limits available    $\phi$

B. Cumulative binomial tables available    $\phi$    Solve (10.32), (10.33)

C. $y \geqslant 10$ and $n - y \geqslant 10$    $\phi$    Use (10.34), (10.35)

D. $y_1$, $y_2$ about 10 or more    $\phi_1 - \phi_2$    Use (10.41)

III. Poisson population

A. Tables of confidence limits available    $\mu$

B. Cumulative Poisson tables available    $\mu$    Use (10.36), (10.37)

C. $y > 12$    $\mu$    Use (10.39), (10.40)

D. $y_1$, $y_2$ not small    $\mu_1 - \mu_2$    Use (10.42)

$\mu_1 - \mu_2$. Then if the significance test $t$ is 2.68, the null difference of 0 will be relatively near the lower limit; whereas, if $t$ were 15.2, both limits on $\mu_1 - \mu_2$ would be relatively far off from 0.

As an example, consider the data on yarn strength given in Section 9.2.3. We could test $\mu_1 = \mu_2$ versus $\mu_1 \neq \mu_2$, and since we found $\sigma_1 < \sigma_2$, we use (9.9)

$$t = \frac{\bar{y}_1 - \bar{y}_2}{\sqrt{s_1^2/n_1 + s_2^2/n_2}} = \frac{6.83 - 7.48}{\sqrt{1.23^2/1782 + 1.33^2/1914}} = -15.4.$$

This $t$ carries enough degrees of freedom by (9.8) that we may treat $t$ as a normal $u$. By any sensible $\alpha$ level this $-15.4$ is significant. But what can we say positively? That large $t$ or $u$ should give us considerable leverage. Let us set 90% confidence limits on $\mu_1 - \mu_2$ by (10.28) since $\sigma_1 < \sigma_2$. These are

$$6.83 - 7.48 \pm 1.96\sqrt{1.23^2/1782 + 1.33^2/1914} = -.567, \quad -.733 \qquad \text{(oz)}.$$

One need not draw a picture to visualize how relatively far the confidence interval keeps its skirts free from 0. Another nice feature of the confidence interval is that it is a result in the same unit as the original data.

One basic thing is that if the null hypothesis condition, whatever it is, lies inside the $1 - \alpha$ confidence interval, we would not reject the null hypothesis; but if it is not in the interval, we would reject it. A second point is that if we are concerned with a one-way significance test, for example, $\mu = \mu_0$ versus $\mu > \mu_0$, we would find this related to a one-way confidence interval setting a $1 - \alpha$ lower limit to $\mu$. Then if this $\mu_L < \mu_0$, the null hypothesis is tenable.

There are some writers who would almost like to do away with significance testing on parameters altogether. But the author believes that they are complementary techniques, both of large practical use, and each useful in its place. One other point against extreme emphasis upon interval estimation to the exclusion of significance testing is that one needs a preliminary test of significance between $s_1$ and $s_2$ to decide whether to use (10.26) or (10.28).

As a matter of fact the author rather leans toward the use of both approaches, first testing at some convincing significance level, such as $\alpha = .05, .02, .01,$ or even .001. Then having established that the null hypothesis is untenable, to set confidence limits on the parameter, or function of parameters, for example, $\mu_1 - \mu_2$ or $\sigma_1/\sigma_2$ at 90% confidence. The reason for going down to only 90% confidence is that the higher percentages tend to give such wide intervals. And 90% is really quite a usefully high confidence. Of course what confidence coefficient to actually use, and whether to use one-way or two-way limits depend on many practical considerations.

An excellent discussion of some of the points in this section is given by Natrella [8].

## 10.12. SUMMARY

The present chapter has been concerned with estimation of population parameters from sample data (as distinct from attempting to determine the *type* of population). On the one hand we had point estimates which are one-number "best" guesses at a parameter. We can never even hope to hit the population parameter *exactly*, but we can choose as estimates those with desirable properties, such as "unbiasedness" (shooting in the right direction) and "efficiency" (shooting with the best statistic to minimize the scatter). The point estimates we use in general have these properties, and also the other two as well: "consistency" and "sufficiency."

The other form of estimation is that of setting from the data at hand an interval which, prior to drawing the sample, had a $1 - \alpha$ probability of containing the parameter. Or if two samples and populations are involved, we may set confidence limits on the difference of the two corresponding parameters, or on $\sigma_1/\sigma_2$ possibly. The interpretation of such intervals is that we may be "$100(1 - \alpha)\%$ confident" of the parameter, or parameter function, lying within the interval. The exact meaning of this is that, prior to sampling, the probability was $1 - \alpha$. But since the parameter values are unknown, and likely always will be, we can for practical purposes treat the $1 - \alpha$ confidence as an ordinary probability.* See Table 10.1, pp. 274–275.

REFERENCES

1. R. M. Brugger, A note on unbiased estimation of the standard deviation. *Amer. Statist.* **23** (no. 4), 32 (1969).
2. "ASTM Manual on Quality Control of Materials." American Soc. for Testing Materials, Philadelphia, Pennsylvania, 1951.
3. R. F. Tate and G. W. Klett, Optimal confidence intervals for the variance of a normal distribution. *J. Amer. Statist. Assoc.* **54**, 674–682 (1959).
4. D. Mainland, L. Herrera, and M. I. Sutcliffe, "Statistical Tables for Use with

* Only when we can obtain an *a priori* distribution of the parameter values, would the author like to modify this statement.

Binomial Samples, Contingency Tests, Confidence Limits, and Sample Size Estimates." N.Y. Univ. College of Medicine, New York, 1956.

5. E. L. Crow and R. S. Gardner, Confidence intervals for the expectation of a Poisson variable. *Biometrika* **46**, 441–453 (1959).
6. E. L. Crow, Confidence intervals for a proportion. *Biometrika* **43**, Parts 3, 4, 423–435 (1956).
7. J. Pachares, Tables of confidence limits for the binomial distribution. *J. Amer. Statist. Assoc.* **55**, 521–533 (1960).
8. M. G. Natrella, Relation between confidence intervals and tests of significance— A teaching aid. *Amer. Statist.* **14**, 20–22 (1960).
9. P. R. Weyl, A redetermination of the constant of gravitation. *J. Res. Nat. Bur. Standards* **5**, 1243–1290 (1930).

ADDITIONAL REFERENCES

C. J. Clopper and E. S. Pearson, The use of confidence or fiducial limits. *Biometrika* **26**, 404–413 (1934).
E. E. Cureton, Unbiased estimation of the standard deviation. *Amer. Statist.* **22**, 22 (1968).
F. A. Graybill, Sample size for a specified width confidence interval on the variance of a normal distribution. *Biometrics* **16**, 636–641 (1960).
W. C. Guenther, Shortest confidence intervals. *Amer. Statist.* **23**, 22–25 (1969).
F. E. Satterthwaite, Binomial and Poisson confidence limits. *Indust. Quality Control* **13** (no. 11), 56–59 (1957).

# Problems

**10.1.** Values of a gravitational constant [9] in units of $10^{-8}$ cm$^3$/g sec$^2$ using gold balls were 6.683, 6.681, 6.676, 6.678, 6.679, and 6.672. Set 90% confidence limits on $\mu$ and on $\sigma$ and interpret the practical meaning of each.

**10.2.** Values of a gravitational constant [9] in units of $10^{-8}$ cm$^3$/g sec$^2$ using platinum balls were 6.661, 6.661, 6.667, 6.667, and 6.664. Set 90% confidence limits on $\mu$ and on $\sigma$ and interpret the practical meaning of each.

**10.3.** Rate coefficients of ethylene oxide absorption in 1b. moles/hr ft$^3$ atm $K_g a$ were determined on samples of 100 ml of gas as follows: $n = 6$, $\bar{y} = 4.285$, $s = .403$. Set 95% confidence limits on $\mu$ and $\sigma$ and interpret the practical meaning of each.

**10.4.** Five determinations of the "water equivalent" at a 2°C rise in calorimeter temperature by standard benzoic acid method were 451.2, 452.1, 446.3, 446.1, and 448.8 g. Set 90% confidence limits on $\mu$ and on $\sigma$ and interpret the practical meaning of each.

10.5. Five determinations of the heat of explosion of 2,2-dinitropropane gave $\bar{y} = 3307.84$ and $s = 5.01$ in calories per gram. Set 90% confidence limits on the true heat of explosion of this compound. Also set 90% confidence limits on the true standard deviation $\sigma$ of measurement error for the technique.

10.6. For the data in Problem 8.2, the diameter of a "low-speed plug" had $s = .000692$ in. for $n = 160$. Set 90% confidence limits on $\sigma$ and interpret.

10.7. In Problem 8.3, 500 observations of percent carbon gave $\bar{y} = .7380\%$, $s = .01883\%$. Set 95% confidence limits on $\mu$ and on $\sigma$, and interpret.

10.8. In Problem 8.4, 450 analyses of blast furnace pig iron gave $\bar{y} = 1.031\%$ and $s = .150\%$. Set 90% confidence limits on $\mu$ and on $\sigma$ and interpret.

10.9. For 25 ceramic disk capacitors fired at top level $\bar{y}_1 = 10,953$ and $s_1 = 845$, while 25 at the bottom level gave $\bar{y}_2 = 10,537$ and $s_2 = 640$, the unit being microfarads of capacitance. Set 90% confidence limits on the true mean difference between top and bottom levels under present conditions.

10.10. The data in Problem 9.2 are on the effect of "air entrainment" on the breaking stress of concrete. Set a 95% confidence limit on the minimum gain in breaking stress using air entrainment. What is the point estimate?

10.11. Data given in Section 9.3.2 on yield of steel from charged metallics were analyzed for significance of differences. Set 90 and 99% confidence limits on $\mu_1 - \mu_2$ from the data:
  (1)   standard practice, $\bar{y}_1 = 76.23\%$, $s_1^2 = 3.325(\%)^2$;
  (2)   proposed practice, $\bar{y}_2 = 79.43\%$, $s_2^2 = 2.225(\%)^2$
        $(n_1 = n_2 = 10)$.

10.12. For the data in Problem 9.3 on a dog's temperature, set 90% confidence limits on the increase in temperature during stormy weather. What is the estimated increase?

10.13. For random independent samples of 300 piston rings (1) before and (2) after, the "tension" was measured. (Tension is the force required to close the gap to the specified distance as it will be in a piston.) The data in pounds follow: $\bar{y}_1 = 4.95$, $s_1 = .465$; $\bar{y}_2 = 5.48$, $s_2 = .459$. Set 95% confidence limits on $\mu_1 - \mu_2$ and interpret the result.

10.14. For the data on strength of chains given in Problem 9.1, set 95% confidence limits on $\mu_1 - \mu_2$. Do they include 0? How does the confidence interval approach compare with results of the significance test?

10.15. Set 90% confidence limits on the true gain in breaking strength of bonded mats of fiberglass, as a result of a change in binder formula. Use data of Problem 9.7.

10.16. For the large sample data on thickness of cork disks in Problem 9.8, set 95% confidence limits on $\mu_1 - \mu_2$, and compare with the significance test results.

10.17. The length of nozzle jets for carburators for autos is quite critical. A processing change was made to improve the consistency of length. This cut the variability in half. Samples of 125 jets before and after the change yielded $s_1 = 1.002$ and $s_2 = .505$ in .001 in. units. The point estimate of $\sigma_1/\sigma_2$ is .50. Set 90% confidence limits on the ratio by using $F$ ratios for $\nu_1 = \nu_2 = 120$.

10.18. Set 90% confidence limits on $\mu_1 - \mu_2$ for weights of individual granules of a molding powder under two methods: granulation (1) and granulation (2). Data in respective samples each of 50 granules: $\bar{y}_1 = .02121$, $s_1 = .00182$; $\bar{y}_2 = .01697$, $s_2 = .00315$.

10.19. In the manufacture of ceramic disk capacitors a kiln is used for silver firing at about $1500°F$. The firing treatment affects the resultant value of capacitance. To compare position effects 25 capacitors were measured from (1) the top level and (2) the bottom level. The following results in microfarads were observed: $\bar{y}_1 = 11,003$, $s_1 = 408$ and $\bar{y}_2 = 10,289$, $s_2 = 688$. Set 95% confidence limits on $\sigma_1/\sigma_2$ and $\mu_1 - \mu_2$.

10.20. Using the data of Problem 9.15 on forces in a blanking operation, set 95% confidence limits on $\mu_{\text{after}} - \mu_{\text{before}}$, and interpret your results.

10.21. Set 99% confidence limits on the relative bias between gages in measuring thickness of cork disks, using data of Problem 9.16. Interpret results. (Also compare with the significance test in that problem if it was assigned.)

10.22. Thickness of coating of tin on steel sheets is measured in pounds per base box (equivalent weight to both sides of 112 sheets $14 \times 20$ in.²). A test strip across the sheet is periodically measured.

| Strip | 1 | 2 | 3 | 4 | 5 | 6 | 7 | 8 | 9 | 10 |
|---|---|---|---|---|---|---|---|---|---|---|
| Drive edge | 76 | 72 | 70 | 74 | 74 | 75 | 74 | 77 | 76 | 66 |
| Center | 71 | 68 | 62 | 65 | 65 | 69 | 67 | 67 | 68 | 59 |

Set 90% confidence limits on the true difference in coating weight between the two positions.

10.23.  Set 95% confidence limits on $\mu_1 - \mu_2$, the true difference in percentages of residual solvent in a plastic, between inner layer 1 and outer layer 2:

| Test sample | 1 | 2 | 3 | 4 | 5 | 6 | 7 | 8 | 9 | 10 |
|---|---|---|---|---|---|---|---|---|---|---|
| $y_{1j}$ | 9.04 | 8.99 | 9.12 | 7.67 | 9.49 | 9.46 | 9.85 | 9.70 | 9.83 | 8.56 |
| $y_{2j}$ | 6.34 | 5.82 | 7.37 | 5.97 | 8.58 | 6.30 | 6.62 | 7.06 | 7.29 | 7.24 |

Interpret your results. If you were to (erroneously) treat these as two random *independent* samples, do you think the limits would be wider or narrower?

10.24.  The diameter of each pin was measured before and after a heat treating process. One hundred pins were measured (1) "before" and (2) "afterward," giving 100 differences, $d_i = y_{1i} - y_{2i}$. The distribution of $d_i$'s is as follows:

| $d_i$ (in.) | 0 | +.0001 | +.0002 | +.0003 | +.0004 | +.0005 | +.0006 |
|---|---|---|---|---|---|---|---|
| Frequency | 1 | 9 | 27 | 26 | 26 | 6 | 5 |

Set 95% limits on the true loss in diameter during heat treating. (Average diameters were about .2430 in. before heat treating.)

10.25.  Set 95% confidence limits on the true relative bias between operators 1 and 2 in reading lamp current in microamperes, using data of Problem 9.13. Comment.

10.26.  Records were kept on the percentage of "menders" (that is, sheets needing recoating) in manufacture of tin-coated steel sheet. For 89 shifts on stack 13, $\bar{y} = 9.92$ and $s = 5.20$, while for 89 shifts on stack 14, $\bar{y} = 9.15$ and $s = 4.03$. Set 90% confidence limits on $\mu_1 - \mu_2$. (Note that here we have treated attribute data, that is, fraction of sheet menders, as though they were measurements.) What do you think of the assumption of normality here?

10.27.  Set 90% confidence limits on $\mu_1 - \mu_2$ for the blood sugar determination results of the example in Section 9.3.3, and interpret your results. Also set 99% limits. Set 90% confidence limits on $\sigma_1/\sigma_2$.

10.28.  What does a random sample of $n = 200$ parts from a large lot of parts tell you about the lot, if you find 12 defective ones in the 200?

10.29.  What does a random sample of $n = 500$ parts from a large lot of parts tell you about the lot, if you find 25 defectives in the 500?

10.30.  Ten carburetors out of 300 for autos were defective at final inspection. Set 95% confidence limits on $\phi$ for this process and interpret.

10.31.  During 18 days, 841 auto radiators were tested for outlet leaks, 116 showing such leaks. Set 90% confidence limits on $\phi$ for the current manufacturing conditions and interpret.

10.32.  Set 90% confidence limits on the true proportion of times a head will come up when spinning your nickel the way you did for Problem 8.24.

10.33.  Of 150 stud-diffusers tested, 18 were rejected. Give a point estimate of $\phi$ and 90% confidence limits, for true fraction rejectable.

10.34.  One hundred percent inspection of 1320 malleable castings by magnaflux showed 76 with cracks. Set 90% confidence limits on $\phi$ for the current conditions from the data and interpret.

10.35.  At the beginning of a study of packing nuts, five samples of 50 each showed defectives as follows: 22, 6, 1, 23, 18, or a total of 70 in 250. Treating this as a single random sample, set 95% confidence limits on the true fraction defective $\phi$ for the process. What do you think of the propriety of treating the data as a homogeneous sample?

10.36.  In a mail-order house 41 samples, each of 100 tickets, were audited, yielding a total of 123 with errors in recording the ordered item and other information. (The 41 samples showed homogeneity.) Set 90% confidence limits on the true $\phi$ for error rate. After intensive work for three months a set of 45 samples each of 100 tickets showed only 30 with errors. Set 90% confidence limits on the new $\phi$.

10.37.  A total of 36 defects was found on 29 pieces of woolen goods (each 100 yd long). Set 90% confidence limits on $\mu$ for 29 lengths. Then use this to set 90% limits on the $\mu$ for *single* lengths under the same conditions.

10.38.  From 30 bales of dry bleached hemlock pulp, a $9 \times 12$ in.$^2$ sheet was taken, and inspected. On the 30 sheets a count was made of the number of specks greater than .1 mm². The total was 580 such specks. Set 90% confidence limits on $\mu$ for 30 sheets and $\mu$ for one sheet, under the prevailing conditions.

10.39. Two different materials were used for tungsten composite contacts which were difficult to make. The first material had 72 cracked out of 163, while for the second material only 20 were cracked out of 105. Set 95% confidence limits on $\phi_1 - \phi_2$ and interpret.

10.40. In a production line making pails, on June 5 there were 48 defective out of 2056, while on June 6 there were 86 defective in 16,835. Set 95% confidence limits on the true difference in fractions defective, $\phi_1$, $\phi_2$, for the respective production conditions. (The last phrase was added because the data were for inspection of the entire output.)

10.41. Two consecutive lots of nuts "3/8–24" gave upon 100% inspection, 387 nonconforming in 3400 and 179 in 3100. Test $\phi_1 = \phi_2$ versus $\phi_1 \neq \phi_2$ at $\alpha = .01$ where $\phi_i$ is for the production process giving rise to the $i$th lot, $i = 1, 2$. Then set 90% confidence limits on $\phi_1 - \phi_2$. Interpret results.

10.42. Set 95% confidence limits on the true difference in proportion of "returns with an order" under the two advertising techniques given in Problem 9.20. Interpret your results to an official in the company.

10.43. For the data on conducting particles on condenser paper given in Problem 9.25, set 95% confidence limits on the difference between $\mu_1$, $\mu_2$, the respective true averages of particles on $n = 400$ ft² on the company's paper and the vendor's paper. Reduce your limits to a per square foot basis.

10.44. For the first 25 large aircraft assemblies the total of all types of defects found was 3898, while for the last 25 in a series, the total was 3468. Test for significance $\mu_1 = \mu_2$ versus $\mu_1 \neq \mu_2$. Then set 90% confidence limits on $\mu_1 - \mu_2$. Do both these where $\mu$ is the total defects for 25 assemblies, then in the latter reduce to limits on $\mu$ for defects on *one* assembly. Interpret results.

10.45. For the second part of the example of Section 9.5, set 90% confidence limits on $\mu_1 - \mu_2$, where $\mu$ is the true average number of missing rivets per aircraft assembly under the prevailing conditions. Data were $8 + 16$ versus $4 + 3$ for the first two assemblies and the last two. Interpret results.

10.46.   For the data in Problem 10.36, set 90% confidence limits on $\phi_1 - \phi_2$ and compare with the separate 90% confidence limits.

10.47.   In Problem 9.24 the first five aircraft subassemblies yielded a total of 854 defects of all kinds, while subassemblies numbered 50–54 yielded a total of only 593. Set 99% confidence limits on $\phi_1 - \phi_2$ where $\phi$ is for the average defects per one subassembly. (Compare with results on Problem 9.24 if it was assigned.)

10.48.   On a May 1, 84 defects were found on 30 farm implements at final inspection. Set 95% limits on the true $\mu$ for total defects on 30 implements under those conditions. On May 2 there were 137 defects on 30 implements. Set 95% limits on this $\mu$. Also set 95% limits on $\mu_1 - \mu_2$. Comment.

10.49.   Each member of the class is to draw three random samples of $n = 5$ $y$'s from a normal population and set 90% confidence limits on $\mu$. (Everyone should be most careful in calculation since the arithmetic will not be checked.) Then the number of intervals actually not containing $\mu$ for the population is counted. Test whether the proportion is significantly different from 10%. A convenient population, fairly normal, may be made by marking chips as follows:

| $y$ | $-5$ | $-4$ | $-3$ | $-2$ | $-1$ | 0 | $+1$ | $+2$ | $+3$ | $+4$ | $+5$ | Total |
|---|---|---|---|---|---|---|---|---|---|---|---|---|
| Frequency | 1 | 3 | 10 | 23 | 39 | 48 | 39 | 23 | 10 | 3 | 1 | 200 |

10.50.   Prove the equivalence of (a) two-way $1 - \alpha$ confidence limits for $\mu$, not containing $\mu_0$, and (b) rejecting $\mu = \mu_0$ in favor of $\mu \neq \mu_0$ at level $\alpha$, with $\sigma$ unknown. Also of (a) one-way lower $1 - \alpha$ confidence limit lying above $\mu_0$ and (b) rejecting $\mu = \mu_0$ in favor of $\mu > \mu_0$.

*Chapter 11*

# SIMPLE REGRESSION

## 11.1. REGRESSION, A STUDY OF RELATIONSHIP

We now change the scene somewhat in this and the next two chapters. Up to this point we have been concerned primarily with a single sample of observations $y$ of a random variable $Y$ or, perhaps, as in Chapter 9, two samples of similar random variables, $Y_1$, $Y_2$. Now we turn our attention to the relationship between a "dependent" random variable $Y$, and one or more "independent" random variables $X$, or $X_j$'s. In all three chapters a "best" equation may be sought. Or in Chapter 12, the independent variable or variables may be purely categorical, for example, people, and we seek to estimate differential "effects" which best explain the observed results. We still have the twin problems of estimation and testing of hypotheses. For example, in all three chapters there is some model proposed which involves random errors $\varepsilon_i$'s, and one of the first tasks after fitting the model is to estimate the strength of the random errors, that is, $\sigma_\varepsilon^2$. This can then be used in testing hypotheses on and interval estimation of such things as coefficients in a mathematical model. Also Chapter 12 considers inferences from a series of samples, that is, a generalization of Chapter 9 where we had just two samples.

The present chapter is concerned with models for $Y$ versus just one "independent variable" $X$. The model is often linear, but later in the chapter we consider nonlinear models.

## 11.2. THE SCATTER DIAGRAM

In studying the relationship between two random variables $X$ and $Y$, we first need data. We take an observation of $Y$, say $y_1$. Then we also measure the *corresponding* value of $X$, called $x_1$. There must be some linkage or tie-up : $x_i$, $y_i$ can be two different characteristics of the same sample of material, or person, or animal, or at the identical time. It would not do to call $x_1$ Paul's height and $y_1$ Helen's weight, and so forth. Nor would we obtain much in studying the thickness of electrolytic tin, $Y$, versus the electrical current $X$, if we obtain $x_1$ at 9 : 10 a.m. and $y_1$ at 11 : 15, and so on. The pair $x_i$, $y_i$ should be for the same person, or at as nearly the same time as possible. Thus we have "correlated" data.

It is very helpful to plot a scatter diagram to help us picture the data. For this we choose a suitable horizontal scale for $X$ and vertical scale for $Y$. Then we plot the points $(x_1, y_1)$, $(x_2, y_2)$, ..., $(x_n, y_n)$.

Let us consider an example of seven samples of road mixes for bituminous surface of highways. For each sample we measure $x_i$, the percent water absorbed, and $y_i$, the percent asphalt absorbed. The former is easier to measure. Can it be used to estimate the latter, and if so, how well? In any case, what sort of a relationship exists? The data follow:

| $X$, % water absorbed | 8.17 | 8.57 | 4.07 | 5.68 | 3.96 | 6.37 | 6.36 |
| $Y$, % asphalt absorbed | 6.95 | 7.94 | 3.44 | 4.14 | 3.02 | 5.86 | 4.88 |

How do the data of Fig. 11.1 look? Do you think $X$ and $Y$ are related? Would you hazard a guess as to what $Y$ is if $X$ is 6%? Or if 4%? Does the trend seem to be along a straight line or a curve? How much error would you expect in estimating $Y$ from a given $X$? These are some of the questions in whose answers we are interested. The author believes that most readers would answer yes to the second question and make respective guesses of about 5 and 3.2%. A straight line relationship seems quite reasonable. A typical error of estimate of $Y$ would appear to be .2–.3%. All of these answers are rather subjective. But they can be of much use; in fact the more closely the two variables are related, the more useful the scatter diagram becomes. The author well remembers a scatter diagram showing the percent fat in a type of meat versus the percent moisture. (Both are subject to legal limits.) The relationship was so surprisingly close and linear that further objective study was almost superfluous.

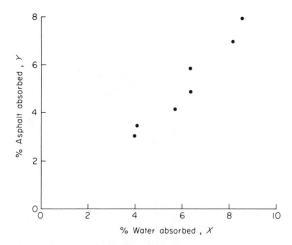

**Fig. 11.1.** Scatter diagram showing percentage of water $X$ and of asphalt $Y$ absorbed in seven samples of road mixes for bituminous surfaces.

Let us summarize by saying that the pattern of points in a scatter diagram may show

1.  how closely or loosely the two variables are related, that is, whether knowledge of the $X$ value enables us to estimate the value of $Y$ relatively accurately or poorly;
2.  whether the relationship seems to be along a straight line (i.e., linear) or along some curved line;
3.  the presence of any extreme or far-off-trend points;
4.  any other irregularities.

If we have only a small number up to 20 or 25 points, we are well advised to plot a scatter diagram to picture the data. But for larger samples we may wish to plot up only a sampling of the $(x_i, y_i)$ points.

## 11.3. LINE OF BEST FIT TO "LINEAR" DATA

We now seek a more objective approach to linear regression involving quantitative results, associated with a statistical model.

### 11.3.1. Least Squares Fitting.

Let us first consider the concept of obtaining a line which "best" fits a set of $X$, $Y$ points. Consider Fig. 11.2. How do you like the short-dashed line $A$? Does it not do a fine job on four of the points? But we

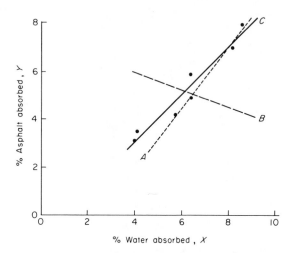

**Fig. 11.2.** The scatter diagram of Fig. 11.1, on percent asphalt and water absorbed by a road mix, with three lines drawn. Line $A$ fits four points well and three not well. Line $B$ does have the algebraic sum of the vertical distances, line to points, totaling zero, but is a *poor* fit. Line $C$ is the least squares line minimizing $\sum (y_i - \hat{y}_i)^2$.

want a line to fit *all* seven of the points as well as possible, in some sense. How about the long-dashed line $B$? It makes the algebraic sum of all of the *vertical* distances, from the line to the points, zero. It also makes the sum of the horizontal distances, from the line to the points, zero. But these properties are enjoyed by every *slanting* line which passes through the mean point $(\bar{x}, \bar{y}) \doteq (6.17, 5.18)$. No one in his right mind would choose $B$.

How then shall we *define* our "best" line and then proceed to find it? There are several ways:

1. Just as for the mean deviation (3.7), we could seek the line which minimizes the total sum of the vertical distances, calling them all plus, or of the horizontal distances.

2. We could seek the line which minimizes the sum of squares of the distances from the line to the points, *at right angles to the line*.

3. We could seek the line which minimizes the sum of the squares of the distances measured vertically from the line to the points, or of the horizontal distances.

4. Other criteria could be given.

We actually choose the first case of 3. Criterion 2 could be used, especially when there is considerable error of measurement in both $X$

and $Y$. Criteria in 1 are messy, requiring successive approximation. But the criterion we choose fits in with the model we shall be using, and is natural for variances and the normal distribution of errors.

The equation of a (nonvertical) straight line takes the form

$$Y = b_0 + b_1 X. \tag{11.1}$$

Our problem is then to find the "best" point estimates $b_0$, $b_1$ from the given data for the population regression values $\beta_0$, $\beta_1$ where the true regression line is

$$\mu_{Y.X} = E(Y \mid X) = \beta_0 + \beta_1 X. \tag{11.2}$$

Thus for any given $X$, the theoretical average value of the population of $Y$'s for this given $X$ is a linear function of $X$: $\beta_0 + \beta_1 X$.*

Now for any given values of $b_0$, $b_1$ leading to a straight line (11.1), we can measure how good a job it does by finding the sum of the squares of the *vertical* distances, line to points. For line $A$, we consider the first point (8.17, 6.95). If the full grid showed in our figure, we might read the estimated line value, for $x_1 = 8.17$, as $7.20 = \hat{y}_1$. Then the vertical distance from the line value 7.20 to the observed $y$ value 6.95 is $-.25$. This we would square, obtaining .0625. Proceeding in this manner we could find $\sum_{i=1}^{7}(y_i - \hat{y}_i)^2$, thus having a measure of how good a job line $A$ does. Similarly, for whatever line determined by the $b_0$, $b_1$ values we choose, we could find $\sum(y_i - \hat{y}_i)^2$. What we want is that combination of $b_0$ and $b_1$ which, for the data at hand, gives the smallest possible total $\sum(y_i - \hat{y}_i)^2$. It turns out in general that there is one such, and only one. But we cannot find it by trying one combination $(b_0, b_1)$ after another. The natural way to proceed is to use calculus (though it can be done purely algebraically, with a little more effort).

Consider the sum of squares of the vertical deviations which we wish to minimize:

$$\sum(y_i - \hat{y}_i)^2 = \sum(y_i - b_0 - b_1 x_i)^2 = S(b_0, b_1). \tag{11.3}$$

The $S(b_0, b_1)$ emphasizes that, *for the data at hand*, the sum of squares

---

* The student should carefully distinguish between the use of $\beta$ as a risk of making an error of the second kind, and the use here introduced by $\beta_0$, $\beta_1$ as exact coefficients in the model. Such subscripts as 0, 1, 2 will indicate the second usage, even though we occasionally use a subscript on risk $\beta$, as, for example, $\beta_\mu$ to show that it is the probability of accepting a null hypothesis when the mean is $\mu$.

is dependent only on $b_0$, $b_1$. Minimizing $S$ leads directly to the following*

$$b_1 = \frac{\sum x_i y_i - [(\sum x_i)(\sum y_i)/n]}{\sum x_i^2 - [(\sum x_i)^2/n]}$$                    (11.4)

$$b_0 = \bar{y} - b_1 \bar{x}.$$                    (11.5)

One should note that the expressions in (11.4) are closely related to the standard deviation (3.11):

$$s = \sqrt{\frac{n \sum y_i^2 - (\sum y_i)^2}{n(n-1)}} = \sqrt{\frac{\sum y_i^2 - [(\sum y_i)^2/n]}{n-1}}.$$                    (11.9)

Although the middle expression here is often more simple for calculations, the latter expression is convenient in our work since it gives

$$\sum y_i^2 - \left[\left(\sum y_i\right)^2/n\right] = \sum (y_i - \bar{y})^2.$$                    (11.10)

Also

$$\sum x_i y_i - \left[\left(\sum x_i\right)\left(\sum y_i\right)/n\right] = \sum (x_i - \bar{x})(y_i - \bar{y}).$$                    (11.11)

### 11.3.2. Calculational Aspects.

The author is well aware that one can readily program a digital computer to do the job of computing $b_0$ and $b_1$ by (11.4) and (11.5) from given data, as well as the quantities needed in subsequent sections. Nevertheless, it seems well worthwhile for the student to learn to calculate such things efficiently via a desk calculator. There are several reasons for this. It helps the student get the "feel" of the data, which aids a good deal in interpretation of results. Then too, many problems are so simply done on a desk calculator that it becomes a real question

---

* By squaring the trinomial in the middle expression of (11.3) and using the laws of summations (3.2)–(3.4), we find

$$S(b_0, b_1) = \sum y_i^2 + b_0^2 n + b_1^2 \sum x_i^2 - 2b_0 \sum y_i - 2b_1 \sum x_i y_i + 2b_0 b_1 \sum x_i.$$                    (11.6)

Differentiating partially with respect to $b_1$ and $b_0$, and setting equal to zero we find

$$b_0 \sum x_i + b_1 \sum x_i^2 = \sum x_i y_i$$                    (11.7)

$$b_0 n + b_1 \sum x_i = \sum y_i.$$                    (11.8)

To find (11.4) from (11.7) and (11.8) we have but to eliminate $b_0$ by multiplying (11.8) by $\sum x_i/n$ and subtracting from (11.7). Then (11.5) comes immediately from (11.8) by solving for $b_0$ in terms of $b_1$. (The only way we can fail to have a unique solution is to have all $x$'s identical.)

whether to bother punching cards and seeking out an appropriate "canned" program. Finally a large-scale computer is not always even available.

For finding $b_0$ and $b_1$ by (11.4) and (11.5) we need $\sum x_i$, $\sum x_i^2$, $\sum y_i$, $\sum x_i y_i$. On most desk calculators it is possible to accumulate all of these, as well as $\sum y_i^2$, in just one single time through the data. To do this we square the number 8170000695. This gives the result 66.7489113.5630048. 3025, where we have placed decimal points appropriately. In this rather "gruesome" square, 66.7489 is $8.17^2$, 48.3025 is $6.95^2$, and 113.5630 is 2(8.17)6.95. Obviously we need a 10-bank machine for this, otherwise the numbers will merge and be lost. Meanwhile 8170000695 will appear in another dial in the machine. Now take the next pair 8.57 and 7.94. Square 8570000794, adding in to the previous large square. One dial now reads 140.1938,249.6546,111.3461, the commas showing the grouping; and the other dial reads 16740001489 (if there are 11 places in it). We are thus forming $\sum x_i^2$, $2\sum x_i y_i$, $\sum y_i^2$ in one dial and $\sum x_i$, $\sum y_i$ in the other. Table 11.1 shows some individual entries and the sums. To obtain $\sum x_i y_i$, we have only to divide the middle entry by 2.

**TABLE 11.1**

Percents Asphalt and Water Absorbed by Seven Samples of Road Mixes for Bituminous Surface. Regression Calculations[a]

| % Water $x$ | % Asphalt $y$ | $x^2$ | $y^2$ | $xy$ |
|---|---|---|---|---|
| 8.17 | 6.95 | 66.7489 | 48.3025 | 56.7815 |
| 8.57 | 7.94 | 73.4449 | 63.0436 | 68.0458 |
| 4.07 | 3.44 | | | |
| 5.68 | 4.14 | | | |
| 3.96 | 3.02 | | | |
| 6.37 | 5.86 | | | |
| 6.36 | 4.88 | | | |
| 43.18 | 36.23 | 285.7292 | 207.5937 | 242.6675 |
| $\sum x_i$ | $\sum y_i$ | $\sum x_i^2$ | $\sum y_i^2$ | $\sum x_i y_i$ |

$$b_1 = \frac{242.6675 - 43.18(36.23)/7}{285.7292 - 43.18^2/7} = \frac{19.1802}{19.3703} = .99018$$

$\bar{x} = 43.18/7 = 6.1686, \quad \hat{y} = 36.23/7 = 5.1757$

$b_0 = 5.1757 - .99018(6.1686) = -.9323$

Estimation equation: $\hat{y} = -.9323 + .9902x$, line $C$, Fig. 11.2

[a] Entries in last three columns should not be recorded, the totals only accumulated, if a desk calculator is available.

It is often possible to code the $x$'s into $v$'s as in (3.12) and $y$'s into $w$'s by $w = (y - B)/d$, perform the calculations, and decode. (See the details in Section 11.7.) Or, if there are three active (varying) digits in the $x$'s and/or $y$'s we can round off the last digit with surprisingly little loss in accuracy. Thus in the data we could have used $x = 8.2, 8.6, ...,$ and $y = 7.0, 7.9, ...$ .

In the techniques to follow there are many shortcuts available, using desk calculators, which should be developed. Space forbids describing them, since they vary a bit from one type of calculator to another. One point, however, is that it is highly desirable to retain many places of precision as we go along, then round off as desirable at the end.

### 11.3.3. The Linear Model and Its Parameters.

Let us now describe the linear model we shall be using in subsequent sections. We assume a population regression as given in (11.2), namely,

$$\mu_{Y.X} = \beta_0 + \beta_1 X. \tag{11.2}$$

This means that for each $X$ there is a distribution of $Y$'s, and that the mean of this distribution is the linear function given above. Secondly, we assume that the variance of each such distribution of $Y$'s is $\sigma_\varepsilon^2$, which therefore does not depend on $X$ at all. This is called "homoscedasticity." Finally we say that the distribution of $Y$'s around the average (11.2) for any one given $X$ is always normal, and the errors $\varepsilon_i$ are independent.

Hence we write

$$Y = \beta_0 + \beta_1 X + \varepsilon, \tag{11.12}$$

where $\varepsilon$ is the vertical error taking us from the straight line to the observed value of $Y$ for a given $X$. Then our assumptions may be summarized by saying*

$$\varepsilon: \text{NID}(0, \sigma_\varepsilon^2), \qquad Y: \text{NID}(\beta_0 + \beta_1 X, \sigma_\varepsilon^2). \tag{11.13}$$

Now what are the parameters of this model? There are three: (1) the slope $\beta_1$, (2) the intercept $\beta_0$, and (3) the error variance $\sigma_\varepsilon^2$. It is our job to learn how to estimate each of these three parameters as well as possible.

Another assumption usually made is that the $x$'s are fixed, that is, we may ask what happens to our three sample estimates $b_0$, $b_1$, and $s_\varepsilon^2$, from sample to sample of $n$ pairs of $(x_i, y_i)$, when we always have the

---

* NID means normally and independently distributed, the first number in the parentheses being the mean, the second the variance.

same set of $x_i$'s. Thus the collection of $x$'s is like a population. As pointed out by Crow [1], this last assumption may be relaxed somewhat without hurting interpretations.

## 11.4. SAMPLING DISTRIBUTIONS FOR ESTIMATES

We have already introduced the estimators for $\beta_0$ and $\beta_1$ in (11.4) and (11.5). We now need an estimator for the error variance $\sigma_\varepsilon^2$ around the true regression line (11.2). The $\varepsilon$'s show up in (11.12) as

$$y_i = \beta_0 + \beta_1 x_i + \varepsilon_i. \tag{11.14}$$

But we do not know $\beta_0$ and $\beta_1$. Hence we cannot know the true error $\varepsilon_i$, even though we know the observed $x_i$ and $y_i$. So we go to

$$y_i = b_0 + b_1 x_i + e_i. \tag{11.15}$$

Compare these equations carefully, $x_i$, $y_i$ are the same in both, but the others are all estimates in (11.15). It was $\sum e_i^2$ which we minimized in the least squares approach to finding estimates $b_0$ and $b_1$.

For finding $\sigma_\varepsilon^2$ we would like an *infinite* collection of $\varepsilon$'s from which to obtain $\sigma_\varepsilon^2$. Instead we must settle for $n$ values of $e_i$'s, and from $\sum_{i=1}^n e_i^2$. Now what shall we divide by? For an estimate of $\sigma_Y^2$, if we had $n$ values of $y_i - \bar{y}$, we would use $\sum(y_i - \bar{y})^2/(n-1)$, the $-1$ being for the one parameter $\mu_Y$ estimated by $\bar{y}$. But for $\sum e_i^2$ there are two parameters being estimated, $\beta_0$ and $\beta_1$ via $b_0$ and $b_1$, since by (11.15)

$$\sum e_i^2 = \sum (y_i - b_0 - b_1 x_i)^2.$$

Hence the appropriate denominator is $n-2$. We therefore define

$$s_\varepsilon^2 = \frac{\sum_{i=1}^n (y_i - b_0 - b_1 x_i)^2}{n-2}. \tag{11.16}$$

This is our estimator of $\sigma_\varepsilon^2$.

It may readily be shown that under our assumptions in Section 11.3.3, the following are the distributions:

$$b_1: N(\beta_1, \sigma_{b_1}^2), \qquad \sigma_{b_1}^2 = \sigma_\varepsilon^2 / \sum (X_i - \bar{X})^2 \tag{11.17}$$

$$b_0: N(\beta_0, \sigma_{b_0}^2), \qquad \sigma_{b_0}^2 = \sigma_\varepsilon^2 \left[ 1/n + \bar{X}^2 / \sum (X_i - \bar{X})^2 \right] \tag{11.18}$$

$$= \sigma_\varepsilon^2 \left[ \sum X_i^2 / n \sum (X_i - \bar{X})^2 \right]$$

$$s_\varepsilon^2: \frac{(n-2) s_\varepsilon^2}{\sigma_\varepsilon^2} = \chi_{n-2}^2 \qquad \text{chi-square with } n-2 \text{ degrees of freedom.} \tag{11.19}$$

We easily note that $E(b_1) = \beta_1$ and $E(b_0) = \beta_0$, and so $b_0$ and $b_1$, respectively, are, unbiased estimators of $\beta_0$ and $\beta_1$. Since it may be shown that $E(\chi_f^2) = f$ (degrees of freedom), we see that $E(s_\varepsilon^2) = \sigma_\varepsilon^2$, so that $s_\varepsilon^2$ is an unbiased estimator of $\sigma_\varepsilon^2$.* These estimators also have the other desirable characteristics of estimators, if our assumptions are met.

Now we have not given the whole story with (11.17)–(11.19). There is the question of interrelationship or independency. It may be shown that $s_\varepsilon^2$ is independent of $b_0$ and $b_1$. However, $b_0$ and $b_1$ are not independent. This is quite apparent, because $b_0 = \bar{y} - b_1\bar{x}$, so that $b_1$ in part determines $b_0$. To achieve independency we may proceed as follows, by writing the regression equation as

$$\mu_{Y.X} = E(Y \mid X) = \mu_Y + \beta_1(X - \bar{X}) \tag{11.20}$$

instead of (11.2). Here $\bar{x}$ could also be called $\mu_X$ because we assume a fixed collection of $X$'s, that is, our population.

The estimated regression line corresponding to (11.20) is easily derived from (11.15) by substituting in (11.5), obtaining

$$y_i = \bar{y} + b_1(x_i - \bar{x}) + e_i, \tag{11.21}$$

or the estimator of $\mu_{Y.X}$ in (11.20) is

$$\hat{y}_x = \bar{y} + b_1(x - \bar{x}). \tag{11.22}$$

Then our estimators for regression line (11.20) are $\bar{y}$, $b_1$, and $s_\varepsilon^2$. All three of these are mutually independent, and (11.17) and (11.19) still hold. But now we have for the third estimator

$$\bar{Y}: N(\mu_Y, \sigma_\varepsilon^2/n). \tag{11.23}$$

Therefore $\bar{Y}$ is an unbiased estimator of $\mu_Y$, and so on. The variance of $\bar{Y}$ is just the first part of that for $b_0$.

In general, we do not know $\sigma_\varepsilon^2$, so that instead of normal-curve tests for $b_0$, $b_1$, or $\bar{y}$, we must use $s_\varepsilon^2$ instead, and this forces us to use $t$ tests. Tests on $s_\varepsilon^2$ are strictly analogous to those for $s^2$ versus $\sigma^2$, using $\chi^2$. Hence from (11.17), (11.18), and (11.23)

$$t_{n-2} = \frac{b_1 - \beta_1}{s_{b_1}} = \frac{b_1 - \beta_1}{s_\varepsilon/\sqrt{\sum (x - \bar{x})^2}} \tag{11.24}$$

---

* Of course it does not follow that $s_\varepsilon$ is an unbiased estimator of $\sigma_\varepsilon$, just as $s$ is not an unbiased estimator of $\sigma$.

$$t_{n-2} = \frac{b_0 - \beta_0}{s_{b_0}} = \frac{b_0 - \beta_0}{s_\varepsilon[\sum x^2/n \sum (x - \bar{x})^2]^{1/2}} \qquad (11.25)$$

$$t_{n-2} = \frac{\bar{y} - \mu_Y}{s_\varepsilon/\sqrt{n}}. \qquad (11.26)$$

The degrees of freedom are $n - 2$ in all three cases, these coming from $s_\varepsilon$.

## 11.5. SIGNIFICANCE TESTS AND CONFIDENCE INTERVALS FOR PARAMETERS IN LINEAR REGRESSION

We are now in a position to draw a variety of inferences about the regression, from our observed data. In fact it is certainly desirable also at this time to carry along some numerical results for illustration, parallel to the general formulas.

In the example we have already obtained three estimates (see Table 11.1):

$$b_1 = .99018, \qquad b_0 = -.9323, \qquad \bar{y} = 5.1757.$$

We next need $s_\varepsilon^2$. Although one could use (11.16), this would involve finding the estimated $\hat{y}_i = b_0 + b_1 x_i$, subtracting these from $y_i$, and finding the sum of squares of these differences. This can be a chore for any size of $n$, especially large ones. Instead we can use

$$s_\varepsilon^2 = \left\{ \sum y_i^2 - \left( \sum y_i \right)^2 / n - \frac{[\sum x_i y_i - (\sum x_i)(\sum y_i)/n]^2}{\sum x_i^2 - (\sum x_i)^2/n} \right\} / (n - 2). \qquad (11.27)$$

(It looks worse but is easier, using summations readily available.) If we let $x_i' = x_i - \bar{x}$, $y_i' = y_i - \bar{y}$, we can write (11.27) compactly as

$$s_\varepsilon^2 = \left[ \sum y_i'^2 - \left( \sum x_i' y_i' \right)^2 / \sum x_i'^2 \right] / (n - 2). \qquad (11.28)$$

[But again (11.28) is not used for computation.]

For our example we have from (11.27) and Table 11.1:

$$s_\varepsilon^2 = \left\{ 207.5937 - (36.23)^2/7 - \frac{[242.6675 - 43.18(36.23)/7]^2}{285.7292 - 43.18^2/7} \right\} / (7 - 2)$$

$$= \left( 20.0776 - \frac{19.1802^2}{19.3703} \right) / 5 = (20.0776 - 18.9920)/5 = .21712.$$

Let us compare this with $s_y^2 = 20.0776/6 = 3.3463$. The last is an estimate of a typical squared deviation of $Y$ from $\mu_Y$, while .21712 is that for squared deviations from the true regression line (11.2), about one-fifteenth as much. Or, if we take the square root, $s_\varepsilon = .4660$ is about one-fourth as great as $s_y = 1.8293$.

### 11.5.1. Slope.

We may test the hypothesis that $\beta_1$ has any given value, say $\beta_{10}$. Such a hypothetical slope may have a value suggested by previous data (to see whether there is a change), by some reasonable physical or other theory, or by common sense. One very important hypothesis is $\beta_1 = 0$. If we cannot disprove that the slope $\beta_1$ is 0, then we have no *reliable* evidence in the data to suppose that $Y$ is linearly related to $X$ at all. Our model might then be simplified to $Y = \beta_0 + \varepsilon$ where $\beta_0$ will be estimated by $\overline{Y}$, unless the scatter diagram shows some nonlinear trend.

Let us test at the $5\%$ significance level two such hypothesis. First

$$\beta_1 = 0 \qquad \text{versus} \qquad \beta_1 \neq 0.$$

Use (11.24) in the form

$$t_5 = \frac{b_1}{s_\varepsilon/\sqrt{\sum x_i^2 - (\sum x_i)^2/n}} = \frac{.99018}{.4660/\sqrt{19.3703}} = 9.35.$$

This is of course extremely significant, and we conclude that $\beta_1 > 0$ and that $Y$ is linearly related to $X$. (This does not rule out all possibility of some nonlinearity too. In fact perhaps most straight line trends are just short pieces of curves!) Next use (11.24) for

$$\beta_1 = 1 \qquad \text{versus} \qquad \beta_1 \neq 1,$$

$$t_5 = \frac{.99018 - 1}{.4660/\sqrt{19.3703}} = \frac{-.00982}{.1059} = -.093.$$

Since the boundaries of the critical region are $\pm 2.571$ we can accept the hypothesis $\beta_1 = 1$.

Next let us set $95\%$ confidence limits for $\beta_1$. We use (11.24) in the same way as we set confidence limits for $\mu_Y$ [see (10.17)]. We have

$$b_1 \pm (t_{\alpha/2,\,\mathrm{df}\,n-2})\, s_\varepsilon/\sqrt{\sum (x_1 - \bar{x})^2}, \qquad 1 - \alpha \quad \text{limits for } \beta_1. \qquad (11.29)$$

Thus we find $95\%$ confidence limits for $\beta$ as

$$.99018 \pm 2.571(.1059) = .718, \quad 1.262.$$

These are relatively close together and include $\beta_1 = 1$, but exclude $\beta_1 = 0$ by a wide margin (the fruit of such a big $t$ as 9.35 before).

### 11.5.2. Intercept.

Often the intercept is of rather slight interest, especially if it is far removed from the data at hand. We can, however, do things analogous to those for the slope $\beta_1$. All we need is to use (11.25). For it we find

$$s_{b_0} = .4660\sqrt{285.7292/[7(19.3703)]} = .6764.$$

One hypothetical value of the intercept $\beta_0$ of potential practical interost is zero. If the hypothesis $\beta_0 = 0$ is tenable, then it might be desirable to fit the regression equation $Y = \beta X + \varepsilon$. (See Section 11.8). Let us make the test from (11.25):

$$t_5 = \frac{b_0}{s_{b_0}} = \frac{-.9323}{.6764} = -1.38.$$

Comparing again with $\pm 2.571$ we find $\beta_0 = 0$ to be tenable. But since the $Y$ percentages seem to be running about $1\%$ below the $X$'s it seems quite reasonable to retain the calculated $b_0$ in the equation. (If, however, we were to adopt the model $Y = \beta X + \varepsilon$, we would find a new $b_1$ estimate of .8493.)

For $1 - \alpha$ confidence limits on $\beta_0$ we have

$$b_0 \pm (t_{\alpha/2,\,\mathrm{df}n-2})\, s_{b_0},$$

from which we obtain

$$-.9323 \pm 2.571(.6764) = -2.671 + .807,$$

with $95\%$ confidence that $\beta_0$ lies between. Note that $\beta_0 = 0$ lies between the limits.

### 11.5.3. Mean of Y's: $\mu_Y$.

The same approaches may be used for $\overline{Y}$ versus $\mu_Y$ via (11.26). We shall only illustrate the most useful one here. We have $95\%$ confidence that $\mu_Y$ lies between

$$\bar{y} \pm (t_{\alpha/2,\,\mathrm{df}n-2})\, s_\varepsilon/\sqrt{n} = 5.1757 \pm 2.571(.4660)/\sqrt{7}$$
$$= 4.72,\quad 5.63.$$

### 11.5.4. Error Variance, $\sigma_\varepsilon{}^2$, and $\sigma_\varepsilon$.

For significance tests on $\sigma_\varepsilon{}^2$ we may proceed to use (11.19) exactly as in Section 8.8. However, reasonable hypothetical values of $\sigma^2$ are not commonly known, and so a more common approach is to use confidence intervals for $\sigma_\varepsilon^2$ and/or $\sigma_\varepsilon$. For these we proceed as in Section 10.7 and (10.20). Thus $1 - \alpha$ confidence limits on $\sigma_\varepsilon$ are

$$s_\varepsilon \sqrt{(n-2)/\chi^2_{1-(\alpha/2) \text{ below, } n-2}},$$

$$s_\varepsilon \sqrt{(n-2)/\chi^2_{\alpha/2 \text{ below, } n-2}}. \qquad (11.30)$$

We find for 95% limits on $\sigma_\varepsilon$:

$$.4660\sqrt{5/12.833} = .291, \qquad .4660\sqrt{5/.831} = 1.143.$$

These may seem relatively quite far apart but this is because of the small sample size. For limits on $\sigma_\varepsilon^2$, we have but to use the squares of the limits for $\sigma_\varepsilon$.

### 11.5.5. Regression Line Mean: $\mu_{Y.X} = \mu_Y + \beta_1(X - \bar{X})$.

We may sometimes wish to set confidence limits on the regression line mean, of the population of $Y$'s for a given $X$. The theoretical variance is $\sigma_{\hat{Y}_X}^2 = \sigma_\varepsilon^2[1/n + (X - \bar{X})^2/\sum(X - \bar{X})^2]$. And $\hat{y}_x = \bar{y} + b_1 (x - \bar{x})$ is an unbiased estimate of $\mu_{Y.X} = \mu_Y + \beta(X - \bar{X})$, and is normally distributed. Since $\sigma_\varepsilon^2$ is not known we use the $t$ distribution:

$$t_{n-2} = \frac{\bar{y} + b(x - \bar{x}) - \mu_{Y.X}}{s_\varepsilon \sqrt{1/n + (x - \bar{x})^2/\sum (x - \bar{x})^2}}. \qquad (11.31)$$

Then $1 - \alpha$ confidence limits on $\mu_{Y.X}$, the true regression mean, are

$$\bar{y} + b_1(x - \bar{x}) \pm (t_{\alpha/2, n-2\text{df}})(\sqrt{1/n + (x - \bar{x})^2/\sum (x - \bar{x})^2})\, s_\varepsilon. \qquad (11.32)$$

It is readily seen that such limits are closest together when $x = \bar{x}$, and that they grow farther apart as $x$ recedes from $\bar{x}$ in either direction. This is natural because the further away we go from $\bar{x}$, the more magnified any error of $b_1$, as an estimate of $\beta_1$, becomes.

One could use (11.32) to test a hypothesis on $\mu_{Y.X}$, but this is seldom done. Also by adding a 1 under the radical in (11.31) and (11.32) one can set confidence limits for a single $y$ for a given $x$. But this seems to the author to be not especially useful in applications, though it does say something about the variation of individual $y$'s around the estimated $\hat{y}$.

### 11.6. CORRELATIONAL ASPECTS

As a supplement to the foregoing regression approach we now discuss the correlation coefficient. This coefficient is often used to describe how relatively good a fit to data is provided by the linear model.

The definition of the sample correlation coefficient $r$ is given most simply by using $x_i' = x_i - \bar{x}$, $y_i' = y_i - \bar{y}$:

$$r_{xy} = \frac{\sum x_i' y_i'}{\sqrt{\sum x_i'^2 \sum y_i'^2}} \tag{11.33}$$

$$r_{xy} = \frac{\sum x_i y_i - (\sum x_i)(\sum y_i)/n}{\sqrt{\sum x_i^2 - [(\sum x_i)^2/n]} \; \sqrt{\sum y_i^2 - [(\sum y_i)^2/n]}} . \tag{11.34}$$

This coefficient is always a dimensionless number, that is, no units. It carries the same sign as does $b_1$. The $r$ always lies between $-1$ and $+1$, the extremes being reached if and only if all points lie exactly on a slanting straight line. (Why not vertical or horizontal?) The simplest interpretation for $r$ comes from $r^2$, namely,

$$\begin{aligned} r^2 &= \text{the proportion of } \sum(y_i - \bar{y})^2 \text{ which is} \\ &\quad \text{linearly relatable to } x \\ &= \text{coefficient of determination.} \end{aligned} \tag{11.35}$$

If we reexamine (11.28), we can see that $\sum(y_i - \bar{y})^2 = \sum y_i'^2$ is the total "variation" in $Y$. Meanwhile that part "explained" by $x$ via the regression line is $(\sum x_i' y_i')^2/\sum x_i'^2$. The difference, which is the part "not explained" by the linear model, is what goes into the error variance $s_\varepsilon^2$. Now if we divide the former $(\sum x_i' y_i')^2/\sum x_i'^2$ by $\sum y_i'^2$ to obtain the proportion explained, we have $(\sum x_i' y_i')^2/\sum x_i'^2 \sum y_i'^2$. But by (11.33), this is $r^2$.

When the points lie very close to a straight line, the explained part of $\sum y_i'^2$ is relatively large and $r^2$ is near 1; meanwhile $s_\varepsilon^2$ is small. But when the points scatter relatively widely from the straight line, the unexplained part of $\sum y_i'^2$ is much larger than the explained, and $r^2$ is small. From the above we may also see that $r^2$ lies between 0 and 1, and thus $r$ ranges from $-1$ to $+1$.

Also we may note that

$$s_\varepsilon^2 = \sum y_i'^2 (1 - r^2)/(n - 2).$$

An interesting interpretation of $r$ is the following, using standardized values of $x$ and $y$, say,

$$u_x = (x - \bar{x})/s_x, \qquad u_y = (y - \bar{y})/s_y .$$

Then we have

$$\hat{u}_y = r u_x .$$

By using $r$ we can make a test exactly identical to the test of the hypothesis $\beta_1 = 0$ versus $\beta_1 \neq 0$. Letting the population correlation coefficient be designated $\rho$, then $\beta_1 = 0$ is identical to $\rho = 0$. Testing this versus $\rho \neq 0$ we would use

$$t_{n-2} = \frac{r}{\sqrt{(1 - r^2)/(n - 2)}}. \tag{11.36}$$

In our example we find from (11.34)

$$r = \frac{19.1802}{\sqrt{19.3703}\,\sqrt{20.0776}} = +.9726.$$

This is a high positive correlation of course. Also the coefficient of determination is

$$r^2 = .9460.$$

Hence nearly 95% of $\sum(y_i - \bar{y})^2$ is linearly relatable to $x$; that is, 95% of the variation in $y$ is thus explained. Let us use (11.36). Then

$$t_5 = \frac{.9726}{\sqrt{1 - .9460}\,/\,\sqrt{5}} = 9.35,$$

just as we had in testing $\beta_1 = 0$.

Another good property enjoyed by $r$ is that if $x$ and $y$ are subjected to any linear transformations, such as coding, with *positive* coefficients, the value of $r$ is unaffected.

### 11.7.* GROUPED BIVARIATE DATA

This is the two-variable analogy of frequency tables such as we studied in Chapter 2. Such a table is especially useful when we have a large number of pairs $(x_i, y_i)$. The first step is to choose a suitable set of class limits for $x$'s and also for $y$'s. See Table 11.2. Only the midvalues rather than class limits are shown there. In tallying such data we take a pair of readings such as $x = 65$, $y = 547$. This would be tallied within the cell having midvalues 64 and 545.5. It would be one of the 13 such pairs occurring in this cell.

The data shown in Table 11.2 were part of a study concerned with the diameters of metal foil condensers. Here $X$ is the total foil thickness. (Two other "independent" variables were studied also, namely, total paper thickness and the number of turns.) Diameters of these condensers were an important characteristic to control.

The table shows the details of coding both $X$ and $Y$. They were coded so that neither of the coded variables $v$ and $w$ were ever negative. This

TABLE 11.2

Condenser Diameters $Y$ in .001 in. Units versus Total Foil Thickness $X$ in .00001 in. Units

| Midvalues $y$ | 52 | 55 | 58 | 61 | 64 | 67 | 70 | 73 | 76 | 79 | 82 | $f_y$ | $w$ |
|---|---|---|---|---|---|---|---|---|---|---|---|---|---|
| 585.5 | — | — | — | — | — | — | — | 1 | 1 | 1 | 1 | 4 | 8 |
| 577.5 | — | — | — | — | 1 | 1 | — | 3 | 4 | 1 | — | 10 | 7 |
| 569.5 | — | — | — | 1 | 1 | 4 | 6 | 4 | 10 | 3 | — | 29 | 6 |
| 561.5 | — | — | — | 3 | 3 | 12 | 11 | 3 | 2 | 1 | 1 | 36 | 5 |
| 553.5 | — | — | 1 | 2 | 12 | 18 | 11 | 7 | 4 | — | — | 55 | 4 |
| 545.5 | — | 2 | — | 2 | 13 | 8 | 2 | 2 | 1 | — | — | 30 | 3 |
| 537.5 | 1 | 4 | — | 1 | — | 1 | — | — | — | — | — | 7 | 2 |
| 529.5 | 1 | — | 1 | 1 | — | 1 | — | — | — | — | — | 4 | 1 |
| 521.5 | — | 1 | 3 | — | — | — | — | — | — | — | — | 4 | 0 |
| $f_x$ | 2 | 7 | 5 | 10 | 30 | 45 | 30 | 20 | 22 | 6 | 2 | 179 | |
| $v$ | 0 | 1 | 2 | 3 | 4 | 5 | 6 | 7 | 8 | 9 | 10 | | |
| $\sum wf_{xy}$ col. | 3 | 14 | 5 | 38 | 115 | 190 | 141 | 102 | 125 | 38 | 13 | 784 | |

For example in the last row, $v = 4$, $\quad \sum wf_{xy} = 3(13) + 4(12) + 5(3) + 6(1) + 7(1) = 115$

$\sum vf_x = 962, \qquad \sum v^2 f_x = 5876, \qquad \sum wf_y = 784, \qquad \sum w^2 f_y = 3872$

$\sum vwf_{xy} = \sum v \sum_{\text{col.}} wf = 0(3) + 1(14) + 2(5) + 3(38) + \cdots + 10(13) = 4580$

Coding equations: $\quad v = (x - A)/c = (x - 52)/3, \qquad x = 52 + 3v$

$\qquad\qquad\qquad\quad w = (y - B)/d = (y - 521.5)/8, \qquad y = 521.5 + 8w$

$\bar{x} = 52 + 3\bar{v} = 52 + 3(962)/179 = 68.123$

$\bar{y} = 521.5 + 8\bar{w} = 521.5 + 8(784)/179 = 556.539$

$$\sum (x - \bar{x})^2 f_x = c^2 \sum (v - \bar{v})^2 f_x = c^2 \left[ \sum v^2 f_x - \frac{(\sum vf_x)^2}{n} \right]$$

$$= 3^2 \left( 5876 - \frac{962^2}{179} \right) = 6353.3$$

$$\sum (y - \bar{y})^2 f_y = d^2 \sum (w - \bar{w})^2 f_y = d^2 \left[ \sum w^2 f_y - \frac{(\sum wf_y)^2}{n} \right]$$

$$= 8^2 \left( 3872 - \frac{784^2}{179} \right) = 28043$$

$$\sum (x - \bar{x})(y - \bar{y}) f_{xy} = cd \sum (v - \bar{v})(w - \bar{w}) f_{xy} = cd \left[ \sum vwf_{xy} - \frac{(\sum vf_x)(\sum wf_y)}{n} \right]$$

$$= 3(8)[4580 - 962(784)/179] = 8797.1$$

facilitates use of desk calculators, so that we can simultaneously cumulate $\Sigma f_x$, $\Sigma v f_x$, $\Sigma v^2 f_x$. This is done by multiplying $v$ placed in the leftmost columns and $v^2$ in the rightmost columns of the keyboard and multiplying by $f_x$, then cumulating. (If long-hand calculation is being used, midvalues toward the center might be used, for example, 67 and 553.5.) Since $\Sigma v w f_{xy}$ involves a product of three numbers it is perhaps easiest to form $\Sigma w f_{xy}$ for each $v$ separately as shown at the bottom of the table and then to cumulate the products with $v$, as shown. The sum of the separate column product sums should be $\Sigma w f_y = 784$ here. Then the calculations are shown for the means $\bar{x}$, $\bar{y}$, and for the sums of squares and of products (of deviations from the means). These are the basic elements needed for the regression analysis. The continuing calculations follow:

$$b_1 = \frac{\Sigma (x - \bar{x})(y - \bar{y}) f_{xy}}{\Sigma (x - \bar{x})^2 f_x} = \frac{8797.1}{6353.3} = 1.3847$$

$$b_0 = \bar{y} - b_1 \bar{x} = 556.539 - 1.3847(68.123) = 462.2$$

$$\hat{y} = 462.2 + 1.3847x.$$

Using (11.27) we find

$$s_\varepsilon^2 = [28043 - (8797.1)^2/6353.3]/177 = 15862/177 = 89.616$$

$$s_\varepsilon = 9.4666.$$

The correlation coefficient is found from (11.34) to be

$$r = \frac{8797.1}{\sqrt{6353.3}\,\sqrt{28043}} = +.6591.$$

We may also find in the usual way:

$$s_y^2 = 28043/178 = 157.54, \qquad s_y = 12.552.$$

Thus we see that our estimate of $\sigma_\varepsilon^2$, the variance around the theoretical straight line of regression, is 89.616, whereas that for all $y$'s (when we have such a collection of $x$'s) is 157.54, which is much larger.

Our $r$ value indicates a substantial positive linear relationship. The coefficient of determination $r^2$ is .4344; that is, about 43 % of the variation in $y$ is linearly relatable to $x$.

We now give some tests and interval estimates that would seem to be of use.

First test the hypothesis $\beta_1 = 0$ versus $\beta_1 \neq 0$ by (11.24) at $\alpha = .05$,

$$t_{177} = \frac{1.3847 - 0}{9.4666/\sqrt{6353.3}} = 11.66.$$

This is to be compared to $\pm 1.96$ from the bottom line of Table I in the Appendix. We thus have very clear evidence of there being linearity of relation between $y$ and $x$. But even with such a very high $t$ we cannot say that there can be no nonlinearity present. We could also have used (11.36), testing $\rho = 0$ versus $\rho \neq 0$:

$$t_{177} = \frac{.6591\sqrt{177}}{\sqrt{1 - .6591^2}} = 11.66$$

as before. The practical problem does not suggest any reasonable hypothetical value of $\beta_1$, but we will set 90% confidence limits for $\beta_1$ by (11.29);

$$1.3847 \pm 1.645(9.4666)/\sqrt{6353.3} = 1.3847 \pm .1954 = 1.19, \quad 1.58.$$

For 90% confidence limits for $\sigma_\varepsilon$, we cannot use (11.30) because $n - 2$ is too large, so we resort to (10.23):

$$\frac{9.4666}{1 + 1.645/\sqrt{358}} = 8.71 \quad \text{and} \quad \frac{9.4666}{1 - 1.645/\sqrt{358}} = 10.37.$$

These are relatively quite close together, due to our large $n$.

The intercept $\beta_0$ has little physical meaning, being of primary interest as a part of the equation for $\hat{y}$ for estimation. But $\bar{y}$ is of greater interest, being at the middle of the data. The 90% confidence limits on $\mu_Y$ are

$$\bar{y} \pm t_{.05,177\mathrm{df}} s_\varepsilon/\sqrt{n} = 556.539 \pm 1.645(9.4666)/\sqrt{179} = 555.38, \quad 557.70.$$

These are at $x = \bar{x}$. For $\mu_{Y.x}$ at other $x$ values, there would be greater spread around $\hat{y} = \bar{y} + b_1(x - \bar{x})$ as (11.32) shows.

One thing an industrial person might wish to know is as to what total foil thickness $x$ to use if he desires $y$ to average not at $\bar{y}$, but at, say, 550 We can substitute this for $y$ in the estimation equation and have

$$550 = 462.2 + 1.3847x$$

or $x = 63.4$. Of course for this to be used, other factors should be held constant. Moreover this assumes some degree of cause and effect which seems reasonable here, but which *cannot be proved* by any statistical analysis.*

As a final word let us say that a bivariate table provides a good picture of the relationship. A calculator may readily be programmed to provide such a table as an output, as well as to make required tests and estimations.

## 11.8. SPECIAL CASE $\mu_{Y.X} = \beta_1 X$

An important special case occurs when $Y$ varies directly as $X$, that is, from some practical or scientific reasons we know the intercept to be zero. Many physical laws are of this form and the main problem is finding the slope $\beta_1$. Proceeding by least squares as in the footnote proof in Section 11.3.1, we readily find the estimate of $\beta_1$ to be

$$b_1 = \sum x_i y_i \big/ \sum x_i^2 \tag{11.37}$$

for

$$\mu_{Y.X} = \beta_1 X. \tag{11.38}$$

We also find

$$\sum (y_i - \hat{y}_i)^2 = \sum (y_i - b_1 x_1)^2 = \sum y_i^2 - \left(\sum x_i y_i\right)^2 \big/ \sum x_i^2 \tag{11.39}$$

and

$$s_\varepsilon^2 = \frac{\sum y_i^2 - (\sum x_i y_i)^2 / \sum x_i^2}{n - 1}, \tag{11.40}$$

there being but one parameter involved and hence but one loss from the original $n$ degrees of freedom. Since we find

$$\sigma_{b_1}^2 = \sigma_\varepsilon^2 \big/ \sum x_i^2, \tag{11.41}$$

we have

$$s_{b_1}^2 = s_\varepsilon^2 \big/ \sum x_i^2,$$

and thus we may use

$$t_{n-1} = \frac{b_1 - \beta_1}{s_\varepsilon / \sqrt{\sum x_i^2}} \tag{11.42}$$

* For example, the length $Y$ and width $X$ of a block of iron would be *positively* correlated when measured at radically different temperatures. But if various large forces were applied to the width, the length would be *inversely* correlated. In neither case is $Y$ a direct result of $X$ as a cause, but their two relationships are due to two different common physical causes.

to test a hypothesis on $\beta_1$, or to set confidence limits on $\beta_1$. We may also use an obvious analogy of (11.30) with $\nu = n - 1$, for limits on $\sigma_\varepsilon$.

Confidence limits for $\mu_{Y.X} = \beta_1 X$ are estimated by $\hat{y} = b_1 x$. Since $\sigma_{\hat{y}} = x\sigma_{b_1}$ we may use

$$t_{n-1} = \frac{y - \mu_{Y.X}}{x s_\varepsilon / \sqrt{\sum x_i^2}}$$

for a significance test or for confidence limits.

### 11.9.* SIGNIFICANCE OF DIFFERENCES BETWEEN TWO SLOPES

One test which may well occur in practice is easily explained from the foregoing. Suppose we have two independent samples of *similar* linear data for each of which we can make the assumptions in (11.13). Moreover suppose that in addition it is reasonable to assume $\sigma_{\varepsilon 1} = \sigma_{\varepsilon 2}$. Then we can prove

$$t_{n_1+n_2-4}$$
$$= \frac{b_{11} - b_{12} - (\beta_{11} - \beta_{12})}{\sqrt{\dfrac{(n_1 - 2) s_1^2 + (n_2 - 2) s_2^2}{n_1 + n_2 - 4} \left( \dfrac{1}{\sum (x_{1i} - \bar{x}_1)^2} + \dfrac{1}{\sum (x_{2i} - \bar{x}_2)^2} \right)}}.$$

This formula is used directly for testing the hypothesis $\beta_{11} = \beta_{12}$ or for setting $1 - \alpha$ confidence limits for $\beta_{11} - \beta_{12}$, just as (11.29) followed from (11.24).

Further analysis of two or more samples of bivariate data is beyond the scope of this book. Much of it comes under the heading of "analysis of covariance."

### 11.10. NONLINEARITY TEST

In Chapter 12, after studying some analysis of variance we shall be in a position to present a test of significance of the departure from linearity. If not significant, then we can be content with the linear fit. But if significant, the indication is that we can *significantly* improve the fit by using some curve. In addition to the linear trend there is *bona fide* nonlinearity. This is given in Chapter 12.

## 11.11.* USE OF LEAST SQUARES FITTING FOR OTHER TRENDS

We now take up briefly some fitting methods for nonlinear relationships. The appropriate method of fitting depends of course on the mathematical relationship between $Y$ and $X$. But also it depends on what assumptions are made about the error terms involved. Relative to the use of least squares, there are three cases: (1) those where a least squares approach may be made directly as in the foregoing fitting of a straight line; (2) cases where, after some sort of transformation, we can make direct use of least squares; and (3) those cases where we cannot obtain normal equations from least squares which can be solved in closed form. The last require successive approximation if we are to use the least squares approach.

The choice of an equation to use in fitting data requires knowledge of the various curves one meets in analytic geometry. It also often requires ingenuity. Choice of the functional forms to fit might well be regarded as an art rather than a science, since sometimes several different curves are all plausible. Scientific laws sometimes supply a guide or even the type of curve.

### 11.11.1.* Functions Linear in the Parameters.

We can always use least squares fitting effectively if (1) the functional form is linear in the parameters, *and* (2) the error term is added to the other terms. Examples are

$$Y = \beta_0 + \beta_1 X + \beta_2 X^2 + \varepsilon \tag{11.43}$$

$$Y = \beta_0 + \beta_1 X + \beta_2 X^2 + \beta_3 X^3 + \varepsilon \tag{11.44}$$

$$Y = \beta_0 + \beta_1 X + \beta_2/X + \varepsilon \tag{11.45}$$

$$Y = \beta_0 + \beta_1 X + \beta_2 2^X + \varepsilon. \tag{11.46}$$

Notice the $+\,\varepsilon$ on the end of each. Applying the least squares approach to a function with, say, three constants will lead to three equations to be solved. Suppose the regression equation is

$$Y_i = \beta_0 + \beta_1 g(X_i) + \beta_2 h(X_i) + \varepsilon_i, \tag{11.47}$$

where the $g(X_i)$ and $h(X_i)$ are known functions and with the $\varepsilon_i$'s normal and independently distributed with mean 0 and variance $\sigma_\varepsilon^2$, that is, NID(0, $\sigma_\varepsilon^2$). Then by minimizing $\sum e_i^2$ for the corresponding estimation equation

$$y_i = b_0 + b_1 g(x_i) + b_2 h(x_i) + e_i, \tag{11.48}$$

we have

$$b_0 n + b_1 \sum g(x_i) + b_2 \sum h(x_i) = \sum y_i \qquad (11.49)$$

$$b_0 \sum g(x_i) + b_1 \sum [g(x_i)]^2 + b_2 \sum g(x_i) h(x_i) = \sum y_i g(x_i) \qquad (11.50)$$

$$b_0 \sum h(x_i) + b_1 \sum g(x_i) h(x_i) + b_2 \sum [h(x_i)]^2 = \sum y_i h(x_i), \qquad (11.51)$$

where the summations are for $i = 1, ..., n$. Since $g(X)$ and $h(X)$ are completely known, all summations are readily calculated, and then the equations must be solved for the estimates $b_0$, $b_1$, and $b_2$. For example, for (11.43), equations (11.49)–(11.51) become

$$b_0 n + b_1 \sum x_i + b_2 \sum x_i^2 = \sum y_i \qquad (11.52)$$

$$b_0 \sum x_i + b_1 \sum x_i^2 + b_2 \sum x_i^3 = \sum x_i y_i \qquad (11.53)$$

$$b_0 \sum x_i^2 + b_1 \sum x_i^3 + b_2 \sum x_i^4 = \sum x_i^2 y_i . \qquad (11.54)$$

In this way we can fit a polynomial of any degree.

If our collection of $x_i$'s form an arithmetic progression, that is, proceed by equal increments, we have an especially easy method available by transforming $x$. For example, if $x_i$'s are 150, 175, 200, 225, 250, and 275°F, then $\bar{x} = 212.5$. Now $x_i - \bar{x}$ are $-62.5$, $-37.5$, $-12.5$, $+12.5$, $+37.5$, and $+62.5$. Dividing by 12.5 gives $-5$, $-3$, $-1$, $+1$, $+3$, and $+5$. Hence if we let $u_i = (x_i - 212.5)/12.5$, these last are the $u_i$ values. Then using (11.52)–(11.54) for $u_i$'s we find four coefficients drop out because $\sum u_i = 0$ and $\sum u_i^3 = 0$. This gives

$$b_0 6 + b_2 70 = \sum y_i$$

$$b_1 70 = \sum u_i y_i$$

$$b_0 70 + b_2 1414 = \sum u_i^2 y_i ,$$

which are easily completed and solved. Whenever $x_i$ is of the form $x_i = A$, $A + \Delta x$, $A + 2\Delta x$, ..., and we have an even number for $n$, we use $u_i = (x_i - \bar{x})/.5 \Delta x$; whereas if $n$ is odd, we use $u_i = (x_i - \bar{x})/\Delta x$. These two transformations always give small whole numbers for the $u_i$, summing to zero. (Experiment a bit.)

Another approach, which is even easier than the last, is to use "orthogonal polynomials." These avoid all solutions of simultaneous equations. See Section 12.3 for a brief introduction to their use.

### 11.11.2.* Least Squares after a Transformation.

In a number of cases where the parameters are not linear, we can still use the least squares approach successfully and easily. For example, two very commonly occurring types of relationships in science are the "power law" and the "exponential" law. They are as follows:

$$Y = \beta_0 X^{\beta_1} \tag{11.55}$$

$$Y = \beta_0 \beta_1{}^X. \tag{11.56}$$

Figure 11.3 shows examples of each. Now although "nature" might be kind enough to give a relationship exactly as one of the forms, usually

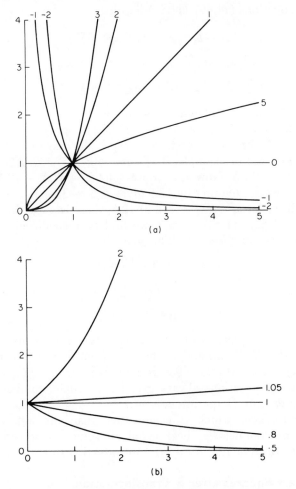

**Fig.11.3.** Two families of curves often used in practice: (a) the power law $Y = \beta_0 X^{\beta_1}$, and (b) the exponential law $Y = \beta_0 \beta_1{}^X$. These are drawn with $\beta_0 = 1$ for convenience, and with $\beta_1$ values as shown at the ends of the curves as drawn.

such an equation is only approximate, and then too there are almost invariably errors of measurement of some consequence. Now how do we put in an error term on, say, $Y = \beta_0 X^{\beta_1}$? One might well wish to use an *additive* error and have $Y = \beta_0 X^{\beta_1} + \varepsilon$. The author would like this model, but unfortunately minimizing $\sum(y_i - b_0 x^{b_1})^2$ leads to equations which can only be solved by laborious successive approximation, preferably by an electronic computer. Instead we assume a model with a *multiplicative* error term as follows:

$$Y = \beta_0 X^{\beta_1} \eta, \tag{11.57}$$

with the $\eta_i$'s such that $\log_{10} \eta_i = \varepsilon_i$ are NID$(0, \sigma_\varepsilon^2)$. Then by taking $\log_{10}$ of both sides of (11.57) we have for $Y_i$ versus $X_i$

$$\log Y_i = \log \beta_0 + \beta_1 \log X_i + \varepsilon_i . \tag{11.58}$$

Now using primes for $\log_{10}$ of a letter:

$$Y_i' = \beta_0' + \beta_1 X_i' + \varepsilon_i . \tag{11.59}$$

On this we can proceed exactly as in Section 11.3. As a matter of fact in not a few cases in scientific work the exact value of $\beta_1$ is known, and we only need to estimate $\beta_0'$, then $\beta_0$. Examples are Boyle's law of physical chemistry $\beta_1 = -1$, volume of similar objects as a function of a linear dimension $\beta_1 = 3$, and the very commonly occurring inverse square law, for example, in gravitation or light intensity, $\beta_1 = -2$.

In like manner the exponential relation

$$Y = \beta_0 \beta_1^X \eta \tag{11.60}$$

using $y_i$ versus $x_i$ becomes after taking $\log_{10}$

$$\log Y_i = \log \beta_0 + X_i \log \beta_1 + \log \eta_i$$

or

$$Y_i' = \beta_0' + X_i \beta_1' + \varepsilon_i . \tag{11.61}$$

Again this relation may be fitted by least squares as in Section 11.3.

Although (11.59) and (11.61) seem quite similar the equations from which they come are radically different. The former says that if we have this law, (11.57), and plot $\log Y$ versus $\log X$, we should have a linear appearing graph. Thus if we plot data directly onto log–log paper and find linearity, we can fit this relation. On the other hand, (11.61) says that if we have this law, (11.60), and we plot $\log Y$ versus $X$, we should have a linear appearing graph. Thus if we plot data directly onto semi-log paper ($Y$ on the log scale) and find linearity, we can fit this relation.

### 11.11.3.* Intrinsically Nonlinear Cases.

Quite distinct from the foregoing cases are those where we cannot obtain least squares normal equations which can be solved, except by resorting to successive approximation. In this class are the following:

$$Y = \beta_0 X^{\beta_1} + \varepsilon \tag{11.62}$$

$$Y = \beta_0 \beta_1{}^X + \varepsilon, \tag{11.63}$$

and also many others where the parameters are not linear and to which function we *add* an error term. Such intractable relationships are called "intrinsically nonlinear," whereas those like (11.57) and (11.60) are called "intrinsically linear" because they may be transformed so as to use linear methods. See Draper and Smith [2] for an excellent discussion on these topics.

## 11.12. APPLICATIONS TO INDUSTRY AND THE LABORATORY

Let us here list a few general types and some specific applications of linear correlation methods. Careful reading should provide some ideas for the reader's applications in his own field of activity.

1. Two general types
   (a) degree of relationship—correlation;
   (b) estimation or prediction—regression.

2. Checking relative importance of process variables and ingredients (percent composition or analysis) versus qualities such as strengths, hardness, taste, odor, color, surface finish, physical properties, length of life. For example, in steel, nonferrous metals, rubber, plastics, glass, pharmaceuticals, candy, and food, comparing dependent or output variables with independent or input variables.

3. Are two tests equivalent? For example, weighing versus acid-dissolving test for thickness of tin coating on sheets.

4. Repeatability of a test method.

5. Relationship of angle of bend to thickness of stock in cold forming.

6. Before and after heat treating.

7. Life of tool versus its hardness, hardness of stock, cutting speed, stock removal, analyses.

8. Tensile strength versus hardness and density.
9. Life of bulb versus quality test on filament.
10. Breaking strength versus thickness of starch film.
11. Calibration of two instruments, or one versus a standard.
12. Percentage of broken paper bags versus height of fall (bags filled with sand).
13. Width of wet lumber versus width of same pieces dry.
14. Production versus quality.
15. Market research: sales versus a variety of potential predictors.
16. Hospital calls versus lost-time accidents.
17. Foundry performance versus atmospheric conditions.
18. Weight of steel ingot versus percent yield.
19. Diameter of condenser coil versus thickness of foil, of paper, number of turns and tension in winding.
20. Thickness of coating of zinc by stripping method and by non-destructive magnetic method.
21. Percentage of defective bottles versus process variables and chemical compositions.
22. Tile finish versus temperature of firing.
23. Inventory and sales versus month and year.
24. Age of food versus bacteria count.
25. Machinability versus percent sulfur in steel.
26. Percent "waste wasters" (nonsalvageable) "menders" (salvageable) in tin-plate production.
27. Production of metal versus heat size, delays, charging time.
28. Viscosity of gelatin versus time since addition, in making capsules.
29. Outside diameter versus time since process was adjusted (tool wear).
30. Electrical properties, for example, resistance, versus inductance of coils.
31. Improvement in number of defects per subassembly over a period of time.
32. Tool life versus hardness of stock, of tool, depth of cut, and rate of stock removal.
33. Production versus quality of raw materials and processing variables.
34. Employee efficiency as predicted from several tests.
35. Moisture content of lumber versus speed of drying, temperature, and dimension of pieces.
36. Weight of coating of electrolytic tin plate versus current, rate of strip, acidity, distance from tin anode.
37. Tensile strength versus density and hardness of stock.

## 11.13. SUMMARY

In this chapter we have been particularly concerned with obtaining linear relationships between a dependent variable $Y$ and some independent variable $X$. Such a linear equation is estimated from pairs of observations, $x_i$, $y_i$, which are made at the same time, or on the same material, or otherwise linked or related. Often $Y$ is the "output" variable and $X$ the "input" variable. Having accomplished the fitting, then a variety of tests or interval estimates may be made, as desired, for example, on slopes, intercepts, standard errors of estimate $\sigma_\varepsilon$, or predicted mean values from a fitted line.

Fitting of linear models is often theoretically justifiable, or if not, then empirically justified by the data seeming to lie along a line. In any case a straight line is commonly a first approximation to a curve, especially if the range of $X$ values is relatively narrow. But when a curved trend clearly appears we should fit some other relationship, such as those in Section 11.11. A graph of $Y$ on $X$ on uniform paper and inspection of Fig. 11.3 may suggest plotting on semilog or log–log paper (or possibly others). If either "straightens out" the graph, then we may proceed as in Section 11.11.2. We may also readily fit polynomials such as a parabola by least squares.

### REFERENCES

1. E. L. Crow, Generality of confidence intervals for a regression function. *J. Amer. Statist. Assoc.* **50**, 850–853 (1955).
2. N. Draper and H. Smith, "Applied Regression Analysis." Wiley, New York, 1966.
3. H. Haug, Reaction of nitric acid with potassium chloride in fused salt solution. Ph.D. Thesis, Purdue Univ., 1963.
4. C. T. Metcalf, A study of the use of sandstone in bituminous surface courses, p. 146. M.S. Thesis, Purdue Univ., 1949.
5. C. L. Dorsey, Some thermodynamic properties of the nitroparaffins. Ph.D. Thesis, Purdue Univ., 1949.
6. V. C. Kashyap, Solubility of nitrogen in liquid iron and iron alloys. Ph.D. Thesis, Purdue Univ., 1956.
7. F. R. Moulton, "An Introduction to Celestial Mechanics," p. 82. Macmillan, New York, 1931.

### ADDITIONAL REFERENCES

G. A. Baker, Linear regression when the standard deviations of arrays are not all equal. *J. Amer. Statist. Assoc.* **36**, 500–506 (1941).
M. S. Bartlett, Fitting a straight line when both variables are subject to error. *Biometrics* **5**, 207–212 (1949).

J. Berkson, Are there two regressions? *J. Amer. Statist. Assoc.* **45**, 164–180 (1950).

G. E. P. Box, Use and abuse of regression. *Technometrics* **8**, 625–629 (1966).

G. E. P. Box and G. S. Watson, Robustness to nonnormality of regression tests. *Biometrika* **49**, 93–106 (1962).

F. D. Carlson, E. Sobel, and G. S. Watson, Linear relationships between variables affected by errors. *Biometrics* **22**, 252–267 (1966).

S. M. Goldfeld and R. E. Quandt, Some tests for homoscedasticity. *J. Amer. Statist. Assoc.* **60**, 539–547 (1965).

J. D. Hinchen, Correlation analysis in batch process control. *Indust. Quality Control* **12**, (no. 11) 54–59 (1956).

P. G. Hoel, On testing the degree of a polynomial. *Technometrics* **10**, 757–767 (1968).

J. E. Kerrich, Fitting the line $y = ax$ when errors of observation are present in both variables. *Amer. Statist.* **20**, 24 (1966).

J. R. Kittrell, R. Mezaki, and C. C. Watson, Estimation of parameters for nonlinear least squares analysis. *Indust. Engrg. Chem.* **57**, (no. 12) 18–27 (1965).

E. G. Olds and F. S. Acton, Mathematics for engineers I: Fitting functions to engineering data. *Electrical Engrg.* **67**, 988–993 (1948).

C. H. Richardson, "Statistical Analysis." Harcourt, New York, 1934.

M. F. Shakun, Nonlinear regression analysis. *Indust. Quality Control* **23**, (no. 1) 11–13 (1966).

C. T. Shewell, Design patterns for slopes and intercepts. *Anal. Chem.* **32**, 1535 (1960).

W. J. Youden, Statistical design method for fitting lines to data illustrates statistical procedures for testing hypotheses. *Indust. Engrg. Chem.* **48**, 107A–109A (1956).

## Problems

11.1. Density of molten salt mixtures $Yg/cc$ was measured at various temperatures $X$ in degrees centigrade. Fit a straight line to the data coding $X$ for simplicity so that $\sum v = 0$. Test the hypothesis $\beta_1 = 0$ versus $\beta_1 \neq 0$, at $\alpha = .05$. Set 90% confidence limits on $\beta_1$, $\mu_{Y.300}$, and $\sigma_\varepsilon$. Estimate $\hat{y}$ at $x = 230°$. (Actual value was 1.972.) Data [3]:

| $x$ | 250 | 275 | 300 | 325 | 350 |
|---|---|---|---|---|---|
| $y$ | 1.955 | 1.934 | 1.916 | 1.898 | 1.878 |

11.2. Fit a straight line to the following data [4], on $Y$, the percent voids in total mix for bituminous course for highway versus $X$, percent asphalt cement:

| $x$ | 5 | 6 | 7 | 8 | 9 | 10 | 11 | 12 | 13 |
|---|---|---|---|---|---|---|---|---|---|
| $y$ | 20.1 | 17.7 | 16.1 | 13.8 | 12.4 | 10.5 | 8.9 | 8.2 | 7.0 |

Do the fitting for $y = \bar{y} + b_1 x'$ where $x' = x - \bar{x}$. This simplifies (11.4), (11.5). Test hypothesis $\beta_1 = 0$ at .01 level versus $\beta_1 \neq 0$. Set 90% confidence limits on $\beta_1$, $\mu_{Y.X}$, $\sigma_\varepsilon$. Find $\hat{y}$ at $x = 14$. (Actual value was 5.5.)

11.3.  Under certain conditions in a laboratory experiment $Y$ is grams of a dye, and $X$ is grams of sodium acetate.

| $x$ | 0 | .5 | 1.0 | 1.5 | 2.0 | 2.5 |
|---|---|---|---|---|---|---|
| $y$ | 3.83 | 3.95 | 3.79 | 3.72 | 3.72 | 3.79 |

Find equation $\hat{y} = \bar{y} + b_1(x - \bar{x})$ (using coding on $x$), $r$, and test $\rho = 0$ versus $\rho \neq 0$ at $\alpha = .05$. Set 95% confidence limits on $\beta_1$ and $\mu_Y$.

11.4.  A small amount of data are here given on "tension" for piston rings, that is, the force required to close the gap to a specified size. Here $Y$ is pounds tension of a ring *after* "ferroxing," a high-temperature steam treatment to give a blue oxide coating; $X$ is pounds tension *before*, for the *same* ring. Test hypotheses $\beta_1 = 0$ versus $\beta_1 \neq 0$ and $\beta_1 = 1$ versus $\beta_1 \neq 1$ at $\alpha = .05$. Find estimation equation, $r$, $r^2$, $s_\varepsilon$, and interpret. Estimate $\hat{y}$ if $x = 4.4$. 5.4, (If no calculator is available, coding $x$ and $y$ will help materially.)

| $x$ | 4.3 | 5.4 | 4.9 | 5.5 | 4.8 | 4.4 |
|---|---|---|---|---|---|---|
| $y$ | 4.8 | 5.9 | 5.5 | 6.0 | 5.4 | 5.1 |

11.5.  The bore hardness of cylinder blocks was a characteristic of importance to be controlled. To measure it the block would have to be cut into two pieces, destroying the block. However, the hardness on the top deck of a block could be measured without destroying the block. Accordingly a study of how well the latter $X$ would predict the former $Y$ was made. An average of five Rockwell B hardness readings on one block was used for $x$, and an average of five top deck Brinell hardness readings was $y$. Twenty-five blocks were thus measured. Coding was used: $v = x - 200$, $w = y - 88.8$. The following were obtained: $\Sigma v = +3$, $\Sigma v^2 = 2821$, $\Sigma w = 82.9$, $\Sigma w^2 = 326.11$, $\Sigma vw = 310.4$. Find $\bar{x}$, $\bar{y}$, $b_0$, $b_1$, $s_\varepsilon$, $r$, $r^2$. Test hypothesis $\beta_1 = 0$ versus $\beta_1 > 0$ at $\alpha = .01$. Make a point estimate of $\hat{y}$ if $x = 187$. To what amount of error is this subject?

11.6. In a study of steel rejection, percent sulfur was one variable studied. Here $Y$ is tons rejected in a "heat" (about 200 tons); $X$ is .001 % of sulfur. Ten heats gave the following:

| x | 38 | 26 | 36 | 33 | 30 | 29 | 31 | 29 | 32 | 32 |
|---|----|----|----|----|----|----|----|----|----|----|
| y | .0 | .3 | 1.1 | 6.5 | 5.8 | 8.1 | 3.4 | 3.1 | .9 | 7.0 |

Draw a scatter diagram. Find $b_0$, $b_1$, $s_\epsilon$, $r^2$, $r$. Interpret the results.

11.7. The porosity in road mix for sandstone samples $Y$, was studied in relation to percent sandstone $X$ retained on number 12 sieve. A random sample of 10 pairs of observations was chosen. Fit a straight line to the data and show on a scatter diagram. Test the hypothesis $\beta_1 = 0$ versus $\beta_1 \neq 0$ at $\alpha = .05$. Find $r$ and similarly test $\rho = 0$. Set 90% confidence limits on $\beta_1$. Data follow, rounded to the nearest whole unit [4].

| x | 74 | 86 | 56 | 62 | 29 | 14 | 73 | 74 | 55 | 66 |
|---|----|----|----|----|----|----|----|----|----|----|
| y | 20 | 11 | 19 | 12 | 24 | 24 | 16 | 17 | 20 | 14 |

11.8. In a study of thermodynamic properties of nitroparaffins, $Y$ is the heat of combustion in calories per gram, $X$ is the oxygen balance, which equals $-(2a + .5b - c)1600/M$, where $a$ is carbon atoms, $b$ is hydrogen atoms, $c$ is oxygen atoms in the molecule, and $M$ is the molecular weight of compound [5]. Seven compounds were here chosen at random from the 17 studied. Fit a straight line and set 95% confidence limits on $\beta_1$, $\mu_Y$, and $\sigma_\epsilon$. Test $\beta_1 = 0$ versus $\beta_1 \neq 0$ at $\alpha = .05$. Find $r$ and $r^2$. Estimate $\hat{y}$ for $x = -134.6$. (Actual value was 5361.)

| x | −95.8 | −163.0 | −59.6 | −59.6 | −127.1 | +4.9 | −20.2 |
|---|-------|--------|-------|-------|--------|------|-------|
| y | 4340 | 6182 | 3243 | 3308 | 5110 | 1777 | 2460 |

11.9. A study of solubility of nitrogen in iron was made [6]. Here $Y$ is the cubic centimeters of nitrogen dissolved per 100 g of iron, and $X$ is the temperature in degrees centigrade.

| y | 33.5 | 34.6 | 34.7 | 34.8 | 34.7 | 34.7 | 33.0 | 34.0 | 37.8 |
|---|------|------|------|------|------|------|------|------|------|
| x | 1543 | 1612 | 1696 | 1754 | 1696 | 1622 | 1539 | 1630 | 1842 |

Find the estimating equation $\hat{y} = b_0 + b_1 x$ and $r$, $r^2$. Test the hypothesis $\beta_1 = 0$ versus $\beta_1 \neq 0$ at $\alpha = .01$. Draw a scatter diagram and show the line. Set 90% confidence limits on $\beta_1$ and on $\mu_{Y.1600}$.

11.10.  Under certain conditions it was conjectured that $Y$, the cubic centimeters of nitrogen dissolved in 100 g of liquid iron, would vary directly as the square root of the pressure of $N_2 = X$ [6]. Using $Y = \beta_1 X + \varepsilon$ as model, estimate $\beta_1$ and $\sigma_\varepsilon$. Show a scatter diagram. Portion of data:

| $x = \sqrt{P(N_2)}$ | 14.14 | 20.0 | 24.49 | 27.22 | 17.32 | 22.36 | 26.46 | 27.33 |
|---|---|---|---|---|---|---|---|---|
| $y$ | 14.3 | 25.8 | 31.4 | 30.9 | 19.2 | 25.8 | 30.4 | 31.5 |

11.11.  In fiberglass manufacture, recovered thickness $Y$ is important to control. This is related to thickness at the machine $X$. Both are in inches. The table below shows the data for 50 tests. Find $b_0$, $b_1$, $s_\varepsilon$, $r$, and test the hypothesis $\rho = 0$ versus $\rho > 0$ at $\alpha = .01$. Also test the hypothesis $\beta_1 = 1$ versus $\beta_1 \neq 1$ at $\alpha = .01$. Set 95% limits on $\beta_1$ and $\mu_{Y.x}$ at $x = .80$, 1.30, and 2.20. About how much drop is there from machine thickness to recovered thickness?

| Recovered thickness $y$ (in.) | Machine thickness $x$ (in.) | | | | | | | | | |
|---|---|---|---|---|---|---|---|---|---|---|
| | .595 | .795 | .995 | 1.195 | 1.395 | 1.595 | 1.795 | 1.995 | 2.195 | 2.395 |
| 1.995 | — | — | — | — | — | — | — | — | — | 3 |
| 1.795 | — | — | — | — | — | — | — | — | 2 | — |
| 1.595 | — | — | — | — | — | — | 5 | 2 | — | — |
| 1.395 | — | — | — | — | — | — | 3 | — | — | — |
| 1.195 | — | — | — | — | 2 | — | — | — | — | — |
| .995 | — | — | — | 14 | 1 | — | — | — | — | — |
| .795 | — | 1 | 6 | — | — | — | — | — | — | — |
| .595 | 4 | 6 | 1 | — | — | — | — | — | — | — |

11.12.  Using the original ungrouped data which were tabulated for Problem 11.11, we have the following: $\sum x = 65.89$, $\sum x^2 = 98.9733$, $\sum y = 53.60$, $\sum y^2 = 66.6798$, $\sum xy = 81.1754$, $n = 50$. Answer the same questions as in Problem 11.11, and compare with answers for that problem.

11.13.  In manufacturing a stamping, the angle between two sides was required to be held to $90 \pm (1/3)°$. The factory reported that this angular tolerance could not be held due to the variation in the thickness of the stock. A number of stampings were checked for both angle and thickness of stock giving the data shown. Test $\beta_1 = 0$ versus $\beta_1 \neq 0$ at $\alpha = .01$. Find estimating equation and $r$. Estimate $\hat{y}$ for $x = 77, 80$ mils. What $x$ would give $\hat{y} = 90°$? What conclusions can you draw?

| Angle $y$ (degrees from 90) | $x$ Thickness of stock [in mils (.001 in.)] | | | | | | | | | | |
|---|---|---|---|---|---|---|---|---|---|---|---|
| | 76.0 | 76.5 | 77.0 | 77.5 | 78.0 | 78.5 | 79.0 | 79.5 | 80.0 | 80.5 | 81.0 |
| +.1 | 1 | 1 | — | — | — | — | — | — | — | — | — |
| .0 | — | 2 | 3 | 1 | 1 | — | — | — | — | — | — |
| −.1 | — | 2 | 6 | 10 | 6 | 3 | 1 | — | — | — | — |
| −.2 | — | — | 5 | 6 | 22 | 11 | 8 | 3 | — | — | — |
| −.3 | — | — | 1 | 1 | 5 | 3 | 8 | 6 | 1 | — | — |
| −.4 | — | — | — | — | 2 | — | 3 | 1 | 1 | 1 | — |
| −.5 | — | — | — | — | — | — | — | — | 2 | — | 1 |

11.14. A study was made on the amount of out-of-roundness of a type of glassware at two places. (Out-of-roundness is maximum minus minimum diameters.) Here $Y$ is "heel" out-of-roundness, and $X$ is that at the "shoulder." Find estimation equation, $r$, $s_\varepsilon$, and test hypothesis $\rho = 0$ versus $\rho \neq 0$ at $\alpha = .01$. Set 90% confidence limits on $\beta_1$, $\mu_{Y, .007}$, and $\sigma_\varepsilon$.

| | $x$ | | | | | | |
|---|---|---|---|---|---|---|---|
| $y$ | .006 | .007 | .008 | .009 | .010 | .011 | .012 |
| .0215 | — | — | — | — | — | 1 | — |
| .0195 | — | — | 1 | — | 2 | 1 | 1 |
| .0175 | — | 1 | 1 | 4 | 2 | — | — |
| .0155 | — | — | 3 | 5 | 1 | — | — |
| .0135 | — | 2 | 6 | 7 | 3 | — | — |
| .0115 | 2 | 3 | 4 | 1 | 1 | 2 | — |
| .0095 | 2 | 2 | 2 | 1 | — | — | — |
| .0075 | 1 | — | — | — | — | — | — |

11.15. A study was made of the dry versus wet width of 1/6 in-thick poplar wood, nominally 99 in. across the grain. Forty-nine original pairs of observations were made to 1/16 in. (49 pieces). The table shows the data grouped in 2 in. classes. Test hypotheses $\beta_1 = 0$ versus $\beta_1 \neq 0$, and $\beta_1 = 1$ versus $\beta_1 \neq 1$, both at $\alpha = .05$. Set 95% confidence limits on $\beta_1$ and $\mu_{Y.X}$ at $X = 107, 113, 121$. Find $r^2$ and interpret results.

| Dry width | Wet width $x$ (in.) | | | | | | | |
|---|---|---|---|---|---|---|---|---|
| $y$ (in.) | 107 | 109 | 111 | 113 | 115 | 117 | 119 | 121 |
| 114 | — | — | — | — | — | — | 1 | 1 |
| 112 | — | — | — | — | — | — | — | — |
| 110 | — | — | — | — | 1 | 4 | — | — |
| 108 | — | — | — | 2 | 2 | — | — | — |
| 106 | — | — | 3 | 6 | 3 | 2 | — | — |
| 104 | — | 1 | 8 | 3 | — | — | — | — |
| 102 | 2 | — | 5 | 3 | — | — | — | — |
| 100 | — | 2 | — | — | — | — | — | — |

**11.16.** Class experiment: Each member of the class is to generate 10 pairs $(x_i, y_i)$ as follows: use three pennies and two nickels, toss all five coins, counting the number of heads $x$. Then the two nickels are picked up, well shaken, and retossed, and a count made of the number of heads on *all five* coins $y$. For $(x_2, y_2)$, all five coins are again shaken and dropped giving $x_2$, and so forth. Each student finds $r$, and tests $\rho = 0$ versus $\rho \neq 0$ at $\alpha = .05$. Also test $\beta_1 = 1$ versus $\beta_1 \neq 1$ at $\alpha = .05$. (All data may be pooled into bivariate correlation table and calculations repeated.) (Actually $\rho = .6$.)

**11.17.** Derive (11.36) from (11.24) if $\beta_1 = 0$.

**11.18.** Using (7.18) and (10.7) find an unbiased point estimate of $\sigma_\varepsilon$ from $s_\varepsilon$.

**11.19.** Prove that for any nonvertical line which goes through the point $(\bar{x}, \bar{y})$, $\sum(y_i - \hat{y}_i) = 0$, using $\hat{y}_i - \bar{y} = k(x_i - \bar{x})$. Thus show that $\sum[y_i - \bar{y} - k(x_i - \bar{x})] = 0$.

**11.20.** Carry out the details of the proof of (11.4) and (11.5) by the least squares approach.

**11.21.** Using the model $Y = \beta X + \varepsilon$ and the least squares approach, derive (11.37) and (11.39).

**11.22.** Water content $Y$, in mole percent, was measured at various temperatures $X$, in degrees centigrade. Plot a scatter diagram. Would a straight line fit well? Fit an exponential curve and draw it on the diagram. In fitting, code $x$ by $v = (x - \bar{x})/25$.
Data [3] are as follows:

| $X$ | 199.5[a] | 250 | 300 | 350 |
|---|---|---|---|---|
| $Y$ | 11.7 | 4.8 | 2.4 | 1.37 |

[a] Call it 200.

11.23.  The "activity coefficient" of water $Y$ at 1 atm in mixed sodium and potassium nitrates versus $X$ temperature in degrees centigrade gave the following data [3]. Plot a scatter diagram. Would a straight line fit well? Fit an exponential curve and draw it on the diagram. In fitting, code $x$ by $v = (x - \bar{x})/25$.

| $X$ | 199.5[a] | 250 | 300 | 350 |
|-----|---------|-----|-----|-----|
| $Y$ | .62 | .63 | .66 | .72 |

[a] Call it 200.

11.24.  Fit an exponential equation (11.60) to the following data on population in millions for the U.S., after plotting on semi-log paper.

| Population | 3.93 | 5.31 | 7.24 | 9.64 | 12.87 | 17.07 | 23.19 | 31.44 |
|-----------|------|------|------|------|-------|-------|-------|-------|
| Year | 1790 | 1800 | 1810 | 1820 | 1830 | 1840 | 1850 | 1860 |

11.25.  Plot the following data on ordinary graph paper and on log–log paper. Here $X$ represents the miles driven in a year, and $Y$ is the cost in cents per mile, including fixed charges, depreciation, insurance, licenses, taxes, as well as operation. Fit an appropriate curve.

| $X$ | 7500 | 10,000 | 12,500 | 14,500 | 20,000 |
|-----|------|--------|--------|--------|--------|
| $Y$ | 14.19 | 11.43 | 9.78 | 8.87 | 7.29 |

11.26.  Let $Y$ be the solubility of nitrogen in cubic centimeters per 100 g of iron under certain conditions, and $X$ the pressure of nitrogen in millimeters of mercury. According to Sievert's law $Y = \beta_0 X^{.5}$. Use the model $Y = \beta_0 X^{\beta_1}\eta$, fit appropriately to find $\beta_0$ and $\beta_1$, and test the hypothesis $\beta_1 = .5$ at $\alpha = .10$. Plot on log–log paper. Data [6] are as follows:

| $X$ | 50 | 100 | 150 | 200 | 250 | 300 | 350 |
|-----|-----|-----|------|------|------|------|------|
| $Y$ | 63.2 | 89.4 | 109.1 | 128.9 | 145.0 | 157.6 | 172.8 |

11.27.  Given below are the semi-major axes of the elliptical orbits of the nine planets of our solar system in millions of miles, and also the periodic time in days.

| Planet | Semi-major axis ($10^6$ miles) | Periodic time (Earth days) |
|---|---|---|
| Mercury | 43.36 | 88.0 |
| Venus | 67.65 | 224.7 |
| Earth | 94.45 | 365.3 |
| Mars | 154.8 | 687.0 |
| Jupiter | 506.7 | 4333 |
| Saturn | 935.6 | 10,759 |
| Uranus | 1867 | 30,686 |
| Neptune | 2817 | 60,188 |
| Pluto | 4600 | 90,737 |

Plot on log–log paper, then fit period $= Y = \beta_0 X^{\beta_1} \eta$, where $X$ is the semi-major axis. Thus verify Kepler's third law [7].

11.28. Conversion of nitric acid $Y$ as a function of the time in minutes $X$. $Y = Cl^-/(Cl^- + HNO_3)$ [3]. Plot data in a scatter diagram. Fit a straight line and draw it on the diagram. Are the departures random ? Fit an appropriate curve.

| $y$ | .396 | .358 | .300 | .234 | .148 |
|---|---|---|---|---|---|
| $x$ | 5 | 9 | 13 | 17 | 22 |

11.29. Fit a straight line to $Y$, the unit weight of total mix for bituminous road course in pounds per cubic foot, versus $X$, the percent asphalt cement [4]. Plot scatter diagram and line. Do fitting in terms of $x' = x - \bar{x}$. Fit $Y = \beta_0 + \beta_1 x' + \beta_2 x'^2$ which simplifies (11.52)–(11.54). Make a point estimate of $y$ for $x = 14$. (Actual valve was 125.4.)

| $x$ | 5 | 6 | 7 | 8 | 9 | 10 | 11 | 12 | 13 |
|---|---|---|---|---|---|---|---|---|---|
| $y$ | 118.1 | 120.1 | 120.9 | 122.8 | 123.3 | 124.5 | 125.4 | 125.0 | 125.2 |

11.30. We are given that $P$ is the vapor pressure in millimeters of mercury of a compound at $T$ absolute degrees Kelvin. Under certain conditions the theoretical equation to be fitted is $\log P_i = \beta_0/T_i + \beta_1 \log T_i + \beta_2 + \varepsilon_i$. (a) How would you go about fitting such an equation ? (b) What form does $P_i$ take ?

11.31. In the work of Haug [3] appear data which closely follow $\log Y = \beta_0 + \beta_1(1/T) + \varepsilon$, where $Y$ is the mole percent of water in salt and $T$ is the absolute temperature. How would you estimate $\beta_0$ and $\beta_1$ ?

11.32. Write out the normal equations we need to solve in fitting (11.44). If the $x$'s are uniformly spaced with $\sum x_i = 0$, what form do these take?

11.33. What form do the normal equations take which we would need to solve in fitting (11.45)?

# Chapter 12

# SIMPLE ANALYSIS OF VARIANCE

## 12.1. GENERAL CONCEPT OF ANALYSIS OF VARIANCE

As pointed out at the beginning of Chapter 11, the present chapter is also concerned with the relationship between variables. In fact there is a rather artificial way of making all analysis of variance problems into special cases of regression. But the natural approach is to say that we gather a collection of $N$ observations $y$ in such a way that we can break the total sum of squares $\sum_{i=1}^{N} (y_i - \bar{y})^2$ into parts. The experiment is so designed that each such part reflects the effect of just one input or independent variable factor, as well as the inevitable random error due to chance. Then all such parts are made comparable by dividing by the appropriate degrees of freedom, and tests made. In this way we can decide whether or not the apparent effect of any one input factor is readily explainable purely by chance.

The simplest case of analysis of variance is that in which we have a collection of samples of data. *Within* each such sample, all $n_i$ observations are taken under the same conditions. Thus the variation between these $n_i$ observations in the first sample is due to chance causes only. Then we take a second sample of $n_2$ observations under a new set of conditions, but these conditions are held constant while the $n_2$ repeated observations are taken. And we thus continue until we have $a$ samples.

In the first part of this chapter we only vary one of the experimental conditions, and thus have a "one-factor" experiment. Subsequently we vary more than one factor among the experimental conditions, in various combinations. Then we can study their independent and also their joint "effects." Such joint effects are called "interactions." Thus analysis of variance opens up a whole field of powerful experimental designs and corresponding analyses.

## 12.2. ONE-FACTOR ANALYSIS OF VARIANCE

In one-factor experiments we will be studying the effect of some *one* factor. It will be permitted to vary from one sample to another, but will be held constant while all of the $n_i$ observations are taken. Also insofar as possible all *other* conditions will be held constant throughout the entire experiment. Thus variation of observations $y$ within any one sample will be due to chance causes only, whereas variation between samples will include not only chance variation but also any real effect due to the factor under study.

Typical factors are temperature of treatment, pressure for a run, hardness of a steel, or individual people, animals, formulations, catalysts, or sources of material. The first three of these examples are numerical factors, the "levels" of which are stated in numbers. On the other hand, the last five examples are purely categories and might be listed as $A_1$, $A_2$, ..., $A_a$. But they are not characterized by a number. Nevertheless the different categories are often referred to as "levels."

Now let us consider the model being assumed, the partition of the total sum of squares, and the test for significance of the factor.

### 12.2.1. The Model.

Let us now describe the model we shall assume in one-factor analysis of variance. It is as follows:

$$Y_{ij} = \mu + \alpha_i + \varepsilon_{ij}, \qquad i = 1, ..., a, \qquad j = 1, ..., n \qquad (12.1)$$

$$\varepsilon_{ij} \text{ NID}(0, \sigma_\varepsilon^2) \qquad (12.2)$$

$$\sum_{i=1}^{a} \alpha_i = 0. \qquad (12.3)$$

In this model there are $a$ levels of factor $A$. At each level we take $n$ repeated observations $y_{i1}, ..., y_{in}$. The true grand mean $\mu$ only applies to this precise collection of levels of factor $A$. The true mean for the $i$th level of factor $A$ is $\mu + \alpha_i$. Thus $\alpha_i$ is the "differential" effect of factor $A$ at the $i$th level. The assumption in (12.3) is not a restriction at all, because $\mu$ is defined to be the grand mean over all levels of $A$ in the experiment. (If $\sum \alpha_i$ were not zero, we would simply adjust $\mu$ by $\sum \alpha_i / a$ to make the differential effects total zero.) The $\varepsilon_{ij}$'s are not only independent within any one sample but also from sample to sample; that is, all *na* of them are independent (and also normally distributed). Again, if they did not average zero we could simply adjust $\mu$ to make $\mu_\varepsilon = 0$. Note that $\sigma_\varepsilon$ is the same for each sample.

Let us carry along an example for illustration as we go. Consider data on weights of specimens (lb/ft³) of road course mixes with an asphalt content of 9%, two specimens being made up from each of eight batches. See Table 12.1. The data look somewhat like matched-pair data as in

**TABLE 12.1**

Weights of Specimens of Road Course Mixes with an Asphalt Content of 9%. Two Specimens from Each of Eight Batches[a]

| Batch $i$ | Coded $y_{i1}$ | Weight $y_{i2}$ | Total $T_i$ | Average $\bar{y}_i$ | $\sum_j y_{ij}^2$ | Range $R_i$ |
|-----------|------|------|------|--------|----------|------|
| 1 | 3.30 | 2.55 | 5.85 | 2.925 | 17.3925 | .75 |
| 2 | 4.05 | 4.11 | 8.16 | 4.080 | . | .06 |
| 3 | 3.86 | 3.74 | 7.60 | 3.800 | . | .12 |
| 4 | 4.05 | 2.92 | 6.97 | 3.485 | . | 1.13 |
| 5 | 3.42 | 3.24 | 6.66 | 3.330 | . | .18 |
| 6 | 2.49 | 2.67 | 5.16 | 2.580 | . | .18 |
| 7 | 3.80 | 3.36 | 7.16 | 3.580 | . | .44 |
| 8 | 2.86 | 2.30 | 5.16 | 2.580 | . | .56 |
| | | | 52.72 | 3.295 | 179.2254 | |
| | | | $T$ | $\bar{y}$ | | |

[a] Data shown are of weight $-120$ lb/ft³. See Metcalf [1].

Table 9.1. But here there is no reason for the first observation to be consistently higher or lower than the second. The two from any one batch $i$ are assumed to be two random observations from a normal population of mean $\mu + \alpha_i$ and variance $\sigma_\varepsilon^2$. Now what we wish to test is whether the true batch means $\mu + \alpha_i$ really do differ, or whether instead all the data could readily have come from just one single population of mean $\mu$. In the latter case all $\alpha_i$ would be zero. This is thus analogous to tests on the significance of the difference between two sample means as in Section 9.3.2, except that now we have eight samples instead of two. Our observed evidence lies in the eight $\bar{y}_i$'s as shown. Do they differ more than can readily be attributed to chance variation, as it is estimated from the variation *within* samples?

## 12.2.2. The Formulas and Test.

Now let us consider how we break up the total sum of squares of the deviations around the grand mean, so that one part shall contain random variation only, whereas the other part will include any differences *between* samples, as well as random variation. Looking at Table 12.1 we find 3.30 and 2.55 from batch 1. Surely the difference between them reflects random variation only, not differences between batches. To measure this variation we could use the variance approach (3.8):

$$(3.30 - 2.925)^2 + (2.55 - 2.925)^2 = .281250,$$

which is then divided by degrees of freedom $2 - 1 = 1$. Or we could use the approach as in the denominator of (11.4):

$$3.30^2 + 2.55^2 - 5.85^2/2 = 17.3925 - 17.11125 = .281250.$$

Now how shall we describe in general terms these two ways of finding the variance? The observations are $y_{11}$ and $y_{12}$, giving a total $T_1$ and $\bar{y}_1$ as in the table. Then the two approaches are, respectively ($n = 2$):

$$\sum_{j=1}^{n} (y_{1j} - \bar{y}_1)^2 = \sum_{j=1}^{n} y_{1j}^2 - T_1^2/n.$$

But next we should like to pool all of the *within-sample* variation.* We therefore let the first subscript on the $y$'s vary from 1 to 8 in our example, or 1 to $a$ in general, and add the contribution from each sample. Calling the total SSE (sum of squares error), we have

$$\text{SSE} = \sum_{i=1}^{a} \sum_{j=1}^{n} (y_{ij} - \bar{y}_i)^2 = \sum_{i=1}^{a} \sum_{j=1}^{n} y_{ij}^2 - \left( \sum_{i=1}^{a} T_i^2 \right) \Big/ n, \tag{12.4}$$

where $T_i = \sum_{j=1}^{n} y_{ij}$. The middle expression is the definitional form, which is readily interpreted as the pooled within-sample variation. The last expression in (12.4) is the one we use most often for calculation since it avoids forming all of the differences $y_{ij} - \bar{y}_i$. But in using it we must be very careful to watch out for heavy cancellation. For our problem the calculation goes as follows:

$$3.30^2 + 2.55^2 + \cdots + 2.86^2 + 2.30^2 - (5.85^2 + \cdots + 5.16)^2/2$$
$$= 179.2254 - 356.0214/2 = 179.2254 - 178.0107 = 1.2147.$$

---

* This is justified only if the sample variances are compatible, that is, may be assumed to be estimates of the same, constant, $\sigma_\varepsilon^2$. A test for this is given in Section 12.5.

(We should use a computer to cumulate simultaneously the sum of squares of the $y_{ij}$'s.) The two numbers subtracted are quite similar, and thus slide rule accuracy is not at all sufficient.* Full precision should be carried along, then any rounding off only done at the end.

Next, how do we measure the sample-to-sample variation? The natural way would seem to be to find the variances of the $\bar{y}_i$'s. Basically this is what we do, but we do include a weighting factor $n$ for each $(\bar{y}_i - \bar{y})^2$. This is to place this sum of squares on the same footing as the SSE, because since by (7.2) $\sigma_{\bar{y}}^2 = \sigma_y^2/n$, $\sigma_y^2 = n\sigma_{\bar{y}}^2$. Thus to make a $(\bar{y}_i - \bar{y})^2$ comparable to a $(y_{ij} - \bar{y}_i)^2$ we multiply the former by $n$. Thus to study the effect of factor $A$ we have for the sum of squares SS$A$:

$$\text{SS}A = \sum_{i=1}^{a} n(\bar{y}_i - \bar{y})^2 = n \sum_{i=1}^{a} (\bar{y}_i - \bar{y})^2 \tag{12.5}$$

$$\text{SS}A = \left( \sum_{i=1}^{a} T_i^2 \right) \bigg/ n - T^2/na.^\dagger \tag{12.6}$$

Again, (12.5) is the interpretational expression or definition, while (12.6) is generally used for calculations. In our example we have from Table 12.1:

$$\text{SS}A = (5.85^2 + \cdots + 5.16^2)/2 - 52.72^2/16$$

$$= 178.0107 - 173.7124 = 4.2983.$$

We now might ask whether these two parts SSE and SSA comprise

---

* But if we had not subtracted 120 from each original weight, we should have had 243,232.0254 − 243,230.8107 = 1.2147!

† That (12.5) and (12.6) are equal may be readily shown:

$$\sum_{i=1}^{a} n(\bar{y}_i - \bar{y})^2 = n \sum_{i=1}^{a} (\bar{y}_i^2 - 2\bar{y}\bar{y}_i + \bar{y}^2) = n \sum_{i=1}^{a} \bar{y}_i^2 - 2n\bar{y} \sum_{i=1}^{a} \bar{y}_i + na\bar{y}^2.$$

But $\bar{y} = \sum \bar{y}_i/a$, and also $\bar{y} = T/na$, $T$ being the grand total $\sum_{i=1}^{a} T_i$. Moreover $\bar{y}_i = T_i/n$. Hence the expression becomes

$$\text{SS}A = n \sum_{i=1}^{a} (T_i/n)^2 - 2n(T/na) \left( \sum_{i=1}^{a} T_i/n \right) + na(T/na)^2$$

$$= \sum_{i=1}^{a} T_i^2/n - 2T^2/na + T^2/na$$

$$= \sum_{i=1}^{a} T_i^2/n - T^2/na.$$

the entire total sum of squares (SST) for all the data. This would be by formula

$$SST = \sum_{i=1}^{a} \sum_{j=1}^{n} (y_{ij} - \bar{y})^2. \tag{12.7}$$

But by the same approach which gives the identity between the two forms of (12.4) and (12.5) we would have

$$SST = \sum_{i=1}^{a} \sum_{j=1}^{n} y_{ij}^2 - T^2/na. \tag{12.8}$$

Once more (12.7) is the definitional expression, and (12.8) is for calculation. In our example these are, respectively,

$$SST = (3.30 - 3.295)^2 + \cdots + (2.30 - 3.295)^2 \quad \text{(16 terms)}$$
$$= 3.30^2 + \cdots + 2.30^2 - 52.72^2/16$$
$$= 179.2254 - 173.7124 = 5.5130.$$

By inspection of the last form of (12.4) and (12.6) we easily see that their sum is (12.8), that is

$$SST = SSA + SSE. \tag{12.9}$$

Hence we have a complete partition of (12.7).

Next consider the degrees of freedom for the three sums of squares. For SST, since this involves the sum of squares of $na = N$ numbers around their average, the degrees of freedom would be $na - 1 = N - 1$. If there were no difference in conditions for the samples, this is what we would divide by, for the variance. But since there are differences we consider the breakdown terms. First, take SSE. Within each sample there were $n$ observations $y$. Hence the degrees of freedom for $\sum_{j=1}^{n} (y_{ij} - \bar{y}_i)^2$ are $n - 1$. To estimate $\sigma_e^2$ from this one sample we would divide by $n - 1$. But we have in SSE, $a$ such sums. Therefore the total of degrees of freedom is $a(n - 1) = na - a$. For the pooled estimate of $\sigma_e^2$, we thus use $SSE/(na - a)$. This is called a mean square error (MSE):

$$MSE = SSE/(na - a). \tag{12.10}$$

Next for SSA, we recall that it reflects the variation between $a$ numbers $\bar{y}_1, ..., \bar{y}_a$, or $T_1, ..., T_a$, and thus it carries $a - 1$ degrees of freedom. So

$$MSA = SSA/(a - 1). \tag{12.11}$$

Now if, in truth, factor $A$ is not influential on $Y$, so that all the $\alpha_i$'s are zero, then both MSE and MSA are estimates of the same thing, that is, $\sigma_\varepsilon^2$. But if the $\alpha_i$'s are not zero, but have

$$\sigma_\alpha^2 = \left(\sum_{i=1}^{a} \alpha_i^2\right)\Big/(a-1) > 0, \tag{12.12}$$

then it can be shown that

$$\mathrm{MS}A = \mathrm{est}(\sigma_\varepsilon^2 + n\sigma_\alpha^2), \tag{12.13}$$

whereas it is still true that

$$\mathrm{MSE} = \mathrm{est}\ \sigma_\varepsilon^2. \tag{12.14}$$

Hence if $\sigma_\alpha^2 > 0$, then MSA is estimating something larger than does MSE. To make the test therefore we use $F$ as follows:

$$F_{a-1,\,na-a} = \mathrm{MS}A/\mathrm{MSE}. \tag{12.15}$$

If this observed $F$ lies above $F_{\alpha\,\text{above}}$, we will conclude that $\sigma_\alpha^2 > 0$, that is, that the $\alpha_i$'s are not all zero (even though their total is). But if the observed $F$ does not exceed the $\alpha$ critical value, we will conclude that we have no *reliable* evidence for supposing factor $A$ really does influence $Y$. If it does, we can only prove it by larger sample sizes.*

Now let us carry along our example to a decision. We use an $\alpha$ risk of .05. We have

$$\mathrm{SSE} = 1.2147, \qquad \mathrm{SS}A = 4.2983, \qquad \mathrm{SST} = 5.5130,$$

and put our results into a convenient table:

| Source of variation | Sum of squares | Degrees of freedom | Mean square | $F_{7,8}$ | $F_{7,8,.05\ \text{above}}$ |
|---|---|---|---|---|---|
| Batch ($A$) | SS$A$ = 4.2983 | 7 | .6140 | 4.04 | 3.500 |
| Error (E) | SSE = 1.2147 | 8 | .1518 | | |
| Total (T) | SST = 5.5130 | 15 | | | |

Since the observed $F$ lies above the 5% critical value, 3.500, we have reliable evidence (at the .05 level) that $\sigma_\alpha^2 > 0$. That is, the true batch

* If $\sigma_\alpha^2 = 0$, then (12.15) does have exactly the $F$ distribution as in Table V in the Appendix. But if $\sigma_\alpha^2 > 0$, then (12.15) follows a "noncentral" $F$ distribution, and has a probability greater than $\alpha$ of lying above the critical value found for the *central $F$* distribution in the table. The larger $\sigma_\alpha^2$ is, the greater the chance of detecting it.

means really do differ. Had we been using $\alpha = .01$ we would not have drawn such a conclusion from the data.

**A Short Cut**   Whenever the number $n$ of repeated measurements is 2, then and only then can we use a shortcut. Since $\sum_{j=1}^{2}(y_{ij} - \bar{y}_i)^2 = |y_{i1} - y_{i2}|^2/2 = R_i^2/2$, we have that

$$\text{SSE} = \left(\sum_{i=1}^{a} R_i^2\right)\Big/2 \quad \text{if} \quad n = 2, \tag{12.16}$$

and the degrees of freedom are $a$, that is, 1 for each $R_i$. In our example, using the last column of Table 12.1 yields

$$\text{SSE} = (.75^2 + .06^2 + \cdots + .56^2)/2 = 2.4294/2 = 1.2147.$$

### 12.2.3. The Case of Unequal Sample Sizes.

A less common case than that where the sample sizes are equal, occurs when the sample sizes vary. Animals will die, litter sizes will vary, and sometimes a test tube is lost, for example. Also the amount of material available can vary. Thus we present this slightly messier case. The basic formulas are similar

$$Y_{ij} = \mu + \alpha_i + \varepsilon_{ij}, \quad i = 1, ..., a, \quad j = 1, ..., n_i \tag{12.17}$$

$$\varepsilon_{ij} \text{ NID}(0, \sigma_\varepsilon^2) \tag{12.18}$$

$$\sum_{i=1}^{a} n_i \alpha_i = 0. \tag{12.19}$$

Note carefully the differences between these and (12.1)–(12.3). See Table 12.2, for an example. We also have the following generalizations of (12.4), (12.6), and (12.8):

$$N = n_1 + n_2 + \cdots + n_a \tag{12.20}$$

$$\text{SSE} = \sum_{i=1}^{a} \sum_{j=1}^{n_i} y_{ij}^2 - \sum_{i=1}^{a} (T_i^2/n_i) = \sum_{i=1}^{a} \sum_{j=1}^{n_i} (y_{ij} - \bar{y}_i)^2 \tag{12.21}$$

$$\text{SSA} = \sum_{i=1}^{a} T_i^2/n_i - T^2/N = \sum_{i=1}^{a} n_i(\bar{y}_i - \bar{y})^2 \tag{12.22}$$

$$\text{SST} = \sum_{i=1}^{a} \sum_{j=1}^{n_i} y_{ij}^2 - T^2/N = \sum_{i=1}^{a} \sum_{j=1}^{n_i} (y_{ij} - \bar{y})^2. \tag{12.23}$$

Note carefully the difference between SSE and SST.

**TABLE 12.2**

Experimental Determinations of a Gravitational Constant Using Three Kinds of Small Balls versus a 66-kg Steel Cylinder[a]

| Factor | Observations | | | | |
|---|---|---|---|---|---|
| Platinum | 6.661, | 6.661, | 6.667, | 6.667, | 6.664 |
| Gold | 6.683, | 6.681, | 6.676, | 6.678, | 6.679, | 6.672 |
| Glass | 6.678, | 6.671, | 6.675, | 6.672, | 6.674 |

Coded in .001 unit from 6.670, that is, (Obs. $-6.670$)/.001:

$$\sum_{j=1}^{n_i} y_{ij} = T_i \qquad \sum_{j=1}^{n_i} y_{ij}^2 \qquad n_i \qquad \bar{y}_i = \frac{T_i}{n_i}$$

| | | | | | |
|---|---|---|---|---|---|
| Platinum | $y_{1j} = -9, -9, -3, -3, -6$ | $-30$ | 216 | 5 | $-6.00$ |
| Gold | $y_{2j} = +13, +11, +6, +8, +9, +2$ | $+49$ | 475 | 6 | $+8.17$ |
| Glass | $y_{3j} = +8, +1, +5, +2, +4$ | $+20$ | 110 | 5 | $+4.00$ |
| | | $T = +39$ | 801 | $N = 16$ | $\bar{y} = +2.44$ |

[a] Unit—$10^{-8}$ cc/g sec. See Weyl [2].

For our example in Table 12.2 we find

$$\text{SST} = 801 - 39^2/16 = 801 - 95.06 = 705.94$$
$$\text{SSE} = 801 - (-30)^2/5 - 49^2/6 - 20^2/5 = 801 - 660.17 = 140.83$$
$$\text{SS}A = 660.17 - 39^2/16 = 660.17 - 95.06 = 565.11.$$

Now for degrees of freedom we have in general

$$\text{SST}: N - 1, \qquad \text{SS}A: a - 1, \qquad \text{SSE}: N - a. \qquad (12.24)$$

The analysis of variance table is as follows, using an $\alpha$ risk of .01:

| Source of variation | Sum of squares | Degrees of freedam | Mean square | $F_{2,13}$ | $F_{2,13,.01 \text{ above}}$ |
|---|---|---|---|---|---|
| Type of ball | SS$A$ = 565.11 | 2 | 282.6 | 26.1 | 6.74 |
| Error | SSE = 140.83 | 13 | 10.83 | | |
| Total | SST = 705.94 | 15 | | | |

The analysis thus provides very clear evidence of a real difference in these three sets of experimental results. See Fig. 12.1 for a graphical picture of the results.

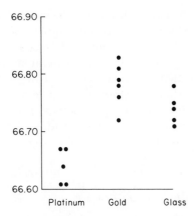

**Fig. 12.1.** Data plotted on determinations of a gravitational constant. Three experiments using different kinds of small balls versus one large steel ball. Plot shows a picture of a significant difference between three sets of sample results, that for the platinum experiment showing no overlapping with the other two sets. Data from Table 12.2 and Weyl [2].

According to the researcher [2], there were found experimental conditions as to why the differences were not attributable to differences in the three materials. In other words, something else besides the kind of balls differed from experiment to experiment. There then becomes a real question as to whether to combine the results and how, which is beyond our present discussion. He did take 6.670 as the combined value.

### 12.2.4.* Orthogonal Contrasts.

After an analysis of variance has indicated that there is definite indication that a factor is significant, that is, is influential on the observed variable, we should not just say the factor is "significant" and let it go at that. We should analyze further. Let us consider this in relation to the first experiment, Table 12.1. There we had a categorical factor: batches. After finding that there are real batch-to-batch differences, one might be tempted to make a $t$ test between each pair of batches, that is 1 versus 2, ..., 1 versus 8, 2 versus 3, ..., 7 versus 8. But how many such comparisons are there? By (4.31) the number is $C(8, 2) = 28$. Now if we were to use an $\alpha = .05$ risk on each such test, we should expect *purely by chance* $.05(28) = 1.4$, that is, one or two "apparently significant" differences when in fact all the data came from one perfectly homogeneous population. Further, we might readily find three or four such spurious differences apparently significant. As a matter of fact, therefore, we really have no right to use such a $t$ test on more than seven such comparisons, because we had only seven degrees of freedom for MS$A$. Moreover we should be making *independent* comparisons or contrasts.

Let us see how this can be done. In general suppose that we have $a$ levels of factor $A$, so that MS$A$ carries $a - 1$ degrees of freedom. Then we may form $a - 1$ distinct contrasts (as a maximum). For simplicity we suppose that the number of replications $n_i$ is constant, say $n$. (If the $n_i$ vary, we may still handle the contrasts readily.) For the first "contrast" $C_1$ we take a set of constants with the following properties

$$C_1 = c_{11}T_1 + c_{21}T_2 + \cdots + c_{a1}T_a . \qquad (12.25a)$$

Then the second to the last contrast are

$$C_2 = c_{12}T_1 + c_{22}T_2 + \cdots + c_{a2}T_a \qquad (12.25b)$$
$$\vdots$$
$$C_{a-1} = c_{1,a-1}T_1 + c_{2,a-1}T_2 + \cdots + c_{a,a-1}T_a . \qquad (12.25c)$$

The constants are to have the following properties:

$$\sum_{i=1}^{a} c_{ij} = 0, \qquad j = 1, 2, ..., a - 1 \qquad (12.26)$$

$$\sum_{i=1}^{a} c_{ij}c_{ik} = 0, \qquad j \neq k, 1, 2, ..., a - 1. \qquad (12.27)$$

The last condition (12.27) is the "orthogonal" property, which guarantees that the $a - 1$ contrasts are independent.

Now how shall we choose the $c_{ij}$ constants? This is commonly not "cut and dried," that is, there are in general several approaches which can be made, depending on what comparisons we wish to study. One possible set for $a = 8$, if the batches were made in a time sequence, is

|  | $C_1$ | $C_2$ | $C_3$ | $C_4$ | $C_5$ | $C_6$ | $C_7$ |
|---|---|---|---|---|---|---|---|
| $i \backslash j$ | 1 | 2 | 3 | 4 | 5 | 6 | 7 |
| 1 | $-1$ | $-1$ | $-1$ | 0 | 0 | 0 | 0 |
| 2 | $-1$ | $-1$ | $+1$ | 0 | 0 | 0 | 0 |
| 3 | $-1$ | $+1$ | 0 | $-1$ | 0 | 0 | 0 |
| 4 | $-1$ | $+1$ | 0 | $+1$ | 0 | 0 | 0 |
| 5 | $+1$ | 0 | 0 | 0 | $-1$ | $-1$ | 0 |
| 6 | $+1$ | 0 | 0 | 0 | $-1$ | $+1$ | 0 |
| 7 | $+1$ | 0 | 0 | 0 | $+1$ | 0 | $-1$ |
| 8 | $+1$ | 0 | 0 | 0 | $+1$ | 0 | $+1$ |

Here $C_1$ will compare the first four totals $T_i$'s (or averages $\bar{y}_i$'s) with the last four. $C_2$ compares the first two with the third and fourth, while $C_3$ simply compares the first with the second. The others are similar. The reader should carefully check that $\sum_{i=1}^{8} c_{ij} c_{ik} = 0$ for all combinations of $j$ and $k$, not equal. That is, the elements of any one column when multiplied by the respective elements in any other column and the products added, always give zero.

Another set could be the following:

| | $C_1$ | $C_2$ | $C_3$ | $C_4$ | $C_5$ | $C_6$ | $C_7$ |
|---|---|---|---|---|---|---|---|
| $i \backslash j$ | 1 | 2 | 3 | 4 | 5 | 6 | 7 |
| 1 | −1 | −2 | −1 | 0 | 0 | 0 | 0 |
| 2 | −1 | −2 | +1 | 0 | 0 | 0 | 0 |
| 3 | −1 | +1 | 0 | −1 | −1 | 0 | 0 |
| 4 | −1 | +1 | 0 | −1 | +1 | 0 | 0 |
| 5 | −1 | +1 | 0 | +1 | 0 | −1 | 0 |
| 6 | −1 | +1 | 0 | +1 | 0 | +1 | 0 |
| 7 | +3 | 0 | 0 | 0 | 0 | 0 | −1 |
| 8 | +3 | 0 | 0 | 0 | 0 | 0 | +1 |

What comparisons does this collection make? For example, $C_1$ compares the average of the first six to that of the last two. Still another set could be

| | $C_1$ | $C_2$ | $C_3$ | $C_4$ | $C_5$ | $C_6$ | $C_7$ |
|---|---|---|---|---|---|---|---|
| $i \backslash j$ | 1 | 2 | 3 | 4 | 5 | 6 | 7 |
| 1 | −5 | −2 | 0 | 0 | 0 | 0 | 0 |
| 2 | −5 | +1 | −1 | 0 | 0 | 0 | 0 |
| 3 | −5 | +1 | +1 | 0 | 0 | 0 | 0 |
| 4 | +3 | 0 | 0 | −4 | 0 | 0 | 0 |
| 5 | +3 | 0 | 0 | +1 | −1 | −1 | 0 |
| 6 | +3 | 0 | 0 | +1 | −1 | +1 | 0 |
| 7 | +3 | 0 | 0 | +1 | +1 | 0 | −1 |
| 8 | +3 | 0 | 0 | +1 | +1 | 0 | +1 |

What comparisons does this make? It is hoped that the reader can begin to see how a set is formulated. He should always be careful to check (12.26) and (12.27) for his prospective contrasts.

Often when the categories are people, animals, materials, or catalysts, say, the desired contrasts may be fairly obvious. But experimentation is the best advice to the reader.

Now how do we make the tests? By using (12.25) for our set of constants we have a set of $a - 1$ contrasts, that is, numbers $C_1$, $C_2$, ..., $C_{a-1}$. Which, if any, are significantly nonzero? (Some ought to be if we had a significant factor $A$, overall.) We have

$$\text{SSC}_m = \frac{C_m{}^2}{n \sum_{i=1}^{a} c_{im}^2}, \qquad m = 1, ..., a - 1. \tag{12.28}$$

Thus for our first table of constants as used on Table 12.1, we find

$$C_1 = -1(5.85 + 8.16 + 7.60 + 6.97) + 1(6.66 + 5.16 + 7.16 + 5.16)$$
$$= -4.44.$$

Likewise $C_2 = +.56$, $C_3 = +2.31$, $C_4 = -.63$, $C_5 = +.50$, $C_6 = -1.50$, $C_7 = -2.00$. To form SSC, by (12.28) we have

$$(-4.44)^2/2 \cdot 8 = 1.2321 = \text{SSC}_1,$$

where the $2 = n$, and 8 is the sum of squares of $c_{i1}$'s, that is, in the $C_1$ column of the table. Again similarly, the other $\text{SSC}_m$, $m = 2$, ..., 7, are .0392, 1.3340, .0992, .0312, .5625, 1.0000. The sum of these seven $\text{SSC}_m$'s should be SSA. It is 4.2982, which agrees to within round-off errors, with SS$A = 4.2983$. Now we may make an $F$ test from each by dividing by the single degree of freedom 1, then by MSE = .1518. We compare such $F$'s with $F_{1,8,.05\,\text{above}} = 5.318$. Carrying out the work we find $C_1$, $C_3$, and $C_7$ significant. The first and last have negative values, indicating significant drops, respectively, between (1) the first four batches and the last four, and (2) the seventh and eighth batch. But $C_3$ indicates a significant increase from the first batch to the second. Thus we have analyzed in this way our partition of SS$A$.

There are many sources of material on this subject, of which we may suggest the work of Ostle [3] and Hicks [4].

## 12.3. ORTHOGONAL POLYNOMIALS AND TESTS

In Section 12.2.4 was presented a method for making various comparisons or contrasts between sample means (or totals). It is especially useful when the factor being studied is categorical rather than quantitative. But the method is general, and can also be applied to quantitative factors, through the form of so-called orthogonal polynomials, the subject of this section. These polynomials are usually only available to use if *both* (1) the jumps or increments in the factor levels are constant, *and* (2) we have the same number of observations at each level. (If one or both of these conditions do not hold, there still exist orthogonal polynomials which could do the desired job, but they are not tabulated.)

What orthogonal polynomials do, is to enable one to fit a polynomial model to data, using the two conditions just mentioned. The fitting is extremely easy, taking full advantage of the equal increments. Then one can test whether the linear component or sum of squares is significant. If so, then the next question is as to whether there is significant non-linearity; that is, having taken out the linear part of SS$A$, is there anything left which is not readily explainable by chance? If so, then we will usually next remove the pure quadratic component and test it for significance, and also the remainder of SS$A$ which is of higher order than second degree.

Let us introduce the analysis, using orthogonal polynomials, through an example. Consider the data of Table 12.3. It is quite obvious that

**TABLE 12.3**

Engler Endpoints—Highest Temperatures Reached without Fractionation in the Column. Sixteen Observations with Various Percents Aviation Solvent Oil Added

| Original | % Aviation solvent oil added, coded by $Y$ = observation − 300 | | |
|---|---|---|---|
| | 0.35 | 0.70 | 1.00 |
| 307 | 320 | 339 | 363 |
| 308 | 317 | 343 | 358 |
| 308 | 318 | 341 | 356 |
| 316 | 318 | 336 | 360 |
| 307 | 331 | 338 | 356 |
| 314 | 320 | 344 | 356 |
| 312 | 322 | 343 | 358 |
| 312 | 318 | 350 | 350 |
| 306 | 322 | 343 | 365 |
| 326 | 315 | 340 | 356 |
| 307 | 316 | 334 | 352 |
| 309 | 322 | 337 | 354 |
| 304 | 318 | 340 | 356 |
| 301 | 322 | 338 | 356 |
| 307 | 315 | 330 | 360 |
| 302 | 315 | 342 | 360 |
| $T_i$ 146 | 309 | 638 | 916 $T = 2009$ |
| $\sum_j Y_{ij}^2$ 1878 | 6237 | 25,758 | 52,658 86,507 |

SST $= 86,507 - 2009^2/64 = 86,507 - 63,063.77 = 23,443.23$
SS$A = (146^2 + \cdots + 916^2)/16 - 2009^2/64 = 85,181.06 - 63,063.77 = 22,117.29$
SSE $= 86,507 - 85,181.06 = 1325.94$

the addition of the solvent oil raises the Engler endpoints. It also seems to be a rather linear relation over this range of the factor: oil addition. But let us make the analysis to see whether the linear model is fully adequate. The table shows the two sums of squares into which SST is partitioned (after subtracting 300 from each observation). The analysis of variance table follows:

| Source of variation | Sum of squares | Degrees of freedom | Mean square | $F_{3,60}$ | $F_{3,60,.05 \text{ above}}$ |
|---|---|---|---|---|---|
| Oil added | 22,117.29 | 3 | 7372.43 | 333.6 | 2.758 |
| Error | 1,325.94 | 60 | 22.10 | | |
| Total | 23,443.23 | 63 | | | |

From this table it is obvious that the amount of oil added is enormously significant in a statistical sense. But is the relation purely linear? At this point we could fit a straight line by the methods of Chapter 11, find $r$, and then the $SSA$(linear) $= r^2 SSA$, and the nonlinear part is $(1 - r^2)$ $SSA$. To use orthogonal polynomials we need to have all increments equal. But $1.00 - .70\%$ oil is not the same as the other increments, that is, $.35\%$. So technically we cannot use ordinary orthogonal polynomials here. However, the gaps are nearly equal, and we will analyze as though the last were $1.05\%$ oil. Then coding the levels of factor $A$ by level $+.35\%)/.35\% = X$, the four levels become $X = 1, 2, 3, 4$. (We can always code levels with equal increments so that they become $1, 2, ..., a$.) Now the orthogonal polynomial values are commonly denoted by that difficult to write Greek letter $\xi$, with one or two subscripts. For four levels such as we have, we can separate out three parts: pure linear, quadratic, and cubic terms, that is, $SSA$ can be broken down into three parts. The $\xi_{ik}$'s are given in the accompanying table. See also Table 12.4.

| $X = i$ | Linear $k = 1, \xi_{i1}$ | Quadratic $k = 2, \xi_{i2}$ | Cubic $k = 3, \xi_{i3}$ |
|---|---|---|---|
| 1 | $-3$ | $+1$ | $-1$ |
| 2 | $-1$ | $-1$ | $+3$ |
| 3 | $+1$ | $-1$ | $-3$ |
| 4 | $+3$ | $+1$ | $+1$ |

Suppose that we wish to fit only a linear model to the $T_i$'s. That is, we will use

$$\hat{T}_i = A_0 + A_1 \xi_{i1}. \tag{12.29}$$

Now $A_0$ is merely $\overline{T}$ over all the data, that is, $T/N = 2009/4 = 502.25$. (Remember we are fitting to $T$'s not $\overline{Y}$'s.) Next for $A_1$ we use

$$A_1 = \frac{\sum_{i=1}^{4} \xi_{i1} T_i}{\sum_{i=1}^{4} \xi_{i1}^2} = \frac{(-3)146 + (-1)309 + (+1)638 + (+3)916}{(-3)^2 + (-1)^2 + (+1)^2 + (+3)^2}$$

$$= \frac{2639}{20} = 131.95.$$

So the fitted linear relationship is

$$\hat{T}_i = 502.25 + 131.95 \xi_{i1}.$$

Using the tabulated values of $\xi_{i1}$ we find the calculated linear values: 106.40, 370.30, 634.20, 898.10. These are shown by the line in Fig. 12.2

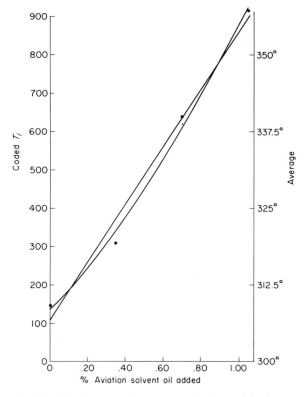

**Fig. 12.2.** Totals of 16 Engler endpoints at each of four levels of percent aviation solvent oil added. Data for 1.00 % were plotted at 1.05 %, as this change was used so as to fit by orthogonal polynomials. Straight line and general parabola shown as fitted to totals $T_i$. Right scale shows averages in original units. Data from Table 12.3.

as well as the actual $T_i$'s. For the sum of squares due to the linear part of $A$ we have

$$\text{SS}A(\text{linear}) = \frac{(\sum_{i=1}^{4} \xi_{i1} T_i)^2}{n \sum_{i=1}^{4} \xi_{i1}^2} = \frac{2639^2}{16(20)} = 21,763.50.$$

| Source of variation | Sum of squares | Degrees of freedom | Mean square | $F_{\text{obs}}$ | $F_{.05 \text{ above}}$ |
|---|---|---|---|---|---|
| $A$, oil added | 22,117.29 | 3 | 7,372.43 | 333.6 | 2.758 |
| $A$ (linear) | 21,763.50 | 1 | 21,763.50 | 984.8 | 4.001 |
| $A$ (nonlinear) | 353.79 | 2 | 176.90 | 8.00 | 3.150 |
| Error | 1,325.94 | 60 | 22.10 | | |
| Total | 23,443.23 | 63 | | | |

The linear part of factor $A$ is of course highly significant, in fact $F$ is much larger than it was for all of factor $A$. (In this way, it occasionally occurs that although a factor, taken as a whole, is not significant, yet the linear component proves to be!) Now we see, however, that the nonlinear component of factor $A$ is also clearly significant with $\alpha = .05$. Thus we now go further and fit a quadratic term. We use the general quadratic model

$$\hat{T}_i = A_0 + A_1 \xi_{i1} + A_2 \xi_{i2}. \tag{12.30}$$

Since the linear and quadratic coefficients are orthogonal, the effects are independent, and so we do not have to start all over again. Our $A_0$ and $A_1$ do not change and we find

$$A_2 = \frac{\sum_{i=1}^{4} \xi_{i2} T_i}{\sum_{i=1}^{4} \xi_{i2}^2} = \frac{(+1)146 + (-1)309 + (-1)638 + (+1)916}{(+1)^2 + (-1)^2 + (-1)^2 + (+1)^2}$$

$$= \frac{115}{4} = 28.75.$$

So the fitted equation of second degree is

$$\hat{T}_i = 502.25 + 131.95 \xi_{i1} + 28.75 \xi_{i2}.$$

Substituting in turn the tabulated values of the $\xi_{ik}$ we find the $T_i$'s: 135.15, 341.55, 605.45, 926.85. These are shown on Fig. 12.2. The departures from the line are entirely due to the $A_2$ term in the equation. The first, second, and fourth points are fitted much better, whereas the parabola misses the third point considerably further.

Let us now see whether the quadratic and residual sums of squares are significant. We have the following table:

| Source of variation | Sum of squares | Degrees of freedom | Mean square | $F_{0 \text{ bs}}$ | $F_{.05 \text{ above}}$ |
|---|---|---|---|---|---|
| $A$, oil added | 22,117.29 | 3 | 7,372.43 | 333.6 | 2.758 |
| $A$ (linear) | 21,763.50 | 1 | 21,763.50 | 984.4 | 4.001 |
| $A$ (quadratic) | 206.64 | 1 | 206.64 | 9.35 | 4.001 |
| $A$ (residual) | 147.15 | 1 | 147.15 | 6.66 | 4.001 |
| Error | 1,325.94 | 60 | 22.10 | | |
| Total | 23,443.23 | 63 | | | |

This table shows that, while the quadratic term is significant, the departures from the best parabola are also significant.

We can go further and fit the pure cubic term for the equation

$$\hat{T}_i = A_0 + A_1\xi_{i1} + A_2\xi_{i2} + A_3\xi_{i3}. \tag{12.31}$$

The other $A$'s are undisturbed, and we find as before

$$A_3 = \frac{-217}{20} = -10.85 \quad \text{and} \quad SSA(\text{cubic}) = \frac{(-217)^2}{16(20)} = 147.15.$$

But note that this is exactly the $SSA$(residual). This must always be, because any four points with unequal $X$'s can always be fitted exactly by some third-degree polynomial.

Now although we believe the departures from the parabola to be real, we cannot assert that the true regression is this cubic. There is naturally some error. We can if desired set confidence limits for the true regression curve at the various points. Let us see how. These are for $T_i$'s. We probably are more interested in the averages at the various levels of percent oil. In uncoded form these are $300° + 146°/16 = 309.1$, 319.3, 339.9, and 357.2°. Each is the average of 16 readings. We proceed exactly as in (10.17), but replace $s$ there by $\sqrt{\text{MSE}}$. Thus $1 - \alpha$ confidence limits for $\mu_i$ are

$$\bar{y}_i \pm t_{\alpha/2} \sqrt{\text{MSE}}/\sqrt{n_i}, \tag{12.32}$$

where the degrees of freedom for $t$ are those for MSE. For 90% confidence limits we then have

$$\bar{y}_i \pm 1.671 \sqrt{22.10}/\sqrt{16} = \bar{y}_i \pm 1.96.$$

These should be thought of in terms of the right-hand vertical scale for average points.

Let us now set out the general formulas and provide a brief table of of the orthogonal multipliers we have been using.

$$\hat{T}_i = \bar{T} + A_1\xi_{i1} + A_2\xi_{i2} + A_3\xi_{i3}, \qquad i = 1, ..., a \qquad \text{(estimation equation)}.$$
$$(12.33)$$

We only go this far (to a cubic, or even only to a quadratic) and would in general call everything else "residual" if $a > 4$ levels of factor $A$:

$$A_j = \sum_{i=1}^{a} \xi_{ij}T_i \Big/ \sum_{i=1}^{a} \xi_{ij}^2 \qquad (12.34)$$

$$SSA(j\text{-degree term}) = \left(\sum_{i=1}^{a} \xi_{ij}T_i\right)^2 \Big/ n \sum_{i=1}^{a} \xi_{ij}^2, \qquad j = 1, 2, 3. \quad (12.35)$$

A few further comments may be useful. To fit data with one $Y$ at each of $a$ equally spaced $X$'s, replace $T_i$ by $Y_i$, and call $n = 1$. Then using (12.34) we will be fitting a straight line, or a general parabola, or a cubic according as we go up to $j = 1, 2,$ or $3$. Secondly, any equation such as (12.33) or of higher degree may be *uniquely* changed into one of the following form

$$\hat{T}_i = a_0 + a_1X_i + \cdots + a_kX_i^k, \qquad k < a, \qquad (12.36)$$

because the $\xi_{ij}$'s are polynomials in $X_i$. For example, if $a = 4$

$$\xi_{i1} = 2X_i - 5, \qquad \xi_{i2} = X_i^2 - 5X_i + 5$$
$$\xi_{i3} = (10X_i^3 - 75X_i^2 + 167X_i - 105)/3.$$

**TABLE 12.4**

Linear, Quadratic, and Cubic Orthogonal Polynomial Values for Two to Eight Levels of $\xi_{ij}$'s[a]

| $a$ | 2 | 3 | | 4 | | | 5 | | | 6 | | | 7 | | | 8 | | |
|---|---|---|---|---|---|---|---|---|---|---|---|---|---|---|---|---|---|---|
| $i \backslash j$ | 1 | 1 | 2 | 1 | 2 | 3 | 1 | 2 | 3 | 1 | 2 | 3 | 1 | 2 | 3 | 1 | 2 | 3 |
| 1 | $-1$ | $-1$ | $+1$ | $-3$ | $+1$ | $-1$ | $-2$ | $+2$ | $-1$ | $-5$ | $+5$ | $-5$ | $-3$ | $+5$ | $-1$ | $-7$ | $+7$ | $-7$ |
| 2 | $+1$ | $0$ | $-2$ | $-1$ | $-1$ | $+3$ | $-1$ | $-1$ | $+2$ | $-3$ | $-1$ | $+7$ | $-2$ | $0$ | $+1$ | $-5$ | $+1$ | $+5$ |
| 3 | | $+1$ | $+1$ | $+1$ | $-1$ | $-3$ | $0$ | $-2$ | $0$ | $-1$ | $-4$ | $+4$ | $-1$ | $-3$ | $+1$ | $-3$ | $-3$ | $+7$ |
| 4 | | | | $+3$ | $+1$ | $+1$ | $+1$ | $-1$ | $-2$ | $+1$ | $-4$ | $-4$ | $0$ | $-4$ | $0$ | $-1$ | $-5$ | $+3$ |
| 5 | | | | | | | $+2$ | $+2$ | $+1$ | $+3$ | $-1$ | $-7$ | $+1$ | $-3$ | $-1$ | $+1$ | $-5$ | $-3$ |
| 6 | | | | | | | | | | $+5$ | $+5$ | $+5$ | $+2$ | $0$ | $-1$ | $+3$ | $-3$ | $-7$ |
| 7 | | | | | | | | | | | | | $+3$ | $+5$ | $+1$ | $+5$ | $+1$ | $-5$ |
| 8 | | | | | | | | | | | | | | | | $+7$ | $+7$ | $+7$ |

[a] $i = 1, 2, ..., a; j = 1, 2, 3.$

Try a few, by substituting in $X_i = 1, 2, 3, 4$. An extensive table of orthogonal polynomial values to $a = 104$ is given by Anderson and Houseman [5], and a useful shorter one by Owen [6]. See Table 12.4.

Finally we might remark that, had we used the correct value of $x = 1.00\%$, then $r^2 = .92336$ and SS(linear) $= 21,646.5$, instead of 21,763.5. This would have left SS(nonlinear) $= 470.8$ instead of 353.8. We would now have to start all over again using (11.52)–(11.54) to fit a parabola. Moral: Space $X$'s with equal increments if at all possible!

## 12.4. A METHOD OF MULTIPLE CONTRASTS

As has been pointed out in Section 12.2.4, we are not free to use an ordinary $t$ test (with $s_y = \sqrt{\text{MSE}}$) to compare *any* two of our $a$ sample means, which the data might suggest would be interesting. For example, we could not make a $t$ test for the highest and lowest $\bar{y}$'s. This is because there are too many possible comparisons of two means (more than our degrees of freedom), the comparisons are not independent, and our $\alpha$ risks for each pair are therefore not applicable. We have just discussed two feasible approaches in Sections 12.2.4 and 12.3, which make independent comparisons. We now discuss a third approach which may be used.

As the title of this section suggests, this is but one of several methods available which have differing characteristics and interpretations. But they may each be justifiably used, *after* the data are obtained, if they are properly interpreted. The literature on these methods is quite extensive and some areas of research are still controversial and developing. Some general material and a lead into the literature may be obtained in references [7–11]. Dunnett [8] and Steel [11] are especially accessible and useful in interpretation.

The multiple comparison test we shall present here and illustrate is described by Duncan [7, p. 26]. It is called the Newman–Keuls test having been developed by D. Newman, 1939, and M. Keuls, 1952. See the references presented by Duncan [7]. It can be called a "multiple range test" because it successively tests a series of ranges of the $\bar{y}_i$'s, for significance. In a sense it tests all of the possible pairs of $\bar{y}_i$'s, but takes account of how many $\bar{y}_i$'s lie between the two being tested at any given moment.

### 12.4.1. The Newman–Keuls Multiple Range Test.

The basic approach is as follows: "the difference between any two means in a set of $k$ means is significant, provided the range of each and every subset which contains the given two means is significant according to an $\alpha$ level test" [7]. In order to accomplish the testing without considering all possible subsets we proceed as follows:

The first step is to find the means of each sample and arrange them in order of magnitude, renaming so that we have

$$\bar{y}_1 > \bar{y}_2 > \cdots > \bar{y}_k .$$

(If there are equal means, an equality could occur.)

Next form a triangular table of all possible differences (ranges) of $\bar{y}_i$'s as follows:

| Rank index | $k$ | $k-1$ | $\cdots$ | 3 | 2 |
|---|---|---|---|---|---|
| 1 | $\bar{y}_1 - \bar{y}_k$ | $\bar{y}_1 - \bar{y}_{k-1}$ | $\cdots$ | $\bar{y}_1 - \bar{y}_3$ | $\bar{y}_1 - \bar{y}_2$ |
| 2 | $\bar{y}_2 - \bar{y}_k$ | $\bar{y}_2 - \bar{y}_{k-1}$ | $\cdots$ | $\bar{y}_2 - \bar{y}_3$ | |
| $\vdots$ | $\vdots$ | $\vdots$ | $\cdots$ | | |
| $k-2$ | $\bar{y}_{k-2} - \bar{y}_k$ | $\bar{y}_{k-2} - \bar{y}_{k-1}$ | | | |
| $k-1$ | $\bar{y}_{k-1} - \bar{y}_k$ | | | | |

To test these ranges for significance, we must find critical ranges for each sample size (number of $\bar{y}_i$'s). We estimate $\sigma_{\bar{y}}$ by

$$s_{\bar{y}} = \sqrt{(MSE)/n}.$$

Then we multiply $s_{\bar{y}}$ by appropriate $q_\alpha$'s from Table VII, in the Appendix, using an $\alpha$ of .05 or .01, as desired. Each part of this table is a double-entry one with the row determined by the degrees of freedom of MSE and the column by the number of means $\bar{y}_i$ . Thus we form critical ranges

$$R_k = q_\alpha(k, \text{df})\, s_{\bar{y}}$$
$$R_{k-1} = q_\alpha(k-1, \text{df})\, s_{\bar{y}}$$
$$\vdots$$
$$R_3 = q_\alpha(3, \text{df})\, s_{\bar{y}}$$
$$R_2 = q_\alpha(2, \text{df})\, s_{\bar{y}} .$$

Then we start in row 1 of the table of ranges, at the left, $\bar{y}_1 - \bar{y}_k$ . If $\bar{y}_1 - \bar{y}_k > R_k$ , we declare $\bar{y}_1$ to be greater than $\bar{y}_k$ but if not, then we stop the test. But if significant, we next compare $\bar{y}_1 - \bar{y}_{k-1}$ with $R_{k-1}$ . Declare $\bar{y}_1 - \bar{y}_{k-1}$ significant if it is greater than $R_{k-1}$ . If not, stop. If so, proceed. Continue this way until the first nonsignificant difference occurs in row 1. Then all differences *to the right of this* first nonsignificant difference in row 1 are also declared nonsignificant. Also all differences *below* each nonsignificant difference in row 1 are also declared to be nonsignificant.

Now starting at the left of row 2, *consider only those columns whose row 1 differences were declared nonsignificant*:

Declare $\bar{y}_2 - \bar{y}_k$ significant if it is greater than $R_{k-1}$. If not, stop. If so, proceed. Continue in this way until the first nonsignificant difference occurs in row 2. All differences *to the right* of this in row 2 are then also declared nonsignificant. All differences *below* these nonsignificant differences in row 2, in the same columns, are also declared nonsignificant.

Proceed similarly in subsequent rows until any remaining differences are ruled nonsignificant.

### 12.4.2. Example.

Now let us illustrate the foregoing. We shall use the following data from Hoefs [12] which give the loss in weight of disks of aluminum bronze suspended in sulfuric acid. The materials studied differ in the addition of metallics, as follows:

|  | None | Silver .36% | Silicon .27% | Silver .87% | Silicon .50% |
|---|---|---|---|---|---|
| $\bar{y}$ | 31.80 | 30.13 | 30.10 | 32.58 | 31.83 |
| $s^2$ | 1.181 | 2.085 | 5.008 | 1.170 | 1.640 |
| $n$ | 10 | 10 | 10 | 10 | 10 |

The unit was mg/dm² day loss from corrosion.

First we rank and name as follows:

$$\bar{y}_1 = 32.58, \quad \bar{y}_2 = 31.83, \quad \bar{y}_3 = 31.80, \quad \bar{y}_4 = 30.13, \quad \bar{y}_5 = 30.10.$$

Next find the critical ranges:

$$\text{MSE} = \sum s_i^2/5 = 11.084/5 = 2.2168.$$

Then $s_{\bar{y}} = \sqrt{\text{MSE}/n} = \sqrt{2.2168/10} = .471$. The appropriate degrees of freedom are for MSE, that is, $9(5) = 45$. Now suppose we use $\alpha = .05$. Then in the first part of Table VII (see the Appendix), we find $q_{.05}(5,45) = 4.02$, $q_{.05}(4,45) = 3.78$, $q_{.05}(3,45) = 3.43$, $q_{.05}(2,45) = 2.85$. Multiplying these by .471 yields $R_5 = 1.89$, $R_4 = 1.78$, $R_3 = 1.62$, $R_2 = 1.34$.

Now form the table of differences:

| Rank index | 5 | 4 | 3 | 2 |
|---|---|---|---|---|
| 1 | $\bar{y}_1 - \bar{y}_5 = 2.48$ | $\bar{y}_1 - \bar{y}_4 = 2.45$ | $\bar{y}_1 - \bar{y}_3 = .78$ | $\bar{y}_1 - \bar{y}_2 = .75$ |
| 2 | $\bar{y}_2 - \bar{y}_5 = 1.73$ | $\bar{y}_2 - \bar{y}_4 = 1.70$ | $\bar{y}_2 - \bar{y}_3 = .03$ | |
| 3 | $\bar{y}_3 - \bar{y}_5 = 1.70$ | $\bar{y}_3 - \bar{y}_4 = 1.67$ | | |
| 4 | $\bar{y}_4 - \bar{y}_5 = .03$ | | | |

Comparing $\bar{y}_1 - \bar{y}_5 = 2.48$ with $R_5 = 1.89$ we declare $\bar{y}_1$ to be significantly above $\bar{y}_5$. Next compare $\bar{y}_1 - \bar{y}_4 = 2.45$ with $R_4 = 1.78$. So $\bar{y}_1$ is significantly above $\bar{y}_4$. But $\bar{y}_1 - \bar{y}_3 = .78 < R_3 = 1.62$. Hence $\bar{y}_1$ is not significantly above $\bar{y}_3$, nor $\bar{y}_2$ either. Moreover $\bar{y}_2$ is, therefore not significantly above $\bar{y}_3$ since the range $\bar{y}_2 - \bar{y}_3$ lies below the range $\bar{y}_1 - \bar{y}_3$, which is nonsignificant. Hence we now go to the second row. Since $\bar{y}_2 - \bar{y}_5 = 1.73 < R_4 = 1.78$ (barely), this difference is not significant. Hence none of the other possible differences in row 2 is significant and hence none in the other rows. Thus all we learn "for sure" is that $\bar{y}_1$ is significantly above $\bar{y}_4$ and $\bar{y}_5$. Thus the corrosion rate is judged to be lower for both the low silicon and low silver alloys than for the high silver alloy. But this is all we can say, using $\alpha = .05$, with the Newman–Keuls test.

Some find it useful to summarize with underlining. For example, here we would have

$$\bar{y}_1 = 32.58, \qquad \bar{y}_2 = 31.83, \qquad \bar{y}_3 = 31.80, \qquad \bar{y}_4 = 30.13, \qquad \bar{y}_5 = 30.10$$

The upper line indicates that we cannot surely separate $\bar{y}_1$, $\bar{y}_2$, $\bar{y}_3$, while the lower line says we cannot surely separate $\bar{y}_2$, $\bar{y}_3$, $\bar{y}_4$, $\bar{y}_5$.

### 12.4.3. Interpretation of Risk $\alpha$.

Now just what is the meaning of the $\alpha$ risk in this technique? This is the probability of committing a type I error on any single difference *selected a priori or at random* that is, concluding $\mu_i \neq \mu_j$ from $|\bar{y}_i - \bar{y}_j|$ when in fact $\mu_i = \mu_j$. Moreover, $\alpha$ is the probability of a type I error on at least one difference, in particular the largest difference, if $\mu_1 = \mu_2 = \cdots = \mu_k$.

A point worth mentioning is that even if the analysis of variance test gives significance at $\alpha = .05$, it can work out in using the Newman–Keuls test that no pair of means shows a significant difference at $\alpha = .05$. The author, to his annoyance, found such a case while seeking an illustrative example for this section.

## 12.5.* TESTING HOMOGENEITY OF VARIANCES

We now discuss a problem of interest in its own right, and also a way of checking the assumption of a constant $\sigma_e^2$ in analysis of variance or regression (homoscedasticity). The available techniques do for variances what one-factor analysis of variance does for means. Namely, we test the hypothesis $\sigma_1^2 = \sigma_2^2 = \cdots = \sigma_k^2$ versus the alternate hypothesis: $\sigma_i^2$'s not all equal. We do this by using observed variances $s_i^2$, $i = 1, \ldots, k$.

We shall assume that all of the sample sizes are equal; that is, $n_i = n$, $i = 1, \ldots, k$. The standard test goes by the name "Bartlett's test," it having been developed by M.S. Bartlett. More complete exposition may be found in other books, including appropriate formulas when $n_i$'s vary. See, for example, Ostle [3]. The formula using equal sample sizes is the following one:

$$K_{k-1}^2 = \left[ k(n-1) \log_e \left( \sum_{i=1}^{k} s_i^2/k \right) - (n-1) \sum_{i=1}^{k} \log_e s_i^2 \right] \Big/ C \qquad (12.37)$$

$$= \frac{k(n-1)}{C} \log_e[\mathrm{AM}(s_i^2)/\mathrm{GM}(s_i^2)] \qquad (12.38)$$

$$C = 1 + \frac{k+1}{3k(n-1)}, \qquad \mathrm{GM}(s_i^2) = (s_1^2 \cdot s_2^2 \cdot \cdots \cdot s_k^2)^{1/k} \qquad (12.39)$$

$$K_{k-1}^2 \text{ approximately } \chi^2 \text{ distribution, } k-1 \text{ df.} \qquad (12.40)$$

Formula (12.37) is the calculational form. In fact one can use base 10 logarithms, by multiplying by $\log_e 10 = 2.3026$ at the end. Formula (12.38) shows that (12.37) is basically a comparison of the arithmetic and geometric means (AM, GM) of the variances, where $\mathrm{GM}(X_1, \ldots, X_k) = (X_1 \cdot X_2 \cdots X_k)^{1/k}$, an average. Now the greater the relative variability in a collection of positive quantities, the greater is the ratio of these two averages. Hence the more the relative variation among the $s_i^2$, the larger is (12.38). Thus to test the hypothesis $\sigma_i^2$ constant versus $\sigma_i^2$ not constant, we use a critical region on the upper tail, of size $\alpha$. We use chi-square tables for the test. If (12.37) lies above the critical point $\chi^2_{\alpha \text{above}, k-1 \text{df}}$, we reject the null hypothesis of constant $\sigma_e^2$, and so on. A rather simpler and to this author considerably more desirable test appears in an unpublished thesis [13]. It is actually linearly related to an earlier test by Brandt [14] and Stevens [15], who only found asymptotic distributions. Foster [13] called it the $Q$ test, where the sample value is found by

$$q = \sum_{i=1}^{k} (s_i^2)^2 \Big/ \left( \sum_{i=1}^{k} s_i^2 \right)^2. \qquad (12.41)$$

We may make a comparison similar to (12.38), remembering (if this is the right word) that the root mean square is RMS $(X_1, ..., X_k) = [\sum_{i=1}^{k} X_i^2/k]^{1/2}$, an average. Thus

$$q = \left[ \frac{\text{RMS}(s_i^2)}{\text{AM}(s_i^2)} \right]^2 \Big/ k. \qquad (12.42)$$

Since RMS $\geqslant$ AM and equal, only if all terms are equal, $q \geqslant 1/k$. The larger $q$ is, the greater the tendency toward heterogeneity, so again we use an upper tail test. Foster [13] and Burr and Foster [16] provide Table IX (see the Appendix), wherein $v = n - 1$ runs from 1 to 10 and $k = 2$ to 10, then through several more $k$'s to 64.

It is quite apparent that (12.41) is very easily calculated from a set of $s_i^2$, by cumulating $\sum s_i^4$ and $\sum s_i^2$ on a calculator.

The $Q$ test would seem to have the following advantages in comparison to Bartlett's test [13]:

1. It is simpler to calculate.
2. It does not make use of an asymptotic distribution. The accuracy of the asymptotic distribution for Barlett's test is somewhat difficult to assess. (But see Pearson and Hartley [17, p. 57].)
3. Bartlett's test is not usually recommended for use with sample $n$'s less than 4 or 5, because of inaccuracy of approximation to $\chi^2$.
4. Bartlett's test is more sensitive than the $Q$ test to a very small $s_i^2$, and less so to a very large $s_i^2$, and yet for testing the assumption of homogeneity of $\sigma_i^2$'s for applying an analysis of variance model, the latter is a much more serious breach of assumption than the former.
5. Bartlett's test blows up completely if any $s_i^2 = 0$, whereas the $Q$ test does not. This can well occur with small $n$'s and rather imprecise measurements.
6. The $Q$ test is apparently quite robust relative to nonnormality of population of $Y$'s.

### 12.5.1. An Example.

Let us now consider an example as illustration. See Table 12.5. Here are repeated measurements on the thicknesses of seven films. The important matter of interest is not as to whather the averages differ; we know they do. What is of interest is whether the variabilities, that is, measurement errors, are homogeneous. At the bottom of the table are given some of the details in making the two tests. We find $q$ by (12.41) as

$$q = .244 \qquad \text{versus} \qquad q_{.05} = .382.$$

**TABLE 12.5**

Thicknesses of Seven Polystyrene Films Calculated from Wavelengths and Spacing of Interference Fringes between 3.6 and 5.0 microns[a]

| Film sample | Run 1 | 2 | 3 | $\bar{y}_{mm}$ | Range $10^4_{mm}$ | $s_i^2 \, 10^8_{mm^2}$ |
|---|---|---|---|---|---|---|
| 1 | .0884 | .0888 | .0882 | .0885 | 6 | 9.333 |
| 2 | .0773 | .0764 | .0770 | .0769 | 9 | 21.00 |
| 3 | .0807 | .0797 | .0821 | .0808 | 24 | 145.33 |
| 4 | .0886 | .0894 | .0897 | .0892 | 11 | 32.33 |
| 5 | .0817 | .0823 | .0813 | .0818 | 10 | 25.33 |
| 6 | .0787 | .0808 | .0788 | .0794 | 21 | 140.33 |
| 7 | .0882 | .0898 | .0894 | .0891 | 16 | 69.33 |

$$\Sigma s_i^2 = 442.983, \qquad \Sigma s_i^4 = 47{,}834.91, \qquad (\Sigma s_i^2)^2 = 196{,}233.9$$
$$q = 47{,}834.91/196{,}233.9 = .244 \text{ versus } q_{.05} = .382$$
$$\Sigma s_i^2/7 = 63.2833, \qquad \log_{10}(\Sigma s_i^2/7) = 1.80129$$
$$\Sigma \log_{10} s_i^2 = 11.35591$$
$$K_6^2 = 2.3026[14(1.80129) - 2(11.35591)]/(1 + 8/42)$$
$$= 4.85 \text{ versus } \chi^2_{6,.05 \text{ above}} = 12.592$$

[a] Repeated observations; thicknesses in millimeters. See Copelin [18].

Hence the hypothesis of homogeneity (i.e., constant $\sigma_\varepsilon^2$) is tenable. Also by using base 10 logarithms in (12.37) we have

$$K_6^2 = 4.85 \qquad \text{versus} \qquad \chi^2_{6,.05 \text{ above}} = 12.592.$$

Again homogeneity is a tenable hypothesis.

### 12.5.2. Q Test with Unequal Degrees of Freedom.

When the sample sizes vary, so that the degrees of freedom $n_i - 1 = \nu_i$ are not equal, then it is suggested by Foster [13] that we define

$$q = \bar{\nu}(\nu_1 s_1^4 + \cdots + \nu_k s_k^4)/(\nu_1 s_1^2 + \cdots + \nu_k s_k^2)^2 \qquad (12.43)$$

and use Table IX (see the Appendix), interpolating on $\bar{\nu} = \sum_{i=1}^k \nu_i/k$.

### 12.5.3. Q Test for Ranges.

When the number of $y$'s in a sample is 2, we use the range in general. Then since when $n = 2$, $s^2 = \sum_{i=1}^2 (\bar{y}_i - \bar{y})^2/1 = R^2/2$, the $Q$ statistic becomes $q = \sum R_i^4/(\sum R_i^2)^2$, the 2's canceling out. Note that $n$ must be 2.

## 12.6. TYPES OF FACTORS

In two-factor analysis of variance designs it is important to distinguish between different types of factors. This is because the kinds of factors assumed make a considerable difference in the model to be assumed, in the tests to make, and in the interpretation of the results. We have already come across one distinction, that is, categorical factors versus numerical ones. Sections 12.2.4 and 12.4 were particularly applicable to the former, and 12.3 to the latter, if there were equal increments between the numerical levels. In both cases we would be assuming that the "levels" of the factors were like a population and that we were only interested in comparing results from precisely these levels. Such a factor in either case is called a "fixed factor." We are studying results from four controlled temperatures, or three pressures, or five lengths of time for an injection cycle, as examples of a *numerical* fixed factor. Or for a *categorical* fixed factor, we could have three particular analysts, our only four test sets, three catalysts being compared, four days of production, or the three shipments just received. These are the only ones under consideration, and we are not trying to extrapolate our results to any other analysts, test sets, catalysts, days, or shipments.

As distinct from the categorical fixed factors, we have categorical "random factors." Suppose that we choose five lots at random from 50 from a supplier; we analyze these and then try to use the results on the five to say something about the variability between the 50 lots. Then we have a random factor. It is from a sample. Or again, we might choose eight operators out of 100 to study, and learn something about the variability of the 100. We might choose a sample of tomato plants or a sample of batches to study. These would be random factors.

In the case of *fixed* factors we will be interested in such comparisons as in Sections 12.2.4, 12.3, and 12.4, and also in the means $\mu + \alpha_i$ as estimated by $\bar{y}_i$ or the differential effects $\alpha_i$ as estimated by $a_i = \bar{y}_i - \bar{y}$. On the other hand, for a random factor such things are of only secondary interest. The main interest is in the variability of the population of levels, that is, $\sigma_\alpha^2$. This is estimated as we shall see by $s_\alpha^2$.

In the next section we shall gain further perspective on factors, models, and interpretations.

We also have "nested factors." Repeated measurements are an example of a nested random factor. Such observations are basically a random sample from a population. In our one-factor design they would be a sample from a population of mean $\mu + \alpha_i$ and variance $\sigma_\varepsilon^2$. Now the order in which we measure the repeated observations and record them is purely happenstance. The errors $\varepsilon_{ij}$ are independent; $y_{11}$ happening to be relatively high so that $\varepsilon_{11}$ is high, has no influence upon $y_{21}$ being high or low. They are uncorrelated, and the various sets of $n$ $\varepsilon_{ij}$'s each are simply random samples.

## 12.7. ANALYSIS OF VARIANCE FOR TWO FACTORS

We shall now go to experimental designs where there are two factors being considered and studied. The factors can be either fixed or random. Say that we have $a$ levels of factor $A$ and $b$ levels of factor $B$. Then there are of course by (4.32) $ab$ distinct combinations of levels of the two factors. Now suppose that for *each* combination of a level of factor $A$ and a level of factor $B$, we take $n$ repeated observations or runs. This then gives us $abn$ observations in all. Such an experimental design is called a "two-factor completely randomized design with replication." The "replication" refers to the repeated (nested) measurements. Let us look at some typical data.

### 12.7.1. An Example.

Table 12.6 shows data which compare ratios of expenditure of energy on a 10-lb task to that for a 5-lb task, for four different subjects (people) at each of four different paces (relative speeds of working). This might seem as though we had three factors: person, pace, and weight. But the factor of weight is eliminated by division: energy expended at 10 lb divided by energy expended at 5 lb. Such *ratios* are the variables under study. This gives 16 different combinations of conditions: person by pace. Thus there were 16 "cells." Within each cell there were two repeated observations, hence 32 observed $y$'s. (The repeated observations are "nested" within the cells.)

Now what kinds of factors do we have? The choice of subjects, if it is a random sample from some population of subjects, would mean that we should treat factor $A$ as random. On the other hand, factor $B$ had a a set of four paces set at convenient levels of interest 60–120%. So it is a fixed factor. When we have one fixed factor and one random factor we have a "mixed" model.

Now let us begin our analysis of the data. The 32 measurements were rounded to the best two decimal places from the original data. For this type of design we use three subscripts on $y$:

$$y_{ijk}, \quad i = 1, ..., 4, \quad \text{factor } A; \quad j = 1, ..., 4, \quad \text{factor } B;$$

$$k = 1, 2, \quad \text{replication.}$$

**TABLE 12.6**

A Two-Factor Mixed Model, Ratios of Energy Expenditures for a Sample of Four Subjects and at Four Paces[a]

| Subject | Pace—% normal | | | | |
|---|---|---|---|---|---|
| | 60 | 80 | 100 | 120 | |
| $i$ | 1 | 2 | 3 | 4 | $T_{i..}$ |
| | 2.70 | 1.38 | 2.35 | 2.26 | |
| 1 | 3.30 (.60) | 1.35 (.03) | 1.95 (.40) | 2.13 (.13) | |
| | 6.00 | 2.73 | 4.30 | 4.39 | 17.42 |
| | 1.70 | 1.74 | 1.67 | 3.41 | |
| 2 | 2.14 (.44) | 1.56 (.18) | 1.50 (.17) | 2.56 (.85) | |
| | 3.84 | 3.30 | 3.17 | 5.97 | 16.28 |
| | 1.90 | 3.14 | 1.63 | 3.17 | |
| 3 | 2.00 (.10) | 2.29 (.85) | 1.05 (.58) | 3.18 (.01) | |
| | 3.90 | 5.43 | 2.68 | 6.35 | 18.36 |
| | 2.72 | 3.51 | 1.39 | 2.22 | |
| 4 | 1.85 (.87) | 3.15 (.36) | 1.72 (.33) | 2.19 (.03) | |
| | 4.57 | 6.66 | 3.11 | 4.41 | 18.75 |
| $T_{.j.}$ | 18.31 | 18.12 | 13.26 | 21.12 | $T = 70.81$ |

$$\sum_i T_{i..}^2 = 1257.1469, \qquad \sum_j T_{.j.}^2 = 1285.4725, \qquad \sum_i \sum_j T_{ij.}^2 = 339.1009$$

$$\sum_i \sum_j \sum_k y_{ijk}^2 = 171.3407, \qquad \sum \sum R_{ij}^2 = 3.5805$$

[a] Expenditure, 10-lb task/expenditure, 5-lb task. There are two repeated runs for each combination. Data are rounded to two decimal places. Numbers in parentheses are ranges. Totals are below the two entries. See Barany [19].

In each cell we have the total of the two $y$'s. For the total for the $i, j$ cell we call it $T_{ij.}$, the "dot" meaning that we have summed out on that subscript. Thus $T_{12.} = y_{121} + y_{122} = 2.73$, and so on. Moreover we have put the range $R_{ij} = |y_{ij1} - y_{ij2}|$ in parenthesis for each cell so that we may use the shortcut formula (12.16) for SSE, since $n = 2$. We may then find the within-cell sum of squares error by

$$\text{SSE} = \sum_i \sum_j R_{ij}^2/2 = (.60^2 + .03^2 + \cdots + .03^2)/2 = 3.5805/2 = 1.79025.$$

Next we find the total for all of the data for subject 1. This is 6.00 +

$2.73 + 4.30 + 4.39 = 17.42$. Now what is a good name for this since we first summed on $k$ for the cell total then summed on $j$ across the table? The answer is $T_{1..}$. We thus have four totals $T_{i..}$ for factor $A$. Similarly, for factor $B$ we have four grand totals for the four columns, namely $T_{.j.}$, $T_{.1.} = 18.31$, and so on. Finally we have the grand total for 32 observations in the entire table, namely $T_{...}$, or more simply $T$. We now give the general formulas in form analogous to (12.6) and (12.8) followed by the numerical illustration:

$$\mathrm{SS}A = \left( \sum_{i=1}^{a} T_{i..} \right) \Big/ nb - T^2/nab, \qquad (12.44)$$

$$\mathrm{SS}A = (17.42^2 + 16.28^2 + 18.36^2 + 18.75^2)/2(4) - 70.81^2/32$$
$$= 1257.1469/8 - 156.68925 = 157.14336 - 156.68925 = .45411;$$

$$\mathrm{SS}B = \left( \sum_{j=1}^{b} T_{.j.}^2 \right) \Big/ na - T^2/nab, \qquad (12.45)$$

$$\mathrm{SS}B = (18.31^2 + 18.12^2 + 13.26^2 + 21.12^2)/2(4) - 70.81^2/32$$
$$= 1285.4725/8 - 156.68925 = 160.68406 - 156.68925 = 3.99481.$$

The student should note the large similarity of these with what we did for one-factor completely randomized designs. It is as though we had eight observations per level, which in fact we do.

Now the new feature here lies in looking toward the joint effect of the two factors, which we call "interaction of $A$ by $B$." To this end we first find the cell-to-cell sum of squares. This is precisely similar to a one-factor experiment with 16 samples of two each:

$$\mathrm{SS}(A, B \text{ cells}) = \left( \sum_{i=1}^{a} \sum_{j=1}^{b} T_{ij.}^2 \right) \Big/ n - T^2/nab, \qquad (12.46)$$

$$\mathrm{SS}(A, B \text{ cells}) = (6.00^2 + 2.73^2 + \cdots + 4.41^2)/2 - 70.81^2/32$$
$$= 339.1009/2 - 156.68925 = 169.55045 - 156.68925$$
$$= 12.86120.$$

Now, although it may not be obvious, $\mathrm{SS}A$ and $\mathrm{SS}B$ are contained within $\mathrm{SS}(A, B \text{ cells})$.* The next is to find the "interaction sum of squares $A$ by $B$":

$$\mathrm{SS}AB = \mathrm{SS}(A, B \text{ cells}) - \mathrm{SS}A - \mathrm{SS}B \qquad (12.47)$$

$$\mathrm{SS}AB = 12.86120 - .45411 - 3.99481 = 8.41228.^{\dagger}$$

* In fact if one ever obtains $\mathrm{SS}A + \mathrm{SS}B > \mathrm{SS}(A, B \text{ cells})$, then he simply starts over again in his calculation.

† The reader may well be frightened by the number of "significant" figures we have here carried. But it is not hard to carry many when using a desk calculator. And then we can be sure that round-off errors will not cause trouble.

The general method for finding the sum of squares *within* cells, commonly used only when $n > 2$, is

$$\text{SSE} = \sum_{i=1}^{a} \sum_{j=1}^{b} \sum_{k=1}^{n} y_{ijk}^2 - \sum_{i=1}^{a} \sum_{j=1}^{b} T_{ij.}^2/n, \qquad (12.48)$$

$$\text{SSE} = 171.3407 - 169.55045 = 1.79025.$$

Note that this result checks perfectly with that from ranges using

$$\text{SSE} = \sum_{i=1}^{a} \sum_{j=1}^{b} R_{ij}^2/2 \quad \text{ranges} \qquad \text{for} \quad n = 2. \qquad (12.49)$$

We now need to find the various degrees of freedom. For the whole set of *nab* observations in a table, there are $nab - 1$ degrees of freedom, or in our example, 31. How are these distributed? There are *ab* cells, and the sum of squares of the *n* deviations within a single cell carries $n - 1$ degrees of freedom. Hence, altogether for SSE we have $ab(n - 1) = nab - ab$. In our example this is $4(4)(2 - 1) = 16$. Next for factor *A* we had *a* totals, so the degrees of freedom are $a - 1$, and similarly for factor *B*, $b - 1$. Thus we have three in each. Finally for SS(*A, B* cells), this is based on *ab* totals and thus has $ab - 1$ degrees of freedom. This leaves $ab - 1 - (a - 1) - (b - 1) = ab - a - b + 1 = (a - 1)(b - 1)$ degrees of freedom for SS*AB*. Or in the example it is $3(3) = 9$.

We now put everything together in an analysis of variance table:

| Source of variation | Sum of squares (SS) | Degrees of freedom (df) | Mean square (MS) | $F$ | $F_{.05}$ | Expected mean square (EMS) |
|---|---|---|---|---|---|---|
| Subject, $A$ | SS$A$ = .4541 | 3 | .1514 | 1.35 | 3.239 | $\sigma_\varepsilon^2 + 8\sigma_\alpha^2$ |
| Pace, $B$ | SS$B$ = 3.9948 | 3 | 1.3316 | 1.42 | 3.863 | $\sigma_\varepsilon^2 + 2\sigma_{\alpha\beta}^2 + 8\sigma_\beta^2$ |
| Interaction $AB$ | SS$AB$ = 8.4123 | 9 | .9347 | 8.35 | 2.538 | $\sigma_\varepsilon^2 + 2\sigma_{\alpha\beta}^2$ |
| Cells $A$, $B$ | SS($A$, $B$) = 12.8612 | 15 | — | | | |
| Error | SSE = 1.7902 | 16 | .1119 | | | $\sigma_\varepsilon^2$ |
| Total | SST = 14.6514 | 31 | | | | |

At first glance the $F$ column seems mixed up. But this is because of the *kinds of factors* assumed in our mixed model. Look at the last column which gives the expected mean squares for this model. These are the long-run averages of the four mean squares if we were to repeat the whole experiment of 32 measurements, time after time: selecting four subjects at random from the population, and putting them through the same four paces twice each, in random order. It is assumed that the four variances in the last column are fixed, thus determining the long run averages as shown. But $\sigma_\alpha^2$ is the variance for the differential effects of the whole population of subjects, the collection of four $\alpha_i$'s varying from experiment to experiment. On the other hand, $\sigma_\beta^2$ is concerned only with the same four paces 60, 80, 100, and 120%, the corresponding differential effects $\beta_1$, $\beta_2$, $\beta_3$, $\beta_4$ being always the same from experiment to experiment. We define $\sigma_\beta^2 = \sum_{i=1}^{4} \beta_i^2/(4-1)$.

Now to test for the presence of $\sigma_{\alpha\beta}$ we divide as follows:

$$\frac{\text{MS}AB}{\text{MSE}} = \frac{\text{est}(\sigma_\varepsilon^2 + 2\sigma_{\alpha\beta}^2)}{\text{est } \sigma_\varepsilon^2} = \frac{.9347}{.1119} = 8.35 = F.$$

This is distributed as $F$ only if $\sigma_{\alpha\beta}^2 = 0$. It carries degrees of freedom $\nu_1 = 9$, $\nu_2 = 16$. If $\sigma_{\alpha\beta}^2 > 0$, then the numerator will tend to be larger than the denominator, and the observed ratio $F$ will have an enhanced chance of being larger than $F_{9,16,.05}$. Here this observed $F$ of 8.35 is far above the critical value, so we conclude that there is an "interaction." Let us defer the interpretation of this for the moment.

Consider next the test for the presence of the random factor $A$, that is, the differences between subjects. To test the hypothesis $\sigma_\alpha^2 = 0$ versus $\sigma_\alpha^2 > 0$, we use

$$\frac{\text{MS}A}{\text{MSE}} = \frac{\text{est}(\sigma_\varepsilon^2 + 8\sigma_\alpha^2)}{\text{est } \sigma_\varepsilon^2} = \frac{.1514}{.1119} = 1.35 = F.$$

This $F$ carries degrees of freedom $\nu_1 = 3$, $\nu_2 = 16$. Interpolating in Table V (see the Appendix), we find $F_{3,16,.05}$ to be 3.239. Since the observed value is well below the critical value, we have no reliable evidence of $\sigma_\alpha^2 > 0$, that is, of individual differences of subjects (over these four paces). If there *are* differences, it will take more data to establish their presence.

Next consider the hypothesis $\sigma_\beta^2 = 0$. Suppose we were *erroneously* to compare MS$B$ with MSE. We would have

$$\frac{\text{MS}B}{\text{MSE}} = \frac{\text{est}(\sigma_\varepsilon^2 + 2\sigma_{\alpha\beta}^2 + 8\sigma_\beta^2)}{\text{est } \sigma_\varepsilon^2} = \frac{1.3316}{.1119} = 11.90.$$

Now suppose that this ratio carrying 3 and 16 degrees of freedom were to lie above the corresponding $F_{.05}$, as it does. We would then have reliable evidence that $2\sigma_{\alpha\beta}^2 + 8\sigma_\beta^2 > 0$. But is this because $\sigma_{\alpha\beta}^2 > 0$, or

$\sigma_\beta{}^2 > 0$, or are both true? We cannot say. The two effects are "confounded." But looking at the EMS column we see that we can use

$$\frac{\text{MS}B}{\text{MS}AB} = \frac{\text{est}(\sigma_\varepsilon{}^2 + 2\sigma_{\alpha\beta}^2 + 8\sigma_\beta{}^2)}{\text{est}(\sigma_\varepsilon{}^2 + 2\sigma_{\alpha\beta}^2)} = \frac{1.3316}{.9347} = 1.42.$$

This carries $\nu_1 = 3$, $\nu_2 = 9$ degrees of freedom so that $F_{3,9,.05} = 3.863$. And thus we have no significance for factor $B$. If in fact $\sigma_\beta{}^2 > 0$, we shall need more evidence to establish the fact. We thus see how basic it is to know the kinds of factors involved and to pay close attention to the EMS column.*

Now let us consider the meaning of the so-called interaction. This effect being significant means that we have reliable evidence that the *pattern of ratios* for the different subjects over the four paces used differ. See Fig. 12.3. There we have plotted the cell averages of the four subjects over the four paces. Thus looking at Table 12.6 we find $\bar{y}_{11\cdot} = 6.00/2$, $\bar{y}_{12\cdot} = 2.73/2 = 1.365$, and so on. These are plotted and connected for the observed pattern for subject 1. Then for subject 2 we plot $\bar{y}_{21\cdot} = 3.84/2 = 1.92$, and so on. We thus have a graph for the energy ratios for each of the four subjects across the four paces. Now such patterns will always vary somewhat. But $\sigma_{\alpha\beta}^2$ being nonzero assures that there is something behind the differences in shapes between these

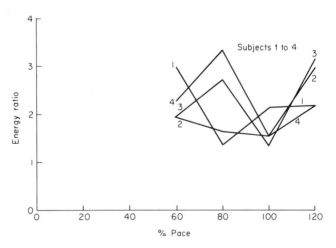

**Fig. 12.3.** Patterns of energy ratios (10lb to 5 lb) for four subjects at paces 60, 80, 100, and 120 %. Data from Table 12.6, plotted because there was a significant interaction.

---

* It might be well to remind the reader that we are here concerned with conclusions on the *ratio* of energies at 10- and 5-lb tasks, rather than either separately.

patterns. Note for example that over 60, 80, and 100%, the pattern of subjects 3 and 4 is just opposite to that for subject 1. Meanwhile 2 is fairly uniform. Then 2 and 3 end high while 1 and 4 end low at 120%. It is up to the experimenter to explain the physiological meaning of these varying reactions of the subjects to the tasks. The statistics simply say that there is something present in the model behind such differing patterns.

## 12.7.2. The Models and Assumptions.

Under the general classification of "two factors with replication" there are really three distinct models:

(1)  fixed: both factors $A$ and $B$ fixed;
(2)  random: both factors $A$ and $B$ random;
(3)  mixed: of factors $A$ and $B$, one is a fixed factor and the other random.

We have just seen an example of a mixed model.

To describe these models it is convenient to give three subscripts $i$, $j$, $k$ to each observed $y$. Subscript $i$ reflects the level of factor $A$, $j$ the level of factor $B$, and $k$ the number of the repeated measurement in the $(i, j)$ cell. Thus we have

$$Y_{ijk} = \mu + \alpha_i + \beta_j + (\alpha\beta)_{ij} + \varepsilon_{k(ij)}, \qquad i = 1, ..., a; \qquad j = 1, ..., b;$$
$$k = 1, ..., n. \qquad (12.50)$$

Here $\mu$ is the grand mean for all data from which we have a sample. Now if factor $A$ is a fixed factor, we have as in (12.3)

$$\sum_{i=1}^{a} \alpha_i = 0, \qquad \text{factor } A \text{ fixed.} \qquad (12.3)$$

On the other hand, if factor $A$ is random, then

$$\alpha_i\text{'s: NID}(0, \sigma_\alpha^2), \qquad \text{factor } A \text{ random;} \qquad (12.51)$$

that is, the $\alpha_i$'s are normally and independently distributed with mean zero and variance $\sigma_\alpha^2$. In this case the actually obtained $\alpha_i$'s will not in general add to zero. Assumptions for $\beta_j$'s are identical. As for $(\alpha\beta)_{ij}$, the interaction term, it is a differential effect which varies from cell to cell. If the true cell mean were dependent only on the two main factor effects $\alpha_i$ and $\beta_j$, it would be $\mu + \alpha_i + \beta_j$. But if there is an interaction effect between the two factors, then such a formula would not give the true cell mean, and for the true $(i, j)$ cell mean we would need $(\alpha\beta)_{ij}$ added, that is,

$$\mu_{ij} = \mu + \alpha_i + \beta_j + (\alpha\beta)_{ij}. \qquad (12.52)$$

If both factors $A$ and $B$ are fixed, we would commonly assume $(\alpha\beta)_{ij}$ to satisfy

$$\sum_j (\alpha\beta)_{ij} = 0 \quad \text{for all } i; \quad \sum_i (\alpha\beta)_{ij} = 0 \quad \text{for all } j. \quad (12.53)$$

For random and mixed models there are several possibilities. For a random model we generally assume

$$(\alpha\beta)_{ij}: \text{NID}(0, \sigma_{\alpha\beta}^2). \quad (12.54)$$

[It is not necessary in our analysis to assume normality in (12.52) or (12.54).] Finally the collection of $abn$ errors $\varepsilon_{k(ij)}$ are assumed NID$(0, \sigma_\varepsilon^2)$. The subscript $k(ij)$ means the $k$th repeated measurement under the conditions of the $(i, j)$ cell. Better reread this subsection!

### 12.7.3. Expected Mean Squares and Significance Tests.

The general formulas for finding the sums of squares reflecting the various "effects" we have just been considering are (12.44)–(12.49). They are appropriate no matter which of the three models of Section 12.7.2 we are using. Let us now give the three corresponding analysis of variance tables and describe the appropriate tests of the hypotheses $\sigma_\alpha^2 = 0$, $\sigma_\beta^2 = 0$, $\sigma_{\alpha\beta}^2 = 0$.

(1)  Fixed model (factors $A$, $B$ fixed):

| Source of variation | Sum of squares | Degrees of freedom | $MS = \dfrac{SS_i}{\nu_i}$ | $F = \dfrac{MS_i}{MS_j}$ | Critical $F_\alpha$ | Expected mean square |
|---|---|---|---|---|---|---|
| Factor $A$ | SS$A$ | $a - 1 = \nu_1$ | MS$A$ | $F_A = \dfrac{MSA}{MSE}$ | $F_{\alpha,\nu_1,\nu_4}$ | $\sigma_\varepsilon^2 + bn\sigma_\alpha^2$ |
| Factor $B$ | SS$B$ | $b - 1 = \nu_2$ | MS$B$ | $F_B = \dfrac{MSB}{MSE}$ | $F_{\alpha,\nu_2,\nu_4}$ | $\sigma_\varepsilon^2 + an\sigma_\beta^2$ |
| Interaction | SS$AB$ | $(a-1)(b-1) = \nu_3$ | MS$AB$ | $F_{AB} = \dfrac{MSAB}{MSE}$ | $F_{\alpha,\nu_3,\nu_4}$ | $\sigma_\varepsilon^2 + n\sigma_{\alpha\beta}^2$ |
| Error | SSE | $ab(n-1) = \nu_4$ | MSE | | | $\sigma_\varepsilon^2$ |
| Total | SST | $abn - 1$ | | | | |

Here the definitions of variances are

$$\sigma_\alpha^2 = \frac{\sum_{i=1}^a \alpha_i^2}{a - 1}, \quad \sigma_\beta^2 = \frac{\sum_{j=1}^b \beta_j^2}{b - 1}, \quad \sigma_{\alpha\beta}^2 = \frac{\sum_{i=1}^a \sum_{j=1}^b (\alpha\beta)_{ij}^2}{(a - 1)(b - 1)}.$$

Particularly note that MSE is the denominator for each of the three $F$ tests of the respective null hypotheses.

(2)  Random model (factors $A$, $B$ random):

| Source of variation | Sum of squares | Degrees of freedom | MS = $\dfrac{SS_i}{\nu_i}$ | $F = \dfrac{MS_i}{MS_j}$ | Critical $F_\alpha$ | Expected mean squares |
|---|---|---|---|---|---|---|
| Factor $A$ | SS$A$ | $a - 1 = \nu_1$ | MS$A$ | $F_A = \dfrac{MSA}{MSAB}$ | $F_{\alpha,\nu_1,\nu_3}$ | $\sigma_\varepsilon^2 + n\sigma_{\alpha\beta}^2 + nb\sigma_\alpha^2$ |
| Factor $B$ | SS$B$ | $b - 1 = \nu_2$ | MS$B$ | $F_B = \dfrac{MSB}{MSAB}$ | $F_{\alpha,\nu_2,\nu_3}$ | $\sigma_\varepsilon^2 + n\sigma_{\alpha\beta}^2 + na\sigma_\beta^2$ |
| Interaction | SS$AB$ | $(a - 1)(b - 1) = \nu_3$ | MS$AB$ | $F_{AB} = \dfrac{MSAB}{MSE}$ | $F_{\alpha,\nu_3,\nu_4}$ | $\sigma_\varepsilon^2 + n\sigma_{\alpha\beta}^2$ |
| Error | SSE | $ab(n - 1) = \nu_4$ | MSE | | | $\sigma_\varepsilon^2$ |
| Total | SST | $abn - 1$ | | | | |

In this case, the definitions of the variances are for the respective normal distributions for $\alpha_i$'s, $\beta_j$'s, $\alpha\beta_{ij}$'s. The null hypotheses are that these are, respectively, zero. As in our example of Section 12.7.1, to obtain a specific test for the presence or absence of $\sigma_\alpha^2$ or $\sigma_\beta^2$, we use MS$AB$ as the denominator of $F$. If we were to use MSE, and we found significance, we would not know whether it was due to the presence of $\sigma_\alpha^2$, or $\sigma_{\alpha\beta}^2$, or both, and so on.

(3)  Mixed model (factor $A$ random, $B$ fixed):

| Source of variation | Sum of squares | Degrees of freedom | MS = $\dfrac{SS_i}{\nu_i}$ | $F = \dfrac{MS_i}{MS_j}$ | Critical $F_\alpha$ | Expected mean squares |
|---|---|---|---|---|---|---|
| Factor $A$ | SS$A$ | $a - 1 = \nu_1$ | MS$A$ | $F_A = \dfrac{MSA}{MSE}$ | $F_{\alpha,\nu_1,\nu_4}$ | $\sigma_\varepsilon^2 + nb\sigma_\alpha^2$ |
| Factor $B$ | SS$B$ | $b - 1 = \nu_2$ | MS$B$ | $F_B = \dfrac{MSB}{MSAB}$ | $F_{\alpha,\nu_2,\nu_3}$ | $\sigma_\varepsilon^2 + n\sigma_{\alpha\beta}^2 + na\sigma_\beta^2$ |
| Interaction | SS$AB$ | $(a - 1)(b - 1) = \nu_3$ | MS$AB$ | $F_{AB} = \dfrac{MSAB}{MSE}$ | $F_{\alpha,\nu_3,\nu_4}$ | $\sigma_\varepsilon^2 + n\sigma_{\alpha\beta}^2$ |
| Error | SSE | $ab(n - 1) = \nu_4$ | MSE | | | $\sigma_\varepsilon^2$ |
| Total | SST | $abn - 1$ | | | | |

This is the general form of our example of Section 12.7.1, a mixed model. To test for $\sigma_\alpha^2$ versus $\sigma_\alpha^2 > 0$ for the random factor we use MSE in the denominator of $F$; while to test for $\sigma_\beta^2 = 0$ versus $\sigma_\beta^2 > 0$ for the fixed factor we use MS$AB$ in the denominator of $F$. This seems just the opposite from what we might expect if we look over the tables for the fixed and random models. But theoretical considerations leading to the given expected mean squares determine these tests. It seems almost unnecessary to point out that if $A$ is fixed and $B$ random, we merely interchange the analysis.

### 12.7.4. Interpretation of Significant Factors.

Whenever we obtain a significant $F$ test, indicating the presence of a factor effect, we seek further interpretation. For example, if factors $A$ and $B$ are random and we have evidence of $\sigma_\alpha^2 > 0$, then we can estimate $\sigma_\alpha^2$ by

$$\text{MS}A = \text{unbiased estimate of } \sigma_\varepsilon^2 + n\sigma_{\alpha\beta}^2 + nb\sigma_\alpha^2$$

$$\text{MS}AB = \text{unbiased estimate of } \sigma_\varepsilon^2 + n\sigma_{\alpha\beta}^2$$

$$\text{MS}A - \text{MS}AB = \text{unbiased estimate of } nb\sigma_\alpha^2.$$

Thus an unbiased estimate of $\sigma_\alpha^2$ is $(\text{MS}A - \text{MS}AB)/nb$. We can always use this approach, being careful to use the two mean squares employed in the $F$ test used to establish $\sigma_\alpha^2 > 0$.

Now if we have a fixed factor, the situation is somewhat different. Interest centers on the $\beta_j$'s rather than $\sigma_\beta^2$. To estimate $\beta_j$ we use $\bar{y}_{.j.} = T_{.j.}/an = \sum_{i=1}^a \sum_{k=1}^n y_{ijk}/an$, as follows:

$$b_j = \hat{\beta}_j = \bar{y}_{.j.} - \bar{y}_{...}.$$

But this may well be not enough for our purposes. We can then use the method of Section 12.3 if factor $B$ is a fixed quantitative factor with equally spaced levels. But if factor $B$ does not have equally spaced quantitative levels, or if it is a fixed categorical factor, then we may employ Section 12.4 or 12.2.4. In all cases we must be careful to use the MS in the denominator for the $F$ test for $\sigma_\beta^2 > 0$, as MSE, for example, in (12.32) or in Section 12.4. We also mention that either the $b_j$'s or the $\bar{y}_{.j.}$'s can be used in such analyses.

### 12.7.5. The Case of Unreplicated Two-Factor Experiments.

Often an experimenter will run each of $ab$ possible combinations of factor $A$ and factor $B$ just once. (He should be careful to perform the $ab$

experiments in random order to avoid biases.) Thus we have the case $n = 1$ in Sections 12.7.2 and 12.7.3.

At the outset let us say that the analysis is easier than when replication (repeated measurements) is present. But there are disadvantages. For one thing we cannot obtain separate estimates of $\sigma_\varepsilon^2$ and $\sigma_{\alpha\beta}^2$. These will be tied together or confounded. For another we have fewer degrees of freedom in the error mean square for the denominator of the $F$ tests. Third, we do not have a collection of cell variances $s_{ij}^2$ from which we can test for homogeneity of cell variance through using Section 12.5. For these reasons it is desirable whenever feasible to have replication, even if only two measurements per cell.

Let us look at the breakdown of SST for the observations $y_{ij}$ :

$$\text{SST} = \sum_{i=1}^{a} \sum_{j=1}^{b} y_{ij}^2 - T_{..}^2/ab, \qquad \text{df} = ab - 1, \qquad (12.55)$$

$$\text{SS}A = \sum_{i=1}^{a} T_{i.}^2/b - T_{..}^2/ab, \qquad \text{df} = a - 1, \qquad (12.56)$$

$$\text{SS}B = \sum_{j=1}^{b} T_{.j}^2/a - T_{..}^2/ab, \qquad \text{df} = b - 1, \qquad (12.57)$$

$$\text{SS}AB = \text{SST} - \text{SS}A - \text{SS}B, \qquad \text{df} = (a-1)(b-1). \qquad (12.58)$$

The last is often called SSE, especially when this case of $n = 1$ is presented in books before the more general replicated case. But note that $\text{MS}AB = \text{SS}AB/(a-1)(b-1)$ is an unbiased estimate of $\sigma_\varepsilon^2 + \sigma_{\alpha\beta}^2$, not $\sigma_\varepsilon^2$ for which there is no available estimate.

Let us first suppose that we have an $ab$ random model unreplicated, that is, like model (2) of Section 12.7.3, but with $n = 1$. The three expected mean squares are

$$\text{EMS}A = \sigma_\varepsilon^2 + \sigma_{\alpha\beta}^2 + b\sigma_\alpha^2$$

$$\text{EMS}B = \sigma_\varepsilon^2 + \sigma_{\alpha\beta}^2 + a\sigma_\beta^2$$

$$\text{EMS}AB = \sigma_\varepsilon^2 + \sigma_{\alpha\beta}^2 .$$

Thus to test $\sigma_\alpha^2 = 0$ versus $\sigma_\alpha^2 > 0$ we use

$$F_{a-1,(a-1)(b-1)} = \text{MS}A/\text{MS}AB$$

and compare with $F_\alpha$ for these degrees of freedom. The test on $\sigma_\beta^2$ is similar.

So far so good. Now consider the fixed model. See (1) of Section 12.7.3. To test $\sigma_\alpha{}^2 = 0$ versus $\sigma_\alpha{}^2 > 0$, that is, all $\alpha_i$'s equal versus some not equal, we see that we need an unbiased estimate of $\sigma_\varepsilon{}^2$ for the denominator. But all we have is one for $\sigma_\varepsilon{}^2 + \sigma_{\alpha\beta}^2$. In this dilemma the usual recourse is to assume $\sigma_{\alpha\beta}^2$ to be zero, or at least negligible. Then we can make the test $F = \mathrm{MS}A/\mathrm{MS}AB$, and so on. The same assumption permits testing for $\sigma_\beta{}^2$.

Finally consider the mixed model. It should now be easily seen from (3) of Section 12.7.3, the mixed model, that we have the exact test available for $\sigma_\beta{}^2 = 0$ versus $\sigma_\beta{}^2 > 0$ for fixed factor $B$, by using

$$F = \mathrm{MS}B/\mathrm{MS}AB,$$

but in order to test $\sigma_\alpha{}^2 = 0$ versus $\sigma_\alpha{}^2 > 0$, we need to assume $\sigma_{\alpha\beta}^2 = 0$, then use for the random factor

$$F = \mathrm{MS}A/\mathrm{MS}AB.$$

Commonly, those who use such unreplicated two-factor designs assume that there is no interaction present. But as we have just seen, this is not always necessary.

### 12.7.6. Example of an Unreplicated Two-Factor Completely Randomized Design.

Let us now use the anova* approach to an earlier problem we considered, namely, Table 9.1 of Section 9.6. There the data were treated by the matched-pair technique. That is, we took the $2 \times 6$ table of data, found differences, and thus reduced the two-factor table of data to a one-factor table, that is, six differences as shown in the last column of Table 9.1. Then we proceeded to use the one-sample significance test of $\mu_d = 0$ versus $\mu_d < 0$.

In the context of the present chapter, we treat the data as a two-factor unreplicated experiment. In this way we can isolate evidence of differences between soil types, and of differences between types of steel pipe.

No type of coding seems likely to help much so we use the data directly as in Table 12.7. The calculations in this table are direct applications of (12.55)–(12.58). We then have the following anova table to summarize the results. In it we assume a fixed model and therefore need to assume $\sigma_{\alpha\beta}^2 = 0$.

---

\* A convenient abbreviation for analysis of variance is "anova."

**TABLE 12.7**

Corrosion Measured in Weight Loss for Pairs of Steel Pipe of Two Kinds, Buried about Eight Years, in Various Soils[a]

| Factor A: soil type | i | Factor B | | $T_{i.}$ |
|---|---|---|---|---|
| | | Lead-coated steel $j = 1$ | Bare steel $j = 2$ | |
| Very fine sandy loam | 1 | .18 | 1.70 | $T_{1.} = 1.88$ |
| Gravelly sandy loam | 2 | .08 | .21 | $T_{2.} = .29$ |
| Muck | 3 | .61 | 1.21 | $T_{3.} = 1.82$ |
| Clay | 4 | .44 | .89 | $T_{4.} = 1.33$ |
| Tidal marsh | 5 | .77 | .86 | $T_{5.} = 1.63$ |
| Alkali | 6 | 1.27 | 2.64 | $T_{6.} = 3.91$ |
| $T_{.j}$ | | $T_{.1} = 3.35$ | $T_{.2} = 7.51$ | $T = 10.86$ |
| $\sum_{i=1}^{6} y_{ij}^2$ | | 2.8103 | 12.8995 | 15.7098 |

$$\text{SST} = 15.7098 - 10.86^2/12 = 15.7098 - 9.8283 = 5.8815$$
$$\text{SS}A = (1.88^2 + .29^2 + \cdots + 3.91^2)/2 - 10.86^2/12$$
$$= 26.6448/2 - 9.8283 = 13.3224 - 9.8283 = 3.4941$$
$$\text{SS}B = (3.35^2 + 7.51^2)/6 - 10.86^2/12 = 11.2704 - 9.8283 = 1.4421$$
$$\text{SS}AB = 5.8815 - 3.4941 - 1.4421 = .9453$$

[a] Units oz/ft² yr. See Logan and Ewing [7], Chap. 9.

| Source | S | df | MS | F | $F_{.05,\nu_1,\nu_2}$ | EMS |
|---|---|---|---|---|---|---|
| Soil type, A | 3.4941 | 5 | .6988 | 4.44 | $F_{.05,5,6} = 4.39$ | $\sigma_\varepsilon^2 + 2\sigma_\alpha^2$ |
| Pipe coating, B | 1.4421 | 1 | 1.442 | 9.16 | $F_{.05,1,6} = 5.99$ | $\sigma_\varepsilon^2 + 6\sigma_\beta^2$ |
| Error (interaction) | .9453 | 6 | .1575 | | | $\sigma_\varepsilon^2$ |
| Total | 5.8815 | | | | | |

We therefore see that both factors are significant. It was already anticipated that there would be differences in corrosion under differing soil types. The main question was as to the different pipe coatings. There proves to be a difference. The results say that we are justified in estimating $\beta_1$ and $\beta_2$ by $\hat{\beta}_1 = b_1 = \bar{y}_{.1} - \bar{y} = 3.35/6 - 10.86/12 = -.347$, $\hat{\beta}_2 = b_2 = +.347$, or an average difference of .694. Further we could estimate the $\alpha_i$'s, for example, $\hat{\alpha}_1 = a_1 = \bar{y}_{1.} - \bar{y} = 1.88/2 - 10.86/12 = +.035$. (But take note that such differential effects are only good for this precise collection of soil types, length of burial, and pipe coatings, that is, the present experimental conditions.)

Another point to be noted is that we assumed no interaction, that is, that the true difference in corrosion rates between the pipe coatings was constant from one soil type to another. This might be true here, although the first and last soil types seem to have differences (see Table 9.1, last column) greater than the others. But how would the reader feel about the last line of the matched-pair data in Problem 9.14?

### 12.8. Other Models

Many models other than those we have considered could be given. Two-factor, completely randomized designs with and without replication are special cases of "factorial designs," in which all combinations of all levels of each factor are used once, or replicated $n$ times. Thus we might have factor $A$ at five levels, $B$ three, $C$ two, $D$ three, and two measurements per cell. This could be called a $5 \times 3 \times 2 \times 3 \times 2$ factorial replicated design. Note that as the number of factors and/or levels increases, the number of measurements builds up very rapidly. The design above would have 180 measurements $y$. But we could study the four main effects of the factors individually, $C(4, 2) = 6$ two-factor interactions, $C(4, 3) = 4$ three-factor interactions, one four-factor interaction. This provides much useful information.

Because of the rapid buildup of the number of combinations, experimental designs have been devised to obtain information on the main effects at the cost of assuming some of the interactions to be negligible. Thus with six factors each at two levels we should have $2^6 = 64$ separate observations, if not replicating. But we can cut this to 32 or even 16 if we are prepared to assume some interactions to be negligible. Such interactions are deliberately confounded with some main effect. Along this general line we might have three factors at four levels each. We should have $4^3 = 64$ observations. But these can be cut to 16 by using a Latin-square design. Or four factors at four levels each ($4^4 = 256$) can be handled in a Greco–Latin square design with only 16 observations.

Then too there are designs with nested factors. For example, we have five lots of a bulk product, from each lot we choose randomly four samples throughout the lot, then make two measurements of some characteristic for each sample. Then the samples are a nested factor. Nesting (or "subsampling") may occur in a factorial completely randomized design too.

Finally we may mention designs and analyses which use both analysis of variance and regression. Such designs are called "analysis of covariance." One use is to remove or eliminate the effect of some uncontrolled factor in an experiment.

For further information on such designs and their analysis, the reader is referred to the literature [3, 4, 20–22].

## 12.9. Summary

In this chapter we have seen how we can gather data in such a way as to be able to break up the total sum of squares of the deviations of the measurements from the grand mean into separate parts. These separate parts enable us to study the main "effects" of input factors, and also their joint "effects" or interactions. Further the within-cell variances which may be used to form the mean square error, can also be used to test for homogeneity of variances, that is, whether $\sigma_\varepsilon^2$ is constant or not, via the $Q$ test or Bartlett's test. Thus we have the natural generalizations of the $t$ test for $\bar{y}_1$ versus $\bar{y}_2$, and the $F$ test for $s_1^2$ versus $s_2^2$. Our introduction to analysis of variance opens the door to other and more complicated experimental designs and their analysis.

It is well worth mentioning here that many general computer programs are available which will carry out the calculations for analysis of variance problems. They are often written to handle 10 or even more factors, to make orthogonal polynomial analyses of numerical factors, to print out cell means, and so on. The reader is well advised to find out what is available in his organization. But working out some simple designs with or without coding and with or without using a desk calculator is a highly useful aid to one's understanding and interpretation of analysis of variance.

## REFERENCES

1. C. T. Metcalf, A study of the use of sandstone in bituminous surface courses, p. 140. M.S. Thesis, Purdue Univ., Lafayette, Indiana, 1949.
2. P. R. Weyl, A redetermination of the constant of gravitation. *J. Res. Nat. Bur. Standards* **5**, 1243–1290 (1930).
3. B. Ostle, "Statistics in Research." Iowa State Univ. Press, Ames, 1963.
4. C. R. Hicks, "Fundamental Concepts in the Design of Experiments." Holt, New York, 1973.
5. R. L. Anderson and E. E. Houseman, Tables of orthogonal polynomial values extended to N = 104. Res. Bull. 297. Agr. Exp. Station, Iowa State Univ., Ames, 1942.
6. D. B. Owen, "Handbook of Statistical Tables." Addison-Wesley, Reading, Massachusetts, 1962.
7. D. B. Duncan, Multiple range and multiple F tests. *Biometrics* **11**, 1–42 (1955).
8. C. W. Dunnett, Multiple comparison tests. *Biometrics* **26**, 139–141 (1970).
9. H. L. Harter, "Order Statistics and Their Use in Testing and Estimation, Vol. 1,

Tests Based on Range and Studentized Range of Sample from a Normal Population."
Aerospace Res. Labs., Wright-Patterson Air Force Base, Ohio, 1969.

10. P. Seeger, "Variance Analysis of Complete Designs," Pt. III, "The Problem of Multiple Comparisons." Almqvist & Wiksell, Stockholm, 1966.

11. R. G. D. Steel, Error rates in multiple comparisons. *Biometrics* **17**, 326–328 (1961).

12. R. H. Hoefs, The effect of minor alloy additions on the corrosion resistance of alpha aluminum bronze. Ph.D. Thesis, Purdue Univ., Lafayette, Indiana, 1953.

13. L. A. Foster, Testing equality of variances. Ph.D. Thesis, Purdue Univ., Lafayette, Indiana, 1964, unpublished.

14. A. E. Brandt, Ph.D. Thesis, Iowa State Univ., Ames, 1932.

15. W. L. Stevens, Heterogeneity of a set of variances. *J. Genet.* **33**, 383 ff, 1936.

16. I. W. Burr and L. A. Foster, A test for equality of variances. Mimeograph Series, No. 282, Statistics Dept., Purdue Univ., Lafayette, Indiana, 1972.

17. E. S. Pearson and H. O. Hartley, "Biometrika Tables for Statisticians," Vol. I. Cambridge Univ. Press, London and New York, 1954.

18. E. C. Copelin, Attempts to establish performance tests for infrared spectrophotometers. Ph.D. Thesis, Purdue Univ., Lafayette, Indiana, 1958.

19. J. W. Barany, A metabolic investigation of the effect of pace on the weight handled secondary adjustment. M.S. Thesis, Purdue Univ., Lafayette, Indiana, 1958.

20. C. A. Bennett and N. L. Franklin, "Statistical Analysis in Chemistry and Chemical Industry." Wiley, New York, 1961.

21. O. L. Davies, "The Design and Analysis of Industrial Experiments." Oliver & Boyd, Edinburgh, 1954.

22. G. W. Snedecor and W. G. Cochran, "Statistical Methods." Iowa State Univ. Press, Ames, 1967.

23. J. T. McCall, Probability of fatigue failure of concrete. Ph.D. Thesis, Purdue Univ., Lafayette, Indiana, 1956.

24. W. B. Seefeldt, Pressure drop in packed columns. M.S. Thesis, Purdue Univ., Lafayette, Indiana, 1948.

25. E. W. Engerer, An optimum procedure for the determination of the dimensional capabilities of a machine tool. M.S. Thesis, Purdue Univ., Lafayette, Indiana, 1950.

26. R. W. Guard, Calcium and magnesium as steel deoxidizers. Ph.D. Thesis, Purdue Univ., Lafayette, Indiana, 1952.

27. I. W. Burr and W. R. Weaver, Stratification control charts. *Indust. Quality Control* **5** (no. 5), 10–15 (1949).

ADDITIONAL REFERENCES

O. J. Dunn, Multiple comparisons among means. *J. Amer. Statist. Assoc.* **56**, 52–64 (1961).

H. L. Harter, Error rates and sample sizes for range tests in multiple comparisons. *Biometrics* **13**, 511–536 (1957).

C. Y. Kramer, Extension of multiple range tests to group means with unequal numbers of replications. *Biometrics* **12**, 307–310 (1956).

R. G. Miller, Jr., "Simultaneous Statistical Inference." McGraw-Hill, New York, 1966.

D. L. Wallace, Multiple comparisons in the analysis of variance. *Nat. Conv. Trans. Amer. Soc. Quality Control* pp. 279–285 (1957).

## Problems

**12.1.** For the data of Section 12.5.1, run a one-factor analysis of variance test (use $\alpha = .05$). (MSE is immediately available as in that section. SS$A$ can be found as usual by converting the $\bar{y}_i$'s to $T_i$'s.)

**12.2.** Data were collected on weights of steel ingots as they come from the molds prior to being rolled into bars (5300 lb was the desired average). Test whether there are significant day-to-day differences in average, using $\alpha = .01$. Calculations can be simplified by coding.

| April | Weight | | | |
|-------|------|------|------|------|
| 1  | 5500 | 5800 | 5740 | 5710 |
| 2  | 5440 | 5680 | 5240 | 5600 |
| 4  | 5400 | 5410 | 5430 | 5400 |
| 9  | 5640 | 5700 | 5660 | 5700 |
| 10 | 5610 | 5760 | 5610 | 5400 |

**12.3.** Do Problem 9.13 as a two-factor unreplicated design, using $\alpha = .01$. If done as matched-pair data, compare your results.

**12.4.** Use the data of Problem 9.16 as a two-factor unreplicated design, and analyze with $\alpha = .01$. Treating as a mixed design, estimate $\sigma^2$ for disk thickness, and the difference between gage readings, if justified by your results.

**12.5.** For the data of Problem 9.1, make an analysis of variance test with $a = 2$, $n = 3$. Compare with the approach used in that problem which yields $t$. To compare, square the observed and critical $t$ values to see whether they give the observed and critical $F$ values, respectively.

**12.6.** For the data of Problem 9.3, make an analysis of variance test with $a = 2$, $n = 3$. Compare with the approach used in that problem which yields $t$. To compare, square the observed and critical $t$ values to see whether they give the observed and critical $F$ values, respectively.

**12.7.** Class experiment: One-factor, fixed factor experiment. For the model $y_{ij} = \mu + \alpha_i + \varepsilon_{ij}$, let $\mu = 10$, $\alpha_1 = -4$, $\alpha_2 = +1$, $\alpha_3 = +3$, $\alpha_4 = 0$. Then $\mu_1 = 10 - 4 = 6$, and so on. To each such cell mean is added an error $\varepsilon_{ij}$ chosen randomly from the distribution of Problem 10.49. Five such errors added to each $\mu_i$ give five $y_{ij}$'s, $j = 1, ..., 5$. Each student collects his own data, namely, 20 $y_{ij}$, and carries through the calculation *carefully*.—Mean square errors are estimates of 2.94, since the error distribution is approximately N(0, 1.715$^2$) from Problem

10.49. Of what is MS$A$ an estimate? Average all results and compare with the expected values.

12.8. Class experiment: One-factor random factor experiment. Just the same as Problem 12.7 but instead each of the four $\alpha_i$'s is a drawing from the population of Problem 10.49.

12.9. Test the data of Table 12.5 for differences between film averages, at $\alpha = .05$, by making a multiple comparison test. See Problem 12.1.

12.10. The following data give cycles at which failure occurred on beams, rounded to 10 for five batches of concrete [23]:

| 1 | 2 | 3 | 4 | 5 |
|-----|-----|-----|-----|-----|
| 790 | 840 | 800 | 640 | 830 |
| 590 | 800 | 870 | 760 | 940 |
| 750 | 950 | 870 | 830 | — |

Test for homogeneity of means at $\alpha = .05$ and interpret your results.

12.11. The following data give results of testing lamps for luminous flux (lumens per watt):

| Company | Measurements | | | | | |
|---------|------|------|------|------|------|------|
| 1 | 9.47, | 9.00, | 9.12, | 9.27, | 9.27, | 9.25 |
| 2 | 10.80, | 11.28, | 11.15 | | | |
| 3 | 10.37, | 10.42, | 10.28 | | | |
| 4 | 10.65, | 10.33 | | | | |
| 1 held for year | 9.54, | 8.62 | | | | |

Round data to nearest .1 (upward if 9.25 or 11.15) for simplicity. Coding also helps. Then test for company differences by analysis of variance using $\alpha = .01$. Interpret your results.

12.12. The following are blood sugar determinations in rats (1) normal, and after injections (2) of pitressin and (3) of pitocin:

(1)   93, 111, 118, 98, 114, 107, 108, 120, 106, 96, 114, 125
(2)   134, 129, 121, 127, 138, 126, 134, 130, 147, 135
(3)   110, 106, 108, 102, 109, 108, 104, 108

Make an analysis of variance test of the data using $\alpha = .01$. Since in advance of data collection comparison of each injection with normal would be desired, make the appropriate $t$ tests, at $\alpha = .01$, and interpret.

12.13. Data given below [24] give the pressure drops under different kinds of packing in columns. Test for significance by analysis of variance. What type of factor do we have here? Interpret your results.

| Packing | Drops measured per run | | | | |
|---------|------|------|------|------|------|
| Random $B$ | 46 | 29 | 48 | | |
| Random $C$ | 37 | 35 | 50 | 39 | |
| Random $D$ | 29 | 16 | 54 | 24 | |
| Dense $E$ | 36 | 27 | 25 | 28 | 23 |
| Dense $F$ | 16 | 23 | 29 | | |
| Random $H$ | 21 | 29 | | | |

12.14. Test the ranges in Table 12.6 for homogeneity using the $Q$ test at the 5% level and interpret the result. See Section 12.5.3.

12.15. For the variances given in Section 12.4.2, test for homogeneity at the 5% level, using the $Q$ test, and interpret your results.

12.16. Test the blood sugar data of Problem 12.12 for homogeneity by the $Q$ test, using $\alpha = .01$, and interpret your result.

12.17. For the lamp data of Problem 12.11, test for homogeneity of variability by the $Q$ test, using $\alpha = .01$, and interpret your results.

12.18. The data shown are for measurements of the pounds "load" for five different piston rings (for automobile cylinders) by five inspectors. We of course expect the rings to differ, but do the inspectors have differing averages, that is, biases relative to each other? Use $\alpha = .05$ If so, use a multiple range test to analyze the inspectors further, $\alpha = .05$. What is the average measurement or repeatability error ($\sqrt{\text{MSE}}$ after decoding, if coding was used)?

| Ring | Inspector | | | | |
|------|-----|-----|-----|-----|-----|
| | 1 | 2 | 3 | 4 | 5 |
| 1 | 9.4 | 9.6 | 9.3 | 9.5 | 9.5 |
| 2 | 9.1 | 9.0 | 8.9 | 9.0 | 9.0 |
| 3 | 8.6 | 8.4 | 8.4 | 8.4 | 8.5 |
| 4 | 10.1 | 10.3 | 10.0 | 10.1 | 10.2 |
| 5 | 9.9 | 9.9 | 9.8 | 9.9 | 10.0 |

12.19.    Same as Problem 12.18, but with the following data:

|        | Inspector |     |     |     |     |
|--------|-----------|-----|-----|-----|-----|
| Ring   | 1         | 2   | 3   | 4   | 5   |
| 1      | 3.8       | 3.7 | 3.8 | 3.8 | 3.8 |
| 2      | 4.2       | 4.3 | 4.1 | 4.2 | 4.1 |
| 3      | 3.9       | 4.0 | 3.9 | 3.9 | 3.9 |
| 4      | 4.6       | 4.6 | 4.5 | 4.6 | 4.5 |
| 5      | 4.4       | 4.3 | 4.4 | 4.3 | 4.3 |

12.20.    The data shown here are for a piece of experimental equipment for measuring the volume of fuel metered out in unit time from four nozzles. Volumes are in cubic centimeters. This is a mixed model.

|       | Nozzles |      |      |      |
|-------|---------|------|------|------|
| Run   | 1       | 2    | 3    | 4    |
| 1     | 96.6    | 96.6 | 94.0 | 97.0 |
| 2     | 97.2    | 96.4 | 95.0 | 96.0 |
| 3     | 96.4    | 97.0 | 94.4 | 95.0 |
| 4     | 97.4    | 96.2 | 94.4 | 95.8 |
| 5     | 97.8    | 96.8 | 89.8 | 97.0 |

After considering what to do about the 89.8 value, test for significance between nozzles and between runs, at the .05 level.

12.21.    Data shown are for the radiation intensity measured in millivolts, at five doors in an open hearth furnace. Door 1 is nearest the flame, 5 farthest.

|        | Door |    |    |    |    |
|--------|------|----|----|----|----|
| Time   | 1    | 2  | 3  | 4  | 5  |
| 1      | 55   | 52 | 43 | 43 | 40 |
| 2      | 55   | 59 | 48 | 44 | 42 |
| 3      | 58   | 57 | 52 | 47 | 42 |
| 4      | 56   | 58 | 49 | 46 | 43 |
| 5      | 57   | 59 | 52 | 48 | 41 |
| 6      | 55   | 57 | 52 | 49 | 44 |

Test for time-to-time and door-to-door differences. For the latter use multiple comparison. Use $\alpha = .05$ on all tests.

12.22. In a study [25] of the process variability the following data were gathered. Ten pieces of 4 in. stock were machined on a lathe and measurements of diameter made at five places along each piece. Data are here given for just four pieces of the 10. Analyze for position and piece-to-piece differences at 5% significance. If either or both factors prove significant, test for a linearity effect or use multiple comparisons.

| Piece | Position | | | | |
|---|---|---|---|---|---|
| | 1 | 2 | 3 | 4 | 5 |
| 1 | 3.9672 in. | 71 | 73 | 70 | 68 |
| 4 | 76 | 72 | 70 | 71 | 70 |
| 7 | 76 | 73 | 72 | 72 | 72 |
| 10 | 75 | 75 | 73 | 73 | 73 |

12.23. Same as Problem 12.22, but using 3/4 in. stock.

| Piece | Position | | | | |
|---|---|---|---|---|---|
| | 1 | 2 | 3 | 4 | 5 |
| 1 | .6607 in. | 7 | 7 | 6 | 6 |
| 4 | 9 | 9 | 9 | 8 | 8 |
| 7 | 11 | 10 | 10 | 9 | 12 |
| 10 | 13 | 13 | 14 | 13 | 13 |

12.24. A fixed, two-factor experiment with replication gave the following impact results. (This is a part of a three-factor experiment.) [26]

| Ca$_2$ Si | | Steel Oxidizer Mg$_2$ Si | | Al | |
|---|---|---|---|---|---|
| Transverse | Longitudinal | Transverse | Longitudinal | Transverse | Longitudinal |
| 8 | 11 | 5 | 7 | 10 | 16 |
| 8 | 10 | 6 | 7 | 10 | 14 |
| 8 | 12 | 6 | 8 | 12 | 15 |

Analyze the data and interpret your results using $\alpha = .05$.

12.25. This is part of an industrial study on paper-making, measuring the finish on coated paper. Three paper machines $A_1$, $A_2$, $A_3$ used four different coatings for paper $B_1$, ..., $B_4$. Two repeat measurements were made per combination.

| Machines | Coatings | | | | | | | |
|---|---|---|---|---|---|---|---|---|
| | $B_1$ | | $B_2$ | | $B_3$ | | $B_4$ | |
| $A_1$ | 42.5 | 42.6 | 42.0 | 42.2 | 43.9 | 43.6 | 42.2 | 42.5 |
| $A_2$ | 42.1 | 42.3 | 41.7 | 41.5 | 43.1 | 43.0 | 41.5 | 41.6 |
| $A_3$ | 43.6 | 43.8 | 43.6 | 43.2 | 44.1 | 44.2 | 42.9 | 43.0 |

Analyze the data and interpret your results, using $\alpha = .05$.

12.26.   Data given here are for percent moisture content of sand for foundry for each of two mullers (mixers) on each of five days, five measurements per day. Take as a fixed model using $\alpha = .05$, analyze, and interpret results. (Coding in .1 % around 4.6 % gives small numbers.)

| Monday Muller | | Tuesday Muller | | Wednesday Muller | | Thursday Muller | | Friday Muller | |
|---|---|---|---|---|---|---|---|---|---|
| E | W | E | W | E | W | E | W | E | W |
| 4.6 | 4.7 | 4.4 | 4.6 | 4.7 | 4.6 | 4.5 | 5.0 | 4.7 | 4.8 |
| 4.6 | 4.7 | 4.4 | 4.5 | 4.7 | 4.2 | 4.6 | 4.6 | 4.6 | 4.7 |
| 4.6 | 4.6 | 4.6 | 4.6 | 4.6 | 4.7 | 4.8 | 5.0 | 4.4 | 4.6 |
| 4.6 | 4.6 | 4.6 | 4.6 | 4.7 | 4.7 | 4.8 | 4.6 | 4.6 | 4.6 |
| 4.6 | 4.6 | 4.5 | 4.6 | 4.6 | 4.7 | 4.8 | 4.8 | 4.6 | 4.7 |

12.27.   Data given below give measurements at three places across a strip of tin-plated steel sheet (in pounds of tin per base box, equivalent weight of tin on 112 sheets $14 \times 20$ in$^2$ both sides) [27].

| Place | Time | | | | | | | | | |
|---|---|---|---|---|---|---|---|---|---|---|
| | 1 | 2 | 3 | 4 | 5 | 6 | 7 | 8 | 9 | 10 |
| Drive side | 71 | 71 | 71 | 74 | 74 | 70 | 71 | 71 | 72 | 74 |
| Center | 61 | 64 | 64 | 64 | 64 | 62 | 63 | 65 | 65 | 61 |
| Operator side | 70 | 71 | 71 | 75 | 75 | 73 | 72 | 74 | 73 | 74 |

Are the time-to-time differences significant, and are the position differences significant? Use $\alpha = .05$. For position make orthogonal tests to compare center with average of the other two, and to compare the two sides.

12.28.   Data collected within the class, a mixed model with replication: Take $\mu = 10$, $\alpha_1 = -4$, $\alpha_2 = -2$, $\alpha_3 = 0$, $\alpha_4 = +2$, and $\alpha_5 = +4$, then $\beta_j$'s as four drawings from the distribution of Problem 10.49. Let $\alpha\beta_{ij} = 0$ throughout, and each $\varepsilon_{ijk}$ be a new drawing from the population of Problem 10.49; use $n = 3$ observations per cell. Analyze the data carefully, including linear orthogonal term for factor $A$, and multiple comparison on factor $B$ if significant. Use $\alpha = .05$.

12.29. The following are data on yields in percent, of a chemical process (hypothetical data, but quite illustrative). Analyze this fixed model at 5% significance level. Interpret interaction.

|  | Catalyst | | |
|---|---|---|---|
| Temperature | 1 | 2 | 3 |
| 1 | 91 | 88 | 93 |
|  | 93 | 86 | 92 |
| 2 | 90 | 95 | 90 |
|  | 94 | 94 | 87 |

12.30. Prove shortcut formula using ranges for SSE, namely, that $\sum_{i=1}^{k} s_i^2 = \sum_{i=1}^{k} R_i^2/2$, when the sample size is $n = 2$. How many degrees of freedom does such an SSE carry?

12.31. Point out some of the similarities between one-factor analysis of variance and linear regression, such as tests, error variance, breakdown of sums of squares, and use of orthogonal polynomials.

*Chapter 13*

# MULTIPLE REGRESSION

## 13.1. INTRODUCTION

This chapter is a perfectly natural extension and generalization of Chapter 11. In fact we have already met many of the ideas and concepts in some models of Chapter 11 where we fitted functions of $y$ which were *linear in the parameters* $\beta_0$, $\beta_1$, $\beta_2$, and so on. Such, for example, are

$$y_i = \beta_0 + \beta_1 x_i + \beta_2 x_i^2 + \beta_3 x_i^3 + \varepsilon_i, \quad \varepsilon_i: \text{NID}(0, \sigma^2) \quad (13.1)$$

$$y_i = \beta_0 + \beta_1 x_i + \beta_2/x_i + \varepsilon_i, \quad \varepsilon_i: \text{NID}(0, \sigma^2). \quad (13.2)$$

Use of least squares fitting for these would lead to the problem of solving four equations for $b_0$, $b_1$, $b_2$, $b_3$ for (13.1), and three equations for $b_0$, $b_1$, $b_2$ for (13.2). In both cases the $b_j$'s are estimates of $\beta_j$'s.

The generalization now comes with removal of the above-mentioned restriction that the random variables whose coefficients were $\beta_j$'s were all functions of the very same random variable $x_i$. Thus we are no longer doing the curve fitting of $y_i$ versus $x_i$. Instead we use models such as

$$y_i = \beta_0 + \beta_1 x_{1i} + \beta_2 x_{2i} + \beta_3 x_{3i} + \varepsilon_i, \quad \varepsilon_i: \text{NID}(0, \sigma^2). \quad (13.3)$$

This type of model includes those above, for there is nothing to prevent $x_{2i}$ from being $x_{1i}^2$, $x_{3i}$ as $x_{1i}^3$. However, $x_{2i}$ is usually not functionally related to $x_{1i}$, nor the others.

Now to what kinds of application may we apply such a model as (13.3)? We may at least try to use it whenever the "dependent variable" or "output variable" may be expected to be related to two or more "independent" or "input variables." Such situations are very numerous

372

in all phases of science and industry. The linearity or first-degree character of (13.3) need not bother us, because for almost any case it can be called a first approximation, and if we want to put in a "curved surface" model, we may add terms such as $\beta_4 x_{1i}^2$ and $\beta_5 x_{1i} x_{2i}$. Typical examples are as follows:

1. Tool life versus hardness of stock to be machined, hardness of tool, depth of cut, rate of stock removal.
2. Productivity versus quality of raw materials and variables in the processing.
3. Employee efficiency as related to employment tests.
4. Sales estimated from economic factors such as gross national income, percentage of equipment five years old or older, and price of semifinished goods.
5. Diameter of roll condenser versus thickness of paper and of foil, number of turns and tension of winding.
6. Weight of coating of electrolytic tin plate versus current, rate of travel of strip, acidity, and distance from tin anode.
7. Tensile strength versus density and hardness of stock.
8. Yield of chemical process (to be maximized) versus amount of catalyst, temperature of reactor, rate of throughput, and ratio of ingredients.
9. Process variables and chemical compositions used in estimating quality of steel, nonferrous metals, rubber, plastics, glass, or food products. Also useful in determining which of many input variables it will be worthwhile to control more closely.
10. Other examples were already given in Section 11.12.

Some of the objectives of multiple regression are (1) to obtain the best available prediction equation for the model chosen; (2) to test whether the model can be justifiably used at all; (3) to test whether all the variables are worth retaining or alternatively whether little will be lost if certain variables are omitted; (4) to determine how closely $y$ may be estimated from a set of $x_j$'s; (5) to test whether the model may be desirably modified; (6) to rank the variables $x_j$ roughly in terms of their usefulness in prediction.

We present the approach to multiple regression in two forms, the first of which does not make use of modern algebra, and is thus somewhat more accessible to less sophisticated readers. Also this approach is likely to aid the understanding of *all* readers. The second approach makes use of vectors and matrices and is intimately related to computers and their output. The equations appear simpler, and more nearly resemble, at least superficially, those of linear regression.

## 13.2. FIRST APPROACH

Observations for multiple regression should be taken carefully if results are to be of maximum value. Insofar as possible the observation of dependent variable $y_1$ and independent variables $x_{11}$, $x_{21}$, ..., $x_{k1}$ should be taken at the same time, or on the same material, or for the same person, or at the same place. Then when $y_2$ and $x_{12}$, ..., $x_{k2}$ are taken, it can be at a different time or place, and so forth. There should be something linking or correlating the observations of the $k+1$ variables.

As a counterexample, the author saw some poorly gathered data, collected for example 6 of Section 13.1. The four independent variables were measured over a period of up to an hour around the time at which the weight of tin coating was measured. This gave a large chance for each variable to be off from the value it did have just at the time the strip, yielding $y_i$, was being plated. Hence the multiple regression relation was very feeble and of negligible value. When the five variables were measured nearly simultaneously the strength of relation and predictability considerably improved.

### 13.2.1. Data Table Format.

Typical data can be placed in a table of the following rectangular form, where each row represents one composite observation:

| Observation | $Y$ | $X_1$ | $X_2$ | $\cdots$ | $X_j$ | $\cdots$ | $X_k$ |
|---|---|---|---|---|---|---|---|
| 1 | $y_1$ | $x_{11}$ | $x_{21}$ | | $x_{j1}$ | | $x_{k1}$ |
| 2 | $y_2$ | $x_{12}$ | $x_{22}$ | | $x_{j2}$ | | $x_{k2}$ |
| $\vdots$ | | | | | | | |
| $i$ | $y_i$ | $x_{1i}$ | $x_{2i}$ | | $x_{ji}$ | | $x_{ki}$ |
| $\vdots$ | | | | | | | |
| $n$ | $y_n$ | $x_{1n}$ | $x_{2n}$ | | $x_{jn}$ | | $x_{kn}$ |

The first subscript on each $x$ tells which $X$ random variable we have an observation of, while the second subscript tells which of the $n$ composite observations this is a part of.

### 13.2.2. Fitting the Equation.

To seek the true "regression equation"

$$\mathrm{E}(Y) = \beta_0 + \beta_1 X_1 + \cdots + \beta_k X_k , \qquad (13.4)$$

or for the $i$th composite observation ($i$th row in the table),

$$\mathrm{E}(y_i) = \beta_0 + \beta_1 x_{1i} + \cdots + \beta_k x_{ki} , \qquad (13.5)$$

we must find estimates for the $\beta_j$'s, namely $b_j$'s. For this we use

$$y_i = b_0' + b_1'x_{1i} + \cdots + b_k'x_{ki} + e_i, \tag{13.6}$$

and then seek that collection of $b_0'$, $b_1'$, ..., $b_k'$ which, *for the data at hand,* minimizes $\sum_{i=1}^{n} e_i^2$. This will be the "least squares" fitting. Proceeding, we form, using $e_i$ from (13.6),

$$T = \sum_{i=1}^{n} e_i^2$$

$$= \sum_{i=1}^{n} (y_i - b_0' - b_1'x_{1i} - \cdots - b_k'x_{ki})^2. \tag{13.7}$$

Now $T$ is a function of $b_0'$, $b_1'$, ..., $b_k'$ only, since all the observed data are held constant. Thus to minimize $T$, we differentiate in turn with respect to each $b_j'$, and set the results equal to zero. That is, $\partial T/\partial b_j' = 0$. We just use laws of differentiation, such as the derivative of a sum is the sum of the derivatives, derivative of a square, and so on. Then, for example,

$$\frac{\partial T}{\partial b_1'} = \frac{\partial}{\partial b_1'} \sum_{i=1}^{n} (y_i - b_0' - b_1'x_{1i} - \cdots - b_k'x_{ki})^2$$

$$= \sum_{i=1}^{n} \frac{\partial}{\partial b_1'} (y_i - b_0' - b_1'x_{1i} - \cdots - b_k'x_{ki})^2$$

$$= \sum_{i=1}^{n} 2(y_i - b_0' - b_1'x_{1i} - \cdots - b_k'x_{ki})(-x_{1i}).$$

Using the laws of summations (3.4) and (3.5), we find

$$2\left[ -\sum_{i=1}^{n} x_{1i}y_i + b_0' \sum_{i=1}^{n} x_{1i} + b_1' \sum_{i=1}^{n} x_{1i}^2 + \cdots + b_k' \sum_{i=1}^{n} x_{1i}x_{ki} \right].$$

We then set such expressions equal to zero, divide through by 2, and then remove the summation involving $y_i$ to the right side, giving, for example,

$$b_0' \sum_{i=1}^{n} x_{1i} + b_1' \sum_{i=1}^{n} x_{1i}^2 + \cdots + b_k' \sum_{i=1}^{n} x_{1i}x_{ki} = \sum_{i=1}^{n} x_{1i}y_i.$$

Now, omitting the limits $i = 1$ to $n$ on each $\sum$, the full set of equations to be solved becomes

$$b_0 n + b_1 \sum x_{1i} + b_2 \sum x_{2i} + \cdots + b_k \sum x_{ki} = \sum y_i$$

$$b_0 \sum x_{1i} + b_1 \sum x_{1i}^2 + b_2 \sum x_{1i}x_{2i} + \cdots + b_k \sum x_{1i}x_{ki} = \sum x_{1i}y_i$$

$$b_0 \sum x_{2i} + b_1 \sum x_{2i}x_{1i} + b_2 \sum x_{2i}^2 + \cdots + b_k \sum x_{2i}x_{ki} = \sum x_{2i}y_i \qquad (13.8)$$

$$\vdots$$

$$b_0 \sum x_{ki} + b_1 \sum x_{ki}x_{1i} + b_2 \sum x_{ki}x_{2i} + \cdots + b_k \sum x_{ki}^2 = \sum x_{ki}y_i .$$

We have omitted the primes from the $b_j$'s since the solution of these equations is the *particular* set of $b_j$'s which minimizes $T = \sum e_i^2$, not just any set of $b_j'$'s that give some value of $T$. The $b_j$'s are actually unbiased estimates of the respective $\beta_j$'s.

The coefficients of $b_j$'s are numbers determined by the data at hand. It can be quite a chore to solve a set of equations such as (13.8) if $k$ is at all sizable. To simplify the solution as much as possible, systematic methods have been developed. These take advantage of the symmetry in (13.8); for example, the coefficient of $b_1$ in the first equation is the same as that of $b_0$ in the second, of $b_2$ in the second is the same as that of $b_1$ in the third, and so on.

Equations (13.8) are often called "normal equations." Their solution provides the fitted regression equation as an estimate of (13.5), namely

$$\hat{y}_i = b_0 + b_1 x_{1i} + b_2 x_{2i} + \cdots + b_k x_{ki} . \qquad (13.9)$$

We shall be interested in how well (13.9) estimates (13.5). Note that the unit in each $b_j$ is the $y$ unit divided by the unit of $x_j$, since when $b_j$ is multiplied by $x_{ji}$ it must give a result in $y$ units.

### 13.2.3. Alternative Forms of Normal Equations and Regression.

With a computer it is commonly desirable to solve the system of normal equations (13.8) just as it is, unless the values of the random variables are radically different in size, in which case we probably should code the data to make the variables become somewhat the same size, before computing. This avoids round-off troubles. However, for some purposes we may wish to transform the equations.

Consider the first equation of (13.8). Solving for $b_0$ one easily finds

$$b_0 = \bar{y} - b_1 \bar{x}_1 - b_2 \bar{x}_2 - \cdots - b_k \bar{x}_k , \qquad (13.10)$$

where, as usual, $\bar{y} = \sum y_i/n$, $\bar{x}_1 = \sum x_{1i}/n$, .... Substituting into (13.9) easily yields the regression equation

$$\hat{y}_i = \bar{y} + b_1(x_{1i} - \bar{x}_1) + b_2(x_{2i} - \bar{x}_2) + \cdots + b_k(x_{ki} - \bar{x}_k). \quad (13.11)$$

From this it is easily seen that if each $x_{ji}$ is at the respective mean $\bar{x}_j$, then the estimated $\hat{y}_i$ is at $\bar{y}$. Now if $k = 2$, then (13.9) and (13.11) represent the best fitting plane in three-dimensional space, namely, that of $X_1$, $X_2$, and $Y$. See Fig. 13.1. But if $k > 2$, then we cannot visualize the "hyperplane" (13.9) or (13.11). But in either case the fitted plane or hyperplane passes through the point where each random variable is at its mean.

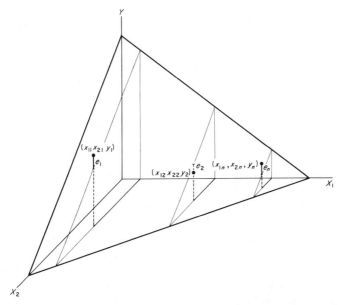

**Fig. 13.1.** Regression plane as in (13.9) for $k = 2$. Three points shown, first, second, and $n$th, by large dots. The vertical distance to such points is the observed $y_i$. Also shown, by a horizontal dash, are the points where the vertical line from $(x_{1i}, x_{2i})$ on the $X_1$, $X_2$ plane hits the fitted plane. Such vertical distances are $y_i$'s. The differences $y_i - \hat{y}_i = e_i$ are also shown, being sometimes positive and sometimes negative.

One advantage of using (13.10) is that (13.8) may be reduced to one fewer equation to solve. In fact if we substitute (13.10) into each equation of (13.8), we may find the following (see Problem 13.13):

$$b_1 \sum x'^2_{1i} + b_2 \sum x'_{1i}x'_{2i} + \cdots + b_k \sum x'_{1i}x'_{ki} = \sum x'_{1i}y'_i$$

$$b_1 \sum x'_{2i}x'_{1i} + b_2 \sum x'^2_{2i} + \cdots + b_k \sum x'_{2i}x'_{ki} = \sum x'_{2i}y'_i \qquad (13.12)$$

$$\vdots$$

$$b_1 \sum x'_{ki}x'_{1i} + b_2 \sum x'_{ki}x'_{2i} + \cdots + b_k \sum x'^2_{ki} = \sum x'_{ki}y'_i,$$

where, for example, $x'_{1i} = x_{1i} - \bar{x}_1$, ..., and the $b_j$'s are identical to those before.

It is possible to transform the system of equations (13.12) by introducing the standardized coefficients

$$b_1{}^* = b_1 s_1/s_y, \qquad b_2{}^* = b_2 s_2/s_y, \qquad ..., \qquad b_k{}^* = b_k s_k/s_y \qquad (13.13)$$

whose units are pure numbers. Using (3.8) for terms like $\sum x'^2_{ji} = (n-1)s_j{}^2$, and (11.33), for example, $\sum x'_{1i}x'_{2i} = (n-1)s_1 s_2 r_{12}$, we may find the system of equations for $b_j{}^*$'s, with coefficients $r$ as follows:

$$1b_1{}^* + r_{12}b_2{}^* + \cdots + r_{1k}b_k{}^* = r_{1y}$$

$$r_{12}b_1{}^* + 1b_2{}^* + \cdots + r_{2k}b_k{}^* = r_{2y} \qquad (13.14)$$

$$\vdots$$

$$r_{1k}b_1{}^* + r_{2k}b_2{}^* + \cdots + 1b_k{}^* = r_{ky}.$$

This system has the desirable feature that regardless of the size of the various values of the random variables, all coefficients of $b^*$'s in (13.14) will be between $-1$ and $+1$. This facilitates the solution for $b^*$'s. On the other hand, (13.8) and (13.12) may have enormous differences in the size of the coefficients of the $b$'s. The solutions of all three systems are algebraically equivalent, and have the same least squares solution.

There is a direct usage of the $b^*$'s in a regression equation. If we standardize the variables by $u_{ji} = (x_{ji} - \bar{x}_j)/s_j$ and $v_i = (y_i - \bar{y}_i)/s_y$, we have the regression

$$\hat{v} = b_1{}^*u_1 + b_2{}^*u_2 + \cdots + b_k{}^*u_k, \qquad (13.15)$$

there being no constant term. From this it may be seen that

$b_j{}^* =$ number of standard deviations by which the estimated $y$ *changes*, when $x_j$ is *increased* by one of its standard deviations $s_j$, while all other $x$'s *remain fixed*.

### 13.2.4. Describing Goodness of Fit.

Almost the first thing to come to mind after fitting a regression is the question: How good a job does it do? There are several ways of describing this goodness of fit.

One way to describe the relative goodness of fit of the multiple regression equation (13.9) is the coefficient of determination. It is very analogous to that for linear regression as given in (11.35). Thus using 0 in the subscript to denote $y$:

$$R^2_{0.12...k} = \text{the proportion of } \sum (y_i - \bar{y})^2 \text{ which is}$$
$$\text{linearly relatable to the } x_j\text{'s via (13.9)}$$
$$= \text{coefficient of multiple determination.} \qquad (13.16)$$

The subscript for $R^2$ gives the number of the dependent variable before the dot, and the numbers of the independent variables involved following the dot. (What is $R^2_{0.2}$ for example?)

$R^2_{0.12...k}$ is always between 0 and 1, and thus tells us what part of the total variation in $y$ is "explained" by the set of $x$'s.

$$R_{0.12...k} = \text{multiple correlation coefficient,} \qquad (13.17)$$

being always defined to be positive. (Recall that linear correlation coefficients can be of either sign.) One interpretation of $R_{0.12...k}$ is that if we were to use (13.9) to obtain $n$ estimates $\hat{y}_i$, then

$$R_{0.12...k} = r_{y_i, \hat{y}_i};$$

that is, it is the simple correlation between the observed value of $Y$ and that estimated.

Another way of describing the goodness of fit of the regression equation (13.9) is through making an estimate of $\sigma^2$ in (13.3), that is, of the variance around the population regression. This is the "unexplained" variance. Define

$$s^2_{0.12...k} = \frac{\sum (y_i - \hat{y}_i)^2}{n - k - 1}, \qquad (13.18)$$

where $\hat{y}_i$ is the estimated $y_i$ from (13.9), and the degrees of freedom are $n - k - 1$, because we fitted $k + 1$ constants, that is, $b_0$, $b_1$, ..., $b_k$. Definition (13.18) looks as though we would need to make $n$ estimates $\hat{y}_i$ and subtract these from $y_i$'s, and so on. But as usual in statistics there are shorter methods. We can use $R_{0.12...k}$ for this as follows:

$$s^2_{0.12...k} = \frac{\sum (y_i - \bar{y})^2 (1 - R^2_{0.12...k})}{n - k - 1}. \qquad (13.19)$$

The numerator is thus the "unexplained variation" and the whole fraction the "unexplained variance." Although $R^2_{0.12...k}$ is a unitless number, $s^2_{0.12...k}$ carries the square of the $Y$ unit, and is therefore an absolute measure of goodness of fit.

Now to use (13.19), we need $R^2_{0.12...k}$. How do we find it? This comes directly by use of

$$R^2_{0.12...k} = b_1{}^*r_{1y} + b_2{}^*r_{2y} + \cdots + b_k{}^*r_{ky}.$$ (13.20)

So now we need to be able to find the $b_j{}^*$'s.

### 13.2.5. Systematic Solution of the Normal Equations.

We shall illustrate with actual data a systematic method of solving the system (13.14). The same method could also solve systems (13.8) or (13.12). This method, the "compact Doolittle,"* and others are fully described by Dwyer [1].

The data were obtained from $n = 357$ "heats" of steel, made in an open hearth furnace. Each such heat yields about 200 tons of steel. The variables being studied were

$Y = X_0$, tons per melting hour;
$X_1$, percent carbon at time of melt;
$X_2$, the difference in percent carbon between time of melt and time of tapping (pouring ingots);
$X_3$, additions of limestone;
$X_4$, the amount of ore fed into furnace.

The linear correlation coefficients are shown at the top of Table 13.1 which shows the calculational steps. Following the listing of the correlation coefficients, at the top of the table, we have the coefficients of the four equations of system (13.14), $k$ being 4. For example, line 1:

$$1b_1{}^* + .5731b_2{}^* + .2761b_3{}^* + .3685b_4{}^* = +.3011;$$

and line 4:

$$.3685b_1{}^* + .7114b_2{}^* + .1185b_3{}^* + 1b_4{}^* = -.0433.$$

In the latter the first three coefficients are not written because they are symmetric, that is, equal to the coefficients of $b_4{}^*$ in lines 1, 2, and 3.

Lines 1–4: Fill in correlation coefficients as indicated by row and column headings. Check sums are formed as if the lower left-hand half

* Named for M. H. Doolittle.

TABLE 13.1

Abbreviated Doolittle Calculation

$r_{01} = +.3011,$ $\quad r_{02} = -.2271,$ $\quad r_{03} = -.2589,$ $\quad r_{04} = -.0433,$ $\quad r_{12} = +.5731$
$r_{13} = +.2761,$ $\quad r_{14} = +.3685,$ $\quad r_{23} = +.2444,$ $\quad r_{24} = +.7114,$ $\quad r_{34} = +.1185$

| Line | | $x_1$ | $x_2$ | $x_3$ | $x_4$ | $x_0$ | Check sums |
|---|---|---|---|---|---|---|---|
| 1 | $x_1$ | 1.0000 | +.5731 | +.2761 | +.3685 | +.3011 | 2.5188 |
| 2 | $x_2$ | | 1.0000 | +.2444 | +.7114 | −.2271 | 2.3018 |
| 3 | $x_3$ | | | 1.0000 | +.1185 | −.2589 | 1.3801 |
| 4 | $x_4$ | | | | 1.0000 | −.0433 | 2.1551 |
| 5 | $A_1$ | 1.0000 | +.5731 | +.2761 | +.3685 | +.3011 | 2.5188 |
| 6 | $B_1$ | 1.0000 | +.5731 | +.2761 | +.3685 | +.3011 | 2.5188 |
| 7 | $A_2$ | | +.6716 | +.0862 | +.5002 | −.3997 | 0.8583 |
| 8 | $B_2$ | | 1.0000 | +.1284 | +.7448 | −.5951 | 1.2780 |
| 9 | $A_3$ | | | +.9127 | −.0474 | −.2907 | 0.5745 |
| 10 | $B_3$ | | | 1.0000 | −.0519 | −.3185 | 0.6295 |
| 11 | $A_4$ | | | | +.4892 | +.1283 | 0.6175 |
| 12 | $B_4$ | | | | 1.0000 | +.2623 | 1.2623 |
| 13 | | | | 1.0000 | | −.3049 | |
| 14 | | | 1.0000 | | | −.7513 | |
| 15 | | 1.0000 | | | | +.7192 | |
| 16 | | +.7192 | −.7513 | −.3049 | +.2623 | | |

of the matrix were filled in, adding down a column to the diagonal and then across the remainder of the row. For example,

$$2.1551 = .3685 + .7114 + .1185 + 1.0000 - .0433.$$

Line 5: Copy of line 1.
Line 6: Copy of line 5. (Actually line 5 divided by $x_1$ item in line 5, i.e., 1.)
Line 7: Combination of lines 2, 5, and 6 as follows:

| Line | 2 | 5 | 6 | 7 |
|---|---|---|---|---|
| | $1.0000 - (+.5731)(+.5731) = +.6716$ | | | |
| | $+.2444 - (+.5731)(+.2761) = +.0862$ | | | |
| | $+.7114 - (+.5731)(+.3685) = +.5002$ | | | |
| | $-.2271 - (+.5731)(+.3011) = -.3997$ | | | |
| | | | | $+.8583$ |

Check sum: $2.3018 - (+.5731)(2.5188) = .8583 =$ sum of line 7 as check.

Line 8: Line 7 entries, each divided by .6716, that is, $(+.0862)/(+.6716) = +.1284$, $(+.5002)/(+.6716) = +.7448$, ..., $(+.8583)/(+.6716) = 1.2780 =$ sum of line 8 to within rounding errors.

Line 9: Combination of lines 3, 5, 6, 7, 8, as follows:

| Line | 3 | 5 | 6 | 7 | 8 | 9 |
|---|---|---|---|---|---|---|
| | $1.0000 - (+.2761)(+.2761) - (+.0862)(+.1284) = +.9127$ | | | | | |
| | $+.1185 - (+.2761)(+.3685) - (+.0862)(+.7448) = -.0474$ | | | | | |
| | $-.2589 - (+.2761)(+.3011) - (+.0862)(-.5951) = -.2907$ | | | | | |
| | | | | | | $+.5746$ |

Check sum: $1.3801 - (+.2761)(2.5188) - (+.0862)(1.2780) = +.5745$ = sum of line 9 to within rounding errors.

Line 10: Line 9 entries, divided by $(+.9127)$, that is, $(-.0474)/(+.9127) = -.0519$, $(-.2907)/(+.9127) = .3185$, $(+.5745)/(+.9127) = +.6295 =$ sum of line 10 to within rounding errors.

Line 11: Combination of lines 4–10 as follows:

| Line | 4 | 5 | 6 | 7 | 8 | 9 | 10 |
|---|---|---|---|---|---|---|---|
| | $1.0000 - (+.3685)(+.3685) - (+.5002)(+.7448) - (-.0474)(-.0519) = +.4892$ | | | | | | |
| | $-.0433 - (+.3685)(+.3011) - (+.5002)(-.5951) - (-.0474)(-.3185) = +.1283$ | | | | | | |

Check sum: $2.1551 - (+.3685)(2.5188) - (+.5002)(1.2780) - (-.0474)(.6295) = .6175 =$ sum of line 11.

Line 12: Line 11 entries divided by $+.4892$, that is, $(+.1283)/(+.4892) = +.2623$, $(+.6175)/(+.4872) = 1.2623 =$ sum of line 12. The entry in line 12, $x_0$ column, is $b_4^*$. To get the other $b^*$'s we successively solve preceding equations. We find $b_3^*$ from line 10, which contains the coefficients for the following equation:

$$1.0000b_3^* + (-.0519)\,b_4^* = -.3185.$$

Hence $b_3^* = -.3185 - (-.0519)(+.2623) = -.3049$

In order to facilitate the back solutions for the $b^*$'s, we fill them in row 16, as they are obtained. Also, they are listed in the $x_0$ column, rows 12–15. We have from line 8,

$$1.0000b_2^* + (+.1284)\,b_3^* + (+.7448)\,b_4^* = -.5951$$

or

$$b_2{}^* = -.5951 - (+.1284)(-.3049) - (+.7448)(+.2623) = -.7513.$$

Similarly, from line 6:

$$b_1{}^* = +.3011 - (+.5731)(-.7513) - (+.2761)(-.3049) - (+.3685)(+.2623)$$
$$= +.7192.$$

As a final check, we substitute the $b*$'s into the normal equations which must be satisfied:

$$1(+.7192) + (+.5731)(-.7513) + (+.2761)(-.3049)$$
$$+ (.3685)(+.2623) = +.3011$$
$$(+.5731)(+.7192) + 1(-.7513) + (+.2444)(-.3049)$$
$$+ (+.7114)(+.2623) = -.2270$$
$$(+.2761)(+.7192) + (+.2444)(-.7513) + 1(-.3049)$$
$$+ (+.1185)(+.2623) = -.2589$$
$$(.3685)(+.7192) + (+.7114)(-.7513)$$
$$+ (+.1185)(-.3049) + 1(+.2623) = -.0433.$$

So we might write the regression equation in standardized variable form as (13.15)

$$\hat{v} = +.7192u_1 - .7513u_2 - .3049u_3 + .2623u_4 .$$

Finally

$$R^2_{0.1234} = b_1{}^*r_{01} + b_2{}^*r_{02} + b_3{}^*r_{03} + b_4{}^*r_{04}$$
$$= (+.7192)(+.3011) + (-.7513)(-.2271) + (-.3049)(-.2589)$$
$$+ (+.2623)(-.0433)$$
$$= +.4548.$$

Or, if the $b_i{}^*$ are not needed, but only the multiple $R$ is desired:

$$R^2_{0.1234} = \sum A_i B_i \qquad \text{from the } x_0 \text{ column}$$
$$= (+.3011)(+.3011) + (-.3997)(-.5951)$$
$$+ (-.2907)(-.3185) + (+.1283)(+.2623)$$
$$= .4548$$

$$R_{0.1234} = \sqrt{.4548} = .6744.$$

It is also true that from the $x_0$ column partial sums of $A_iB_i$ give

$$R_{0.1}^2 = r_{01}^2 = (+.3011)(+.3011) = .0907$$

$$R_{0.12}^2 = (+.3011)(+.3011) + (-.3997)(-.5951) = .3285$$

$$R_{0.123}^2 = (+.3011)(+.3011) + (-.3997)(-.5951) + (-.2907)(-.3185) = .4211.$$

Therefore, if one may want the multiple $R$ (and/or the regression equation) for a certain set of independent variables smaller than the total set, these variables should be placed first (i.e., in the left-hand columns and the top rows) in the matrix for the solution.

Such calculations as this are not prohibitively difficult on a desk calculator, by cumulating sums of products (of two factors) algebraically.

### 13.2.6. Significance Tests on the Explained Variation.

A basic test on the multiple correlation coefficient is concerned with whether there is any reliable evidence *at all* of $Y$ being linearly related to the set of $X$'s. Under the model, an example of which is (13.3), we can use the $F$ distribution for an exact test as to whether the population multiple regression is zero or not. The null hypothesis is $\rho_{0.12...k}^2 = 0$ versus $\rho_{0.12...k}^2 > 0$. The test statistic is*

$$F_{k,n-k-1} = \frac{R_{0.12...k}^2/k}{(1 - R_{0.12...k}^2)/(n - k - 1)}. \tag{13.21}$$

Hence to test we have but to evaluate and compare with the appropriate $F$ value in the table.

If such a test indicates the null hypothesis to be tenable, this simply means that we have no reliable evidence of any *linear* relation of $Y$ to the $X$'s. But this does not mean that no other model could fit, because another model, perhaps nonlinear, might still fit well.

The foregoing test is for all of the $X$'s. But we often want to know whether a single additional variable or a set of them adds significantly to the predictability, that is, significantly cuts down the unexplained variation. The $F$ test is directly applicable to this test too, granted the assumption of the model in (13.3). Suppose we wish to compare the $R^2$ for all $k$ $X$'s with the $R^2$ for a subset of $h$ of the $X$'s. If these are numbered 1, 2, ..., $h$, then we have $R_{0.12...k}^2$ versus $R_{0.12...h}^2$. It may be shown that always

$$R_{0.12...k}^2 \geqslant R_{0.12...h}^2, \qquad h < k. \tag{13.22}$$

---

* If numerator and denominator were each multiplied by $s_{0.12...k}^2/\sigma^2$, then each would be distributed as an ordinary $s^2/\sigma^2$, under the null hypothesis, and the ratio is thus of the $F$ distribution (7.27). But such multipliers cancel out, leaving (13.21).

Then the test of $\rho^2_{0.12...k} = \rho^2_{0.12...h}$ versus $\rho^2_{0.12...k} > \rho^2_{0.12...h}$ is by

$$F_{k-h,n-k-1} = \frac{(R^2_{0.12...k} - R^2_{0.12...h})/(k-h)}{(1 - R^2_{0.12...k})/(n-k-1)}. \tag{13.23}$$

If $F$ is above the $\alpha$ level $F$, then the addition of variables $X_{h+1},...,X_k$ significantly improves the predictability (that is, decreases significantly the unexplained variance). Then we might well add these $X$'s to the "team" unless the gain, even though *statistically* significant, is of negligible *practical* significance.

There are programs written for computers which tell which of the $X$'s not yet chosen for the model will give the greatest increase in $R^2$. Moreover, there are programs written which decide which, of all the $k$ $X$'s in the prediction team, we may drop, with the least loss in $R^2$. Both types of progams also test whether the gain or loss in $R^2$ is significant in a statistical sense. The aim is to obtain as small a team of independent variables or predictors of $y$ as we can, and still do an "adequate" job of prediction. What is "adequate" is a practical matter.

### 13.2.7. Simple Example.

A simple example of a multiple regression was given by Shewhart [2], wherein the aim was to predict the tensile strength of a metal product by using both the hardness and density of the product, the latter two characteristics being much less expensive to measure than tensile strength.

The following results were obtained for $n = 60$ sample pieces:

$X_0 =$ tensile strength in lb/in.$^2 = Y$
$X_1 =$ density in g/cc
$X_2 =$ hardness, Rockwell E
$\bar{x}_0 = 31{,}870,$   $s_0 = 3996$ lbs/in.$^2,$   $r_{01} = .657$
$\bar{x}_1 = 2.679,$   $s_1 = .0994$ g/cc   $r_{02} = .683$
$\bar{x}_2 = 69.83,$   $s_2 = 11.87$ E unit,   $r_{12} = .616.$

The nine results above are from the sample of 60 composite observations, such as the table in Section 13.2.1, and the basic ingredients of the regression study. They summarize the observed data.

When $k = 2$ we have the following solutions for the system (13.14):

$$b_1{}^* = \frac{r_{01} - r_{12}r_{02}}{1 - r^2_{12}} \tag{13.24}$$

$$b_2{}^* = \frac{r_{02} - r_{12}r_{01}}{1 - r^2_{12}}. \tag{13.25}$$

Upon substituting the three $r_{ij}$'s, these become

$$b_1{}^* = .381, \qquad b_2{}^* = .448,$$

both of which are dimensionless numbers, since the $r$'s are.

Then we have the following

$$b_1 = b_1{}^* \frac{s_0}{s_1} = .381 \frac{3996}{.0994} = 15,300 \qquad (\text{lb/in.}^2 \text{ g/cc})$$

$$b_2 = b_2{}^* \frac{s_0}{s_2} = .448 \frac{3996}{11.87} = 151 \qquad (\text{lb/in.}^2 \text{ E unit})$$

$$b_0 = \bar{x}_0 - b_1\bar{x}_1 - b_2\bar{x}_2 = -19,700 \qquad (\text{lb/in.}^2).$$

Thus the estimation equation is

$$y = 15,300x_1 + 151x_2 - 19,700.$$

Next we wish to know whether we are justified in using such an estimation equation at all, and if so, how good a job it does. For this we need the multiple $R_{0.12}^2$. Using (13.20)

$$R_{0.12}^2 = b_1{}^*r_{01} + b_2{}^*r_{02} = .381(.657) + .448(.683) = .556$$

and

$$R_{0.12} = .746.$$

Thus 55.6% of $\sum(y - \bar{y})^2$, which is the variation in tensile strength, is linearly relatable to density and hardness.

This compares with the following by (11.35)

$$r_{01}^2 = .657^2 = 43.2\% \qquad \left[\text{of } \sum (y - \bar{y})^2 \text{ explained by density}\right]$$

$$r_{02}^2 = .683^2 = 46.7\% \qquad \left[\text{of } \sum (y - \bar{y})^2 \text{ explained by hardness}\right].$$

Thus the two predictors together do about 8% more "explaining" of $\sum(y - \bar{y})^2$ than the better of the two does separately. This 55.6% is not the sum of 43.2% and 46.7%, since the two predictors are not independent ($r_{12} = .616$) and thus, here, the story each tells separately overlaps with what the other can tell about strength.

Let us use the tests of Section 13.2.6. Is the multiple $R_{0.12}^2$ significantly above zero, using $\alpha = .01$. For this we use (13.21):

$$F_{2,57} = \frac{.556/2}{(1 - .556)/57} = 35.7.$$

This is far above the nearest 1% value which has degrees of freedom 2 and 60, namely 4.977. Hence we have clear evidence of some linearity between $Y$ and $X_1$, $X_2$. Next we may ask whether the gain from using $X_2$ alone and using both is significant. Use (13.23) for this with $h = 1$, $k = 2$:

$$F_{1,57} = \frac{(.556 - .467)/(2 - 1)}{(1 - .556)/57} = 11.43.$$

This is compared with the 1% $F$ of 7.077 for 1, 60 degrees of freedom. Therefore the addition of $X_1$ to $X_2$ significantly improves the predictability in a statistical sense. It is up to the worker involved to decide whether the gain in predictability is worth the additional time, effort, and cost.

By (13.19)

$$s_{0.12}^2 = (n - 1) s_0^2 (1 - R_{0.12}^2)/(n - k - 1) = 59(3996^2)(1 - .556)/57$$
$$= 7,339,000 \quad (\text{lb}/\text{in.}^2)^2$$

or

$$s_{0.12} = 2710 \quad \text{lb}/\text{in.}^2 \qquad \text{versus} \qquad s_{0.2} = 2940 \quad \text{lb}/\text{in.}^2$$

found similarly.

### 13.2.8. Second Example.

Let us now reconsider the example given in Section 13.2.5 illustrating the solution of the normal equations. First we might look at $r_{01} = +.3011$ and $r_{02} = -.2271$. The first, when squared, indicates that $X_1$ explains only 9.07% of $\Sigma(y - \bar{y})^2$, while the latter squared gives 5.16%. From these figures one might naively suppose that together the best one could hope for in amount of explained variation is about 14.23%. But now let us see what $R_{0.12}^2$ turns out to be. We have from Section 13.2.5,

$$R_{0.12}^2 = .3285.$$

Therefore, as a team, $X_1$ and $X_2$ explain 32.85% of the variation in $Y$. How does this come about? It simply means that $X_2$, the carbon drop, adds very materially to the picture painted by $X_1$, the melt carbon. So together they explain a considerable amount of $\Sigma(y - \bar{y})^2$. Another way of looking at it is that $r_{01}$ being positive says that $\hat{y}$ goes up as $x_1$ does, while $r_{02}$ being negative says that $\hat{y}$ goes down as $x_2$ goes up. These are the separate effects. They suggest that $x_1$ and $x_2$ ought to be inversely related. But are they? No; $r_{12} = +.5731$, *not negative*. Hence we see from this and the preceding example that the proportion of the variation $\Sigma(y - \bar{y})^2$ explained by $X_1$, $X_2$ together may be less than or more than the total explained separately, $r_{01}^2 + r_{02}^2$. Only when $r_{12} = 0$ will that sum be $R_{0.12}^2$.

With such a large $n$ as 357 we are quite likely to have statistical significance, even if there is not a worthwhile gain in predictability in a practical sense. But let us apply the tests of Section 13.2.6 to this example.

First test $\rho_{0.1234}^2 = 0$ versus $\rho_{0.1234}^2 \neq 0$, at $\alpha = .01$. By (13.21)

$$F_{4,352} = \frac{.4548/4}{(1 - .4548)/352} = 73.4,$$

which is far beyond the $1\%$ $F$ value. Next, does the inclusion of $X_4$ to $X_1$, $X_2$, $X_3$ add significantly?

$$F_{1,352} = \frac{(R_{0.1234}^2 - R_{0.123}^2)/1}{(1 - R_{0.1234}^2)/352} = \frac{(.4548 - .4211)}{(1 - .4548)/352} = 21.8.$$

This is of course indicative of a significant gain in predictability, by the addition of $X_4$ to the "team."

Finally the addition of $X_3$ and $X_4$ to $X_1$ and $X_2$ gives

$$F_{2,352} = \frac{(R_{0.1234}^2 - R_{0.12}^2)/2}{(1 - R_{0.1234}^2)/352} = \frac{(.4548 - .3285)/2}{(1 - .4548)/352} = 40.8,$$

which is again significant.

### 13.3. SECOND APPROACH

The present section is not really a different approach, but rather a reworking of the previous section, making use of vectors and matrices. In this way the formulas and calculations are more in line with those used in modern calculators. Output from most computers uses the format of vectors and matrices. Moreover with this approach we can go further more easily to additional tests. A most helpful reference in this section as well as this entire chapter is the work of Draper and Smith [3].

### 13.3.1. Vectors and Matrices.

A very brief sketch is given here to serve more as a refresher to those who have studied the subject, rather than to teach those who have not.

Basically a "matrix" is simply a rectangular table of numbers, and a "vector" is a special case of a matrix, in which there is but one row or one column. These are, respectively, row or column vectors. We shall use boldface type for matrices and vectors, and represent the number of rows and columns by a subscript. Thus two notations are

$$
\mathbf{A}_{p \cdot q} =
\begin{bmatrix}
a_{11} & a_{12} & \cdots & a_{1q} \\
a_{21} & a_{22} & \cdots & a_{2q} \\
\vdots & \vdots & & \vdots \\
a_{p1} & a_{p2} & \cdots & a_{pq}
\end{bmatrix}
= (a_{ij}), \quad
\mathbf{B}_{p \cdot q} =
\begin{bmatrix}
b_{11} & b_{12} & \cdots & b_{1q} \\
b_{21} & b_{22} & \cdots & b_{2q} \\
\vdots & \vdots & & \vdots \\
b_{p1} & b_{p2} & \cdots & p_{pq}
\end{bmatrix}
= (b_{ij}).
$$

(13.26)

The parentheses enclose a typical term, namely that in the $i$th row and $j$th column, where $i = 1, 2, ..., p$ and $j = 1, 2, ..., q$. The sum of $\mathbf{A}$ and $\mathbf{B}$ is

$$
\mathbf{A} + \mathbf{B} = (a_{ij} + b_{ij});
$$

(13.27)

that is, a $p \cdot q$ matrix with the element in the $i$th row and $j$th column the sum of those for $\mathbf{A}$, $\mathbf{B}$. This is an operation which can only be performed if the two matrices are the same size ($p \cdot q$ in the above). Subtraction is similar.

To define multiplication of matrices let us define

$$
\mathbf{C}_{q \cdot r} =
\begin{bmatrix}
c_{11} & c_{12} & \cdots & c_{1r} \\
c_{21} & c_{22} & \cdots & c_{2r} \\
\vdots & \vdots & & \vdots \\
c_{q1} & c_{q2} & \cdots & c_{qr}
\end{bmatrix}
= (c_{jk}).
$$

(13.28)

Then the product $\mathbf{AC} = \mathbf{D}$ is a $p \cdot r$ matrix whose terms are the result of multiplying elements in the $i$th row of $\mathbf{A}$ by those of the $k$th column of $\mathbf{C}$, respectively. For example, the element in the second row and third column of $\mathbf{D}$ is $d_{23} = a_{21}c_{13} + a_{22}c_{23} + \cdots + a_{2q}c_{q3} = \sum_{j=1}^{q} a_{2j}c_{j3}$. Hence

$$
\mathbf{A}_{p \cdot q} \mathbf{C}_{q \cdot r} = (a_{ij})(c_{jk}) = \left( \sum_{j=1}^{q} a_{ij}c_{jk} \right) = (d_{ij}) = \mathbf{D}_{p \cdot r} .
$$

(13.29)

For the operation to be possible the first factor, here $\mathbf{A}$, must have the same number of columns as the second factor, here $\mathbf{C}$, has rows. In general $\mathbf{AC}$ is not $\mathbf{CA}$; in fact, the latter may not be possible even though the former is.

Two matrices are equal if and only if corresponding elements throughout are equal; that is, $\mathbf{A}_{p \cdot q} = \mathbf{B}_{p \cdot q}$ if $a_{ij} = b_{ij}$ for all $i$, $j$ combinations.

A vector is also denoted by boldface type and may be a row, for example, $(a_1, a_2, ..., a_n) = \mathbf{A}_{1 \cdot n}$ or a column

$$\mathbf{E}_{n \cdot 1} = \begin{bmatrix} e_1 \\ e_2 \\ \vdots \\ e_n \end{bmatrix}.$$

Consider the two possible products of these vectors:

$$\mathbf{A}_{1 \cdot n}\mathbf{E}_{n \cdot 1} = \mathbf{F}_{1 \cdot 1} = \sum_{i=1}^{n} a_i e_i. \tag{13.30}$$

Here we see that we have a $1 \cdot 1$ product, or just a single real number, often called a "scalar." On the other hand, $\mathbf{EA}$ is possible too:

$$\mathbf{E}_{n \cdot 1}\mathbf{A}_{1 \cdot n} = \mathbf{G}_{n \cdot n} = (e_i a_j) = (g_{ij}); \tag{13.31}$$

that is, the product is an $n$ by $n$ matrix with simple products of respective elements of $\mathbf{E}$ and $\mathbf{A}$.

For the "transpose" of a matrix we merely interchange the rows and columns, and designate by a prime. Thus

$$\mathbf{A}'_{q \cdot p} = \text{transpose of } \mathbf{A}_{p \cdot q} = \begin{bmatrix} a_{11} & a_{21} & \cdots & a_{p1} \\ a_{12} & a_{22} & \cdots & a_{p2} \\ \vdots & \vdots & & \vdots \\ a_{1q} & a_{2q} & \cdots & a_{pq} \end{bmatrix}, \quad \begin{bmatrix} 3 & 1 & 5 \\ 2 & 6 & 7 \end{bmatrix}' = \begin{bmatrix} 3 & 2 \\ 1 & 6 \\ 5 & 7 \end{bmatrix}. \tag{13.32}$$

The transpose of a column vector is a row vector, and vice versa.

If a matrix is square, then it is said to have a determinant. This is designated by vertical bars. For example,

$$\text{determinant of } \mathbf{A}_{n \cdot n} = |\mathbf{A}| = |(a_{ij})|. \tag{13.33}$$

Evaluation of determinants is given in high school algebra texts. A way of reducing the evaluation of an $n \cdot n$ determinant to $(n-1) \cdot (n-1)$ determinants is through the use of "cofactors." Consider an $n \cdot n$ determinant $|\mathbf{A}_{n \cdot n}| = |(a_{ij})|$. Then the cofactor corresponding to the element $a_{ij}$ uses the $(n-1) \cdot (n-1)$ determinant, obtained from $\mathbf{A}$ by deleting the $i$th row and $j$th column of $\mathbf{A}$ then taking the determinant of the resulting $(n-1) \cdot (n-1)$ matrix. Finally we must multiply such a determinant by $(-1)^{i+j}$ to give the cofactor of $a_{ij}$, namely $A(ij)$. Then an expansion giving $|\mathbf{A}|$ via elements in the $i$th row is

$$|\mathbf{A}| = a_{i1}A(i1) + a_{i2}A(i2) + \cdots + a_{in}A(in). \tag{13.34}$$

But if we take

$$a_{i1}A(k1) + a_{i2}A(k2) + \cdots + a_{in}A(kn) = 0 \quad \text{if } i \neq k, \quad (13.35)$$

because this would be the expansion of a determinant with two identical rows.

Let us now define the "identity" matrix and the inverse of a matrix. Both are square matrices.

$$\mathbf{I}_{n \cdot n} = \begin{bmatrix} 1 & 0 & \cdots & 0 \\ 0 & 1 & \cdots & 0 \\ \vdots & \vdots & & \vdots \\ 0 & 0 & \cdots & 1 \end{bmatrix} = \text{identity } n \cdot n \text{ matrix.} \quad (13.36)$$

It is easily shown that

$$\mathbf{I}_{n \cdot n}\mathbf{A}_{n \cdot n} = \mathbf{A}_{n \cdot n} = \mathbf{A}_{n \cdot n}\mathbf{I}_{n \cdot n}. \quad (13.37)$$

So in matrix algebra $I$ behaves like 1 in arithmetic. The "inverse" of a square matrix $\mathbf{A}_{n \cdot n}$ which is "nonsingular," that is, whose determinant is nonzero, is written as $\mathbf{A}^{-1}$ and defined by

$$\mathbf{A}_{n \cdot n}\mathbf{A}_{n \cdot n}^{-1} = \mathbf{I}_{n \cdot n}. \quad (13.38)$$

It is easily shown that the reverse order also gives $I$, namely

$$\mathbf{A}_{n \cdot n}^{-1}\mathbf{A}_{n \cdot n} = \mathbf{I}_{n \cdot n}. \quad (13.39)$$

Letting the *nonzero* determinant of $\mathbf{A}$ be $|\mathbf{A}|$, indicating a real number, we use cofactors as follows:

$$\mathbf{A}_{n \cdot n}^{-1} = (a_{ij}^{-1}) = \left(\frac{A(ji)}{|\mathbf{A}|}\right). \quad (13.40)$$

Thus the 3, 4 element of the inverse $\mathbf{A}_{n \cdot n}^{-1}$ is the cofactor of $a_{4,3}$ divided by $A$.

Finding the inverse of a square matrix is a rather considerable arithmetical chore. Accordingly there are many computer programs written for the job. It can, however, be done through an extension of the systematic method presented in Section 13.2.5.

### 13.3.2. The Matrix Approach.

We now define a number of matrices and vectors and go over the regression approach of Section 13.2 again.

We place the observed data in a vector and a matrix:

$$
\mathbf{Y}_{n\cdot1} = \begin{bmatrix} y_1 \\ y_2 \\ \vdots \\ y_n \end{bmatrix}, \qquad
\mathbf{X}_{n\cdot(k+1)} = \begin{bmatrix} 1 & x_{11} & x_{21} & \cdots & x_{k1} \\ 1 & x_{12} & x_{22} & \cdots & x_{k2} \\ \vdots & \vdots & \vdots & & \vdots \\ 1 & x_{1n} & x_{2n} & \cdots & x_{kn} \end{bmatrix}. \qquad (13.41)
$$

The second matrix requires a bit of explanation. The column of 1's is to be associated with the constant term $\beta_0$ or $b_0$. The 1's in a sense mean that we always have $\beta_0$ or $b_0$ present in $y$. The second column contains the observed data for random variable $X_1$, ..., to the $k+1$ column, those for $X_k$. Note also that $x_{21}$ is not the element of $X$ in the second row and first column as is often the case in matrix notations. Next we have four column vectors

$$
\boldsymbol{\beta}_{(k+1)\cdot1} = \begin{bmatrix} \beta_0 \\ \beta_1 \\ \vdots \\ \beta_k \end{bmatrix}, \qquad
\mathbf{b}_{(k+1)\cdot1} = \begin{bmatrix} b_0 \\ b_1 \\ \vdots \\ b_k \end{bmatrix}, \qquad
\boldsymbol{\varepsilon}_{n\cdot1} = \begin{bmatrix} \varepsilon_1 \\ \varepsilon_2 \\ \vdots \\ \varepsilon_n \end{bmatrix}, \qquad
\mathbf{e}_{n\cdot1} = \begin{bmatrix} e_1 \\ e_2 \\ \vdots \\ e_n \end{bmatrix}.
$$

$$(13.42)$$

The theoretical model for the linear regression is now expressible:

$$
\mathbf{Y}_{n\cdot1} = \mathbf{X}_{n\cdot(k+1)}\boldsymbol{\beta}_{(k+1)\cdot1} + \boldsymbol{\varepsilon}_{n\cdot1}. \qquad (13.43)
$$

Each of the three terms $\mathbf{Y}$, $\mathbf{X\beta}$, and $\boldsymbol{\varepsilon}$ is an $n \cdot 1$ column matrix, and the equality means really $n$ equations such as

$$
y_i = \beta_0 + \beta_1 x_{1i} + \beta_2 x_{2i} + \cdots + \beta_k x_{ki} + \varepsilon_i, \qquad i = 1, ..., n, \quad (13.44)
$$

that is, like (13.3). Moreover the matrix equation

$$
\mathbf{Y} = \mathbf{X\beta} + \boldsymbol{\varepsilon} \qquad (13.45)
$$

is very similar in form to the one-variable linear model (11.12), since $\boldsymbol{\beta}$ is composite, not single. The observed errors and $b$'s come in similarly

$$
\mathbf{Y}_{n\cdot1} = \mathbf{X}_{n\cdot(k+1)}\mathbf{b}_{(k+1)\cdot1} + \mathbf{e}_{n\cdot1}, \qquad (13.46)
$$

or like (13.6) and (11.15)

$$
\mathbf{Y} = \mathbf{Xb} + \mathbf{e}. \qquad (13.47)
$$

Next consider the following, the $\sum$ being for $i = 1, ..., n$:

$$\mathbf{X}'_{(k+1) \cdot n} \mathbf{X}_{n \cdot (k+1)}$$

$$= \begin{bmatrix} 1 & 1 & \cdots & 1 \\ x_{11} & x_{12} & \cdots & x_{1n} \\ x_{21} & x_{22} & \cdots & x_{2n} \\ \vdots & \vdots & & \vdots \\ x_{k1} & x_{k2} & \cdots & x_{kn} \end{bmatrix} \begin{bmatrix} 1 & x_{11} & x_{21} & \cdots & x_{k1} \\ 1 & x_{12} & x_{22} & \cdots & x_{k2} \\ \vdots & \vdots & \vdots & & \vdots \\ 1 & x_{1n} & x_{2n} & \cdots & x_{kn} \end{bmatrix} = (\mathbf{X}'\mathbf{X})_{(k+1) \cdot (k+1)}$$

$$= \begin{bmatrix} n & \sum x_{1i} & \sum x_{2i} & \cdots & \sum x_{ki} \\ \sum x_{1i} & \sum x_{1i}^2 & \sum x_{1i} x_{2i} & \cdots & \sum x_{1i} x_{ki} \\ \sum x_{2i} & \sum x_{2i} x_{1i} & \sum x_{2i}^2 & \cdots & \sum x_{2i} x_{ki} \\ \vdots & \vdots & \vdots & & \vdots \\ \sum x_{ki} & \sum x_{ki} x_{1i} & \sum x_{ki} x_{2i} & \cdots & \sum x_{ki}^2 \end{bmatrix}. \qquad (13.48)$$

Note that these are exactly the coefficients of the $b$'s in the normal equations (13.8). Therefore

$$(\mathbf{X}'\mathbf{X})_{(k+1) \cdot (k+1)} \mathbf{b}_{(k+1) \cdot 1} = (\mathbf{X}'\mathbf{X}\mathbf{b})_{(k+1) \cdot 1} \qquad (13.49)$$

gives exactly the $k + 1$ left sides of (13.8), for example,

$$b_0 \sum x_{1i} + b_1 \sum x_{1i}^2 + b_2 \sum x_{1i} x_{2i} + \cdots + b_k \sum x_{1i} x_{ki}. \qquad (13.50)$$

On the other hand,

$$\mathbf{X}'_{(k+1) \cdot n} \mathbf{Y}_{n \cdot 1} = (\mathbf{X}'\mathbf{Y})_{(k+1) \cdot 1} = \begin{bmatrix} 1 & 1 & \cdots & 1 \\ x_{11} & x_{12} & \cdots & x_{1n} \\ x_{21} & x_{22} & \cdots & x_{2n} \\ \vdots & \vdots & & \vdots \\ x_{k1} & x_{k2} & \cdots & x_{kn} \end{bmatrix} \begin{bmatrix} y_1 \\ y_2 \\ y_3 \\ \vdots \\ y_n \end{bmatrix}$$

$$= \begin{bmatrix} \sum y_i \\ \sum x_{1i} y_i \\ \sum x_{2i} y_i \\ \vdots \\ \sum x_{ki} y_i \end{bmatrix}, \qquad (13.51)$$

which are the $k + 1$ right-hand sides of (13.8). Thus equating, the $k + 1$ normal equations are all contained in

$$\mathbf{X}'\mathbf{X}\mathbf{b} = \mathbf{X}'\mathbf{Y}. \qquad (13.52)$$

These $k + 1$ equations are to be solved for $b_0$, $b_1$, ..., $b_k$, that is, for $b$. But this is algebraically easy. All we need do is multiply both sides on the left by $(\mathbf{X'X})^{-1}*$:

$$(\mathbf{X'X})^{-1}(\mathbf{X'X})\mathbf{b} = (\mathbf{X'X})^{-1}\mathbf{X'Y},$$

or since $\mathbf{Ib} = \mathbf{b}$:

$$\mathbf{b}_{(k+1)\cdot 1} = [(\mathbf{X'X})^{-1}\mathbf{X'Y}]_{(k+1)\cdot 1}. \tag{13.53}$$

Both sides are $(k + 1) \cdot 1$ column vectors.

The **b** vector found in this manner is the estimation vector for the parameter vector $\beta$. In fact under our assumed normal model the $b$'s are normally distributed with means the respective $\beta$'s. This comes about because they are linear functions of the $y$'s. Moreover the $b$'s are maximum likelihood estimates of the $\beta$'s, respectively.

The $n$ fitted or estimated values of the random variable $Y$, namely, elements in $\hat{\mathbf{Y}}$, are obtained by

$$\hat{\mathbf{Y}} = \mathbf{Xb}. \tag{13.54}$$

Many computer programs will print out the vector of $\hat{y}_1$, $\hat{y}_2$, ..., $\hat{y}_n$ and/or the $n$ residuals

$$\mathbf{e} = \mathbf{Y} - \hat{\mathbf{Y}} = \mathbf{Y} - \mathbf{Xb}. \tag{13.55}$$

The latter are often checked for normality (see Chapter 14), or tested for outliers, that is, extreme isolated values. We can use (13.54) for predictions, or if we have a particular set of $X_1$, $X_2$, ... $X_k$ for which we wish to predict $\hat{Y}$, we can substitute into (13.9). In matrix notation this would be by $(1, x_{10}, x_{20}, ..., x_{k0}) = \mathbf{X_0}'$. Then

$$\hat{y}_0 = \mathbf{X_0}'\mathbf{b}. \tag{13.56}$$

Now how do we obtain the multiple correlation coefficient $R_{0.12...k}$, using the present approach? Following (13.16):

$$R^2_{0.12...k} = \frac{\sum(\hat{y}_i - \bar{y})^2}{\sum(y_i - \bar{y})^2} = \frac{\text{explained variation in } Y \text{ by } X_1, ..., X_k}{\text{total variation in } Y}. \tag{13.57}$$

The machine could find this directly, but it is much better for it to use various sums. As usual

$$\sum(y_i - \bar{y})^2 = \sum y_i^2 - \left[\left(\sum y_i\right)^2 / n\right] = \mathbf{Y'Y} - \left[\left(\sum y_i\right)^2 / n\right]. \tag{13.58}$$

---

* The matrix $\mathbf{X'X}$ will always have an inverse unless two or more of the $x_{ji}$, $x_{ki}$ vectors are mathematically dependent.

For the numerator we can use

$$\mathbf{b'X'Y} - \left[\left(\sum y_i\right)^2 \big/ n\right] = \sum (\hat{y}_i - \bar{y})^2. \qquad (13.59)$$

Now what is $\mathbf{b'}_{1 \cdot (k+1)} \mathbf{X'}_{(k+1) \cdot n} \mathbf{Y}_{n \cdot 1}$? We have met $\mathbf{X'Y}$ in (13.51) and $\mathbf{b'}_{1 \cdot (k+1)} = (b_0, b_1, ..., b_k)$, so

$$\mathbf{b'X'Y} = b_0 \sum y_i + b_1 \sum x_{1i} y_i + \cdots + b_k \sum x_{ki} y_i. \qquad (13.60)$$

And hence

$$\sum (\hat{y}_i - \bar{y})^2 = b_0 \sum y_i + b_1 \sum x_{1i} y_i + \cdots + b_k \sum x_{ki} y_i - \left[\left(\sum y_i\right)^2 \big/ n\right]. \quad (13.61)$$

Thus we may find $R^2_{0.12...k}$ by the quotient of (13.59) or (13.61) by (13.58). Then (13.19) provides $s^2_{0.12...k}$.

An important use of the inverse matrix $(\mathbf{X'X})_{(k+1) \cdot (k+1)}$ is the following. By definition

$$\text{variance } (b_j) = \sigma^2_{b_j} = E[(b_j - \beta_j)^2], \qquad j = 0, 1, ..., k \qquad (13.62)$$

$$\text{covariance } (b_h, b_j) = \text{cov}(b_h, b_j) = E[(b_h - \beta_h)(b_j - \beta_j)], \qquad h \neq j. \quad (13.63)$$

Then the square matrix $\sigma^2(\mathbf{X'X})^{-1}$, in which each element in $(\mathbf{X'X})^{-1}$ is multiplied by $\sigma^2$, gives on the principal diagonal $\sigma^2_{b_j}$ and off the principal diagonal the covariances. That is, the "variance–covariance matrix" of $b$'s is

$$\sigma^2(\mathbf{X'X}) = \begin{bmatrix} \sigma^2_{b_0} & \text{cov}(b_0, b_1) & \cdots & \text{cov}(b_0, b_k) \\ \text{cov}(b_0, b_1) & \sigma^2_{b_1} & \cdots & \text{cov}(b_1, b_k) \\ \vdots & & & \vdots \\ \text{cov}(b_0, b_k) & \text{cov}(b_1, b_k) & \cdots & \sigma^2_{b_k} \end{bmatrix}. \qquad (13.64)$$

From this we may test the hypotheses, for example, $\beta_j = 0$ versus $\beta_j \neq 0$, by

$$u = \frac{b_j - 0}{\sigma_{b_j}} \qquad (u \text{ normal}) \qquad (13.65)$$

if $\sigma$ in (13.3) is somehow known, or if unknown, substitute $s^2_{0.12...k}$ from (13.19) for $\sigma^2$ in (13.64) and use the $t$ distribution statistic

$$t_{n-k-1} = \frac{b_j - 0}{s_{b_j}}. \qquad (13.66)$$

Calculator programs often provide the "standard errors of $b_j$'s" so that the hypotheses just given may be tested for each $\beta_j$, $j = 0, ..., k$. Moreover the correlation coefficients are available:

$$r(b_h, b_j) = \text{cov}(b_h, b_j)/\sigma_{b_h}\sigma_{b_j} = \rho(b_h, b_j), \tag{13.67}$$

where $\rho$ can be used since we are treating our collection of $X$'s as though they were a population, that is, constant. [It might be mentioned in passing that if an $r$ in (13.67) should happen to be 1, then $\mathbf{X}'\mathbf{X}$ has a zero determinant, and thus fails to have an inverse. The same is true if some multiple correlation of $b$'s is 1.]

If it is desired to set confidence limits on the true regression mean $E(Y \mid x_{10}, x_{20}, ..., x_{k0})$, that is, at a point of interest $(x_{10}, x_{20}, ..., x_{k0})$, we define the row vector $\mathbf{X_0}' = (1, x_{10}, x_{20}, ..., x_{k0})$. The point estimate of this mean value is $\hat{y} = b_0 + b_1 x_{10} + \cdots + b_k x_{k0} = \mathbf{b}'\mathbf{X_0} = \mathbf{X_0}'\mathbf{b}$. Then we have as $1 - \alpha$ confidence limits on $E(Y)$:

$$\hat{y} \pm t[n - k - 1, \alpha/2]s_{0.12...k} \sqrt{\mathbf{X_0}'(\mathbf{X'X})^{-1}\mathbf{X_0}} = 1 - \alpha \quad \text{limits on} \quad E(Y). \tag{13.68}$$

### 13.3.3. Selection of a Set of Predictors.

As already mentioned in Section 13.2.6, there are rather complicated computer programs for seeking a set of $X$'s less than the full collection of $k$ of them, which will do the job of prediction adequately. One is the "buildup" technique which selects at the start that variable $X_h$ which has the largest $r_{0h}^2$, if it is significantly nonzero, (11.24) or (11.36). Then it hunts through the remaining $X$'s to find that $X_j$ which in the presence of $X_h$ will make the greatest gain $R_{0.hj}^2 - r_{0h}^2$. If the gain is significant at a chosen $\alpha$ risk, by (13.23), it seeks out another $X$ among those remaining which will give the largest jump above $R_{0.hj}$, tests for significance, and so on. One trouble is that if $n$ is large, each jump is often significant up through all $k$ $X$'s. Another trouble is that two $X$'s can give a much larger jump together than might be expected from the separate jumps. We saw an example of this with $X_1$, $X_2$ in Section 13.2.8. In fact it is possible for the separate jumps to be both nonsignificant, but for the combined jump to be clearly significant. And following this further we have no guarantee that say five $X$'s chosen this way from 10 $X$'s will provide the highest $R^2$ for any five chosen from the 10. It probably will, however.

There are also "tear-down" programs which obtain the full $R_{0.12...k}^2$, then seek out that $X_h$ which will cause the smallest loss in $R_{0.12...k}^2$. If the loss by (13.23) is not statistically significant, then it seeks out the next suspected $X_h$ for deletion, and so on. In this way a subset of $X$'s is sought. Again there is no guarantee of the chosen set being "best" in any absolute sense, for reasons such as those just discussed in the previous paragraph.

Then too there is the question of whether such gains or losses in predictability (or explained variation) are of practical importance. This question shows up especially when $n$ is large.

The problem of seeking "best" sets of predictors is still an open question to some extent, and is being researched further.

### 13.3.4. Calculation of an Inverse.

This job can readily be done systematically when the matrix of coefficients is symmetric as in (13.8), (13.12), or (13.14). Consider $(\mathbf{X'X})^{-1}{}_{(k+1)\cdot(k+1)} \cdot (\mathbf{X'X})_{(k+1)\cdot(k+1)} = I_{(k+1)\cdot(k+1)}$ . The procedure of finding $(\mathbf{X'X})$ simply builds upon the compact Doolittle technique as illustrated in Section 13.2.5. Instead of having just one column of right-hand sides of the normal equation, namely, the $x_0$ column there, we add $k + 1$ more columns in the form of $I$, that is,

$$\begin{bmatrix} 1 & 0 & 0 & \cdots & 0 \\ 0 & 1 & 0 & \cdots & 0 \\ 0 & 0 & 1 & \cdots & 1 \\ \vdots & \vdots & \vdots & & \vdots \\ 0 & 0 & 0 & \cdots & 1 \end{bmatrix} = \mathbf{I}.$$

Everything in the $x_1$, $x_2$, $x_3$, $x_4$ columns through line 12 is exactly the same. The back solution for the $b^*$'s from the $x_0$ column is also the same. But now we have $k + 1$ back solutions instead of just one, one for each column of $I$. Some of these duplicate others because, since $\mathbf{X'X}$ is symmetric, so is $(\mathbf{X'X})^{-1}$. The elements of the first column or row of $(\mathbf{X'X})^{-1}$ come from the back solution of the $(1, 0, 0, ..., 0)$ column just the same as the $b^*$'s came from the back solution using the $x_0$ column. Then the elements of the second column or row of $(\mathbf{X'X})^{-1}$ come from the $(0, 1, 0, ... 0)$ column of $I$, and so forth. This can be done quite compactly. See Dwyer [1, p. 191].

### 13.4. SUMMARY OF APPROACH

We summarize by listing the following available steps:

1. Start by observed data in the form of Section 13.2.1, which form $\mathbf{X}$ and $\mathbf{Y}$ matrices.
2. Model considered is $\mathbf{Y} = \mathbf{X}\boldsymbol{\beta} + \boldsymbol{\varepsilon}$, like (13.3).

3. Find $b$'s as estimates of the $\beta$'s by $\mathbf{b} = (\mathbf{X'X})^{-1}\mathbf{X'Y}$, or by solving normal equations (13.8), (13.12), or (13.14) by Section 13.2.5.

4. Find $R^2_{0.12\ldots k}$ by (13.57) through (13.61), or as in (13.20) and Section 13.2.5.

5. Find $s^2_{0.12\ldots k}$ by (13.19) from $R^2$.

6. Can test the significance of $b_h$ by the hypothesis $\beta_h = 0$ versus $\beta_h \neq 0$ through using (13.66). This test is not independent of the team of $X_j$'s chosen.

7. Can find the variance–covariance matrix for the $b$'s: $\sigma^2(\mathbf{X'X})^{-1}$, or as estimated by $s^2_{0.12\ldots k}(\mathbf{X'X})^{-1}$.

8. Can find the estimated values of $Y$ which correspond to the $n$ observed values of $Y$: $y_1, \ldots, y_n$. These are $\hat{\mathbf{Y}} = \mathbf{Xb}$ or $b_0 + b_1 x_{1i} + b_2 x_{2i} + \cdots + b_k x_{ki}$.

9. Can find the $n$ residuals $\mathbf{e} = \mathbf{Y} - \hat{\mathbf{Y}}$ or $e_i = y_i - b_0 - b_1 x_{1i} - \cdots - b_k x_{ki}$ for all $i$. These residuals may be tested for normality, for example, as in Chapter 14. Or the most extreme positive and negative $e_i$'s may be tested for significance as outliers.

10. Can use (13.68) to set confidence limits on the true regression $E(Y)$ at an $X$ point of interest.

## 13.5. ADEQUACY OF REGRESSION MODEL

This topic has its analogy in linear regression, where the straight line regression of $Y$ versus $X$ is questioned. In Chapter 11, $s_e^2$ includes not only the pure error variance around the true regression, which might be a curve rather than a straight line, but also the departure of this curve from the straight line. It is possible to "purify" $s_e^2$ by measuring it from the best fitting curve instead of the line. The new $s^2_{e\,\text{from curve}}$, is then smaller than $s^2_{e\,\text{from line}}$, and the difference can be tested for significance.

In the case of multiple regression, the error variance $s^2_{0.12\ldots k}$ is under study. There are two cases. In one case, which may be rare in practice, we might have a variance $s_e^2$ from previous or similar studies. If it is deemed to be from adequate models and therefore pure error, we can form

$$F_{\nu_1,\nu_2} = \frac{s^2_{0.12\ldots k}}{s_e^2}, \tag{13.69}$$

where $\nu_1$ and $\nu_2$ are the corresponding degrees of freedom. If significant, then this indicates that $s^2_{0.12\ldots k}$ contains some "lack of fit" and that the proposed model is not adequate.

The second type of case is that in which we use only the data in the current study. There are two subcases here. If some whole sets of $X$ variables are duplicated, so that we have more than one $y$ at a given $x$ point, then the sum of squares at each such point $\sum_{j=1}^{n_i}(y_{ij} - \bar{y}_i)^2$ can be found and pooled. Then the pooled "within" sum of squares is divided by the total degrees of freedom $\sum(n_i - 1) = n_e$. This forms $s_e^2$. The breakdown then is as follows:

| Source | Sum of squares | Degrees of freedom | Mean square |
|---|---|---|---|
| Total | $(n - k - 1) s_{0.12\ldots k}^2$ | $(n - k - 1)$ | $s_{0.12\ldots k}^2$ |
| Pure error | $\sum\sum (y_{ij} - \bar{y}_i)^2$ | $n_e$ | $s_e^2$ |
| Lack of fit | $(n - k - 1) s^2 - \sum\sum (y_{ij} - \bar{y}_i)^2$ | $n - k - 1 - n_e$ | Ratio |

Then the test is

$$F_{n-k-1-n_e,n_e} = \frac{[(n - k - 1)s_{0.12\ldots k}^2 - \sum\sum (y_{ij} - \bar{y}_i)^2]/(n - k - 1 - n_e)}{\sum\sum (y_{ij} - \bar{y}_i)^2/n_e}.$$

$$(13.70)$$

The repetition of a whole set of $X$'s will not in general "just happen" (unless $k$ is say only 2 and $n$ large) if we are just taking data as they occur. But it is possible to control the $X$'s in a systematically designed experiment to include duplicate runs yielding replicated $y$'s.

The second subcase occurs where we already have, say, $X_1$, $X_2$, $X_3$, $X_4$ in the model. Some curvilinear tendency shows up in the scatter diagram of $Y$ versus $X_1$, and versus $X_2$. So we may try adding three more terms $X_5 = X_1^2$, $X_6 = X_1X_2$, and $X_7 = X_2^2$. Then we may use the approach already discussed in Section 13.2.6 and use the test in (13.23) which does the same thing as (13.70).

## 13.6. COMMENTS AND PRECAUTIONS

The following may prove useful as supplements to the preceding sections:

1. Applications often stem from the need to be able to estimate $Y$ from several $X$'s. Using several $X$'s gives better estimates on the average than using just one $X$. Another type of application can readily come from a study of the $b_j$*'s, that is, the standardized regression coefficients. If none of the $X$'s are very strongly correlated, then the $b_j$*'s will tend to be somewhere between $-1$

and $+1$.* Those $X$'s having the largest $b_j$*'s, in size, will be the input variables most worth controlling further to control $y$ more closely. At what levels to control such variables can be judged by study of the regression equation.

2. If the $X$'s are greatly different in size, coding them prior to calculation (by hand or machine) is helpful. Results can be decoded if desired, at the end.

3. A great many significant figures have to be carried if the matrix of $r_{hj}$'s is near zero.

4. In general $R_{0.12...k}$ is not going to exceed the reliability of $Y$, that is, how well $Y$ correlates with itself in repeated measures. That is, the $X$'s ought not to be expected to predict $Y$ better than it can predict itself. For example, no battery of criteria can predict college freshman grades, which are always subject to unreliability, better than their own reliability.

5. Suppose that we choose the best four predictors, $X_j$'s, out of, say, 50 available. Such variables cannot be expected to do quite as well on new data of the same type. This is because those chosen from the large group of $X$'s are likely to be those for which luck or chance favored them. Others might well do better in the long run.

6. The range of the $X$'s in the sample has much to do with $r$'s and $R$'s. Relatively narrow ranges tend to give small $r$'s and $R$'s.

7. Very high $b_j$*'s tend to have high standard errors.

8. Planning of data gathering can make most efficient use of multiple regression. For example, we can actually make the $X$ variables entirely independent as in two-way analysis of variance designs (with or without replication). The same is true in designs seeking optimum conditions to maximize or minimize $E(Y)$. In fact, through rather arbitrary and somewhat artificial definitions of variables, analysis of variance can be handled by multiple regression techniques.

REFERENCES

1. P. S. Dwyer, "Linear Computations." Wiley, New York, 1951.
2. W. A. Shewhart, "Economic Control of Quality of Manufactured Product." Van Nostrand-Reinhold, Princeton, New Jersey, 1931.
3. N. R. Draper and H. Smith, "Applied Regression Analysis." Wiley, New York, 1966.
4. F. R. Lloyd, The caustic fusion of a 6-methyl-2-sodium naphthalenesulphonate, p. 50. Ph.D. Thesis, Purdue Univ., Lafayette, Indiana, 1949.

* But if a couple of $X$'s are very strongly correlated as might happen if $X_6 = X_5^2$, then $b_5$* and $b_6$* could be perhaps $-20$ and $+22$, which makes meaningless the interpretation of $b_j$* given after (13.15).

## Problems

13.1. In a blooming mill a study was made of the effect of length, width, and thickness of ingots being rolled, upon the yield of usable steel for a purpose. Thirty-four ($n = 34$) composite observations were made. Given $Y = X_0 =$ yield, $X_1 =$ length, $X_2 =$ width, $X_3 =$ thickness, the following linear correlation coefficients were obtained: $r_{01} = +.5600$, $r_{02} = -.4524$, $r_{03} = -.3429$, $r_{12} = -.8043$, $r_{13} = -.7364$, $r_{23} = +.2373$. Use (13.24) and (13.25) to find the coefficients $b_1^*$ and $b_2^*$, and then $R_{0.12}^2$. Interpret its meaning in words. Test the null hypothesis $\rho_{0.12} = 0$ versus $\rho_{0.12} > 0$ at the 1% significance level. Is the gain in going from $X_1$ to $X_1$ and $X_2$ significant at the 1% level?

13.2. Using all three independent variables given in Problem 13.1, find by the compact Doolittle technique $R_{0.123}^2$ and test $\rho_{0.123} = 0$ versus $\rho_{0.123} > 0$ at $\alpha = .01$. Interpret the meaning of $R_{0.123}^2$. Test at $\alpha = .01$ whether the gain in adding $X_3$ to the predictors is significant.

13.3. In a study of "flush" heats in an open hearth furnace the operating rate for $n = 100$ heats was observed. The variables were as follows:

$Y = X_0 =$ tons per hour, $\quad \bar{x}_0 = 13.63$ tons/hr, $\quad s_0 = 1.57$ tons/hr

$X_1 =$ heat size in tons, $\quad \bar{x}_1 = 205$ tons, $\quad s_1 = 8.07$ tons

$X_2 =$ charging time in hours, $\quad \bar{x}_2 = 3.59$ hr, $\quad s_2 = .942$ hr

$X_3 =$ gallons of oil per ton of steel, $\quad \bar{x}_3 = 27.96$ gal/ton, $\quad s_3 = 3.74$ gal/ton

$r_{01} = +.424$, $\quad r_{02} = -.446$, $\quad r_{03} = -.791$, $\quad r_{12} = -.258$,

$\qquad r_{13} = -.413$, $\quad r_{23} = +.324$.

(a) Use (13.24) and (13.25) to find the standardized coefficients $b_2^*$ and $b_3^*$.

(b) Find $R_{0.23}^2$ and interpret its meaning in words.

(c) Find the regression equation for $X_0$ versus $X_2$ and $X_3$ and $s_{0.23}$.

(d) Test $\rho_{0.23} = 0$ versus $\rho_{0.23} > 0$ at $\alpha = .01$.

(e) Test at $\alpha = .01$ whether the gain from $r_{03}^2$ to $R_{0.23}^2$ is significant.

13.4. For the data of Problem 13.3:

(a) Find $R_{0.123}^2$.

(b) Find the regression equation for $X_0$ versus $X_1$, $X_2$, $X_3$, and $s_{0.123}$.

(c) Interpret the meaning of the results in the previous parts.

(d) Test at $\alpha = .01$, $\rho_{0.123} = 0$ versus $\rho_{0.123} > 0$.

(e)  Test whether the gain from $r_{03}^2$ to $R_{0.123}^2$ is significant at $\alpha = .01$.

13.5.  For a study of factors influencing the diameter of a roll condenser for automobiles the following were the results. Note that the actual data were coded so as to give numbers roughly the same size.

$Y = X_0 =$ diameter of condenser in .001 in.,
$\quad\ X_1 =$ total thickness of foil in .00001 in.,
$\quad\ X_2 =$ total thickness of paper in .00001 in.,
$\quad\ X_3 =$ number of turns in condenser;

| | | | | | |
|---|---|---|---|---|---|
| $\bar{X}_0 =$ | 556.5, | $s_0 =$ | 12.56, | $n = 179$ | |
| $\bar{X}_1 =$ | 68.12, | $s_1 =$ | 5.975 | | |
| $\bar{X}_2 =$ | 155.3, | $s_2 =$ | 4.787 | | |
| $\bar{X}_3 =$ | 96.82, | $s_3 =$ | 2.018 | | |
| $r_{01} =$ | .6591, | $r_{02} =$ | .1295, | $r_{03} =$ | .3824 |
| $r_{12} =$ | $-.1876$, | $r_{13} =$ | .0229, | $r_{23} =$ | $-.0107$ |

(a)  For $X_0$ versus $X_1$ and $X_2$, find $b_1{}^*$ and $b_2{}^*$, the regression equation, $R_{0.12}^2$, and $s_{0.12}$.
(b)  Test at $\alpha = .01$, $\rho_{0.12} = 0$ versus $\rho_{0.12} > 0$.
(c)  Test at $\alpha = .01$, whether the gain from $r_{01}^2$ to $R_{0.12}^2$ is significant.

13.6  For the data of Problem 13.5:
(a)  Find $R_{0.123}^2$.
(b)  Find the $b^*$'s and regression equation for $X_0$ versus $X_1$, $X_2$, $X_3$, and $s_{0.123}$.
(c)  Interpret the meaning of the results in the previous parts.
(d)  Test at $\alpha = .01$, $\rho_{0.123} = 0$ versus $\rho_{0.123} > 0$.
(e)  Test whether the gain from $r_{01}^2$ to $R_{0.123}^2$ is significant.
(f)  Which variable is most influential in predicting $X_0$?
(g)  How does $r_{01}^2 + r_{02}^2 + r_{03}^2$ compare with $R_{0.123}^2$?

13.7.  A study of blast furnace production in the making of pig iron yielded the following results for $n = 120$ days:

$Y = X_0 =$ average daily tonnage of blast furnace,
$\quad\ X_1 =$ average cubic feet per minute of wind blown,
$\quad\ X_2 =$ percent $Al_2O_3$ in slag,
$\quad\ X_3 =$ volatile material in coke times free carbon,
$\quad\ X_4 =$ moisture in grains;
$\quad\ \bar{X}_0 = 781.6, \quad \bar{X}_1 = 50.16, \quad \bar{X}_2 = 13.23, \quad \bar{X}_3 = .84,$
$\quad\quad \bar{X}_4 = 3.45$
$\quad\ s_0 = 45.83, \quad s_1 = 2.015, \quad s_2 = 1.127, \quad s_3 = .165,$
$\quad\quad s_4 = 1.713.$

Correlation coefficients:

$$r_{01} = +.410, \quad r_{02} = +.376, \quad r_{03} = +.301, \quad r_{04} = -.169$$
$$r_{12} = -.239, \quad r_{13} = +.051, \quad r_{14} = +.195, \quad r_{23} = +.124$$
$$r_{24} = +.023, \quad r_{34} = +.251.$$

(a)  Use the abbreviated Doolittle technique to find the $b^*$'s and $R^2_{0.1234}$ .

(b)  Test at $\alpha = .01$, $\rho_{0.1234} = 0$ versus $\rho_{0.1234} > 0$.

(c)  Find the regression equation for $X_0$ versus $X_1$ to $X_4$ , and $s_{0.1234}$ .

(d)  Test for significance, the addition to explained variance as each of $X_2$ , $X_3$ , $X_4$ is added, using $\alpha = .01$.

(e)  Do any assigned tasks for subsamples of $X_1$ to $X_4$ .

13.8.  Excessive rejection of cast glass insulators occasioned the following study, the object being to predict and control the tank temperature of the molten glass from the surface temperature and height of glass. This was done for the only furnace with a thermocouple installed to measure tank temperature directly. The other two furnaces were not so equipped. The variables were $X_0$ , the tank temperature in degrees centigrade inside (by thermocouple); $X_1$ , the surface temperature of glass in degrees centigrade; and $X_2$ , the height of glass in inches. The observed data follow:

| $X_0$ | $X_1$ | $X_2$ |
|---|---|---|
| 1165 | 1115 | 12.50 |
| 1180 | 1131 | 12.50 |
| 1195 | 1130 | 10.50 |
| 1230 | 1185 | 7.50 |
| 1260 | 1225 | 4.50 |
| 1120 | 1080 | 13.00 |
| 1160 | 1105 | 11.50 |
| 1225 | 1165 | 10.50 |
| 1220 | 1167 | 8.00 |
| 1240 | 1195 | 5.00 |
| 1150 | 1103 | 11.25 |
| 1190 | 1140 | 9.50 |
| 1195 | 1150 | 6.25 |
| 1200 | 1150 | 5.00 |

(a)  Find the required $R^2_{0.12}$ , the regression equation, and $s_{0.12}$ .

(b)  Test $R^2_{0.12}$ for significance.

(c)  Estimate $x_0$ from $x_1 = 1115$ and $x_2 = 12.50$, the first reading. *Within the observed data*, what is the "average" error in such estimates, $x_0 - \hat{x}_0$ ?

Estimates were considered useful and a nomogram constructed for the estimation of $X_0$.

13.9.  An alloy was being developed to have desired percent elongation under stress. Accordingly for 24 melts this dependent variable was measured along with the percentages of five elements. The following data were obtained ordered according to elongation.

| Melt No. | $X_0$ | $X_1$ | $X_2$ | $X_3$ | $X_4$ | $X_5$ |
|---|---|---|---|---|---|---|
| 16 | 11.3 | .50 | 1.3 | .4 | 3.4 | .010 |
| 20 | 10.0 | .47 | 1.2 | .3 | 3.6 | .012 |
| 8 | 9.8 | .48 | 3.1 | .7 | 4.3 | .000 |
| 12 | 8.8 | .54 | 2.6 | .7 | 4.0 | .022 |
| 17 | 7.8 | .45 | 2.8 | .7 | 4.2 | .000 |
| 18 | 7.4 | .41 | 3.2 | .7 | 4.7 | .000 |
| 11 | 6.7 | .62 | 3.0 | .6 | 4.7 | .026 |
| 1 | 6.3 | .53 | 4.1 | .9 | 4.6 | .035 |
| 2 | 6.3 | .57 | 3.7 | .8 | 4.6 | .000 |
| 9 | 6.3 | .67 | 2.7 | .6 | 4.8 | .013 |
| 25 | 6.0 | .54 | 3.1 | .7 | 4.2 | .000 |
| 6 | 6.0 | .42 | 3.1 | .7 | 4.4 | .000 |
| 24 | 5.8 | .33 | 2.6 | .6 | 4.7 | .008 |
| 3 | 5.5 | .51 | 3.9 | .9 | 4.4 | .000 |
| 4 | 5.5 | .54 | 3.1 | .7 | 4.2 | .000 |
| 15 | 4.7 | .48 | 4.0 | 1.1 | 3.7 | .024 |
| 5 | 4.1 | .38 | 3.3 | .8 | 4.1 | .000 |
| 22 | 4.1 | .39 | 3.2 | .7 | 4.6 | .016 |
| 26 | 3.9 | .60 | 2.9 | .7 | 4.3 | .025 |
| 21 | 3.5 | .54 | 3.2 | .7 | 4.9 | .022 |
| 14 | 3.1 | .33 | 2.9 | 2.9 | 1.0 | .063 |
| 13 | 1.6 | .40 | 3.2 | 3.2 | 1.0 | .059 |
| 7 | 1.1 | .64 | 2.5 | .7 | 3.8 | .018 |
| 23 | 0.6 | .34 | 5.0 | 1.3 | 3.9 | .044 |

(a)  Find the matrix of correlation coefficients, $R^2_{0.12345}$, the regression equation, and $s_{0.12345}$.

(b)  Interpret $R^2$ and the regression equation.

(c)  Find the standardized regression coefficients $b_j^*$ and interpret them. Do any appear insignificant?

(d)  Test the significance, at $\alpha = .01$, of $R_{0.12345}$.

(e)  Does the observed range of the levels of the independent variables $X_1, \ldots, X_5$ have anything to do with the significance of the $b_j^*$'s? How?

13.10.  A study of C 1120 steel from 15 heats in an open hearth

furnace gave the following results, the objective being to find relations with percent rejection. The heats are in order of percent rejection, so as to facilitate subjective comparison with the other variables.
Code the data suitably, and use a computer program for the calculations.

| % Rejections, $X_0$ | % Ladle manganese, $X_1$ | % Ladle sulfur, $X_2$ | Pouring temperature (°F) $X_3$ | Transit Time (hr, min), $X_4$ |
|---|---|---|---|---|
| 1 | .90 | .115 | 2890 | 2 20 |
| 3 | .94 | .115 | 2870 | 1 50 |
| 3 | .83 | .100 | 2875 | 1 45 |
| 4 | .87 | .115 | 2880 | 2 10 |
| 6 | .82 | .090 | 2885 | 2 30 |
| 7 | .80 | .110 | 2870 | 2 15 |
| 8 | .83 | .095 | 2890 | 2 50 |
| 8 | .80 | .110 | 2875 | 2 10 |
| 10 | .79 | .100 | 2895 | 2 40 |
| 12 | .75 | .090 | 2875 | 2 20 |
| 13 | .82 | .105 | 2885 | 2 45 |
| 14 | .80 | .095 | 2880 | 2 30 |
| 15 | .72 | .080 | 2865 | 2 20 |
| 15 | .78 | .095 | 2880 | 2 40 |
| 16 | .70 | .085 | 2885 | 3 10 |

(a) Test the significance of $R_{0.1234}$ at $\alpha = .01$.
(b) Give regression equation and $s_{0.1234}$ for original data.
(c) Are any regression coefficients not significant?
(d) Which predictor seems most important?

13.11. The following data are from a designed experiment.* The dependent variable under study was percent conversion as a function of time and temperature. The data follow:

| Conversion percent, $X_0$ | Time in hours, $X_1$ | Average fusion temperature (°C), $X_2$ |
|---|---|---|
| 62.7 | 3 | 297.5 |
| 76.2 | 3 | 322.5 |
| 80.8 | 3 | 347.5 |
| 80.8 | 6 | 297.5 |
| 89.2 | 6 | 322.5 |
| 78.6 | 6 | 347.5 |
| 90.1 | 9 | 297.5 |
| 88.0 | 9 | 322.5 |
| 76.1 | 9 | 347.5 |

* See Lloyd [4].

Analyze the data by multiple regression. Note that in this designed experiment $r_{12} = 0$, and hence $R_{0.12}^2 = r_{01}^2 + r_{02}^2$. The calculations can also be made very easily by coding $X_1$ and $X_2$ to $-1$, $0$, $+1$. Find $R_{0.12}^2$ and $s_{0.12}$ and test the former for significance.

13.12.   Analyze the data in Problem 13.11 by analysis of variance (a two-way unreplicated design). Compare results with the previous problem. Does it seem (subjectively) that there might be an interaction in the data, which of course cannot be separated from the pure error term in this unreplicated design?

13.13.   Substitute $b_0$ from (13.10) into (13.8) and reduce to (13.12).

13.14.   Substitute $b*_j$'s from (13.13) into (13.12) and reduce to (13.14).

13.15.   Consider the model $v_i = b_0' + b_1' u_{1i} + \cdots + b_k' u_{ki} + e_i'$ where, for example, $u_{ji} = (x_{ji} - \bar{x}_j)/s_j$ are the standardized variables. Fit by minimizing $\sum_{i=1}^{n}(e_i')^2$, thus deriving (13.14) directly by least squares.

# Chapter 14

# GOODNESS OF FIT TESTS, CONTINGENCY TABLES

## 14.1. INTRODUCTION

This chapter presents a number of useful tests, concerned primarily with the comparison between observed and theoretical frequencies. The latter are determined according to some hypothesis or model. The question then becomes that of deciding whether the observed frequencies are compatible with the hypothesis or model. As is usual in statistics we can make two errors: (a) rejecting the model when in fact it is correct, type I error; (b) failing to reject the model when in fact it is in some degree false, type II error. The probability of a type I error is $\alpha$, and of a type II error $\beta$, where $\beta$ depends on the actual condition or model. Power is $1 - \beta$.

Two closely related types of tests are (a) those for testing the goodness of fit of a theoretical frequency curve *type* with possibly a parameter or two specified, (b) those for testing independence of classification, between two or more factors. In the latter case the question is whether the "input factors" act independently, or instead interact.

We shall be involved largely with the use of the chi-square distribution in these problems, although there are other approaches, and indeed the subject of this chapter is still being actively researched. This use of the chi-square distribution is quite distinct from that for variances from a normal population, (7.11), and associated tests.

## 14.2. THE CHI-SQUARE TEST FOR CELL FREQUENCIES, OBSERVED VERSUS THEORETICAL

The general problem considered is that in which (a) we have $k$ classes into which an observation may fall; (b) the probability of an observation falling in each of the $k$ classes or cells is known, for example, $\phi_i$, and $\sum_{i=1}^{k} \phi_i = 1$; (c) we take $n$ independent observations, counting the observed frequency $f_i$ in the $i$th cell for all $i$, $\sum_{i=1}^{k} f_i = n$; (d) the expected frequency $n\phi_i = F_i$ is to be compared with $f_i$ to see whether the whole set of discrepancies $f_i - F_i$ is explainable by chance. This problem was attacked and solved in an early paper by Pearson [1].

The criterion developed for goodness of fit was called "chi-square," defined by

$$\chi^2 = \sum_{i=1}^{k} (f_i - F_i)^2 / F_i, \tag{14.1}$$

where for the $k$ classes and $n$ observations

$$\phi_i = \text{probability of an observation in the } i\text{th class} \tag{14.2}$$

$$F_i = n\phi_i = \text{expected or calculated frequency in the } i\text{th class} \tag{14.3}$$

$$f_i = \text{observed frequency in the } i\text{th class} \tag{14.4}$$

$$\sum_{i=1}^{k} \phi_i = 1, \qquad \sum_{i=1}^{k} F_i = n, \qquad \sum_{i=1}^{k} f_i = n. \tag{14.5}$$

The distribution of the criterion of (14.1) has been shown to approach (7.12) as a limit as $n$ becomes infinite, with

$$\text{degrees of freedom} = \nu = k - 1 - p, \tag{14.6}$$

where $p$ is the number of independent parameters estimated from the data, if any.

Now why does the criterion of goodness of fit in (14.1) make sense? In the first place to compare $f_i$ with $F_i$, we want a measure based on the whole set of discrepancies $f_i - F_i$. Their sum $\sum(f_i - F_i)$ cannot be used since it is always zero, whether or not the fit is good. We might then, as in variance, consider $\sum(f_i - F_i)^2$. But this does not take into account the *relative* amount of discrepancy. For example, if $f_i = 8$, $F_i = 5, f_i - F_i = 3$, this is a much more serious discrepancy than that for $f_i = 38$ and $F_i = 35$. Then why not use relative discrepancies squared and have $\sum[(f_i - F_i)/F_i]^2$. This is better, but it gives equal importance to the two such cases as $f_1 = 7$, $F_1 = 5$, and $f_5 = 70$,

$F_5 = 50$. For both of these $(f_i - F_i)/F_i = .4$, but the latter is a much more serious departure.* We thus find it logical to use a weighting of, say, $F_i$ to $[(f_i - F_i)/F_i]^2$, obtaining

$$\sum_{i=1}^{k} F_i[(f_i - F_i)/F_i]^2 = \sum_{i=1}^{k} (f_i - F_i)^2/F_i$$

as in (14.1).

The use of $k - 1$ degrees of freedom is tied in with the fact that the $f_i$'s are not wholly independent because in every running of such an experiment their sum must always be $n$.

It must be obvious from (14.1) that relatively poor fits give relatively large values of the criterion $\chi^2$, and vice versa.

### 14.2.1. An Example.

Let us give an example of the use of chi-square to test the goodness of fit. In a table of random numbers, there are the digits 0, 1, ..., 9 in random order. See Table X in the Appendix. At any given point in the table, the probability for each of the 10 digits to appear next is supposed to be precisely .10, as in Section 5.6. The author chose independently a starting point in a table of random numbers, and tallied the next $200 = n$ digits. The following results occurred:

| $i$ | 0 | 1 | 2 | 3 | 4 | 5 | 6 | 7 | 8 | 9 | |
|---|---|---|---|---|---|---|---|---|---|---|---|
| $f_i$ | 21 | 23 | 23 | 16 | 19 | 19 | 21 | 20 | 14 | 24 | |
| $F_i$ | 20 | 20 | 20 | 20 | 20 | 20 | 20 | 20 | 20 | 20 | |
| $(f_i - F_i)^2/F_i$ | .05 | .45 | .45 | .80 | .05 | .05 | .05 | .00 | 1.80 | .80 | 4.50 |

The observed value of $\chi^2$ thus proved to be 4.50. Is this high or low? Such a $\chi^2$ value carries $10 - 1 = 9$ degrees of freedom. Since $E(\chi^2) = \nu = 9$ here, our observed $\chi^2$ value is well *below* average and not even close to $\chi^2_{9df,\alpha=.05\,above} = 16.919$. Thus the agreement is better than average, under the null hypothesis.

## 14.3. TESTING THE GOODNESS OF FIT OF THEORETICAL DISTRIBUTIONS

A typical example of the tests in this section is the test that the population from which $n$ observations were drawn is normal. This test enables us to give more substance to the assumption of normality, which

---

* This is tied in with the fact that as $\phi_i$ increases the expected frequency $n\phi_i$ is proportional to $\phi_i$, whereas $\sigma_f = \sqrt{n\phi_i(1 - \phi_i)}$ goes up approximately as $\sqrt{\phi_i}$ by (5.33), (5.34).

we so often made in Chapters 8–13. We can also test whether the population was one or another of those given in Chapters 5 and 6. In Section 14.2.1 we tested whether the data could reasonably have come from a *completely specified* uniform distribution as in Section 5.6.

There are two cases to consider. In one the population is completely specified; that is, all the *parameters* are *known*. In many cases, however, one or more *parameters* of the population are *unknown* and must be estimated from the sample. This fact lowers the degrees of freedom, by the number of parameters $p$ being estimated.

There is another complication. Because the derivation of the chi-square distribution involves letting $n$ become infinite, early workers required all calculated or expected frequencies to be at least 10. To meet this criterion small adjacent tail frequencies were lumped together in a single class till $F_i \geqslant 10$. Subsequently this was reduced to 5. Currently (see Cochran [2], p. 420) the recommendation is to group at each tail so that the two minimum $F_i$'s are at least 1.0.*

Let us give the following instruction for testing the goodness of fit between observed and calculated frequencies for some distribution type:

1. Set up classes as in Section 2.3. (If discrete, the class interval is often 1.)
2. Using the population *type* being hypothesized, calculate the theoretical or expected frequencies $f_i$ for a sample of size $n$. This may require that one or more parameters of the population are estimated from the data at hand. (For example, $\mu$ and $\sigma$ for a normal population of $\bar{y}$ and $s$.)
3. Lump small expected frequencies $F_i$ at either tail, adding to adjacent ones, so as to have at least 1.0 for each tail $F_i$.
4. Let the number of classes *after lumping* be $k$.
5. Then the degrees of freedom for $\chi^2$ are $\nu = k - 1 - p$, where $p$ is the number of parameters of the population estimated from the sample.
6. Calculate $\chi^2$ by (14.1) and compare with $\chi^2_{\nu,\alpha\text{above}}$.
   (a) Obs$\chi^2 > \chi^2_{\nu,\alpha\text{above}}$: reject hypothesis that this more or less completely specified population is the one in question.
   (b) Obs$\chi^2 < \chi^2_{\nu,\alpha\text{above}}$: "accept" hypothesis that this more or less completely specified population is the one. (This means only that this could readily be the population.)

---

* Such lumping together avoids some small class giving a huge contribution to $\chi^2$. For example, if $f_i = 1$, $F_i = .1$, then the contribution is $(1 - .1)^2/.1 = 8.1$. But with $F_i = .1$, $f_i$ has to be 0, 1, or 2.

### 14.3.1. A Binomial Example.

In the work of Glover and Carver [3] are given some dice data (probably gathered by Weldon). Each throw consisted of a roll of 12 dice. A count was made on each throw of the number of "successes," where a success was defined as the occurrence of 4, 5, or 6 showing. Thus there could be 0, 1, 2, ..., 12 successes on one throw. A total of 4096 such throws was made. The observed frequencies $f_i$, $i = 0, 1, ..., 12$, are shown in the second column of Table 14.1.

**A. Test of hypothesis $H_0$:** The population is binomial (5.32) with $\phi = .5$, $n = 12$. Under this hypothesis $F_i = 4096\ C(12, i)(.5^i)(.5)^{12-i}$ $i = 0, 1, 2, ..., 12$. These $F_i$ are given in the third column of Table 14.1. All are whole numbers, since $4096(.5)^{12} = 1$. Comparison of $f_i$ with $F_i$

**TABLE 14.1**

Results of 4096 Throws, Each of 12 Dice[a]

| Number of successes | Observed frequencies, $f_i$ | Expected frequencies $F_i$, $\phi = .5$ | Contributions to $\chi^2$ | Expected frequencies $F_i$, $\hat{\phi} = .511576$ | Contribution to $\chi^2$ |
|---|---|---|---|---|---|
| 0 | 0 | 1 | 1.0 | 1 | 1.0 |
| 1 | 7 | 12 | 2.1 | 9 | .4 |
| 2 | 60 | 66 | .5 | 55 | .5 |
| 3 | 198 | 220 | 2.2 | 191 | .3 |
| 4 | 430 | 495 | 8.5 | 450 | .9 |
| 5 | 731 | 792 | 4.7 | 754 | .7 |
| 6 | 948 | 924 | .6 | 921 | .8 |
| 7 | 847 | 792 | 3.8 | 827 | .5 |
| 8 | 536 | 495 | 3.4 | 541 | .0 |
| 9 | 257 | 220 | 6.2 | 252 | .1 |
| 10 | 71 | 66 | .4 | 79 | .8 |
| 11 | 11 | 12 | .1 | 15 | .1 |
| 12 | 0 | 1 | 1.0 | 1 | 1.0 |
| | 4096 | 4096 | $\chi^2_{12} = 34.5$ | 4096 | $\chi^2_{11} = 7.1$ |

[a] Success is the occurrence of 4, 5, or 6 showing. Expected frequencies calculated by the binomial distribution: $4096[\phi + (1 - \phi)]^{12}$. The $F_i$ found by $\phi = .5$ and $\hat{\phi} = .511576$. See Glover and Carver [3, p. 53].

seems to show rather good agreement. The fourth column shows the contributions to $\chi^2$ for each of the 13 classes. There was no need to lump classes at the tails since the minimum $F_0$ and $F_{12}$ were each 1. Thus the degrees of freedom are $\nu = 13 - 1 - 0 = 12$, since no parameter was estimated from the data. The observed $\chi^2 = 34.5$, which is far above the critical value with $\alpha = .05$. namely, $\chi^2_{12df,.05above} = 21.026$, $\alpha = .05$ being a level frequently used in such tests. In fact the departures are clearly significant at $\alpha = .005$. Hence we conclude that the distribution is *not binomial with $\phi = .5$.* Is it possible that the data follow some other binomial law with $\phi \neq .5$? This is suggested by looking at the $f_i$ and the first $F_i$ columns, where $f_i$ is mostly below $F_i$ at small numbers of successes and mostly above for larger $i$'s. This suggests that the dice are biased with $\phi > 5$.

**B.    Test of hypothesis $H_0'$:**    The population is binomial (5.32). Now we have only a hypothesis about the *type* of population, *but no parameter value for $\phi$ is given.* We therefore estimate $\phi$ from the data; $f_i$'s:

$$\hat{\phi} = \frac{0(0) + 1(7) + 2(60) + \cdots + 11(11) + 12(0)}{4096(12)} = \frac{25,145}{49,152} = .511,576.$$

Calculation of the binomial probabilities is most easily accomplished by use of the recursion relation (5.36). This gives the fifth column of Table 14.1, rounded to integers. The contributions to $\chi^2$ are shown in the last column, and are far smaller. The total is $\chi^2_{11} = 7.1$, where now $p$ in (14.6) is 1, so that there are only 11 degrees of freedom. We compare the observed $\chi^2$ with $\chi^2_{11df,.05above} = 19.675$. Hence we conclude that the population could very well be binomial, with the biased probability of success a bit above .5. (We of course have not proved $\phi$ to be .511,576!) The bias comes about because the faces 1, 2, and 3 are a little heavier because of fewer holes for the markings, and hence their opposite faces 6, 5, and 4 have slightly greater probability of showing.

### 14.3.2. Examples of Tests of Normality.

As a first example consider Table 14.2, on the dimension of an electrical contact. The observed frequencies look quite normal. The reader might well draw a frequency polygon for them. The next step in testing for normality is to find expected frequencies $F_i$. This is done as shown in the table. For use of a desk calculator the coded variable $v$ was started at the lowest midvalue, .412 in. The class boundaries for $v$ appear in the fourth column. These are then converted to standardized values $u$, by the linear formula at the bottom of the table. (The $u$'s can then be found continuously on a desk calculator.) The next column gives the cumulative normal probability $\Phi(u)$ for each $u$, interpolating, for example, by $\Phi(-2.166) = .4\Phi(-2.160) + .6\Phi(-2.170)$. The differences between these $\Phi$ values give the class probabilities, which when multiplied by

TABLE 14.2

Dimension of an Electrical Contact, Tested for Normality by $\chi^2$

| Dimension (in.) | Observed frequency, $f_i$ | Coded variable, $v_i$ | Class boundary | Standard variable $u = \dfrac{v - \bar{v}}{s_v}$ | $\Phi(u)$ | $F_i = [\Delta\Phi(u)]n$ | Contribution to $\chi^2 = \dfrac{(f_i - F_i)^2}{F_i}$ |
|---|---|---|---|---|---|---|---|
| .410–.414 | 6 | 0 | | | | 15.2 | 5.6 |
| | | | .5 | −2.166 | .0152 | | |
| .415–.419 | 34 | 1 | | | | 40.9 | 1.2 |
| | | | 1.5 | −1.588 | .0561 | | |
| .420–.424 | 132 | 2 | | | | 100.1 | 10.1 |
| | | | 2.5 | −1.010 | .1562 | | |
| .425–.429 | 179 | 3 | | | | 176.7 | .0 |
| | | | 3.5 | −.432 | .3329 | | |
| .430–.434 | 218 | 4 | | | | 225.1 | .2 |
| | | | 4.5 | +.146 | .5580 | | |
| .435–.439 | 183 | 5 | | | | 207.5 | 2.9 |
| | | | 5.5 | +.724 | .7655 | | |
| .440–.444 | 146 | 6 | | | | 138.0 | .5 |
| | | | 6.5 | +1.302 | .9035 | | |
| .445–.449 | 69 | 7 | | | | 66.4 | .1 |
| | | | 7.5 | +1.880 | .9699 | | |
| .450–.454 | 30 | 8 | | | | 23.1 | 2.1 |
| | | | 8.5 | +2.458 | .9930 | | |
| .455–.459 | 3 | 9 | | | | 7.0 | 2.3 |
| | 1000 | | | | | | $\chi^2 = 25.0$ |

$$\sum f_i v_i = 4248, \qquad \sum f_i v_i^2 = 21{,}036, \qquad \bar{v} = 4.248,$$

$$s_v = \sqrt{\frac{1000(21{,}036) - 4248^2}{1000(999)}} = 1.73017$$

$$u = \frac{v - 4.248}{1.73017} = .57798v - 2.4553, \qquad \text{degrees of freedom} = \nu = 10 - 1 - 2 = 7$$

$$\text{Also} \qquad a_3 = .1415, \qquad a_4 = 2.5059$$

$n = 1000$ give the expected class frequencies $F_i$. Then the contributions to $\chi^2$ are as shown in the last column. (It might be mentioned that $F_0 = 15.2$ is the expected frequency for all cases below the boundary .4145 in., and similarly for $F_9 = 7.0$: all above .4545 in.)

Now the chi-square value carries degrees of freedom $\nu = 10 - 1 - 2 = 7$ because two parameters $\mu$ and $\sigma$ were estimated from the data at hand.

The observed $\chi_7^2 = 25.0$. But since $\chi_{7df,.05above}^2 = 14.067$, the hypothesis that the population is normal is refuted.

One might say that since there is skewness of .1415 we might try a gamma distribution. Proceeding similarly to above with $\alpha_3 = .14$ yields $\chi_6^2 = 20.7$, still significant. Note the drop in degrees of freedom $\nu$ by 1 because now $\mu$, $\sigma$, $\alpha_3$ were the parameters estimated. The real trouble would seem to be the rather abrupt endings of the frequencies. This is suggested by $a_4 = 2.5059$. One could go further and fit using this fact and Section 6.7 or 6.10.

A second example is from a distribution in the literature [4]. It involves weights of coating of tin in ounces per square foot for sheets.

| $Y$ | 1.30 | 1.35 | 1.40 | 1.45 | 1.50 | 1.55 | 1.60 | 1.65 | 1.70 | 1.75 | Total |
|---|---|---|---|---|---|---|---|---|---|---|---|
| $f_i$ | 2 | 6 | 7 | 14 | 14 | 22 | 17 | 10 | 3 | 5 | 100 |
| $F_i$ | 2.3 | 4.1 | 8.4 | 13.7 | 17.9 | 18.7 | 15.6 | 10.4 | 5.5 | 3.4 | 100 |
| $\chi^2$ | .0 | .9 | .2 | .0 | .8 | .6 | .1 | .0 | 1.1 | .8 | $4.5 = \chi_7^2$ |

Here there are seven degrees of freedom, having used $\hat{\mu} = \bar{y} = 1.5345$ in., $\hat{\sigma} = s = .1046$ in. The critical value is $\chi_{7df,.05above}^2 = 14.067$. Thus the fit is perfectly satisfactory and the hypothesis that the population is a normal one is tenable.

It would be instructive to make a histogram for each of these two examples to compare the $f_i$ with $F_i$ in each case. The second does not "look" much better than the first. But the much larger $n$ in the first gives the test more power, and thus greater likelihood of rejecting the hypothesis of normality when departures from normality are not very great.

### 14.3.3.* Example of a Gamma Distribution Fit.

If we fitted the strongly skewed data of Table 2.18 by a normal curve, we should of course find a significant departure, especially so with $n = 2700$. We might therefore try a gamma distribution (Section 6.6). Proceeding similarly to Table 14.2, but using $\hat{\alpha}_3 = a_3 = .966$, we find for the eccentricities (off-center measurements) in .0001 in.:

| $Y$ | 4.5 | 14.5 | 24.5 | 34.5 | 44.5 | 54.5 | 64.5 | 74.5 | 84.5 | 94.5 | 104.5 |
|---|---|---|---|---|---|---|---|---|---|---|---|
| $f_i$ | 171 | 477 | 561 | 505 | 379 | 289 | 150 | 75 | 43 | 30 | 10 |
| $F_i$ | 185.0 | 450.7 | 569.9 | 517.4 | 386.0 | 254.7 | 154.2 | 87.2 | 47.0 | 24.4 | 12.2 |
| $\chi^2$ | 1.1 | 1.5 | .1 | .3 | .1 | 4.6 | .1 | 1.7 | .3 | 1.3 | .4 |

| $Y$ | 114.5 | 124.5 | 134.5 | 144.5 | |
|---|---|---|---|---|---|
| $f_i$ | 1 | 5 | 3 | 1 | 2700 |
| $F_i$ | 6.0 | 2.9 | 1.3 | 1.1 | 2700 |
| $\chi^2$ | 4.2 | 1.5 | 2.2 | .0 | 19.4 |

The degrees of freedom are $\nu = 15 - 1 - 3 = 11$. The critical value at $\alpha = .05$ is $\chi^2_{11df,.05\,\text{above}} = 19.675$. Thus the fit is satisfactory. This is unusual for such a large sample to be fitted satisfactorily by *any* theoretical distribution. Most of the agreements $f_i$ to $F_i$ are excellent with contributions below 1.0. Most of the larger contributions are at the tails. The population could well be a gamma distribution.

### 14.3.4. Example of a Poisson Distribution.

Let us consider the following data as observed and recorded on 54 large aircraft assemblies. A record was kept on many types of defects, one of which was points needing adjustment of some kind. The accompanying table gives the data and calculations:

| Number of adjustments, $i$ | Observed frequency, $f_i$ | Total adjustment, $i \cdot f_i$ | Probability $p(i)$ | Expected frequency $F_i = 53\,p(i)$ | Contribution to $\chi^2$ |
|---|---|---|---|---|---|
| 0 | 19 | 0 | .2524 | 13.4 | 2.3 |
| 1 | 15 | 15 | .3474 | 18.4 | .6 |
| 2 | 9 | 18 | .2393 | 12.7 | 1.1 |
| 3 | 4 | 12 | .1098 | 5.8 | .6 |
| 4 | 3 | 12 | .0378 | 2.0 | |
| 5 | 2 ⎫ 6 | 10 | .0104 | .55 ⎫ 2.7 | 4.0 |
| 6 | 1 ⎬ | 6 | .0024 | .13 ⎬ | |
| 7 | 0 ⎭ | 0 | .0005 | .03 ⎭ | |
| ⋮ | ⋮ | ⋮ | | | $\chi^2 = 8.6$ |
| 10 | 0 | 0 | | | |
| 11 | 1 | 11 | | | |
| | 54 | 84 | | | |

$\hat{\mu} = 84/54 = 1.556$, revised $\hat{\mu} = 73/53 = 1.377$

Now what shall we do about the extreme number of 11? If we have a Poisson distribution with $\mu = 1.56$, the probability of $i \geqslant 11$ is only .000,000,8 [5]. Hence we reject this outlier as being not a part of the distribution. Revising our estimated mean we find $\hat{\mu} = 1.377$. Using this and interpolating in [5] we find $p(i)$'s then $F_i$ by multiplying by $n$. The contributions to chi-square are found after lumping the small $F_i$'s for high $i$'s, until the sum is at least 1.0. Then at that class the contribution is $(6 - 2.7)^2/2.7$. Now, is the observed $\chi^2 = 8.6$ significant at, say, $\alpha = .05$?

The degrees of freedom are, by (14.6),

$$\nu = 5 - 1 - 1 = 3,$$

the second 1 being because we estimated $\mu$ from the data. Since

$$\chi^2_{3\mathrm{df}, .05\,\mathrm{above}} = 7.815,$$

we can say that the hypothesis that the population is Poisson is untenable and that some other population is called for.

If we examine the data carefully, we see that there are too many zeros and too few ones in the data, as well as somewhat more tailing out toward high numbers (even after eliminating the 11). This is rather a common departure from a Poisson distribution. It means that there are quite a few instances of zero adjustments, but that when conditions are such as to produce one adjustment, there is an enhanced chance of other such. This is rather aptly described as a "contagious" tendency. It means that the adjustments or occurrences are not acting wholly independently, contrary to condition 4 of Section 5.4.6.

### 14.4.* OTHER GOODNESS OF FIT TESTS

A good many other tests for goodness of fit occur in the literature. This is especially true of the important problem of testing for normality. See Shapiro *et al.* [6] for a comparative study of some of them. These authors have developed one such test [7], which seems to possess a fair amount more power to reject the hypothesis of normality, when it is not true in several different ways. (There is a great multiplicity, of course, in the ways the population can be nonnormal.) But the method is quite a chore if $n$ is sizable. Another widely suggested test is the Kolmogorov–Smirnov test. Owen [8] gives a good description and useful tables. This test requires (a) a continuous distribution for the random variable, such as those Chapter 6; (b) that we can and do draw the curve of $F(y)$; (c) that we put together the empirical distribution function (Section 2.6), though usually done with staircase steps; and (d) that we find the largest *absolute* discrepancy between the distribution function in (b) and (c). Then standard tables such as those presented by Owen [8] provide critical values for the discrepancy, which, if exceeded, reject the null hypothesis that the proposed distribution is the one from which the sample came.

Since the $\chi^2$ test seems to have reasonably good power, wide applicability, is easy to interpret, and quite easy to use, we recommend it for general use.

## 14.5. CONTINGENCY TABLES

An important class of problems, to which we can apply the chi-square approach, goes by the name of "contingency tables." These in general are in the form of a rectangular table of frequencies. Commonly there are two different types of factors: those associated with the columns and those with the rows. These can be time, people, localities, samples of material, formulations, or methods. Or, three categories could be "good," "seconds," and "scrap," as in industrial production. As such we have tables much like a two-way analysis of variance without replication, as in Section 12.7.5. Only now we have observed frequencies in the cells instead of measurements $y_{ij}$. In a contingency table the hypothesis being tested suffices to enable us to calculate expected frequencies $F_{ij}$'s which must then be compared with the observed cell frequencies $f_{ij}$.

Commonly, the hypothesis being tested is that the two factors are independent. This is often called "independence of classification." The alternative hypothesis is that the two factors do not act independently; that is, that they interact. (In fact testing a contingency table for significance is commonly analogous to testing whether a two-factor replicated analysis of variance experiment has a significant interaction.)

### 14.5.1. An Example.

Let us consider a simple example for illustration and then give the general picture. Thirty-six caps for food jars were taken from each of three types of ring construction. They were subjected to certain rigorous test conditions, and checked as to whether they showed an undesirable condition called "hiked." The accompanying table shows the data. Thus subjecting a random sample to ring construction of type $A$ showed two

| Ring construction | Hiked | Not hiked | Total |
|---|---|---|---|
| Type $A$ | 2 (7.3) | 34 (28.7) | 36 |
| Type $B$ | 5 (7.3) | 31 (28.7) | 36 |
| Type $C$ | 15 (7.3) | 21 (28.7) | 36 |
| Total | 22 | 86 | 108 |

hiked caps and 34 not hiked, out of 36 tested, and so on. Now if the proportion hiked were independent of the type, that is, all three have the same probabilty of yielding a hiked cap under test, then we would *expect* the same number to be hiked in each sample of 36. Our best estimate under such a hypothesis is $22/108 = .204$. Taking this fraction of 36 yields the expected frequency $F$ of 7.3 for the number hiked in each type, shown in parentheses in the table. Meanwhile the expected number not hiked is $36 - 7.3 = 28.7$, or $(1 - .204)36$. Now how do the observed and expected frequencies compare?

$$\chi^2 = \frac{(2 - 7.3)^2}{7.3} + \frac{(5 - 7.3)^2}{7.3} + \cdots + \frac{(21 - 28.7)^2}{28.7}$$
$$= 3.9 + .7 + 8.1 + 1.0 + .2 + 2.1 = 16.0.$$

The degrees of freedom are the number of frequencies we can fill in arbitrarily ("freely") while maintaining the marginal totals around the bottom and right side of the table. We could place in (type $A$, hiked), any frequency from 0 to 22. When this frequency is set, then (type $A$, not hiked) is 36 minus this frequency. Next we can fill in (type $B$, hiked) with any frequency which does not make the column total exceed 22. Then (type $B$, not hiked) has the complementary frequency filled in. The last two frequencies for type $C$ must now make the column totals 22 and 86. We therefore have but two degrees of freedom for this six-cell contingency table. We have

$$\text{observed } \chi_2{}^2 = 16.0, \qquad \text{critical } \chi^2_{2\,df,\,.05\,above} = 5.991.$$

Therefore the hypothesis of independence of factors is rejected. Hence the proportion hiked does depend on the type of ring construction. We note that this test has statistically proven significance between the three proportions 2/36, 5/36, and 15/36. The proportions did "look" significant, but the samples were rather small. Similarly a $2 \cdot 2$ contingency table can test for significance of difference between two proportions as in Section 9.4.

### 14.5.2. The General Setup of a Contingency Table.

Table 14.3 gives a general setup for an $a \cdot b$ contingency table. The observed frequencies are $f_{ij}$, $i = 1, ..., a; j = 1, ..., b$. The population cell probabilities are $\phi_{ij}$, and their marginal totals are $\phi_{i.}$ and $\phi_{.j}$. The former is, for example, P(factor $A$ at level $i$). Now under the null hypothesis

$$H_0: \text{factor } A \text{ and factor } B \text{ independent}, \qquad (14.7)$$

we have by (4.22)

$$\phi_{ij} = P(A \text{ at } i \cap B \text{ at } j) = P(A \text{ at } i)\,P(B \text{ at } j) = \phi_{i.}\phi_{.j}. \qquad (14.8)$$

Now how do we estimate $\phi_{i.}$? This is by $\hat{\phi}_{i.} = n_{i.}/n$, and similarly $\hat{\phi}_{.j} = n_{.j}/n$. Accordingly we have

$$\hat{\phi}_{ij} = n_{i.}(n_{.j})/n^2. \qquad (14.9)$$

TABLE 14.3

A General Contingency Table, $a \cdot b$

| Factor $A$ | Factor $B$ | | | | | Total |
|---|---|---|---|---|---|---|
| | 1 | 2 $\cdots$ | $j$ | $\cdots$ $b$ | | |
| 1 | $f_{11}$ | $f_{12}$ | $f_{1j}$ | $f_{1b}$ | $n_{1.}$ | |
| | $\phi_{11}$ | $\phi_{12}$ | $\phi_{1j}$ | $\phi_{1b}$ | | $\phi_{1.}$ |
| 2 | $f_{21}$ | $f_{22}$ | $f_{2j}$ | $f_{2b}$ | $n_{2.}$ | |
| | $\phi_{21}$ | $\phi_{22}$ | $\phi_{2j}$ | $\phi_{2b}$ | | $\phi_{2.}$ |
| $\vdots$ | | | | | $\vdots$ | |
| $i$ | $f_{i1}$ | $f_{i2}$ | $f_{ij}$ | $f_{ib}$ | $n_{i.}$ | |
| | $\phi_{i1}$ | $\phi_{i2}$ | $\phi_{ij}$ | $\phi_{ib}$ | | $\phi_{i.}$ |
| $\vdots$ | | | | | $\vdots$ | |
| $a$ | $f_{a1}$ | $f_{a2}$ | $f_{aj}$ | $f_{ab}$ | $n_{a.}$ | |
| | $\phi_{a1}$ | $\phi_{a2}$ | $\phi_{aj}$ | $\phi_{ab}$ | | $\phi_{a.}$ |
| Total | $n_{.1}$ | $n_{.2}$ $\cdots$ | $n_{.j}$ | $\cdots$ $n_{.b}$ | $n$ | |
| | $\phi_{.1}$ | $\phi_{.2}$ $\cdots$ | $\phi_{.j}$ | $\cdots$ $\phi_{.b}$ | | 1 |

For the expected frequency in the $i, j$ cell we have

$$F_{ij} = n\hat{\phi}_{ij} = (n_{i.})(n_{.j})/n. \tag{14.10}$$

Note that

$$\sum_{i=1}^{a} F_{ij} = n_{.j}, \qquad \sum_{j=1}^{b} F_{ij} = n_{i.}, \qquad \sum\sum F_{ij} = n. \tag{14.11}$$

We now have $F_{ij}$'s to compare with the $f_{ij}$'s by

$$\chi^2 = \sum_{i=1}^{a} \sum_{j=1}^{b} (f_{ij} - F_{ij})^2/F_{ij}. \tag{14.12}$$

This carries degrees of freedom

$$\nu = (a - 1)(b - 1). \tag{14.13}$$

The formula for $\nu$ may be justified as in the previous section (preserving the marginal totals). Or we may use (14.6) as follows, $k = ab$. For $p$ we first note that the parameters $\phi_{1.}$, $\phi_{2.}$, ..., $\phi_{(a-1).}$ are independently estimated by $n_{1.}/n$, $n_{2.}/n$, ..., $n_{(a-1).}/n$. But $\phi_{a.}$ is not independent since $\sum\phi_{i.} = 1$. Thus there are $a - 1$ ratios for rows, and similarly $b - 1$ for columns. Therefore by (14.6),

$$\nu = ab - 1 - (a - 1) - (b - 1) = (a - 1)(b - 1).$$

Also compare this with unreplicated two-factor analysis of variance (12.58) for the interaction $SSAB$.

### 14.5.3. A 2 · 2 Contingency Table.

An example of a $2 \cdot 2$ contingency table is given in Table 14.4. Following (14.10), we find the expected frequencies, for example,

TABLE 14.4

Contamination of Tanks versus Technicians and Operators[a]

|  | Number of tanks harvested for which samples were | | |
|---|---|---|---|
| Samples taken by | Contaminated | Not contaminated | Total |
| Technicians | 17 (27.0) | 157 (147.0) | 174 |
| Operators | 56 (46.0) | 240 (250.0) | 296 |
|  | 73 | 397 | 470 |

[a] Data from Noel and Brumbaugh [9].

$F_{11} = 174(73)/470 = 27.0$. The others may be found similarly, or by subtraction. Then using (14.12):

$$\chi_1^2 = \frac{(17 - 27.0)^2}{27.0} + \frac{(157 - 147.0)^2}{147.0} + \frac{(56 - 46.0)^2}{46.0} + \frac{(240 - 250.0)^2}{250.0}$$
$$= 6.96.$$

Note that all four differences are the same (because of the marginal totals for $f_{ij}$, $F_{ij}$ being the same). Hence the calculation can be done by the square of the difference times $\sum\sum(1/F_{ij})$. The degrees of freedom are 1 by (14.13). We may therefore compare the observed $\chi^2$ with $\chi_{1df,.05above}^2 = 3.841$.

However, in the $2 \cdot 2$ contingency tables being tested for independence it is considered desirable to apply a correction for continuity. This is to take

$$(|f_{ij} - F_{ij}| - .5)^2 \quad \text{(continuity correction for 2 · 2 table)} \quad (14.14)$$

for the numerators, still using $F_{ij}$'s for the denominators. Using this for the above we find

$$\chi_1^2 = 6.28.$$

This is significant at the 5% level, but not quite at the 1% level. Without using the continuity correction the hypothesis would have been rejected even at the 1% level also.

This study proved valuable. The results were reported to *both* analysts and technicians and a request for improvement made. In another month the difference had disappeared and by continuing to report data for another three months, the contamination rate of both groups had dropped below 1%.

The present approach is quite comparable with a significance of difference test on $p_1 = y_1/n_1$ versus $p_2 = y_2/n_2$, that is, using (9.11) or (9.13).

There is a so-called exact method when the frequencies are small, developed by Fisher. See most any edition of the work of Fisher [10]. It is based upon hypergeometric calculations on the probability of as bad as or a worse agreement with the hypothesis. Useful tables of many cases are given by Mainland [11].

### 14.5.4. Case of a 2·b Contingency Table.

In such a table we have a series of $b$ columns, in each of which there are just two cells, for example, "success" or "failure." The question then becomes that of determining whether the observed ratios are compatible with the hypothesis that they are all random samples from a single proportion $\phi$ to $1 - \phi$. Calculations could be made as in the general case, figuring all the squares of deviations, and so forth. But a simpler approach is available.

"Recovering" versus "not recovering" was under study for a series of length classes as shown in the accompanying table.

| | 8.5 in. | 9.5 in. | 10.5 in. | 11.5 in. | 12.5 in. | 13.5 in. | 14.5 in. | 15.5 in. | Totals |
|---|---|---|---|---|---|---|---|---|---|
| Recovering, $f_{1j}$ | 6 | 39 | 112 | 168 | 183 | 84 | 38 | 8 | $638 = n_1.$ |
| Not recovering, $f_{2j}$ | 10 | 56 | 138 | 142 | 153 | 80 | 19 | 11 | $609 = n_2.$ |
| Totals, $n_{.j}$ | 16 | 95 | 250 | 310 | 336 | 164 | 57 | 19 | $1247 = n$ |
| Ratio, $f_{1j}/n_{.j}$ | .37500 | .41053 | .44800 | .54194 | .54464 | .51220 | .66667 | .42105 | $.51163 = \bar{p}$ |

Then $\chi^2$ may be figured as follows [10]:

$$\chi^2_{b-1} = \frac{1}{\bar{p}(1-\bar{p})} \left[ \sum_{j=1}^{b} f_{1j} \left( \frac{f_{1j}}{n_{.j}} \right) - n_1 \cdot \bar{p} \right]. \tag{14.15}$$

Here

$$\chi^2_{b-1} = \frac{1}{.51163(.48837)} [6(.37500) + \cdots + 8(.42105) - 638(.51163)]$$

$$\chi^2_7 = \frac{4.45843}{.24986} = 17.84.$$

This compares with $\chi^2_{7df,.05above} = 14.067$. Thus the hypothesis of independence is refuted, and the ratio of recovery does vary among the different lengths.

The present case of a $2 \cdot b$ contingency table is often used to determine whether two observed frequency distributions could well have come from the same population. If refuted, it could be because the *type* of distribution is different and/or the $\mu_i$'s or $\sigma_i$'s differ. Any one or more of these can cause the hypothesis to be rejected.

## 14.6. THE SIGN TEST

We close this chapter with a simple and useful test, which is somewhat related to chi-square tests in that it is concerned with frequencies. Consider again the matched-pair type of experiment as discussed in Section 9.6. Let us take as a simple example the matched-pair data of Problem 9.14. The null hypothesis is that there is no difference in weight loss from corrosion at any given site (set of conditions). The reasonable alternative hypothesis is that the weight loss in lead-covered pipe is a consistent amount less than that for bare pipe. Now if the null hypothesis were true, then positive and negative deviations, $d_i$, ought to be equally likely. There are $n = 14$ deviations, $d_i$'s. But if the alternative hypothesis is true, we should expect more minus $d_i$'s than plus. Therefore counting the number of positives and negatives, the rejection region will be some small number of positives on down to zero.

Therefore, by noting whether the difference is positive or negative, we now have a binomial situation with null hypothesis $\phi_+ = .5 = \phi_0$, versus the alternate hypothesis $\phi_+ < .5$. Inspection of the data reveals 13 negatives and one positive. What is the probability of one or less positives if $\phi_+ = .5$ and $n = 14$? Calculation or tables show it to be .00092. Thus we would reject the null hypothesis at any $\alpha$ risk down to .001!

If the alternate hypothesis were to include differences in mean in either direction, then a two-tail binomial test would be in order.

The sign test has less confining assumptions than the matched-pair test of Section 9.6. It does not require normality, nor equal variability from pair to pair, nor a consistent difference $\Delta_j \equiv \Delta$ in means. Also it is easy to calculate.

## 14.7. SUMMARY

This chapter has been primarily concerned with statistical tests for comparing a set of observed frequencies with corresponding frequencies calculated according to some null hypothesis. The theoretical probabilities for the cells may be completely determined by the hypothesis. Such probabilities, when multiplied by the total sample size $n$, give the expected frequencies. Or, one or more parameters may be estimated from the observed frequencies and used in calculating the estimated frequencies. In either case the relative agreement between the $f_i$'s and $F_i$'s is measured by $\chi^2 = \sum_{i=1}^{k}(f_i - F_i)^2/F_i$, which has degrees of freedom $\nu = k - 1 - p$, where $k$ is the number of classes after combining any small expected frequencies to make each $F_i \geqslant 1$, and $p$ is the number of parameters estimated from the data. The critical region for $\chi^2$ is always in the upper tail.

Cases considered were the following:

1. hypothesis that the data follow a completely specified distribution, curve type, *and* given parameters;
2. hypothesis that the data follow a specified type of distribution (parameters unspecified);
3. $2 \cdot 2$ contingency tables, commonly equivalent to a test of significance of difference between two proportions (but not always);
4. $2 \cdot b$ contingency tables, commonly equivalent to a test of significance between $b$ proportions, and also useful for testing the difference between two observed distributions;
5. general contingency tables for testing whether or not the factors are independent;
6. the sign test for a series of differences. It could be done by $\chi^2$, but usually is done by comparison with a binomial with $\phi = .5$. It can be used on matched-pair data.

REFERENCES

1. K. Pearson, On a criterion that a given system of deviations from the probable in the case of a correlated system of variables is such that it can be reasonably supposed to have arisen in random sampling. *Philos. Mag.* **50**, 157 ff. (1900).
2. W. G. Cochran, Some methods for strengthening the common $\chi^2$ tests. *Biometrics* **10**, 417–451 (1954).

3. J. W. Glover and H. C. Carver, "Introduction to Mathematical Statistics." Edwards, Ann Arbor, Michigan, 1928.
4. "ASTM Manual on Quality Control of Materials." American Soc. for Testing Materials, Philadelphia, Pennsylvania, 1951.
5. T. Kitagawa, "Tables of Poisson Distribution." Baifukan, Tokyo, Japan, 1952.
6. S. S. Shapiro, M. B. Wilk, and H. J. Chen, A comparative study of various tests for normality. *J. Amer. Statist. Assoc.* **63**, 1343–1372 (1968).
7. S. S. Shapiro and M. B. Wilk, An analysis of variance test for normality (complete samples). *Biometrika* **52**, 591–611 (1965).
8. D. B. Owen, "Handbook of Statistical Tables." Addison-Wesley, Reading, Massachusetts, 1962.
9. R. H. Noel and M. A. Brumbaugh, Applications of statistics to drug manufacture. *Indust. Quality Control* **7** (no. 2), 7–14 (1950).
10. R. A. Fisher, "Statistical Methods for Research Workers." Oliver & Boyd, Edinburgh, 1970.
11. D. Mainland, Statistical methods in medical research. *Canad. J. Res. Sect. E* **26**, 1–166 (1948).
12. L. G. Ghering, Refined method of control of cordiness and workability in glass during production. *J. Amer. Ceram. Soc.* **27**, 373–387 (1944).
13. P. R. Rider, "Modern Statistical Methods." Wiley, New York, 1939.

ADDITIONAL REFERENCES

D. Lewis and C. J. Burke, The use and misure of chi-square test. *Psychol. Bull.* **46**, 433–498 (1949).

F. J. Massey, Jr., The Kolmogorov-Smirnov test for goodness of fit. *J. Amer. Statist. Assoc.* **46**, 68–78 (1951).

## Problems

14.1.   Starting at a random point in Table X (see the Appendix) of random numbers, tally 200 consecutive digits, and test the hypothesis that the distribution is uniform, at $\alpha = .05$.

14.2.   A class project in which each member rolled six dice 50 times, counting aces as successes, resulted in the following totals of occurrences:

| Number of aces in 6 | 0 | 1 | 2 | 3 | 4 | 5 |
|---|---|---|---|---|---|---|
| Number of such event | 274 | 351 | 132 | 40 | 2 | 1 |

Test the results against the hypothesis that the dice were unbiased and rolled independently, using $\alpha = .05$.

14.3.   Over a period of time 100 samples, each of 200 cartons, were tested for sealing. The following table gives the observed frequency,

$f_i$ of 0, 1, 2, ..., poorly sealed cartons in 200. The average number of defective cartons was about 2, or 1%. The expected frequencies $F_i$ were found from $100 \cdot (.99 + .01)^{200}$. Keeping in mind that $\phi$ was estimated from the data, test the hypothesis that the distribution is binomial, at $\alpha = .01$.

| $i$ | 0 | 1 | 2 | 3 | 4 | 5 | 6 | Over 6 |
|-----|-----|-----|-----|-----|-----|-----|-----|--------|
| $f_i$ | 19 | 30 | 19 | 14 | 6 | 6 | 4 | 2 |
| $F_i$ | 13.4 | 27.1 | 27.2 | 18.1 | 9.0 | 3.6 | 1.2 | .4 |

14.4.   A record was kept of the number of defective rivets found on 54 large aircraft assemblies.

| Defective rivets | 0 | 1 | 2 | 3 | 4 | 5 | Total |
|------------------|-----|-----|-----|-----|-----|-----|-------|
| Frequency | 28 | 12 | 8 | 2 | 3 | 1 | 54 |

Using $\alpha = .05$. test the hypothesis that the population distribution is Poisson. Comment.

14.5.   A distribution of densities of glass, in grams per cubic centimeter, at 20°C was obtained over three months as follows [12]:

| Density | 2.5012 | 2.5022 | 2.5032 | 2.5042 | 2.5052 | 2.5062 | 2.5072 |
|---------|--------|--------|--------|--------|--------|--------|--------|
| Observed frequency, $f_i$ | 2 | 6 | 25 | 33 | 19 | 10 | 4 |
| Expected frequency, $F_i$ | 2.1 | 8.5 | 21.5 | 30.1 | 23.6 | 10.3 | 2.9 |

Sample estimates of $\mu$ and $\sigma$ were used for $F_i$'s. Test for normality using $\alpha = .05$.

14.6.   For 500 heats of steel in an open hearth furnace the percents carbon gave a distribution as shown, along with estimated frequencies using $\bar{y}$ and $s$.

| % Carbon | .69 | .70 | .71 | .72 | .73 | .74 | .75 | .76 | .77 | .78 | .79 | .80 |
|----------|-----|-----|-----|-----|-----|-----|-----|-----|-----|-----|-----|-----|
| Observed frequency, $f_i$ | 2 | 21 | 32 | 63 | 89 | 106 | 97 | 51 | 33 | 4 | 1 | 1 |
| Expected frequency, $F_i$ | 5 | 13 | 34 | 66 | 96 | 106 | 87 | 55 | 26 | 9 | 2.5 | .5 |

Test the hypothesis of normality at $\alpha = .05$,

14.7.   For a table at the end of Chapter 2, as assigned, test for normality by finding $\bar{y}$ and $s$, finding expected frequencies $F_i$, and completing the test by $\chi^2$. Use $\alpha = .05$ or as assigned.

14.8.   Do Problem 9.18 by a 2 · 2 contingency table and interpret your results.

14.9.   Do Problem 9.17 by a 2 · 2 contingency table and interpret your results.

14.10.   Do Problem 9.19 as a contingency table and interpret your results. (Can be done as a 1 · 2 table since the totals are so large.)

14.11.   Do Problem 9.20 as a 2 · 2 contingency table and interpret your results.

14.12.   A 2 · 2 contingency table is given by Rider [13], in which the degrees of freedom are not 1. The table gives offspring from a cross between two hybrids, with two Mendelian factors to be inherited. If they are inherited independently, that is, no linkage between the two, then the four possible combinations should theoretically occur in the proportion 9 : 3 : 3 : 1. An experiment with flowers shows the following:

|          | Flower |     |
| -------- | ------- | --- |
| Stigma   | Magenta | Red |
| Green    | 120     | 36  |
| Red      | 48      | 13  |

Notice that the theoretical frequencies will be completely determined by $n$ and the given ratio. Hence there are no parameters to estimate from the data and $p$ in (14.6) is zero. Test the data versus the ratio with $\alpha = .05$.

14.13.   The following 3 · 2 contingency table shows a way to group results for a study of relationship which can be quite useful. Carbon percent is divided into categories.

|                             | % Carbon |                          |                |
| --------------------------- | -------- | ------------------------ | -------------- |
| Casts judged as to gas holes | Below 1.155 | Between 1.155 and 1.195 | Above 1.195 |
| Some                        | 28       | 35                       | 37             |
| None                        | 46       | 47                       | 49             |

Test whether percent carbon is related to gas holes.

14.14. A check inspection of 12 samples, each of 200 cartridge cases for visual defects, gave the following total defects per cell (that is, out of 2400):

| Week | Foreman | | |
|------|-----|-----|--------|
|      | Pat | Rex | George |
| 1    | 30  | 60  | 39     |
| 2    | 41  | 37  | 63     |

Test for significance at the $\alpha = .01$ level. Note that although we should figure $\chi^2$ on 12 cells including nondefectives, the Poisson distribution can be assumed, and we can find expected frequencies as in a $3 \cdot 2$ table. It turned out that Rex and his crew worked the day shift in week 1 and George and his crew the day shift in week 2. Comment.

14.15. Auditing of caps for food jars gave the following results for caps with functional defects:

| Shift | Total inspected | Number defective |
|-------|-----------------|------------------|
| 1     | 74,740          | 9                |
| 2     | 86,880          | 5                |
| 3     | 83,680          | 29               |

Test the hypothesis of independence, that is, that the defective ratio is the same for all three shifts, using $\alpha = .01$.

14.16. Do the example in Section 14.5.1 by the short formula of Section 14.5.4, and compare results and work involved.

14.17. Use the sign test for appropriate hypotheses on the matched-pair data of Problem 9.13. Use $\alpha = .05$ and interpret your results.

14.18. Use the sign test for appropriate hypotheses on the matched-pair data of Problem 9.16. Use $\alpha = .05$ and interpret your results.

14.19. Use the sign test for appropriate hypotheses on the matched-pair data of Problem 9.15. Use $\alpha = .05$ and interpret your results.

For the following problems test goodness of fit of chosen type of distribution at $\alpha = .05$, lumping tail frequencies to give $F_i \geq 1$.

14.20. Problem 6.12: Table 2.13 fitted by normal curve.

14.21.  Problem 6.11: Table 2.10 fitted by normal curve.

14.22.  Problem 6.14: Table 2.21 fitted by normal curve.

14.23.  Problem 6.13: Table 2.17 fitted by normal curve.

14.24.  Problem 6.15: Table 2.22 fitted by normal curve.

# APPENDIX

**TABLE I**

Probability under Standard Normal Curve: $\Phi(u) = \int_{-\infty}^{u} \phi(z)\, dz$

| $u$ | .00 | .01 | .02 | .03 | .04 | .05 | .06 | .07 | .08 | .09 |
|---|---|---|---|---|---|---|---|---|---|---|
| +0.0 | .5000 | .5040 | .5080 | .5120 | .5160 | .5199 | .5239 | .5279 | .5319 | .5359 |
| +0.1 | .5398 | .5438 | .5478 | .5517 | .5557 | .5596 | .5636 | .5675 | .5714 | .5753 |
| +0.2 | .5793 | .5832 | .5871 | .5910 | .5948 | .5987 | .6026 | .6064 | .6103 | .6141 |
| +0.3 | .6179 | .6217 | .6255 | .6293 | .6331 | .6368 | .6406 | .6443 | .6480 | .6517 |
| +0.4 | .6554 | .6591 | .6628 | .6664 | .6700 | .6736 | .6772 | .6808 | .6844 | .6879 |
| +0.5 | .6915 | .6950 | .6985 | .7019 | .7054 | .7088 | .7123 | .7157 | .7190 | .7224 |
| +0.6 | .7257 | .7291 | .7324 | .7357 | .7389 | .7422 | .7454 | .7486 | .7517 | .7549 |
| +0.7 | .7580 | .7611 | .7642 | .7673 | .7704 | .7734 | .7764 | .7794 | .7823 | .7852 |
| +0.8 | .7881 | .7910 | .7939 | .7967 | .7995 | .8023 | .8051 | .8078 | .8106 | .8133 |
| +0.9 | .8159 | .8186 | .8212 | .8238 | .8264 | .8289 | .8315 | .8340 | .8365 | .8389 |
| +1.0 | .8413 | .8438 | .8461 | .8485 | .8508 | .8531 | .8554 | .8577 | .8599 | .8621 |
| +1.1 | .8643 | .8665 | .8686 | .8708 | .8729 | .8749 | .8770 | .8790 | .8810 | .8830 |
| +1.2 | .8849 | .8869 | .8888 | .8907 | .8925 | .8944 | .8962 | .8980 | .8997 | .9015 |
| +1.3 | .9032 | .9049 | .9066 | .9082 | .9099 | .9115 | .9131 | .9147 | .9162 | .9177 |
| +1.4 | .9192 | .9207 | .9222 | .9236 | .9251 | .9265 | .9279 | .9292 | .9306 | .9319 |
| +1.5 | .9332 | .9345 | .9357 | .9370 | .9382 | .9394 | .9406 | .9418 | .9429 | .9441 |
| +1.6 | .9452 | .9463 | .9474 | .9484 | .9495 | .9505 | .9515 | .9525 | .9535 | .9545 |
| +1.7 | .9554 | .9564 | .9573 | .9582 | .9591 | .9599 | .9608 | .9616 | .9625 | .9633 |
| +1.8 | .9641 | .9649 | .9656 | .9664 | .9671 | .9678 | .9686 | .9693 | .9699 | .9706 |
| +1.9 | .9713 | .9719 | .9726 | .9732 | .9738 | .9744 | .9750 | .9756 | .9761 | .9767 |
| +2.0 | .9772 | .9778 | .9783 | .9788 | .9793 | .9798 | .9803 | .9808 | .9812 | .9817 |
| +2.1 | .9821 | .9826 | .9830 | .9834 | .9838 | .9842 | .9846 | .9850 | .9854 | .9857 |
| +2.2 | .9861 | .9864 | .9868 | .9871 | .9875 | .9878 | .9881 | .9884 | .9887 | .9890 |
| +2.3 | .9893 | .9896 | .9898 | .9901 | .9904 | .9906 | .9909 | .9911 | .9913 | .9916 |
| +2.4 | .9918 | .9920 | .9922 | .9925 | .9927 | .9929 | .9931 | .9932 | .9934 | .9936 |
| +2.5 | .9938 | .9940 | .9941 | .9943 | .9945 | .9946 | .9948 | .9949 | .9951 | .9952 |
| +2.6 | .9953 | .9955 | .9956 | .9957 | .9959 | .9960 | .9961 | .9962 | .9963 | .9964 |
| +2.7 | .9965 | .9966 | .9967 | .9968 | .9969 | .9970 | .9971 | .9972 | .9973 | .9974 |
| +2.8 | .9974 | .9975 | .9976 | .9977 | .9977 | .9978 | .9979 | .9979 | .9980 | .9981 |
| +2.9 | .9981 | .9982 | .9982 | .9983 | .9984 | .9984 | .9985 | .9985 | .9986 | .9986 |
| +3.0 | .9987 | .9987 | .9987 | .9988 | .9988 | .9989 | .9989 | .9989 | .9990 | .9990 |
| +3.1 | .9990 | .9991 | .9991 | .9991 | .9992 | .9992 | .9992 | .9992 | .9993 | .9993 |
| +3.2 | .9993 | .9993 | .9994 | .9994 | .9994 | .9994 | .9994 | .9995 | .9995 | .9995 |
| +3.3 | .9995 | .9995 | .9995 | .9996 | .9996 | .9996 | .9996 | .9996 | .9996 | .9997 |
| +3.4 | .9997 | .9997 | .9997 | .9997 | .9997 | .9997 | .9997 | .9997 | .9997 | .9998 |

| $\Phi(u)$ | .75 | .80 | .90 | .95 | .975 | .98 | .99 | .995 | .999 |
|---|---|---|---|---|---|---|---|---|---|
| $u$ | .6745 | .8416 | 1.2816 | 1.6449 | 1.9600 | 2.0537 | 2.3263 | 2.5758 | 3.0902 |

**TABLE II**

Critical Values for Chi-Square Distribution $P(\chi^2$ random variable with $\nu$ degrees of freedom $\leqslant$ tabled value) $= P$

| | | | | P | | |
|---|---|---|---|---|---|---|
| $\nu$ | 0.005 | 0.01 | 0.025 | 0.05 | 0.10 | 0.25 |
| 1 | - | - | 0.001 | 0.004 | 0.016 | 0.102 |
| 2 | 0.010 | 0.020 | 0.051 | 0.103 | 0.211 | 0.575 |
| 3 | 0.072 | 0.115 | 0.216 | 0.352 | 0.584 | 1.213 |
| 4 | 0.207 | 0.297 | 0.484 | 0.711 | 1.064 | 1.923 |
| 5 | 0.412 | 0.554 | 0.831 | 1.145 | 1.610 | 2.675 |
| 6 | 0.676 | 0.872 | 1.237 | 1.635 | 2.204 | 3.455 |
| 7 | 0.989 | 1.239 | 1.690 | 2.167 | 2.833 | 4.255 |
| 8 | 1.344 | 1.646 | 2.180 | 2.733 | 3.490 | 5.071 |
| 9 | 1.735 | 2.088 | 2.700 | 3.325 | 4.168 | 5.899 |
| 10 | 2.156 | 2.558 | 3.247 | 3.940 | 4.865 | 6.737 |
| 11 | 2.603 | 3.053 | 3.816 | 4.575 | 5.578 | 7.584 |
| 12 | 3.074 | 3.571 | 4.404 | 5.226 | 6.304 | 8.438 |
| 13 | 3.565 | 4.107 | 5.009 | 5.892 | 7.042 | 9.299 |
| 14 | 4.075 | 4.660 | 5.629 | 6.571 | 7.790 | 10.165 |
| 15 | 4.601 | 5.229 | 6.262 | 7.261 | 8.547 | 11.037 |
| 16 | 5.142 | 5.812 | 6.908 | 7.962 | 9.312 | 11.912 |
| 17 | 5.697 | 6.408 | 7.564 | 8.672 | 10.085 | 12.792 |
| 18 | 6.265 | 7.015 | 8.231 | 9.390 | 10.865 | 13.675 |
| 19 | 6.844 | 7.633 | 8.907 | 10.117 | 11.651 | 14.562 |
| 20 | 7.434 | 8.260 | 9.591 | 10.851 | 12.443 | 15.452 |
| 21 | 8.034 | 8.897 | 10.283 | 11.591 | 13.240 | 16.344 |
| 22 | 8.643 | 9.542 | 10.982 | 12.338 | 14.042 | 17.240 |
| 23 | 9.260 | 10.196 | 11.689 | 13.091 | 14.848 | 18.137 |
| 24 | 9.886 | 10.856 | 12.401 | 13.848 | 15.659 | 19.037 |
| 25 | 10.520 | 11.524 | 13.120 | 14.611 | 16.473 | 19.939 |
| 26 | 11.160 | 12.198 | 13.844 | 15.379 | 17.292 | 20.843 |
| 27 | 11.808 | 12.879 | 14.573 | 16.151 | 18.114 | 21.749 |
| 28 | 12.461 | 13.565 | 15.308 | 16.928 | 18.939 | 22.657 |
| 29 | 13.121 | 14.257 | 16.047 | 17.708 | 19.768 | 23.567 |
| 30 | 13.787 | 14.954 | 16.791 | 18.493 | 20.599 | 24.478 |
| 31 | 14.458 | 15.655 | 17.539 | 19.281 | 21.434 | 25.390 |
| 32 | 15.134 | 16.362 | 18.291 | 20.072 | 22.271 | 26.304 |
| 33 | 15.815 | 17.074 | 19.047 | 20.867 | 23.110 | 27.219 |
| 34 | 16.501 | 17.789 | 19.806 | 21.664 | 23.952 | 28.136 |
| 35 | 17.192 | 18.509 | 20.569 | 22.465 | 24.797 | 29.054 |
| 36 | 17.887 | 19.233 | 21.336 | 23.269 | 25.643 | 29.973 |
| 37 | 18.586 | 19.960 | 22.106 | 24.075 | 26.492 | 30.893 |
| 38 | 19.289 | 20.691 | 22.878 | 24.884 | 27.343 | 31.815 |
| 39 | 19.996 | 21.426 | 23.654 | 25.695 | 28.196 | 32.737 |
| 40 | 20.707 | 22.164 | 24.433 | 26.509 | 29.051 | 33.660 |
| 41 | 21.421 | 22.906 | 25.215 | 27.326 | 29.907 | 34.585 |
| 42 | 22.138 | 23.650 | 25.999 | 28.144 | 30.765 | 35.510 |
| 43 | 22.859 | 24.398 | 26.785 | 28.965 | 31.625 | 36.436 |
| 44 | 23.584 | 25.148 | 27.575 | 29.787 | 32.487 | 37.363 |
| 45 | 24.311 | 25.901 | 28.366 | 30.612 | 33.350 | 38.291 |

REPRODUCED from D. C. Owen, "Handbook of Statistical Tables," Addison-Wesley, Reading, Massachusetts, 1962, with kind permission of author and publisher.

**TABLE II** *(continued)*

| | P | | | | | |
|---|---|---|---|---|---|---|
| $\nu$ | 0.75 | 0.90 | 0.95 | 0.975 | 0.99 | 0.995 |
| 1 | 1.323 | 2.706 | 3.841 | 5.024 | 6.635 | 7.879 |
| 2 | 2.773 | 4.605 | 5.991 | 7.378 | 9.210 | 10.597 |
| 3 | 4.108 | 6.251 | 7.815 | 9.348 | 11.345 | 12.838 |
| 4 | 5.385 | 7.779 | 9.488 | 11.143 | 13.277 | 14.860 |
| 5 | 6.626 | 9.236 | 11.071 | 12.833 | 15.086 | 16.750 |
| 6 | 7.841 | 10.645 | 12.592 | 14.449 | 16.812 | 18.548 |
| 7 | 9.037 | 12.017 | 14.067 | 16.013 | 18.475 | 20.278 |
| 8 | 10.219 | 13.362 | 15.507 | 17.535 | 20.090 | 21.955 |
| 9 | 11.389 | 14.684 | 16.919 | 19.023 | 21.666 | 23.589 |
| 10 | 12.549 | 15.987 | 18.307 | 20.483 | 23.209 | 25.188 |
| 11 | 13.701 | 17.275 | 19.675 | 21.920 | 24.725 | 26.757 |
| 12 | 14.845 | 18.549 | 21.026 | 23.337 | 26.217 | 28.299 |
| 13 | 15.984 | 19.812 | 22.362 | 24.736 | 27.688 | 29.819 |
| 14 | 17.117 | 21.064 | 23.685 | 26.119 | 29.141 | 31.319 |
| 15 | 18.245 | 22.307 | 24.996 | 27.488 | 30.578 | 32.801 |
| 16 | 19.369 | 23.542 | 26.296 | 28.845 | 32.000 | 34.267 |
| 17 | 20.489 | 24.769 | 27.587 | 30.191 | 33.409 | 35.718 |
| 18 | 21.605 | 25.989 | 28.869 | 31.526 | 34.805 | 37.156 |
| 19 | 22.718 | 27.204 | 30.144 | 32.852 | 36.191 | 38.582 |
| 20 | 23.828 | 28.412 | 31.410 | 34.170 | 37.566 | 39.997 |
| 21 | 24.935 | 29.615 | 32.671 | 35.479 | 38.932 | 41.401 |
| 22 | 26.039 | 30.813 | 33.924 | 36.781 | 40.289 | 42.796 |
| 23 | 27.141 | 32.007 | 35.172 | 38.076 | 41.638 | 44.181 |
| 24 | 28.241 | 33.196 | 36.415 | 39.364 | 42.980 | 45.559 |
| 25 | 29.339 | 34.382 | 37.652 | 40.646 | 44.314 | 46.928 |
| 26 | 30.435 | 35.563 | 38.885 | 41.923 | 45.642 | 48.290 |
| 27 | 31.528 | 36.741 | 40.113 | 43.194 | 46.963 | 49.645 |
| 28 | 32.620 | 37.916 | 41.337 | 44.461 | 48.278 | 50.993 |
| 29 | 33.711 | 39.087 | 42.557 | 45.722 | 49.588 | 52.336 |
| 30 | 34.800 | 40.256 | 43.773 | 46.979 | 50.892 | 53.672 |
| 31 | 35.887 | 41.422 | 44.985 | 48.232 | 52.191 | 55.003 |
| 32 | 36.973 | 42.585 | 46.194 | 49.480 | 53.486 | 56.328 |
| 33 | 38.058 | 43.745 | 47.400 | 50.725 | 54.776 | 57.648 |
| 34 | 39.141 | 44.903 | 48.602 | 51.966 | 56.061 | 58.964 |
| 35 | 40.223 | 46.059 | 49.802 | 53.203 | 57.342 | 60.275 |
| 36 | 41.304 | 47.212 | 50.998 | 54.437 | 58.619 | 61.581 |
| 37 | 42.383 | 48.363 | 52.192 | 55.668 | 59.892 | 62.883 |
| 38 | 43.462 | 49.513 | 53.384 | 56.896 | 61.162 | 64.181 |
| 39 | 44.539 | 50.660 | 54.572 | 58.120 | 62.428 | 65.476 |
| 40 | 45.616 | 51.805 | 55.758 | 59.342 | 63.691 | 66.766 |
| 41 | 46.692 | 52.949 | 56.942 | 60.561 | 64.950 | 68.053 |
| 42 | 47.766 | 54.090 | 58.124 | 61.777 | 66.206 | 69.336 |
| 43 | 48.840 | 55.230 | 59.304 | 62.990 | 67.459 | 70.616 |
| 44 | 49.913 | 56.369 | 60.481 | 64.201 | 68.710 | 71.893 |
| 45 | 50.985 | 57.505 | 61.656 | 65.410 | 69.957 | 73.166 |

**TABLE II** *(continued)*

Critical Values for Chi-Square Distribution $P(\chi^2$ random variable with $\nu$ degrees of freedom $\leqslant$ tabled value) $= P$

| | | | | P | | |
|---|---|---|---|---|---|---|
| $\nu$ | 0.005 | 0.01 | 0.025 | 0.05 | 0.10 | 0.25 |
| 46 | 25.041 | 26.657 | 29.160 | 31.439 | 34.215 | 39.220 |
| 47 | 25.775 | 27.416 | 29.956 | 32.268 | 35.081 | 40.149 |
| 48 | 26.511 | 28.177 | 30.755 | 33.098 | 35.949 | 41.079 |
| 49 | 27.249 | 28.941 | 31.555 | 33.930 | 36.818 | 42.010 |
| 50 | 27.991 | 29.707 | 32.357 | 34.764 | 37.689 | 42.942 |
| 51 | 28.735 | 30.475 | 33.162 | 35.600 | 38.560 | 43.874 |
| 52 | 29.481 | 31.246 | 33.968 | 36.437 | 39.433 | 44.808 |
| 53 | 30.230 | 32.018 | 34.776 | 37.276 | 40.308 | 45.741 |
| 54 | 30.981 | 32.793 | 35.586 | 38.116 | 41.183 | 46.676 |
| 55 | 31.735 | 33.570 | 36.398 | 38.958 | 42.060 | 47.610 |
| 56 | 32.490 | 34.350 | 37.212 | 39.801 | 42.937 | 48.546 |
| 57 | 33.248 | 35.131 | 38.027 | 40.646 | 43.816 | 49.482 |
| 58 | 34.008 | 35.913 | 38.844 | 41.492 | 44.696 | 50.419 |
| 59 | 34.770 | 36.698 | 39.662 | 42.339 | 45.577 | 51.356 |
| 60 | 35.534 | 37.485 | 40.482 | 43.188 | 46.459 | 52.294 |
| 61 | 36.300 | 38.273 | 41.303 | 44.038 | 47.342 | 53.232 |
| 62 | 37.068 | 39.063 | 42.126 | 44.889 | 48.226 | 54.171 |
| 63 | 37.838 | 39.855 | 42.950 | 45.741 | 49.111 | 55.110 |
| 64 | 38.610 | 40.649 | 43.776 | 46.595 | 49.996 | 56.050 |
| 65 | 39.383 | 41.444 | 44.603 | 47.450 | 50.883 | 56.990 |
| 66 | 40.158 | 42.240 | 45.431 | 48.305 | 51.770 | 57.931 |
| 67 | 40.935 | 43.038 | 46.261 | 49.162 | 52.659 | 58.872 |
| 68 | 41.713 | 43.838 | 47.092 | 50.020 | 53.548 | 59.814 |
| 69 | 42.494 | 44.639 | 47.924 | 50.879 | 54.438 | 60.756 |
| 70 | 43.275 | 45.442 | 48.758 | 51.739 | 55.329 | 61.698 |
| 71 | 44.058 | 46.246 | 49.592 | 52.600 | 56.221 | 62.641 |
| 72 | 44.843 | 47.051 | 50.428 | 53.462 | 57.113 | 63.585 |
| 73 | 45.629 | 47.858 | 51.265 | 54.325 | 58.006 | 64.528 |
| 74 | 46.417 | 48.666 | 52.103 | 55.189 | 58.900 | 65.472 |
| 75 | 47.206 | 49.475 | 52.942 | 56.054 | 59.795 | 66.417 |
| 76 | 47.997 | 50.286 | 53.782 | 56.920 | 60.690 | 67.362 |
| 77 | 48.788 | 51.097 | 54.623 | 57.786 | 61.586 | 68.307 |
| 78 | 49.582 | 51.910 | 55.466 | 58.654 | 62.483 | 69.252 |
| 79 | 50.376 | 52.725 | 56.309 | 59.522 | 63.380 | 70.198 |
| 80 | 51.172 | 53.540 | 57.153 | 60.391 | 64.278 | 71.145 |
| 81 | 51.969 | 54.357 | 57.998 | 61.261 | 65.176 | 72.091 |
| 82 | 52.767 | 55.174 | 58.845 | 62.132 | 66.076 | 73.038 |
| 83 | 53.567 | 55.993 | 59.692 | 63.004 | 66.976 | 73.985 |
| 84 | 54.368 | 56.813 | 60.540 | 63.876 | 67.876 | 74.933 |
| 85 | 55.170 | 57.634 | 61.389 | 64.749 | 68.777 | 75.881 |
| 86 | 55.973 | 58.456 | 62.239 | 65.623 | 69.679 | 76.829 |
| 87 | 56.777 | 59.279 | 63.089 | 66.498 | 70.581 | 77.777 |
| 88 | 57.582 | 60.103 | 63.941 | 67.373 | 71.484 | 78.726 |
| 89 | 58.389 | 60.928 | 64.793 | 68.249 | 72.387 | 79.675 |
| 90 | 59.196 | 61.754 | 65.647 | 69.126 | 73.291 | 80.625 |

**TABLE II** *(continued)*

| | | | P | | | |
|---|---|---|---|---|---|---|
| ν | 0.75 | 0.90 | 0.95 | 0.975 | 0.99 | 0.995 |
| 46 | 52.056 | 58.641 | 62.830 | 66.617 | 71.201 | 74.437 |
| 47 | 53.127 | 59.774 | 64.001 | 67.821 | 72.443 | 75.704 |
| 48 | 54.196 | 60.907 | 65.171 | 69.023 | 73.683 | 76.969 |
| 49 | 55.265 | 62.038 | 66.339 | 70.222 | 74.919 | 78.231 |
| 50 | 56.334 | 63.167 | 67.505 | 71.420 | 76.154 | 79.490 |
| 51 | 57.401 | 64.295 | 68.669 | 72.616 | 77.386 | 80.747 |
| 52 | 58.468 | 65.422 | 69.832 | 73.810 | 78.616 | 82.001 |
| 53 | 59.534 | 66.548 | 70.993 | 75.002 | 79.843 | 83.253 |
| 54 | 60.600 | 67.673 | 72.153 | 76.192 | 81.069 | 84.502 |
| 55 | 61.665 | 68.796 | 73.311 | 77.380 | 82.292 | 85.749 |
| 56 | 62.729 | 69.919 | 74.468 | 78.567 | 83.513 | 86.994 |
| 57 | 63.793 | 71.040 | 75.624 | 79.752 | 84.733 | 88.236 |
| 58 | 64.857 | 72.160 | 76.778 | 80.936 | 85.950 | 89.477 |
| 59 | 65.919 | 73.279 | 77.931 | 82.117 | 87.166 | 90.715 |
| 60 | 66.981 | 74.397 | 79.082 | 83.298 | 88.379 | 91.952 |
| 61 | 68.043 | 75.514 | 80.232 | 84.476 | 89.591 | 93.186 |
| 62 | 69.104 | 76.630 | 81.381 | 85.654 | 90.802 | 94.419 |
| 63 | 70.165 | 77.745 | 82.529 | 86.830 | 92.010 | 95.649 |
| 64 | 71.225 | 78.860 | 83.675 | 88.004 | 93.217 | 96.878 |
| 65 | 72.285 | 79.973 | 84.821 | 89.177 | 94.422 | 98.105 |
| 66 | 73.344 | 81.085 | 85.965 | 90.349 | 95.626 | 99.330 |
| 67 | 74.403 | 82.197 | 87.108 | 91.519 | 96.828 | 100.554 |
| 68 | 75.461 | 83.308 | 88.250 | 92.689 | 98.028 | 101.776 |
| 69 | 76.519 | 84.418 | 89.391 | 93.856 | 99.228 | 102.996 |
| 70 | 77.577 | 85.527 | 90.531 | 95.023 | 100.425 | 104.215 |
| 71 | 78.634 | 86.635 | 91.670 | 96.189 | 101.621 | 105.432 |
| 72 | 79.690 | 87.743 | 92.808 | 97.353 | 102.816 | 106.648 |
| 73 | 80.747 | 88.850 | 93.945 | 98.516 | 104.010 | 107.862 |
| 74 | 81.803 | 89.956 | 95.081 | 99.678 | 105.202 | 109.074 |
| 75 | 82.858 | 91.061 | 96.217 | 100.839 | 106.393 | 110.286 |
| 76 | 83.913 | 92.166 | 97.351 | 101.999 | 107.583 | 111.495 |
| 77 | 84.968 | 93.270 | 98.484 | 103.158 | 108.771 | 112.704 |
| 78 | 86.022 | 94.374 | 99.617 | 104.316 | 109.958 | 113.911 |
| 79 | 87.077 | 95.476 | 100.749 | 105.473 | 111.144 | 115.117 |
| 80 | 88.130 | 96.578 | 101.879 | 106.629 | 112.329 | 116.321 |
| 81 | 89.184 | 97.680 | 103.010 | 107.783 | 113.512 | 117.524 |
| 82 | 90.237 | 98.780 | 104.139 | 108.937 | 114.695 | 118.726 |
| 83 | 91.289 | 99.880 | 105.267 | 110.090 | 115.876 | 119.927 |
| 84 | 92.342 | 100.980 | 106.395 | 111.242 | 117.057 | 121.126 |
| 85 | 93.394 | 102.079 | 107.522 | 112.393 | 118.236 | 122.325 |
| 86 | 94.446 | 103.177 | 108.648 | 113.544 | 119.414 | 123.522 |
| 87 | 95.497 | 104.275 | 109.773 | 114.693 | 120.591 | 124.718 |
| 88 | 96.548 | 105.372 | 110.898 | 115.841 | 121.767 | 125.913 |
| 89 | 97.599 | 106.469 | 112.022 | 116.989 | 122.942 | 127.106 |
| 90 | 98.650 | 107.565 | 113.145 | 118.136 | 124.116 | 128.299 |

**TABLE III**

Critical Values for Student's $t$ Distribution[a]

| $\nu$ | | | P | | | |
|---|---|---|---|---|---|---|
| | .25 | .10 | .05 | .025 | .01 | .005 |
| 1 | 1.000 | 3.078 | 6.314 | 12.706 | 21.821 | 63.657 |
| 2 | .816 | 1.886 | 2.920 | 4.303 | 6.965 | 9.925 |
| 3 | .765 | 1.638 | 2.353 | 3.182 | 4.541 | 5.841 |
| 4 | .741 | 1.533 | 2.132 | 2.776 | 3.747 | 4.604 |
| 5 | .727 | 1.476 | 2.015 | 2.571 | 3.365 | 4.032 |
| 6 | .718 | 1.440 | 1.943 | 2.447 | 3.143 | 3.707 |
| 7 | .711 | 1.415 | 1.895 | 2.365 | 2.998 | 3.499 |
| 8 | .706 | 1.397 | 1.860 | 2.306 | 2.896 | 3.355 |
| 9 | .703 | 1.383 | 1.833 | 2.262 | 2.821 | 3.250 |
| 10 | .700 | 1.372 | 1.812 | 2.228 | 2.764 | 3.169 |
| 11 | .697 | 1.363 | 1.796 | 2.201 | 2.718 | 3.106 |
| 12 | .695 | 1.356 | 1.782 | 2.179 | 2.681 | 3.055 |
| 13 | .694 | 1.350 | 1.771 | 2.160 | 2.650 | 3.012 |
| 14 | .692 | 1.345 | 1.761 | 2.145 | 2.624 | 2.977 |
| 15 | .691 | 1.341 | 1.753 | 2.131 | 2.602 | 2.947 |
| 16 | .690 | 1.337 | 1.746 | 2.120 | 2.583 | 2.921 |
| 17 | .689 | 1.333 | 1.740 | 2.110 | 2.567 | 2.898 |
| 18 | .688 | 1.330 | 1.734 | 2.101 | 2.552 | 2.878 |
| 19 | .688 | 1.328 | 1.729 | 2.093 | 2.539 | 2.861 |
| 20 | .687 | 1.325 | 1.725 | 2.086 | 2.528 | 2.845 |
| 21 | .686 | 1.323 | 1.721 | 2.080 | 2.518 | 2.831 |
| 22 | .686 | 1.321 | 1.717 | 2.074 | 2.508 | 2.819 |
| 23 | .685 | 1.319 | 1.714 | 2.069 | 2.500 | 2.807 |
| 24 | .685 | 1.318 | 1.711 | 2.064 | 2.492 | 2.797 |
| 25 | .684 | 1.316 | 1.708 | 2.060 | 2.485 | 2.787 |
| 26 | .684 | 1.315 | 1.706 | 2.056 | 2.479 | 2.779 |
| 27 | .684 | 1.314 | 1.703 | 2.052 | 2.473 | 2.771 |
| 28 | .683 | 1.313 | 1.701 | 2.048 | 2.467 | 2.763 |
| 29 | .683 | 1.311 | 1.699 | 2.045 | 2.462 | 2.756 |
| 30 | .683 | 1.310 | 1.697 | 2.042 | 2.457 | 2.750 |
| 35 | .682 | 1.306 | 1.690 | 2.030 | 2.438 | 2.724 |
| 40 | .681 | 1.303 | 1.684 | 2.021 | 2.423 | 2.704 |
| 45 | .680 | 1.301 | 1.679 | 2.014 | 2.412 | 2.690 |
| 50 | .679 | 1.299 | 1.676 | 2.009 | 2.403 | 2.678 |
| 55 | .679 | 1.297 | 1.673 | 2.004 | 2.396 | 2.668 |
| 60 | .679 | 1.296 | 1.671 | 2.000 | 2.390 | 2.660 |
| 65 | .678 | 1.295 | 1.669 | 1.997 | 2.385 | 2.654 |
| 70 | .678 | 1.294 | 1.667 | 1.994 | 2.381 | 2.648 |
| 75 | .678 | 1.293 | 1.665 | 1.992 | 2.377 | 2.643 |
| 80 | .678 | 1.292 | 1.664 | 1.990 | 2.374 | 2.639 |
| 90 | .677 | 1.291 | 1.662 | 1.987 | 2.368 | 2.632 |
| 100 | .677 | 1.290 | 1.660 | 1.984 | 2.364 | 2.626 |
| 200 | .676 | 1.286 | 1.652 | 1.972 | 2.345 | 2.601 |
| 500 | .675 | 1.283 | 1.648 | 1.965 | 2.334 | 2.586 |
| 1000 | .675 | 1.282 | 1.646 | 1.962 | 2.330 | 2.581 |
| $\infty$ | .674 | 1.282 | 1.645 | 1.960 | 2.326 | 2.576 |

[a] $\nu$ = degrees of freedom; P = upper tail probability.
ADAPTED from D. B. Owen, "Handbook of Statistical Tables," Addison-Wesley, Reading, Massachusetts, 1962, with kind permission of author and publisher.

**TABLE IV**

Moment Constants for Distribution of Sample Standard deviation $s^a$

| $n$ | $c_4$ | $c_5$ | $\alpha_{3:s}$ | $n$ | $c_4$ | $c_5$ | $\alpha_{3:s}$ |
|---|---|---|---|---|---|---|---|
| 2 | .7979 | .6028 | .995 | 32 | .9920 | .1265 | .130 |
| 3 | .8862 | .4633 | .631 | 34 | .9925 | .1226 | .125 |
| 4 | .9213 | .3888 | .486 | 36 | .9929 | .1191 | .122 |
| 5 | .9400 | .3412 | .406 | 38 | .9933 | .1158 | .118 |
| 6 | .9515 | .3075 | .354 | 40 | .9936 | .1129 | .115 |
| 7 | .9594 | .2822 | .318 | 42 | .9939 | .1101 | .112 |
| 8 | .9650 | .2621 | .291 | 44 | .9942 | .1075 | .109 |
| 9 | .9693 | .2458 | .269 | 46 | .9945 | .1051 | .107 |
| 10 | .9727 | .2322 | .252 | 48 | .9947 | .1029 | .105 |
| 11 | .9754 | .2207 | .237 | 50 | .9949 | .1008 | .102 |
| 12 | .9776 | .2107 | .225 | 52 | .9951 | .0988 | .100 |
| 13 | .9794 | .2019 | .215 | 54 | .9953 | .0969 | .098 |
| 14 | .9810 | .1942 | .205 | 56 | .9955 | .0951 | .096 |
| 15 | .9823 | .1872 | .197 | 58 | .9956 | .0935 | .095 |
| 16 | .9835 | .1810 | .190 | 60 | .9958 | .0919 | .093 |
| 17 | .9845 | .1753 | .184 | 62 | .9959 | .0903 | .091 |
| 18 | .9854 | .1702 | .178 | 64 | .9960 | .0889 | .090 |
| 19 | .9862 | .1655 | .172 | 66 | .9962 | .0875 | .089 |
| 20 | .9869 | .1611 | .168 | 68 | .9963 | .0862 | .087 |
| 21 | .9876 | .1571 | .163 | 70 | .9964 | .0850 | .086 |
| 22 | .9882 | .1534 | .159 | 72 | .9965 | .0838 | .085 |
| 23 | .9887 | .1499 | .155 | 74 | .9966 | .0826 | .083 |
| 24 | .9892 | .1466 | .151 | 76 | .9967 | .0815 | .082 |
| 25 | .9896 | .1436 | .148 | 78 | .9968 | .0805 | .081 |
| 26 | .9901 | .1407 | .145 | 80 | .9968 | .0794 | .080 |
| 27 | .9904 | .1380 | .142 | 84 | .9970 | .0775 | .078 |
| 28 | .9908 | .1354 | .139 | 88 | .9971 | .0757 | .076 |
| 29 | .9911 | .1330 | .137 | 92 | .9973 | .0740 | .075 |
| 30 | .9914 | .1307 | .134 | 96 | .9974 | .0725 | .073 |
| 31 | .9917 | .1286 | .132 | 100 | .9975 | .0710 | .072 |

$^a$ $E(s) = c_4\sigma$; $\sigma_s = c_5\sigma$; $\alpha_{3:s}$ .

**TABLE V**

Critical Values of the $F$ Distribution
$P(F$ r.v. with $\nu_1$ , $\nu_2$ degrees of freedom $\leqslant$ table value) $=$ P

| Degrees of freedom for denominator, $\nu_2$ | P | Degrees of freedom for numerator, $\nu_1$ | | | | | | | |
|---|---|---|---|---|---|---|---|---|---|
| | | 1 | 2 | 3 | 4 | 5 | 6 | 7 | 8 |
| 1 | .950 | 161.4 | 199.5 | 215.7 | 224.6 | 230.2 | 234.0 | 236.0 | 238.9 |
| | .975 | 648.8 | 799.5 | 864.2 | 899.6 | 921.8 | 937.1 | 948.2 | 956.7 |
| | .990 | 4052 | 4999 | 5403 | 5624 | 5763 | 5859 | 5928 | 5981 |
| | .995 | 16211 | 20000 | 21615 | 22500 | 23056 | 23437 | 23715 | 23925 |
| 2 | .950 | 18.51 | 19.00 | 19.16 | 19.25 | 19.30 | 19.33 | 19.35 | 19.37 |
| | .975 | 38.51 | 39.00 | 39.17 | 39.25 | 39.30 | 39.33 | 39.36 | 39.37 |
| | .990 | 98.50 | 99.00 | 99.17 | 99.25 | 99.30 | 99.33 | 99.36 | 99.37 |
| | .995 | 198.5 | 199.0 | 199.2 | 199.2 | 199.3 | 199.3 | 199.4 | 199.4 |
| 3 | .950 | 10.13 | 9.552 | 9.277 | 9.117 | 9.014 | 8.941 | 8.887 | 8.845 |
| | .975 | 17.44 | 16.04 | 15.44 | 15.10 | 14.88 | 14.73 | 14.62 | 14.54 |
| | .990 | 34.12 | 30.82 | 29.46 | 28.71 | 28.24 | 27.91 | 27.67 | 27.49 |
| | .995 | 55.55 | 49.80 | 47.47 | 46.19 | 45.39 | 44.84 | 44.43 | 44.13 |
| 4 | .950 | 7.709 | 6.944 | 6.591 | 6.388 | 6.256 | 6.163 | 6.094 | 6.041 |
| | .975 | 12.22 | 10.65 | 9.979 | 9.604 | 9.364 | 9.197 | 9.074 | 8.980 |
| | .990 | 21.20 | 18.00 | 16.69 | 15.98 | 15.52 | 15.21 | 14.98 | 14.80 |
| | .995 | 31.33 | 26.28 | 24.26 | 23.15 | 22.46 | 21.97 | 21.62 | 21.35 |
| 5 | .950 | 6.608 | 5.786 | 5.410 | 5.192 | 5.050 | 4.950 | 4.876 | 4.818 |
| | .975 | 10.01 | 8.434 | 7.764 | 7.388 | 7.146 | 6.978 | 6.853 | 6.757 |
| | .990 | 16.26 | 13.27 | 12.06 | 11.39 | 10.97 | 10.67 | 10.46 | 10.29 |
| | .995 | 22.78 | 18.31 | 16.53 | 15.56 | 14.94 | 14.51 | 14.20 | 13.96 |
| 6 | .950 | 5.987 | 5.143 | 4.757 | 4.534 | 4.387 | 4.284 | 4.207 | 4.147 |
| | .975 | 8.813 | 7.260 | 6.599 | 6.227 | 5.988 | 5.820 | 5.696 | 5.600 |
| | .990 | 13.75 | 10.92 | 9.780 | 9.148 | 8.746 | 8.466 | 8.260 | 8.102 |
| | .995 | 18.63 | 14.54 | 12.92 | 12.03 | 11.46 | 11.07 | 10.79 | 10.57 |
| 7 | .950 | 5.591 | 4.737 | 4.347 | 4.120 | 3.972 | 3.866 | 3.787 | 3.726 |
| | .975 | 8.073 | 6.542 | 5.890 | 5.523 | 5.285 | 5.119 | 4.995 | 4.899 |
| | .990 | 12.25 | 9.547 | 8.451 | 7.847 | 7.460 | 7.191 | 6.993 | 6.840 |
| | .995 | 16.24 | 12.40 | 10.88 | 10.05 | 9.522 | 9.155 | 8.885 | 8.678 |
| 8 | .950 | 5.318 | 4.459 | 4.066 | 3.838 | 3.688 | 3.581 | 3.500 | 3.438 |
| | .975 | 7.571 | 6.060 | 5.416 | 5.053 | 4.817 | 4.652 | 4.529 | 4.433 |
| | .990 | 11.26 | 8.649 | 7.591 | 7.006 | 6.632 | 6.371 | 6.178 | 6.029 |
| | .995 | 14.69 | 11.04 | 9.596 | 8.805 | 8.302 | 7.952 | 7.694 | 7.496 |

ADAPTED from D. B. Owen, "Handbook of Statistical Tables," Addison-Wesley, Reading, Massachusetts, 1962, with kind permission of author and publisher.

**TABLE V** *(continued)*

| Degrees of freedom for denominator, $\nu_2$ | P | Degrees of freedom for numerator, $\nu_1$ | | | | | | | |
|---|---|---|---|---|---|---|---|---|---|
| | | 9 | 10 | 12 | 15 | 20 | 30 | 60 | 120 |
| 1 | .950 | 240.5 | 241.9 | 243.9 | 245.9 | 248.0 | 250.1 | 252.2 | 253.3 |
| | .975 | 963.3 | 968.6 | 976.7 | 984.9 | 993.1 | 1001 | 1010 | 1014 |
| | .990 | 6022 | 6056 | 6106 | 6157 | 6209 | 6261 | 6313 | 6339 |
| | .995 | 24091 | 24224 | 24426 | 24630 | 24836 | 25044 | 25253 | 25359 |
| 2 | .950 | 19.38 | 19.40 | 19.41 | 19.43 | 19.45 | 19.46 | 19.48 | 19.49 |
| | .975 | 39.39 | 39.40 | 39.41 | 39.43 | 39.45 | 39.46 | 39.48 | 39.49 |
| | .990 | 99.39 | 99.40 | 99.42 | 99.43 | 99.45 | 99.47 | 99.48 | 99.49 |
| | .995 | 199.4 | 199.4 | 199.4 | 199.4 | 199.4 | 199.5 | 199.5 | 199.5 |
| 3 | .950 | 8.812 | 8.786 | 8.745 | 8.703 | 8.660 | 8.617 | 8.572 | 8.549 |
| | .975 | 14.47 | 14.42 | 14.34 | 14.25 | 14.17 | 14.08 | 13.99 | 13.95 |
| | .990 | 27.35 | 27.23 | 27.05 | 26.87 | 26.69 | 26.50 | 26.32 | 26.22 |
| | .995 | 43.88 | 43.69 | 43.39 | 43.08 | 42.78 | 42.47 | 42.15 | 41.99 |
| 4 | .950 | 5.999 | 5.964 | 5.912 | 5.858 | 5.802 | 5.746 | 5.688 | 5.658 |
| | .975 | 8.905 | 8.844 | 8.751 | 8.656 | 8.560 | 8.461 | 8.360 | 8.309 |
| | .990 | 14.66 | 14.55 | 14.37 | 14.20 | 14.02 | 13.84 | 13.65 | 13.56 |
| | .995 | 21.14 | 20.97 | 20.70 | 20.44 | 20.17 | 19.89 | 19.61 | 19.47 |
| 5 | .950 | 4.772 | 4.735 | 4.678 | 4.619 | 4.558 | 4.496 | 4.431 | 4.398 |
| | .975 | 6.681 | 6.619 | 6.525 | 6.428 | 6.328 | 6.227 | 6.122 | 6.069 |
| | .990 | 10.16 | 10.05 | 9.888 | 9.722 | 9.553 | 9.379 | 9.202 | 9.112 |
| | .995 | 13.77 | 13.62 | 13.38 | 13.15 | 12.90 | 12.66 | 12.40 | 12.27 |
| 6 | .950 | 4.099 | 4.060 | 4.000 | 3.938 | 3.874 | 3.808 | 3.740 | 3.705 |
| | .975 | 5.523 | 5.461 | 5.366 | 5.269 | 5.168 | 5.065 | 4.959 | 4.904 |
| | .990 | 7.976 | 7.874 | 7.718 | 7.559 | 7.396 | 7.228 | 7.057 | 6.969 |
| | .995 | 10.25 | 10.13 | 10.03 | 9.814 | 9.589 | 9.358 | 9.122 | 9.002 |
| 7 | .950 | 3.677 | 3.636 | 3.575 | 3.511 | 3.444 | 3.376 | 3.304 | 3.267 |
| | .975 | 4.823 | 4.761 | 4.666 | 4.568 | 4.467 | 4.362 | 4.254 | 4.199 |
| | .990 | 6.719 | 6.620 | 6.469 | 6.314 | 6.155 | 5.992 | 5.824 | 5.737 |
| | .995 | 8.514 | 8.380 | 8.176 | 7.968 | 7.754 | 7.534 | 7.309 | 7.193 |
| 8 | .950 | 3.388 | 3.347 | 3.284 | 3.218 | 3.150 | 3.079 | 3.005 | 2.967 |
| | .975 | 4.357 | 4.295 | 4.200 | 4.101 | 4.000 | 3.894 | 3.784 | 3.728 |
| | .990 | 5.911 | 5.814 | 5.667 | 5.515 | 5.359 | 5.198 | 5.032 | 4.946 |
| | .995 | 7.339 | 7.211 | 7.015 | 6.814 | 6.608 | 6.396 | 6.177 | 6.065 |

**TABLE V** *(continued)*

Critical Values of the $F$ Distribution

$P(F$ r.v. with $\nu_1$, $\nu_2$ degrees of freedom $\leqslant$ table values) $= P$

| Degrees of freedom for denominator, $\nu_2$ | P | Degrees of freedom for numerator, $\nu_1$ | | | | | | | |
|---|---|---|---|---|---|---|---|---|---|
| | | 1 | 2 | 3 | 4 | 5 | 6 | 7 | 8 |
| 9 | .950 | 5.117 | 4.256 | 3.863 | 3.633 | 3.482 | 3.374 | 3.293 | 3.230 |
| | .975 | 7.209 | 5.715 | 5.078 | 4.718 | 4.484 | 4.320 | 4.197 | 4.102 |
| | .990 | 10.56 | 8.022 | 6.992 | 6.422 | 6.057 | 5.802 | 5.613 | 5.467 |
| | .995 | 13.81 | 10.11 | 8.717 | 7.956 | 7.471 | 7.134 | 6.885 | 6.693 |
| 10 | .950 | 4.965 | 4.103 | 3.708 | 3.478 | 3.326 | 3.217 | 3.136 | 3.072 |
| | .975 | 6.937 | 5.456 | 4.826 | 4.468 | 4.236 | 4.072 | 3.950 | 3.855 |
| | .990 | 10.04 | 7.559 | 6.552 | 5.994 | 5.636 | 5.386 | 5.200 | 5.057 |
| | .995 | 12.83 | 9.427 | 8.081 | 7.343 | 6.872 | 6.545 | 6.302 | 6.116 |
| 12 | .950 | 4.747 | 3.885 | 3.490 | 3.259 | 3.106 | 2.996 | 2.913 | 2.849 |
| | .975 | 6.554 | 3.096 | 4.474 | 4.121 | 3.891 | 3.728 | 3.606 | 3.512 |
| | .990 | 9.330 | 6.927 | 5.953 | 5.412 | 5.064 | 4.821 | 4.640 | 4.499 |
| | .995 | 11.75 | 8.510 | 7.226 | 6.521 | 6.071 | 5.757 | 5.524 | 5.345 |
| 15 | .950 | 4.543 | 3.682 | 3.287 | 3.056 | 2.901 | 2.790 | 2.707 | 2.641 |
| | .975 | 6.200 | 4.765 | 4.153 | 3.804 | 3.576 | 3.415 | 3.293 | 3.199 |
| | .990 | 8.683 | 6.359 | 5.417 | 4.893 | 4.556 | 4.318 | 4.142 | 4.004 |
| | .995 | 10.80 | 7.701 | 6.476 | 5.803 | 5.372 | 5.071 | 4.847 | 4.674 |
| 20 | .950 | 4.351 | 3.493 | 3.098 | 2.866 | 2.711 | 2.599 | 2.514 | 2.447 |
| | .975 | 5.872 | 4.461 | 3.859 | 3.515 | 3.289 | 3.128 | 3.007 | 2.913 |
| | .990 | 8.096 | 5.849 | 4.938 | 4.431 | 4.103 | 3.871 | 3.699 | 3.564 |
| | .995 | 9.944 | 6.986 | 5.818 | 5.174 | 4.762 | 4.472 | 4.257 | 4.090 |
| 30 | .950 | 4.171 | 3.316 | 2.922 | 2.690 | 2.534 | 2.420 | 2.334 | 2.266 |
| | .975 | 5.568 | 4.182 | 3.589 | 3.250 | 3.026 | 2.867 | 2.746 | 2.651 |
| | .990 | 7.562 | 5.390 | 4.510 | 4.018 | 3.699 | 3.474 | 3.304 | 3.173 |
| | .995 | 9.180 | 6.355 | 5.239 | 4.623 | 4.228 | 3.949 | 3.742 | 3.580 |
| 60 | .950 | 4.001 | 3.150 | 2.758 | 2.525 | 2.368 | 2.254 | 2.166 | 2.097 |
| | .975 | 5.286 | 3.925 | 3.342 | 3.008 | 2.786 | 2.627 | 2.507 | 2.412 |
| | .990 | 7.077 | 4.977 | 4.126 | 3.649 | 3.339 | 3.119 | 2.953 | 2.823 |
| | .995 | 8.495 | 5.795 | 4.729 | 4.140 | 3.760 | 3.492 | 3.291 | 3.134 |
| 120 | .950 | 3.920 | 3.072 | 2.680 | 2.447 | 2.290 | 2.175 | 2.087 | 2.016 |
| | .975 | 5.152 | 3.805 | 3.227 | 2.894 | 2.674 | 2.515 | 2.395 | 2.299 |
| | .990 | 6.851 | 4.786 | 3.949 | 3.480 | 3.174 | 2.956 | 2.792 | 2.663 |
| | .995 | 8.179 | 5.539 | 4.497 | 3.921 | 3.548 | 3.285 | 3.087 | 2.933 |

ADAPTED from D. B. Owen, "Handbook of Statistical Tables," Addison-Wesley, Reading, Massachusetts, 1962, with kind permission of author and publisher.

**TABLE V** *(continued)*

| Degrees of freedom for denominator, $\nu_2$ | P | Degrees of freedom for numerator, $\nu_1$ | | | | | | | |
|---|---|---|---|---|---|---|---|---|---|
| | | 9 | 10 | 12 | 15 | 20 | 30 | 60 | 120 |
| 9 | .950 | 3.179 | 3.173 | 3.073 | 3.006 | 2.936 | 2.864 | 2.787 | 2.748 |
| | .975 | 4.026 | 3.964 | 3.868 | 3.769 | 3.667 | 3.560 | 3.449 | 3.392 |
| | .990 | 5.351 | 5.256 | 5.111 | 4.962 | 4.808 | 4.649 | 4.483 | 4.398 |
| | .995 | 6.541 | 6.417 | 6.227 | 6.032 | 5.832 | 5.625 | 5.410 | 5.300 |
| 10 | .950 | 3.020 | 2.978 | 2.913 | 2.845 | 2.774 | 2.700 | 2.621 | 2.580 |
| | .975 | 3.779 | 3.717 | 3.621 | 3.522 | 3.419 | 3.311 | 3.198 | 3.140 |
| | .990 | 4.942 | 4.849 | 4.706 | 4.558 | 4.405 | 4.247 | 4.082 | 3.996 |
| | .995 | 5.968 | 5.847 | 5.661 | 5.471 | 5.274 | 5.070 | 4.859 | 4.750 |
| 12 | .950 | 2.796 | 2.753 | 2.687 | 2.617 | 2.544 | 2.466 | 2.384 | 2.341 |
| | .975 | 3.436 | 3.374 | 3.277 | 3.177 | 3.073 | 2.963 | 2.848 | 2.787 |
| | .990 | 4.388 | 4.296 | 4.155 | 4.010 | 3.858 | 3.701 | 3.536 | 3.449 |
| | .995 | 5.202 | 5.086 | 4.906 | 4.721 | 4.530 | 4.331 | 4.123 | 4.015 |
| 15 | .950 | 2.588 | 2.544 | 2.475 | 2.404 | 2.328 | 2.247 | 2.160 | 2.114 |
| | .975 | 3.123 | 3.060 | 2.963 | 2.862 | 2.756 | 2.644 | 2.524 | 2.461 |
| | .990 | 3.895 | 3.805 | 3.666 | 3.522 | 3.372 | 3.214 | 3.047 | 2.960 |
| | .995 | 4.536 | 4.424 | 4.250 | 4.070 | 3.883 | 3.687 | 3.480 | 3.372 |
| 20 | .950 | 2.393 | 2.348 | 2.278 | 2.203 | 2.124 | 2.039 | 1.946 | 1.896 |
| | .975 | 2.836 | 2.774 | 2.676 | 2.573 | 2.464 | 2.349 | 2.223 | 2.156 |
| | .990 | 3.457 | 3.368 | 3.231 | 3.088 | 2.938 | 2.778 | 2.608 | 2.517 |
| | .995 | 3.956 | 3.847 | 3.678 | 3.502 | 3.318 | 3.123 | 2.916 | 2.806 |
| 30 | .950 | 2.211 | 2.165 | 2.092 | 2.015 | 1.932 | 1.841 | 1.740 | 1.684 |
| | .975 | 2.575 | 2.511 | 2.412 | 2.307 | 2.195 | 2.074 | 1.940 | 1.866 |
| | .990 | 3.066 | 2.979 | 2.843 | 2.700 | 2.549 | 2.386 | 2.208 | 2.111 |
| | .995 | 3.450 | 3.344 | 3.179 | 3.006 | 2.823 | 2.628 | 2.415 | 2.300 |
| 60 | .950 | 2.040 | 1.993 | 1.917 | 1.836 | 1.748 | 1.649 | 1.534 | 1.467 |
| | .975 | 2.334 | 2.270 | 2.169 | 2.061 | 1.944 | 1.815 | 1.667 | 1.581 |
| | .990 | 2.718 | 2.632 | 2.496 | 2.352 | 2.198 | 2.028 | 1.836 | 1.726 |
| | .975 | 3.008 | 2.904 | 2.742 | 2.570 | 2.387 | 2.187 | 1.962 | 1.834 |
| 120 | .950 | 1.959 | 1.910 | 1.834 | 1.750 | 1.659 | 1.554 | 1.429 | 1.352 |
| | .975 | 2.222 | 2.157 | 2.055 | 1.945 | 1.825 | 1.690 | 1.530 | 1.433 |
| | .990 | 2.559 | 2.472 | 2.336 | 2.192 | 2.035 | 1.860 | 1.656 | 1.533 |
| | .995 | 2.808 | 2.705 | 2.544 | 2.373 | 2.188 | 1.984 | 1.747 | 1.606 |

**TABLE VI**

Graphs of Operating Characteristic of Student's $t$ Test for $\alpha = .005^a$

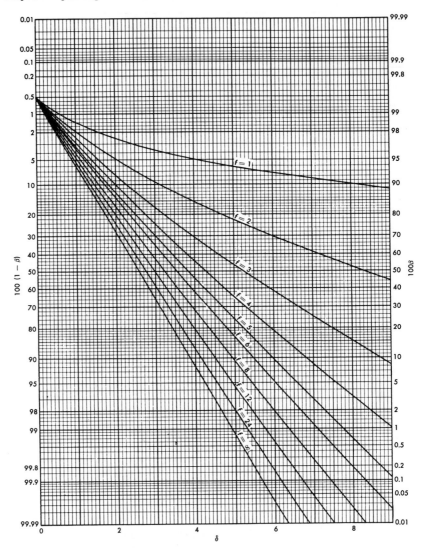

$^a$ $\alpha$ = single tail probability; $\delta = (\mu - \mu_0)\sqrt{n}/\sigma$; $f$ = degrees of freedom = $\nu$.
REPRODUCED from D. B. Owen, "Handbook of Statistical Tables," Addison-Wesley, Reading, Massachusetts, 1962, with kind permission of author and publisher.

**TABLE VI** *(continued)*

Graphs of Operating Characteristic of Student's *t* Test for $\alpha = .01^a$

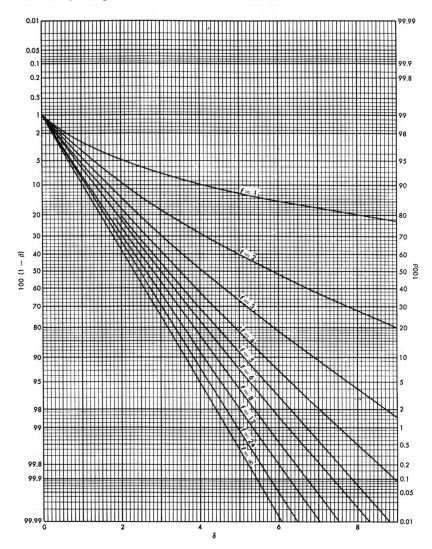

$^a$ $\alpha$ = single tail probability; $\delta = (\mu - \mu_0)\sqrt{\bar{n}}/\sigma$; $f$ = degrees of freedom = $\nu$.

**TABLE VI** *(continued)*

Graphs of Operating Characteristic of Student's $t$ Test for $\alpha = .025^a$

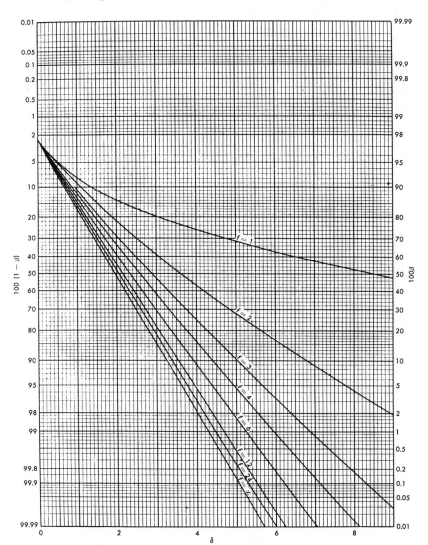

$^a$ $\alpha$ = single tail probability; $\delta = (\mu - \mu_0)\sqrt{n}/\sigma$; $f$ = degrees of freedom = $\nu$.
REPRODUCED from D. B. Owen, "Handbook of Statistical Tables," Addison-Wesley, Reading, Massachusetts, 1962, with kind permission of author and publisher.

**TABLE VI** *(continued)*

Graphs of Operating Characteristic of Student's $t$ Test for $\alpha = .05^a$

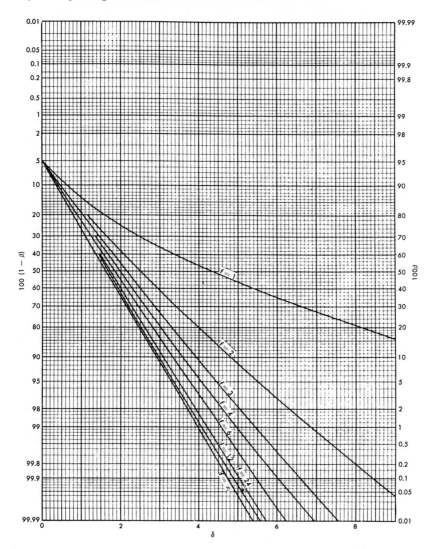

$^a$ $\alpha$ = single-tail probability; $\delta = (\mu - \mu_0)\sqrt{n}/\sigma$; $f$ = degrees of freedom = $\nu$.

TABLE VII

Studentized Range Points $q_\alpha$, $\alpha = .05$[a]

| $\nu$＼n | 2 | 3 | 4 | 5 | 6 | 7 | 8 | 9 | 10 |
|---|---|---|---|---|---|---|---|---|---|
| 1 | 17.97 | 26.98 | 32.82 | 37.08 | 40.41 | 43.12 | 45.40 | 47.36 | 49.07 |
| 2 | 6.085 | 8.331 | 9.798 | 10.88 | 11.74 | 12.44 | 13.03 | 13.54 | 13.99 |
| 3 | 4.501 | 5.910 | 6.825 | 7.502 | 8.037 | 8.478 | 8.853 | 9.177 | 9.462 |
| 4 | 3.927 | 5.040 | 5.757 | 6.287 | 6.707 | 7.053 | 7.347 | 7.602 | 7.826 |
| 5 | 3.635 | 4.602 | 5.218 | 5.673 | 6.033 | 6.330 | 6.582 | 6.802 | 6.995 |
| 6 | 3.461 | 4.339 | 4.896 | 5.305 | 5.628 | 5.895 | 6.122 | 6.319 | 6.493 |
| 7 | 3.344 | 4.165 | 4.681 | 5.060 | 5.359 | 5.606 | 5.815 | 5.998 | 6.158 |
| 8 | 3.261 | 4.041 | 4.529 | 4.886 | 5.167 | 5.399 | 5.597 | 5.767 | 5.918 |
| 9 | 3.199 | 3.949 | 4.415 | 4.756 | 5.024 | 5.244 | 5.432 | 5.595 | 5.739 |
| 10 | 3.151 | 3.877 | 4.327 | 4.654 | 4.912 | 5.124 | 5.305 | 5.461 | 5.599 |
| 11 | 3.113 | 3.820 | 4.256 | 4.574 | 4.823 | 5.028 | 5.202 | 5.353 | 5.487 |
| 12 | 3.082 | 3.773 | 4.199 | 4.508 | 4.751 | 4.950 | 5.119 | 5.265 | 5.395 |
| 13 | 3.055 | 3.735 | 4.151 | 4.453 | 4.690 | 4.885 | 5.049 | 5.192 | 5.318 |
| 14 | 3.033 | 3.702 | 4.111 | 4.407 | 4.639 | 4.829 | 4.990 | 5.131 | 5.254 |
| 15 | 3.014 | 3.674 | 4.076 | 4.367 | 4.595 | 4.782 | 4.940 | 5.077 | 5.198 |
| 16 | 2.998 | 3.649 | 4.046 | 4.333 | 4.557 | 4.741 | 4.897 | 5.031 | 5.150 |
| 17 | 2.984 | 3.628 | 4.020 | 4.303 | 4.524 | 4.705 | 4.858 | 4.991 | 5.108 |
| 18 | 2.971 | 3.609 | 3.997 | 4.277 | 4.495 | 4.673 | 4.824 | 4.956 | 5.071 |
| 19 | 2.960 | 3.593 | 3.977 | 4.253 | 4.469 | 4.645 | 4.794 | 4.924 | 5.038 |
| 20 | 2.950 | 3.578 | 3.958 | 4.232 | 4.445 | 4.620 | 4.768 | 4.896 | 5.008 |
| 24 | 2.919 | 3.532 | 3.901 | 4.166 | 4.373 | 4.541 | 4.684 | 4.807 | 4.915 |
| 30 | 2.888 | 3.486 | 3.845 | 4.102 | 4.302 | 4.464 | 4.602 | 4.720 | 4.824 |
| 40 | 2.858 | 3.442 | 3.791 | 4.039 | 4.232 | 4.389 | 4.521 | 4.635 | 4.735 |
| 60 | 2.829 | 3.399 | 3.737 | 3.977 | 4.163 | 4.314 | 4.441 | 4.550 | 4.646 |
| 120 | 2.800 | 3.356 | 3.685 | 3.917 | 4.096 | 4.241 | 4.363 | 4.468 | 4.560 |
| ∞ | 2.772 | 3.314 | 3.633 | 3.858 | 4.030 | 4.170 | 4.286 | 4.387 | 4.474 |

| $\nu$＼n | 11 | 12 | 13 | 14 | 15 | 16 | 17 | 18 | 19 |
|---|---|---|---|---|---|---|---|---|---|
| 1 | 50.59 | 51.96 | 53.20 | 54.33 | 55.36 | 56.32 | 57.22 | 58.04 | 58.83 |
| 2 | 14.39 | 14.75 | 15.08 | 15.38 | 15.65 | 15.91 | 16.14 | 16.37 | 16.57 |
| 3 | 9.717 | 9.946 | 10.15 | 10.35 | 10.53 | 10.69 | 10.84 | 10.98 | 11.11 |
| 4 | 8.027 | 8.208 | 8.373 | 8.525 | 8.664 | 8.794 | 8.914 | 9.028 | 9.134 |
| 5 | 7.168 | 7.324 | 7.466 | 7.596 | 7.717 | 7.828 | 7.932 | 8.030 | 8.122 |
| 6 | 6.649 | 6.789 | 6.917 | 7.034 | 7.143 | 7.244 | 7.338 | 7.426 | 7.508 |
| 7 | 6.302 | 6.431 | 6.550 | 6.658 | 6.759 | 6.852 | 6.939 | 7.020 | 7.097 |
| 8 | 6.054 | 6.175 | 6.287 | 6.389 | 6.483 | 6.571 | 6.653 | 6.729 | 6.802 |
| 9 | 5.867 | 5.983 | 6.089 | 6.186 | 6.276 | 6.359 | 6.437 | 6.510 | 6.579 |
| 10 | 5.722 | 5.833 | 5.935 | 6.028 | 6.114 | 6.194 | 6.269 | 6.339 | 6.405 |
| 11 | 5.605 | 5.713 | 5.811 | 5.901 | 5.984 | 6.062 | 6.134 | 6.202 | 6.265 |
| 12 | 5.511 | 5.615 | 5.710 | 5.798 | 5.878 | 5.953 | 6.023 | 6.089 | 6.151 |
| 13 | 5.431 | 5.533 | 5.625 | 5.711 | 5.789 | 5.862 | 5.931 | 5.995 | 6.055 |
| 14 | 5.364 | 5.463 | 5.554 | 5.637 | 5.714 | 5.786 | 5.852 | 5.915 | 5.974 |
| 15 | 5.306 | 5.404 | 5.493 | 5.574 | 5.649 | 5.720 | 5.785 | 5.846 | 5.904 |
| 16 | 5.256 | 5.352 | 5.439 | 5.520 | 5.593 | 5.662 | 5.727 | 5.786 | 5.843 |
| 17 | 5.212 | 5.307 | 5.392 | 5.471 | 5.544 | 5.612 | 5.675 | 5.734 | 5.790 |
| 18 | 5.174 | 5.267 | 5.352 | 5.429 | 5.501 | 5.568 | 5.630 | 5.688 | 5.743 |
| 19 | 5.140 | 5.231 | 5.315 | 5.391 | 5.462 | 5.528 | 5.589 | 5.647 | 5.701 |
| 20 | 5.108 | 5.199 | 5.282 | 5.357 | 5.427 | 5.493 | 5.553 | 5.610 | 5.663 |
| 24 | 5.012 | 5.099 | 5.179 | 5.251 | 5.319 | 5.381 | 5.439 | 5.494 | 5.545 |
| 30 | 4.917 | 5.001 | 5.077 | 5.147 | 5.211 | 5.271 | 5.327 | 5.379 | 5.429 |
| 40 | 4.824 | 4.904 | 4.977 | 5.044 | 5.106 | 5.163 | 5.216 | 5.266 | 5.313 |
| 60 | 4.732 | 4.808 | 4.878 | 4.942 | 5.001 | 5.056 | 5.107 | 5.154 | 5.199 |
| 120 | 4.641 | 4.714 | 4.781 | 4.842 | 4.898 | 4.950 | 4.998 | 5.044 | 5.086 |
| ∞ | 4.552 | 4.622 | 4.685 | 4.743 | 4.796 | 4.845 | 4.891 | 4.934 | 4.974 |

[a] Degrees of freedom $\nu$, for $n$ $Y$'s. Probability $= \alpha$ for a range exceeding $q_\alpha$.
REPRODUCED from H. L. Harter, "Order Statistics and Their Use in Testing and Estimation, Vol. 1, Tests Based on Range and Studentized Range of Sample from a Normal Population," Aerospace Res. Labs, Wright-Patterson Air Force Base, Ohio, 1969, with kind permission.

TABLE VII *(continued)*

Studentized Range Points $q_\alpha$ , $\alpha = .01$ [a]

| $\nu \backslash n$ | 2 | 3 | 4 | 5 | 6 | 7 | 8 | 9 | 10 |
|---|---|---|---|---|---|---|---|---|---|
| 1 | 90.03 | 135.0 | 164.3 | 185.6 | 202.2 | 215.8 | 227.2 | 237.0 | 245.6 |
| 2 | 14.04 | 19.02 | 22.29 | 24.72 | 26.63 | 28.20 | 29.53 | 30.68 | 31.69 |
| 3 | 8.261 | 10.62 | 12.17 | 13.33 | 14.24 | 15.00 | 15.64 | 16.20 | 16.69 |
| 4 | 6.512 | 8.120 | 9.173 | 9.958 | 10.58 | 11.10 | 11.55 | 11.93 | 12.27 |
| 5 | 5.702 | 6.976 | 7.804 | 8.421 | 8.913 | 9.321 | 9.669 | 9.972 | 10.24 |
| 6 | 5.243 | 6.331 | 7.033 | 7.556 | 7.973 | 8.318 | 8.613 | 8.869 | 9.097 |
| 7 | 4.949 | 5.919 | 6.543 | 7.005 | 7.373 | 7.679 | 7.939 | 8.166 | 8.368 |
| 8 | 4.746 | 5.635 | 6.204 | 6.625 | 6.960 | 7.237 | 7.474 | 7.681 | 7.863 |
| 9 | 4.596 | 5.428 | 5.957 | 6.348 | 6.658 | 6.915 | 7.134 | 7.325 | 7.495 |
| 10 | 4.482 | 5.270 | 5.769 | 6.136 | 6.428 | 6.669 | 6.875 | 7.055 | 7.213 |
| 11 | 4.392 | 5.146 | 5.621 | 5.970 | 6.247 | 6.476 | 6.672 | 6.842 | 6.992 |
| 12 | 4.320 | 5.046 | 5.502 | 5.836 | 6.101 | 6.321 | 6.507 | 6.670 | 6.814 |
| 13 | 4.260 | 4.964 | 5.404 | 5.727 | 5.981 | 6.192 | 6.372 | 6.528 | 6.667 |
| 14 | 4.210 | 4.895 | 5.322 | 5.634 | 5.881 | 6.085 | 6.258 | 6.409 | 6.543 |
| 15 | 4.168 | 4.836 | 5.252 | 5.556 | 5.796 | 5.994 | 6.162 | 6.309 | 6.439 |
| 16 | 4.131 | 4.786 | 5.192 | 5.489 | 5.722 | 5.915 | 6.079 | 6.222 | 6.349 |
| 17 | 4.099 | 4.742 | 5.140 | 5.430 | 5.659 | 5.847 | 6.007 | 6.147 | 6.270 |
| 18 | 4.071 | 4.703 | 5.094 | 5.379 | 5.603 | 5.788 | 5.944 | 6.081 | 6.201 |
| 19 | 4.046 | 4.670 | 5.054 | 5.334 | 5.554 | 5.735 | 5.889 | 6.022 | 6.141 |
| 20 | 4.024 | 4.639 | 5.018 | 5.294 | 5.510 | 5.688 | 5.839 | 5.970 | 6.087 |
| 24 | 3.956 | 4.546 | 4.907 | 5.168 | 5.374 | 5.542 | 5.685 | 5.809 | 5.919 |
| 30 | 3.889 | 4.455 | 4.799 | 5.048 | 5.242 | 5.401 | 5.536 | 5.653 | 5.756 |
| 40 | 3.825 | 4.367 | 4.696 | 4.931 | 5.114 | 5.265 | 5.392 | 5.502 | 5.599 |
| 60 | 3.762 | 4.282 | 4.595 | 4.818 | 4.991 | 5.133 | 5.253 | 5.356 | 5.447 |
| 120 | 3.702 | 4.200 | 4.497 | 4.709 | 4.872 | 5.005 | 5.118 | 5.214 | 5.299 |
| ∞ | 3.643 | 4.120 | 4.403 | 4.603 | 4.757 | 4.882 | 4.987 | 5.078 | 5.157 |

| $\nu \backslash n$ | 11 | 12 | 13 | 14 | 15 | 16 | 17 | 18 | 19 |
|---|---|---|---|---|---|---|---|---|---|
| 1 | 253.2 | 260.0 | 266.2 | 271.8 | 277.0 | 281.8 | 286.3 | 290.4 | 294.3 |
| 2 | 32.59 | 33.40 | 34.13 | 34.81 | 35.43 | 36.00 | 36.53 | 37.03 | 37.50 |
| 3 | 17.13 | 17.53 | 17.89 | 18.22 | 18.52 | 18.81 | 19.07 | 19.32 | 19.55 |
| 4 | 12.57 | 12.84 | 13.09 | 13.32 | 13.53 | 13.73 | 13.91 | 14.08 | 14.24 |
| 5 | 10.48 | 10.70 | 10.89 | 11.08 | 11.24 | 11.40 | 11.55 | 11.68 | 11.81 |
| 6 | 9.301 | 9.485 | 9.653 | 9.808 | 9.951 | 10.08 | 10.21 | 10.32 | 10.43 |
| 7 | 8.548 | 8.711 | 8.860 | 8.997 | 9.124 | 9.242 | 9.353 | 9.456 | 9.554 |
| 8 | 8.027 | 8.176 | 8.312 | 8.436 | 8.552 | 8.659 | 8.760 | 8.854 | 8.943 |
| 9 | 7.647 | 7.784 | 7.910 | 8.025 | 8.132 | 8.232 | 8.325 | 8.412 | 8.495 |
| 10 | 7.356 | 7.485 | 7.603 | 7.712 | 7.812 | 7.906 | 7.993 | 8.076 | 8.153 |
| 11 | 7.128 | 7.250 | 7.362 | 7.465 | 7.560 | 7.649 | 7.732 | 7.809 | 7.883 |
| 12 | 6.943 | 7.060 | 7.167 | 7.265 | 7.356 | 7.441 | 7.520 | 7.594 | 7.665 |
| 13 | 6.791 | 6.903 | 7.006 | 7.101 | 7.188 | 7.269 | 7.345 | 7.417 | 7.485 |
| 14 | 6.664 | 6.772 | 6.871 | 6.962 | 7.047 | 7.126 | 7.199 | 7.268 | 7.333 |
| 15 | 6.555 | 6.660 | 6.757 | 6.845 | 6.927 | 7.003 | 7.074 | 7.142 | 7.204 |
| 16 | 6.462 | 6.564 | 6.658 | 6.744 | 6.823 | 6.898 | 6.967 | 7.032 | 7.093 |
| 17 | 6.381 | 6.480 | 6.572 | 6.656 | 6.734 | 6.806 | 6.873 | 6.937 | 6.997 |
| 18 | 6.310 | 6.407 | 6.497 | 6.579 | 6.655 | 6.725 | 6.792 | 6.854 | 6.912 |
| 19 | 6.247 | 6.342 | 6.430 | 6.510 | 6.585 | 6.654 | 6.719 | 6.780 | 6.837 |
| 20 | 6.191 | 6.285 | 6.371 | 6.450 | 6.523 | 6.591 | 6.654 | 6.714 | 6.771 |
| 24 | 6.017 | 6.106 | 6.186 | 6.261 | 6.330 | 6.394 | 6.453 | 6.510 | 6.563 |
| 30 | 5.849 | 5.932 | 6.008 | 6.078 | 6.143 | 6.203 | 6.259 | 6.311 | 6.361 |
| 40 | 5.686 | 5.764 | 5.835 | 5.900 | 5.961 | 6.017 | 6.069 | 6.119 | 6.165 |
| 60 | 5.528 | 5.601 | 5.667 | 5.728 | 5.785 | 5.837 | 5.886 | 5.931 | 5.974 |
| 120 | 5.375 | 5.443 | 5.505 | 5.562 | 5.614 | 5.662 | 5.708 | 5.750 | 5.790 |
| ∞ | 5.227 | 5.290 | 5.348 | 5.400 | 5.448 | 5.493 | 5.535 | 5.574 | 5.611 |

[a] Degrees of freedom $\nu$, for $n$ $Y$'s. Probability $= \alpha$ for a range exceeding $q_\alpha$ .

TABLE VIII

Single Sample Test for $\sigma_0 = \sigma_1$ versus $\sigma_0 < \sigma_1$, with $\alpha = \beta$[a]

| Sample size | Ratio of $\sigma_1/\sigma_0$ for $\alpha = \beta$ | | | | Multiplier for $\sigma_0^2$ to get $K$ $\alpha = \beta$ | | | |
|---|---|---|---|---|---|---|---|---|
| | .10 | .05 | .02 | .01 | .10 | .05 | .02 | .01 |
| 2 | 13.1 | 31.3 | 92.6 | 206. | 2.71 | 3.84 | 5.41 | 6.64 |
| 3 | 4.67 | 7.63 | 13.9 | 21.4 | 2.30 | 3.00 | 3.91 | 4.60 |
| 4 | 3.27 | 4.71 | 7.29 | 9.93 | 2.08 | 2.60 | 3.28 | 3.78 |
| 5 | 2.70 | 3.65 | 5.22 | 6.69 | 1.94 | 2.37 | 2.92 | 3.32 |
| 6 | 2.40 | 3.11 | 4.22 | 5.22 | 1.85 | 2.21 | 2.68 | 3.02 |
| 7 | 2.20 | 2.76 | 3.64 | 4.39 | 1.77 | 2.10 | 2.51 | 2.80 |
| 8 | 2.06 | 2.55 | 3.26 | 3.86 | 1.72 | 2.01 | 2.37 | 2.64 |
| 9 | 1.96 | 2.38 | 2.99 | 3.49 | 1.67 | 1.94 | 2.27 | 2.51 |
| 10 | 1.88 | 2.26 | 2.79 | 3.22 | 1.63 | 1.88 | 2.19 | 2.41 |
| 11 | 1.81 | 2.16 | 2.63 | 3.01 | 1.60 | 1.83 | 2.12 | 2.32 |
| 12 | 1.76 | 2.07 | 2.50 | 2.85 | 1.57 | 1.79 | 2.06 | 2.25 |
| 13 | 1.72 | 2.01 | 2.40 | 2.71 | 1.55 | 1.75 | 2.00 | 2.18 |
| 14 | 1.68 | 1.95 | 2.31 | 2.60 | 1.52 | 1.72 | 1.96 | 2.13 |
| 15 | 1.64 | 1.90 | 2.24 | 2.50 | 1.50 | 1.69 | 1.92 | 2.08 |
| 16 | 1.61 | 1.86 | 2.17 | 2.42 | 1.49 | 1.67 | 1.88 | 2.04 |
| 17 | 1.59 | 1.82 | 2.12 | 2.35 | 1.47 | 1.64 | 1.85 | 2.00 |
| 18 | 1.57 | 1.78 | 2.07 | 2.28 | 1.46 | 1.62 | 1.82 | 1.97 |
| 19 | 1.55 | 1.75 | 2.02 | 2.23 | 1.44 | 1.60 | 1.80 | 1.93 |
| 20 | 1.53 | 1.73 | 1.98 | 2.18 | 1.43 | 1.59 | 1.77 | 1.90 |
| 21 | 1.51 | 1.70 | 1.95 | 2.13 | 1.42 | 1.57 | 1.75 | 1.88 |
| 22 | 1.50 | 1.68 | 1.91 | 2.09 | 1.41 | 1.56 | 1.73 | 1.85 |
| 23 | 1.48 | 1.66 | 1.88 | 2.05 | 1.40 | 1.54 | 1.71 | 1.83 |
| 24 | 1.47 | 1.64 | 1.86 | 2.02 | 1.39 | 1.53 | 1.69 | 1.81 |
| 25 | 1.46 | 1.62 | 1.83 | 1.99 | 1.38 | 1.52 | 1.68 | 1.79 |
| 26 | 1.44 | 1.61 | 1.81 | 1.96 | 1.38 | 1.51 | 1.66 | 1.77 |
| 27 | 1.43 | 1.59 | 1.79 | 1.93 | 1.37 | 1.50 | 1.65 | 1.76 |
| 28 | 1.42 | 1.58 | 1.77 | 1.91 | 1.36 | 1.49 | 1.63 | 1.74 |
| 29 | 1.41 | 1.56 | 1.75 | 1.89 | 1.35 | 1.48 | 1.62 | 1.72 |
| 30 | 1.41 | 1.55 | 1.73 | 1.87 | 1.35 | 1.47 | 1.61 | 1.71 |
| 31 | 1.40 | 1.54 | 1.72 | 1.84 | 1.34 | 1.46 | 1.60 | 1.70 |
| 40 | 1.34 | 1.46 | 1.60 | 1.71 | 1.30 | 1.40 | 1.52 | 1.60 |
| 50 | 1.30 | 1.40 | 1.52 | 1.61 | 1.27 | 1.35 | 1.46 | 1.53 |
| 60 | 1.27 | 1.36 | 1.46 | 1.54 | 1.24 | 1.32 | 1.41 | 1.48 |
| 70 | 1.25 | 1.33 | 1.42 | 1.49 | 1.22 | 1.30 | 1.38 | 1.44 |
| 80 | 1.23 | 1.30 | 1.39 | 1.45 | 1.21 | 1.28 | 1.36 | 1.41 |
| 90 | 1.21 | 1.28 | 1.36 | 1.42 | 1.20 | 1.26 | 1.33 | 1.38 |
| 100 | 1.20 | 1.26 | 1.34 | 1.39 | 1.19 | 1.24 | 1.31 | 1.36 |

[a] For $n$, seek entry in columns 2–5, $\leqslant \sigma_1/\sigma_0$, giving $n$. Then for this $n$ and $\alpha = \beta$, find in columns 6–9, the multiplier for $\sigma_0^2$ to give $K$. Then $s^2 \leqslant K$, accept null hypothesis; $s^2 > K$, reject null hypothesis.

**TABLE IX**

$Q$ Test for Homogeneity of Variances[a]

| Number $p$ of samples | Degrees of freedom per sample, $v = n - 1$ | | | | | | | | | |
|---|---|---|---|---|---|---|---|---|---|---|
| | 1 | 2 | 3 | 4 | 5 | 6 | 7 | 8 | 10 | 12 |
| 3 | [b] | .863 | .757 | .684 | .631 | .593 | .562 | .539 | .512 | .486 |
| | .915 | .752 | .657 | .596 | .554 | .524 | .501 | .483 | .456 | .438 |
| 4 | .920 | .720 | .605 | .549 | .498 | .461 | .434 | .413 | .383 | .362 |
| | .799 | .612 | .517 | .461 | .425 | .399 | .379 | .364 | .343 | .328 |
| 5 | .828 | .608 | .512 | .443 | .399 | .368 | .346 | .328 | .303 | .287 |
| | .702 | .511 | .423 | .374 | .342 | .320 | .303 | .291 | .274 | .262 |
| 6 | .744 | .539 | .430 | .369 | .334 | .307 | .288 | .271 | .250 | .236 |
| | .621 | .436 | .356 | .312 | .285 | .266 | .252 | .242 | .227 | .217 |
| 7 | .671 | .469 | .372 | .318 | .284 | .261 | .244 | .230 | .212 | .201 |
| | .555 | .379 | .307 | .267 | .243 | .227 | .215 | .206 | .194 | .185 |
| 8 | .609 | .412 | .325 | .276 | .246 | .226 | .210 | .199 | .184 | .174 |
| | .500 | .334 | .268 | .234 | .212 | .198 | .187 | .180 | .169 | .161 |
| 10 | .528 | .333 | .257 | .218 | .194 | .178 | .165 | .157 | .145 | .137 |
| | .415 | .269 | .214 | .185 | .168 | .157 | .149 | .142 | .134 | .128 |
| 12 | .448 | .276 | .211 | .179 | .159 | | | | | |
| | .352 | .223 | .177 | .153 | .139 | | | | | |
| 15 | .365 | .217 | .165 | .140 | .123 | .113 | .106 | .101 | .094 | .089 |
| | .285 | .177 | .140 | .121 | .110 | .102 | .097 | .093 | .088 | .084 |
| 16 | .343 | .202 | .154 | .130 | .115 | | | | | |
| | .268 | .166 | .131 | .113 | .103 | | | | | |
| 20 | .273 | .158 | .120 | .101 | .090 | .083 | .078 | .074 | .069 | .066 |
| | .214 | .131 | .103 | .089 | .081 | .076 | .072 | .068 | .065 | .063 |
| 24 | .224 | .129 | .098 | .082 | .074 | | | | | |
| | .177 | .108 | .085 | .074 | .067 | | | | | |
| 30 | .176 | .100 | .075 | .064 | .058 | .053 | .050 | .048 | .045 | .043 |
| | .140 | .085 | .067 | .058 | .053 | .050 | .047 | .045 | .043 | .041 |
| 32 | .163 | .093 | .070 | .060 | .054 | | | | | |
| | .131 | .079 | .062 | .054 | .050 | | | | | |
| 40 | .127 | .072 | .055 | .047 | .042 | .039 | .037 | .035 | .033 | .032 |
| | .103 | .062 | .049 | .043 | .039 | .037 | .035 | .034 | .032 | .031 |
| 64 | .074 | .042 | .033 | .028 | .025 | | | | | |
| | .062 | .037 | .030 | .026 | .024 | | | | | |

[a] $Q = (\sum_{i=1}^{p} S_i^4)/(\sum_{i=1}^{p} S_i^2)^2$. Upper entry in cell is for $\alpha = .01$ level test, lower entry for $\alpha = .05$ level test.

[b] Not currently available.

TABLE X

Random Numbers

| | | | | | | | | | |
|---|---|---|---|---|---|---|---|---|---|
| 1368 | 9621 | 9151 | 2066 | 1208 | 2664 | 9822 | 6599 | 6911 | 5112 |
| 5953 | 5936 | 2541 | 4011 | 0408 | 3593 | 3679 | 1378 | 5936 | 2651 |
| 7226 | 9466 | 9553 | 7671 | 8599 | 2119 | 5337 | 5953 | 6355 | 6889 |
| 8883 | 3454 | 6773 | 8207 | 5576 | 6386 | 7487 | 0190 | 0867 | 1298 |
| 7022 | 5281 | 1168 | 4099 | 8069 | 8721 | 8353 | 9952 | 8006 | 9045 |
| | | | | | | | | | |
| 4576 | 1853 | 7884 | 2451 | 3488 | 1286 | 4842 | 7719 | 5795 | 3953 |
| 8715 | 1416 | 7028 | 4616 | 3470 | 9938 | 5703 | 0196 | 3465 | 0034 |
| 4011 | 0408 | 2224 | 7626 | 0643 | 1149 | 8834 | 6429 | 8691 | 0143 |
| 1400 | 3694 | 4482 | 3608 | 1238 | 8221 | 5129 | 6105 | 5314 | 8385 |
| 6370 | 1884 | 0820 | 4854 | 9161 | 6509 | 7123 | 4070 | 6759 | 6113 |
| | | | | | | | | | |
| 4522 | 5749 | 8084 | 3932 | 7678 | 3549 | 0051 | 6761 | 6952 | 7041 |
| 7195 | 6234 | 6426 | 7148 | 9945 | 0358 | 3242 | 0519 | 6550 | 1327 |
| 0054 | 0810 | 2937 | 2040 | 2299 | 4198 | 0846 | 3937 | 3986 | 1019 |
| 5166 | 5433 | 0381 | 9686 | 5670 | 5129 | 2103 | 1125 | 3404 | 8785 |
| 1247 | 3793 | 7415 | 7819 | 1783 | 0506 | 4878 | 7673 | 9840 | 6629 |
| | | | | | | | | | |
| 8529 | 7842 | 7203 | 1844 | 8619 | 7404 | 4215 | 9969 | 6948 | 5643 |
| 8973 | 3440 | 4366 | 9242 | 2151 | 0244 | 0922 | 5887 | 4883 | 1177 |
| 9307 | 2959 | 5904 | 9012 | 4951 | 3695 | 4529 | 7197 | 7179 | 3239 |
| 2923 | 4276 | 9467 | 9868 | 2257 | 1925 | 3382 | 7244 | 1781 | 8037 |
| 6372 | 2808 | 1238 | 8098 | 5509 | 4617 | 4099 | 6705 | 2386 | 2830 |
| | | | | | | | | | |
| 6922 | 1807 | 4900 | 5306 | 0411 | 1828 | 8634 | 2331 | 7247 | 3230 |
| 9862 | 8336 | 6453 | 0545 | 6127 | 2741 | 5967 | 8447 | 3017 | 5709 |
| 3371 | 1530 | 5104 | 3076 | 5506 | 3101 | 4143 | 5845 | 2095 | 6127 |
| 6712 | 9402 | 9588 | 7019 | 9248 | 9192 | 4223 | 6555 | 7947 | 2474 |
| 3071 | 8782 | 7157 | 5941 | 8830 | 8563 | 2252 | 8109 | 5880 | 9912 |
| | | | | | | | | | |
| 4022 | 9734 | 7852 | 9096 | 0051 | 7387 | 7056 | 9331 | 1317 | 7833 |
| 9682 | 8892 | 3577 | 0326 | 5306 | 0050 | 8517 | 4376 | 0788 | 5443 |
| 6705 | 2175 | 9904 | 3743 | 1902 | 5393 | 3032 | 8432 | 0612 | 7972 |
| 1872 | 8292 | 2366 | 8603 | 4288 | 6809 | 4357 | 1072 | 6822 | 5611 |
| 2559 | 7534 | 2281 | 7351 | 2064 | 0611 | 9613 | 2000 | 0327 | 6145 |
| | | | | | | | | | |
| 4399 | 3751 | 9783 | 5399 | 5175 | 8894 | 0296 | 9483 | 0400 | 2272 |
| 6074 | 8827 | 2195 | 2532 | 7680 | 4288 | 6807 | 3101 | 6850 | 6410 |
| 5155 | 7186 | 4722 | 6721 | 0838 | 3632 | 5355 | 9369 | 2006 | 7681 |
| 3193 | 2800 | 6184 | 7891 | 9838 | 6123 | 9397 | 4019 | 8389 | 9508 |
| 8610 | 1880 | 7423 | 3384 | 4625 | 6653 | 2900 | 6290 | 9286 | 2396 |
| | | | | | | | | | |
| 4778 | 8818 | 2992 | 6300 | 4239 | 9595 | 4384 | 0611 | 7687 | 2088 |
| 3987 | 1619 | 4164 | 2542 | 4042 | 7799 | 9084 | 0278 | 8422 | 4330 |
| 2977 | 0248 | 2793 | 3351 | 4922 | 8878 | 5703 | 7421 | 2054 | 4391 |
| 1312 | 2919 | 8220 | 7285 | 5902 | 7882 | 1403 | 5354 | 9913 | 7109 |
| 3890 | 7193 | 7799 | 9190 | 3275 | 7840 | 1872 | 6232 | 5295 | 3148 |
| | | | | | | | | | |
| 0793 | 3468 | 8762 | 2492 | 5854 | 8430 | 8472 | 2264 | 9279 | 2128 |
| 2139 | 4552 | 3444 | 6462 | 2524 | 8601 | 3372 | 1848 | 1472 | 9667 |
| 8277 | 9153 | 2880 | 9053 | 6880 | 4284 | 5044 | 8931 | 0861 | 1517 |
| 2236 | 4778 | 6639 | 0862 | 9509 | 2141 | 0208 | 1450 | 1222 | 5281 |
| 8837 | 7686 | 1771 | 3374 | 2894 | 7314 | 6856 | 0440 | 3766 | 6047 |
| | | | | | | | | | |
| 6605 | 6380 | 4599 | 3333 | 0713 | 8401 | 7146 | 8940 | 2629 | 2006 |
| 8399 | 8175 | 3525 | 1646 | 4019 | 8390 | 4344 | 8975 | 4489 | 3423 |
| 8053 | 3046 | 9102 | 4515 | 2944 | 9763 | 3003 | 3408 | 1199 | 2791 |
| 9837 | 9378 | 3237 | 7016 | 7593 | 5958 | 0068 | 3114 | 0456 | 6840 |
| 2557 | 6395 | 9496 | 1884 | 0612 | 8102 | 4402 | 5498 | 0422 | 3335 |

REPRODUCED from D. B. Owen, "Handbook of Statistical Tables," Addison-Wesley, Reading, Massachusetts, 1962, with kind permission of author and publisher.

TABLE X *(continued)*

Random Numbers

| | | | | | | | | | |
|------|------|------|------|------|------|------|------|------|------|
| 2671 | 4690 | 1550 | 2262 | 2597 | 8034 | 0785 | 2978 | 4409 | 0237 |
| 9111 | 0250 | 3275 | 7519 | 9740 | 4577 | 2064 | 0286 | 3398 | 1348 |
| 0391 | 6035 | 9230 | 4999 | 3332 | 0608 | 6113 | 0391 | 5789 | 9926 |
| 2475 | 2144 | 1886 | 2079 | 3004 | 9686 | 5669 | 4367 | 9306 | 2595 |
| 5336 | 5845 | 2095 | 6446 | 5694 | 3641 | 1085 | 8705 | 5416 | 9066 |
| | | | | | | | | | |
| 6808 | 0423 | 0155 | 1652 | 7897 | 4335 | 3567 | 7109 | 9690 | 3739 |
| 8525 | 0577 | 8940 | 9451 | 6726 | 0876 | 3818 | 7607 | 8854 | 3566 |
| 0398 | 0741 | 8787 | 3043 | 5063 | 0617 | 1770 | 5048 | 7721 | 7032 |
| 3623 | 9636 | 3638 | 1406 | 5731 | 3978 | 8068 | 7238 | 9715 | 3363 |
| 0739 | 2644 | 4917 | 8866 | 3632 | 5399 | 5175 | 7422 | 2476 | 2607 |
| | | | | | | | | | |
| 6713 | 3041 | 8133 | 8749 | 8835 | 6745 | 3597 | 3476 | 3816 | 3455 |
| 7775 | 9315 | 0432 | 8327 | 0861 | 1515 | 2297 | 3375 | 3713 | 9174 |
| 8599 | 2122 | 6842 | 9202 | 0810 | 2936 | 1514 | 2090 | 3067 | 3574 |
| 7955 | 3759 | 5254 | 1126 | 5553 | 4713 | 9605 | 7909 | 1658 | 5490 |
| 4766 | 0070 | 7260 | 6033 | 7997 | 0109 | 5993 | 7592 | 5436 | 1727 |
| | | | | | | | | | |
| 5165 | 1670 | 2534 | 8811 | 8231 | 3721 | 7947 | 5719 | 2640 | 1394 |
| 9111 | 0513 | 2751 | 8256 | 2931 | 7783 | 1281 | 6531 | 7259 | 6993 |
| 1667 | 1084 | 7889 | 8963 | 7018 | 8617 | 6381 | 0723 | 4926 | 4551 |
| 2145 | 4587 | 8585 | 2412 | 5431 | 4667 | 1942 | 7238 | 9613 | 2212 |
| 2739 | 5528 | 1481 | 7528 | 9368 | 1823 | 6979 | 2547 | 7268 | 2467 |
| | | | | | | | | | |
| 8769 | 5480 | 9160 | 5354 | 9700 | 1362 | 2774 | 7980 | 9157 | 8788 |
| 6531 | 9435 | 3422 | 2474 | 1475 | 0159 | 3414 | 5224 | 8399 | 5820 |
| 2937 | 4134 | 7120 | 2206 | 5084 | 9473 | 3958 | 7320 | 9878 | 8609 |
| 1581 | 3285 | 3727 | 8924 | 6204 | 0797 | 0882 | 5945 | 9375 | 9153 |
| 6268 | 1045 | 7076 | 1436 | 4165 | 0143 | 0293 | 4190 | 7171 | 7932 |
| | | | | | | | | | |
| 4293 | 0523 | 8625 | 1961 | 1039 | 2856 | 4889 | 4358 | 1492 | 3804 |
| 6936 | 4213 | 3212 | 7229 | 1230 | 0019 | 5998 | 9206 | 6753 | 3762 |
| 5334 | 7641 | 3258 | 3769 | 1362 | 2771 | 6124 | 9813 | 7915 | 8960 |
| 9373 | 1158 | 4418 | 8826 | 5665 | 5896 | 0358 | 4717 | 8232 | 4859 |
| 6968 | 9428 | 8950 | 5346 | 1741 | 2348 | 8143 | 5377 | 7695 | 0685 |
| | | | | | | | | | |
| 4229 | 0587 | 8794 | 4009 | 9691 | 4579 | 3302 | 7673 | 9629 | 5246 |
| 3807 | 7785 | 7097 | 5701 | 6639 | 0723 | 4819 | 0900 | 2713 | 7650 |
| 4891 | 8829 | 1642 | 2155 | 0796 | 0466 | 2946 | 2970 | 9143 | 6590 |
| 1055 | 2968 | 7911 | 7479 | 8199 | 9735 | 8271 | 5339 | 7058 | 2964 |
| 2983 | 2345 | 0568 | 4125 | 0894 | 8302 | 0506 | 6761 | 7706 | 4310 |
| | | | | | | | | | |
| 4026 | 3129 | 2968 | 8053 | 2797 | 4022 | 9838 | 9611 | 0975 | 2437 |
| 4075 | 0260 | 4256 | 0337 | 2355 | 9371 | 2954 | 6021 | 5783 | 2827 |
| 8488 | 5450 | 1327 | 7358 | 2034 | 8060 | 1788 | 6913 | 6123 | 9405 |
| 1976 | 1749 | 5742 | 4098 | 5887 | 4567 | 6064 | 2777 | 7830 | 5668 |
| 2793 | 4701 | 9466 | 9554 | 8294 | 2160 | 7486 | 1557 | 4769 | 2781 |
| | | | | | | | | | |
| 0916 | 6272 | 6825 | 7188 | 9611 | 1181 | 2301 | 5516 | 5451 | 6832 |
| 5961 | 1149 | 7946 | 1950 | 2010 | 0600 | 5655 | 0796 | 0569 | 4365 |
| 3222 | 4189 | 1891 | 8172 | 8731 | 4769 | 2782 | 1325 | 4238 | 9279 |
| 1176 | 7834 | 4600 | 9992 | 9449 | 5824 | 5344 | 1008 | 6678 | 1921 |
| 2369 | 8971 | 2314 | 4806 | 5071 | 8908 | 8274 | 4936 | 3357 | 4441 |
| | | | | | | | | | |
| 0041 | 4329 | 9265 | 0352 | 4764 | 9070 | 7527 | 7791 | 1094 | 2008 |
| 0803 | 8302 | 6814 | 2422 | 6351 | 0637 | 0514 | 0246 | 1845 | 8594 |
| 9965 | 7804 | 3930 | 8803 | 0268 | 1426 | 3130 | 3613 | 3947 | 8086 |
| 0011 | 2387 | 3148 | 7559 | 4216 | 2946 | 2865 | 6333 | 1916 | 2259 |
| 1767 | 9871 | 3914 | 5790 | 5287 | 7915 | 8959 | 1346 | 5482 | 9251 |

**TABLE XI**

Brief Table of Poisson Distribution

$$P(y) = \sum_{i=0}^{y} p(i \mid \mu) \qquad (5.37)$$

| y | | | | | $\mu$ | | | | |
|---|---|---|---|---|---|---|---|---|---|
| | .1 | .2 | .3 | .4 | .5 | .6 | .7 | .8 | .9 |
| 0 | .90484 | .81873 | .74082 | .67032 | .60653 | .54881 | .49659 | .44933 | .40657 |
| 1 | .99532 | .98248 | .96306 | .93845 | .90980 | .87810 | .84420 | .80879 | .77248 |
| 2 | .99985 | .99885 | .99640 | .99207 | .98561 | .97688 | .96586 | .95258 | .93714 |
| 3 | 1.00000 | .99994 | .99973 | .99922 | .99825 | .99664 | .99425 | .99092 | .98654 |
| 3 | | 1.00000 | .99998 | .99994 | .99983 | .99961 | .99921 | .99859 | .99766 |
| 5 | | | 1.00000 | 1.00000 | .99999 | .99996 | .99991 | .99982 | .99966 |
| 6 | | | | | 1.00000 | 1.00000 | .99999 | .99998 | .99996 |
| 7 | | | | | | | 1.00000 | 1.00000 | 1.00000 |

| y | 1.0 | 1.5 | 2.0 | 2.5 | 3.0 | 3.5 | 4.0 | 4.5 | 5.0 |
|---|---|---|---|---|---|---|---|---|---|
| 0 | .36788 | .22313 | .13534 | .08208 | .04979 | .03020 | .01832 | .01111 | .00674 |
| 1 | .73576 | .55783 | .40601 | .28730 | .19915 | .13589 | .09158 | .06110 | .04043 |
| 2 | .91970 | .80885 | .67668 | .54381 | .42319 | .32085 | .23810 | .17358 | .12465 |
| 3 | .98101 | .93436 | .85712 | .75758 | .64723 | .53663 | .43347 | .34230 | .26503 |
| 4 | .99634 | .98142 | .94735 | .89118 | .81526 | .72544 | .62884 | .53210 | .44049 |
| 5 | .99941 | .99554 | .98344 | .95798 | .91608 | .85761 | .78513 | .70293 | .61596 |
| 6 | .99992 | .99907 | .99547 | .98581 | .96649 | .93471 | .88933 | .83105 | .76218 |
| 7 | .99999 | .99983 | .99890 | .99575 | .98810 | .97326 | .94887 | .91341 | .86663 |
| 8 | 1.00000 | .99997 | .99976 | .99886 | .99620 | .99013 | .97864 | .95974 | .93191 |
| 9 | | 1.00000 | .99995 | .99972 | .99890 | .99669 | .99187 | .98291 | .96817 |
| 10 | | | .99999 | .99994 | .99971 | .99898 | .99716 | .99333 | .98630 |
| 11 | | | 1.00000 | .99999 | .99993 | .99971 | .99908 | .99760 | .99455 |
| 12 | | | | 1.00000 | .99998 | .99992 | .99973 | .99919 | .99798 |
| 13 | | | | | 1.00000 | .99998 | .99992 | .99975 | .99930 |
| 14 | | | | | | 1.00000 | .99998 | .99993 | .99977 |
| 15 | | | | | | | 1.00000 | .99998 | .99993 |
| 16 | | | | | | | | .99999 | .99998 |
| 17 | | | | | | | | 1.00000 | .99999 |
| 18 | | | | | | | | | 1.00000 |

**TABLE XI** *(continued)*

|    | 6.0 | 7.0 | 8.0 | 9.0 | 10.0 | 11.0 | 12.0 | 13.0 | 14.0 |
|----|------|------|------|------|------|------|------|------|------|
| 0  | .00248 | .00091 | .00034 | .00012 | .00005 | .00002 | .00001 | .00000 | .00000 |
| 1  | .01735 | .00730 | .00302 | .00123 | .00050 | .00020 | .00008 | .00003 | .00001 |
| 2  | .06197 | .02964 | .01375 | .00623 | .00277 | .00121 | .00052 | .00022 | .00009 |
| 3  | .15120 | .08177 | .04238 | .02123 | .01034 | .00492 | .00229 | .00105 | .00047 |
| 4  | .28506 | .17299 | .09963 | .05496 | .02925 | .01510 | .00760 | .00374 | .00181 |
| 5  | .44568 | .30071 | .19124 | .11569 | .06709 | .03752 | .02034 | .01073 | .00553 |
| 6  | .60630 | .44971 | .31337 | .20678 | .13014 | .07861 | .04582 | .02589 | .01423 |
| 7  | .74398 | .59871 | .45296 | .32390 | .22022 | .14319 | .08950 | .05403 | .03162 |
| 8  | .84724 | .72909 | .59255 | .45565 | .33282 | .23199 | .15503 | .09976 | .06206 |
| 9  | .91608 | .83050 | .71662 | .58741 | .45793 | .34051 | .24239 | .16581 | .10940 |
| 10 | .95738 | .90148 | .81589 | .70599 | .58304 | .45989 | .34723 | .25168 | .17568 |
| 11 | .97991 | .94665 | .88808 | .80301 | .69678 | .57927 | .46160 | .35316 | .26004 |
| 12 | .99117 | .97300 | .93620 | .87577 | .79156 | .68870 | .57597 | .46310 | .35846 |
| 13 | .99637 | .98719 | .96582 | .92615 | .86446 | .78129 | .68154 | .57304 | .46445 |
| 14 | .99860 | .99428 | .98274 | .95853 | .91654 | .85404 | .77202 | .67513 | .57044 |
| 15 | .99949 | .99759 | .99177 | .97796 | .95126 | .90740 | .84442 | .76361 | .66936 |
| 16 | .99983 | .99904 | .99628 | .98889 | .97296 | .94408 | .89871 | .83549 | .75592 |
| 17 | .99994 | .99964 | .99841 | .99468 | .98572 | .96781 | .93703 | .89046 | .82720 |
| 18 | .99998 | .99987 | .99935 | .99757 | .99281 | .98231 | .96258 | .93017 | .88264 |
| 19 | .99999 | .99996 | .99975 | .99894 | .99655 | .99071 | .97872 | .95733 | .92350 |
| 20 | 1.00000 | .99999 | .99991 | .99956 | .99841 | .99533 | .98840 | .97499 | .95209 |
| 21 |  | 1.00000 | .99997 | .99983 | .99930 | .99775 | .99393 | .98592 | .97116 |
| 22 |  |  | .99999 | .99993 | .99970 | .99896 | .99695 | .99238 | .98329 |
| 23 |  |  | 1.00000 | .99998 | .99988 | .99954 | .99853 | .99603 | .99067 |
| 24 |  |  |  | .99999 | .99995 | .99980 | .99931 | .99801 | .99498 |
| 25 |  |  |  | 1.00000 | .99998 | .99992 | .99969 | .99903 | .99739 |

TABLE XII

Brief Table of Binomial Distribution $P(y) = \sum_{i=0}^{p} p(i \mid n, \phi)$    (5.31), (5.32)

| $n$ | $y$ | $\phi$ | | | | | | | |
|---|---|---|---|---|---|---|---|---|---|
| | | .05 | .10 | .15 | .20 | .25 | .30 | .40 | .50 |
| 5 | 0 | .77378 | .59049 | .44371 | .32768 | .23730 | .16807 | .07776 | .03125 |
| | 1 | .97741 | .91854 | .83521 | .73728 | .63281 | .52822 | .33696 | .18750 |
| | 2 | .99884 | .99144 | .97339 | .94208 | .89648 | .83692 | .68256 | .50000 |
| | 3 | .99997 | .99954 | .99777 | .99328 | .98438 | .96922 | .91296 | .81250 |
| | 4 | 1.00000 | .99999 | .99992 | .99968 | .99902 | .99757 | .98976 | .96875 |
| | 5 | | 1.00000 | 1.00000 | 1.00000 | 1.00000 | 1.00000 | 1.00000 | 1.00000 |
| 10 | 0 | .59874 | .34868 | .19687 | .10737 | .05631 | .02825 | .00605 | .00098 |
| | 1 | .91386 | .73610 | .54430 | .37581 | .24403 | .14931 | .04636 | .01074 |
| | 2 | .98850 | .92981 | .82020 | .67780 | .52559 | .38278 | .16729 | .05469 |
| | 3 | .99897 | .98720 | .95003 | .87913 | .77588 | .64961 | .38228 | .17188 |
| | 4 | .99994 | .99837 | .99013 | .96721 | .92187 | .84973 | .63310 | .37695 |
| | 5 | 1.00000 | .99985 | .99862 | .99363 | .98027 | .95265 | .83376 | .62305 |
| | 6 | | .99999 | .99987 | .99914 | .99649 | .98941 | .94524 | .82812 |
| | 7 | | 1.00000 | .99999 | .99992 | .99958 | .99841 | .98771 | .94531 |
| | 8 | | | 1.00000 | 1.00000 | .99997 | .99986 | .99832 | .98926 |
| | 9 | | | | | 1.00000 | .99999 | .99990 | .99902 |
| | 10 | | | | | | 1.00000 | 1.00000 | 1.00000 |
| 15 | 0 | .46329 | .20589 | .08735 | .03518 | .01336 | .00475 | .00047 | .00003 |
| | 1 | .82905 | .54904 | .31859 | .16713 | .08018 | .03527 | .00517 | .00049 |
| | 2 | .96380 | .81594 | .60423 | .39802 | .23609 | .12683 | .02711 | .00369 |
| | 3 | .99453 | .94444 | .82266 | .64816 | .46129 | .29687 | .09050 | .01758 |
| | 4 | .99939 | .98728 | .93829 | .83577 | .68649 | .51549 | .21728 | .05923 |
| | 5 | .99995 | .99775 | .98319 | .93895 | .85163 | .72162 | .40322 | .15088 |
| | 6 | 1.00000 | .99969 | .99639 | .98194 | .94338 | .86886 | .60981 | .30362 |
| | 7 | | .99997 | .99939 | .99576 | .98270 | .94999 | .78690 | .50000 |
| | 8 | | 1.00000 | .99992 | .99922 | .99581 | .98476 | .90495 | .69638 |
| | 9 | | | .99999 | .99989 | .99921 | .99635 | .96617 | .84912 |
| | 10 | | | 1.00000 | .99999 | .99988 | .99933 | .99065 | .94077 |
| | 11 | | | | 1.00000 | .99999 | .99991 | .99807 | .98242 |
| | 12 | | | | | 1.00000 | .99999 | .99972 | .99631 |
| | 13 | | | | | | 1.00000 | .99997 | .99951 |
| | 14 | | | | | | | 1.00000 | .99997 |
| | 15 | | | | | | | | 1.00000 |
| 20 | 0 | .35849 | .12158 | .03876 | .01153 | .00317 | .00080 | .00004 | .00000 |
| | 1 | .73584 | .39175 | .17556 | .06918 | .02431 | .00764 | .00052 | .00002 |
| | 2 | .92452 | .67693 | .40490 | .20608 | .09126 | .03548 | .00361 | .00020 |
| | 3 | .98410 | .86705 | .64773 | .41145 | .22516 | .10709 | .01596 | .00129 |
| | 4 | .99743 | .95683 | .82985 | .62965 | .41484 | .23751 | .05095 | .00591 |
| | 5 | .99967 | .98875 | .93269 | .80421 | .61717 | .41637 | .12560 | .02069 |
| | 6 | .99997 | .99761 | .97806 | .91331 | .78578 | .60801 | .25001 | .05766 |
| | 7 | 1.00000 | .99958 | .99408 | .96786 | .89819 | .77227 | .41589 | .13159 |
| | 8 | | .99994 | .99867 | .99002 | .95907 | .88667 | .59560 | .25172 |
| | 9 | | .99999 | .99975 | .99741 | .98614 | .95204 | .75534 | .41190 |
| | 10 | | 1.00000 | .99996 | .99944 | .99606 | .98286 | .87248 | .58810 |
| | 11 | | | 1.00000 | .99990 | .99906 | .99486 | .94347 | .74828 |
| | 12 | | | | .99998 | .99982 | .99872 | .97897 | .86841 |
| | 13 | | | | 1.00000 | .99997 | .99974 | .99353 | .94234 |
| | 14 | | | | | 1.00000 | .99996 | .99839 | .97931 |
| | 15 | | | | | | .99999 | .99968 | .99409 |
| | 16 | | | | | | 1.00000 | .99995 | .99871 |
| | 17 | | | | | | | .99999 | .99980 |
| | 18 | | | | | | | 1.00000 | .99998 |
| | 19 | | | | | | | | 1.00000 |

**TABLE XIII**

Parameters for General System of Distributions, (6.52)[a]

| $\alpha_3$ | $\alpha_4$ | $c$ | $k$ | $\mu$ | $\sigma$ |
|---|---|---|---|---|---|
| .00 | 2.8 | 3.939 | 19.865 | .4275 | .1243 |
|  | 3.0 | 4.874 | 6.158 | .6447 | .1620 |
|  | 3.2 | 6.065 | 3.745 | .7673 | .1644 |
|  | 3.4 | 7.696 | 2.701 | .8507 | .1519 |
|  | 3.6 | 10.182 | 2.090 | .9111 | .1300 |
| .10 | 2.8 | 3.520 | 19.606 | .3901 | .1257 |
|  | 3.0 | 4.297 | 6.283 | .6077 | .1714 |
|  | 3.3 | 5.832 | 3.235 | .7830 | .1776 |
|  | 3.6 | 8.298 | 2.164 | .8896 | .1539 |
|  | 3.9 | 13.716 | 1.549 | .9595 | .1097 |
| .20 | 2.8 | 3.071 | 30.547 | .2957 | .1069 |
|  | 3.0 | 3.705 | 7.244 | .5420 | .1737 |
|  | 3.3 | 4.895 | 3.548 | .7347 | .1944 |
|  | 3.6 | 6.640 | 2.353 | .8538 | .1803 |
|  | 3.9 | 9.858 | 1.694 | .9352 | .1447 |
| .30 | 3.0 | 3.140 | 9.872 | .4411 | .1617 |
|  | 3.3 | 4.058 | 4.168 | .6636 | .2054 |
|  | 3.6 | 5.289 | 2.684 | .8000 | .2054 |
|  | 3.9 | 7.227 | 1.933 | .8951 | .1818 |
|  | 4.2 | 11.492 | 1.410 | .9648 | .1353 |
| .40 | 3.0 | 2.625 | 19.800 | .2887 | .1214 |
|  | 3.3 | 3.341 | 5.369 | .5638 | .2039 |
|  | 3.6 | 4.226 | 3.223 | .7240 | .2235 |
|  | 3.9 | 5.449 | 2.289 | .8353 | .2152 |
|  | 4.2 | 7.492 | 1.706 | .9190 | .1863 |
| .50 | 3.2 | 2.538 | 12.523 | .3353 | .1477 |
|  | 3.5 | 3.162 | 4.927 | .5653 | .2172 |
|  | 3.8 | 3.917 | 3.156 | .7134 | .2378 |
|  | 4.1 | 4.925 | 2.306 | .8211 | .2332 |
|  | 4.4 | 6.514 | 1.752 | .9054 | .2093 |
| .60 | 3.2 | 2.066 | 190.007 | .0700 | .0356 |
|  | 3.6 | 2.738 | 5.983 | .4836 | .2086 |
|  | 4.0 | 3.549 | 3.253 | .6859 | .2503 |
|  | 4.4 | 4.673 | 2.219 | .8232 | .2484 |
|  | 4.8 | 6.748 | 1.574 | .9287 | .2141 |
| .70 | 3.5 | 2.065 | 20.670 | .2080 | .1087 |
|  | 3.9 | 2.656 | 5.246 | .5017 | .2258 |
|  | 4.3 | 3.354 | 3.155 | .6815 | .2641 |
|  | 4.7 | 4.292 | 2.238 | .8105 | .2656 |
|  | 5.1 | 5.896 | 1.635 | .9141 | .2386 |
| .80 | 3.8 | 2.005 | 14.088 | .2434 | .1325 |
|  | 4.3 | 2.653 | 4.406 | .5413 | .2494 |
|  | 4.8 | 3.448 | 2.717 | .7260 | .2823 |
|  | 5.3 | 4.651 | 1.897 | .8616 | .2727 |
|  | 5.8 | 8.262 | 1.204 | .9858 | .2044 |

[a] See third additional reference Chapter 6.

**TABLE XIV**

Common Logarithms of Factorials, $n!$, 1–400

|  | 0 | 1 | 2 | 3 | 4 | 5 | 6 | 7 | 8 | 9 |
|---|---|---|---|---|---|---|---|---|---|---|
| 0 | 0.0000 | 0.0000 | .3010 | .7782 | 1.3802 | 2.0792 | 2.8573 | 3.7024 | 4.6055 | 5.5598 |
| 10 | 6.5598 | 7.6012 | 8.6803 | 9.7943 | 10.9404 | 12.1165 | 13.3206 | 14.5511 | 15.8063 | 17.0851 |
| 20 | 18.3861 | 19.7083 | 21.0508 | 22.4125 | 23.7927 | 25.1906 | 26.6056 | 28.0370 | 29.4841 | 30.9465 |
| 30 | 32.4237 | 33.9150 | 35.4202 | 36.9387 | 38.4702 | 40.0142 | 41.5705 | 43.1387 | 44.7185 | 46.3096 |
| 40 | 47.9116 | 49.5244 | 51.1477 | 52.7811 | 54.4246 | 56.0778 | 57.7406 | 59.4127 | 61.0939 | 62.7841 |
| 50 | 64.4831 | 66.1906 | 67.9066 | 69.6309 | 71.3633 | 73.1037 | 74.8519 | 76.6077 | 78.3712 | 80.1420 |
| 60 | 81.9202 | 83.7055 | 85.4979 | 87.2972 | 89.1034 | 90.9163 | 92.7359 | 94.5619 | 96.3945 | 98.2333 |
| 70 | 100.0784 | 101.9297 | 103.7870 | 105.6503 | 107.5196 | 109.3946 | 111.2754 | 113.1619 | 115.0540 | 116.9516 |
| 80 | 118.8547 | 120.7632 | 122.6770 | 124.5961 | 126.5204 | 128.4498 | 130.3843 | 132.3238 | 134.2683 | 136.2177 |
| 90 | 138.1719 | 140.1310 | 142.0948 | 144.0632 | 146.0364 | 148.0141 | 149.9964 | 151.9831 | 153.9744 | 155.9700 |
| 100 | 157.9700 | 159.9743 | 161.9829 | 163.9958 | 166.0128 | 168.0340 | 170.0593 | 172.0887 | 174.1221 | 176.1595 |
| 110 | 178.2009 | 180.2462 | 182.2955 | 184.3485 | 186.4054 | 188.4661 | 190.5306 | 192.5988 | 194.6707 | 196.7462 |
| 120 | 198.8254 | 200.9082 | 202.9945 | 205.0844 | 207.1779 | 209.2748 | 211.3751 | 213.4790 | 215.5862 | 217.6967 |
| 130 | 219.8107 | 221.9280 | 224.0485 | 226.1724 | 228.2995 | 230.4298 | 232.5634 | 234.7001 | 236.8400 | 238.9830 |
| 140 | 241.1291 | 243.2783 | 245.4306 | 247.5860 | 249.7443 | 251.9057 | 254.0700 | 256.2374 | 258.4076 | 260.5808 |
| 150 | 262.7569 | 264.9359 | 267.1177 | 269.3024 | 271.4899 | 273.6803 | 275.8734 | 278.0693 | 280.2679 | 282.4693 |
| 160 | 284.6735 | 286.8803 | 289.0898 | 291.3020 | 293.5168 | 295.7343 | 297.9544 | 300.1771 | 302.4024 | 304.6303 |
| 170 | 306.8608 | 309.0938 | 311.3293 | 313.5674 | 315.8079 | 318.0509 | 320.2965 | 322.5444 | 324.7948 | 327.0477 |
| 180 | 329.3030 | 331.5606 | 333.8207 | 336.0832 | 338.3480 | 340.6152 | 342.8847 | 345.1565 | 347.4307 | 349.7071 |
| 190 | 351.9859 | 354.2669 | 356.5502 | 358.8358 | 361.1236 | 363.4136 | 365.7059 | 368.0003 | 370.2970 | 372.5959 |
| 200 | 374.8969 | 377.2001 | 379.5054 | 381.8129 | 384.1226 | 386.4343 | 388.7482 | 391.0642 | 393.3822 | 395.7024 |
| 210 | 398.0246 | 400.3489 | 402.6752 | 405.0036 | 407.3340 | 409.6664 | 412.0009 | 414.3373 | 416.6758 | 419.0162 |
| 220 | 421.2587 | 423.7031 | 426.0494 | 428.3977 | 430.7480 | 433.1002 | 435.4543 | 437.8103 | 440.1682 | 442.5281 |
| 230 | 444.8898 | 447.2534 | 449.6189 | 451.9862 | 454.3555 | 456.7265 | 459.0994 | 461.4742 | 463.8508 | 466.2292 |
| 240 | 468.6094 | 470.9914 | 473.3752 | 475.7608 | 478.1482 | 480.5374 | 482.9283 | 485.3210 | 487.7154 | 490.1116 |

| | 0 | 1 | 2 | 3 | 4 | 5 | 6 | 7 | 8 | 9 |
|---|---|---|---|---|---|---|---|---|---|---|
| 250 | 492.5096 | 494.9093 | 497.3107 | 499.7138 | 502.1186 | 504.5252 | 506.9334 | 509.3433 | 511.7549 | 514.1682 |
| 260 | 516.5832 | 518.9999 | 521.4182 | 523.8381 | 526.2597 | 528.6830 | 531.1078 | 533.5344 | 535.9625 | 538.3922 |
| 270 | 540.8236 | 543.2566 | 545.6912 | 548.1273 | 550.5651 | 553.0044 | 555.4453 | 557.8878 | 560.3318 | 562.7774 |
| 280 | 565.2246 | 567.6733 | 570.1235 | 572.5753 | 575.0287 | 577.4835 | 579.9399 | 582.3977 | 584.8571 | 587.3180 |
| 290 | 589.7804 | 592.2443 | 594.7097 | 597.1766 | 599.6449 | 602.1147 | 604.5860 | 607.0588 | 609.5330 | 612.0087 |
| 300 | 614.4858 | 616.9644 | 619.4444 | 621.9258 | 624.4087 | 626.8930 | 629.3787 | 631.8659 | 634.3544 | 636.8444 |
| 310 | 639.3357 | 641.8285 | 644.3226 | 646.8182 | 649.3151 | 651.8134 | 654.3131 | 656.8142 | 659.3166 | 661.8204 |
| 320 | 664.3255 | 666.8320 | 669.3399 | 671.8491 | 674.3596 | 676.8715 | 679.3847 | 681.8993 | 684.4152 | 686.9324 |
| 330 | 689.4509 | 691.9707 | 694.4918 | 697.0143 | 699.5380 | 702.0631 | 704.5894 | 707.1170 | 709.6460 | 712.1762 |
| 340 | 714.7076 | 717.2404 | 719.7744 | 722.3097 | 724.8463 | 727.3841 | 729.9232 | 732.4635 | 735.0051 | 737.5479 |
| 350 | 740.0920 | 742.6373 | 745.1838 | 747.7316 | 750.2806 | 752.8308 | 755.3823 | 757.9349 | 760.4888 | 763.0439 |
| 360 | 765.6002 | 768.1577 | 770.7164 | 773.2764 | 775.8375 | 778.3997 | 780.9632 | 783.5279 | 786.0937 | 788.6608 |
| 370 | 791.2290 | 793.7983 | 796.3689 | 798.9406 | 801.5135 | 804.0875 | 806.6627 | 809.2390 | 811.8165 | 814.3952 |
| 380 | 816.9749 | 819.5559 | 822.1379 | 824.7211 | 827.3055 | 829.8909 | 832.4775 | 835.0652 | 837.6540 | 840.2440 |
| 390 | 842.8351 | 845.4272 | 848.0205 | 850.6149 | 853.2104 | 855.8070 | 858.4047 | 861.0035 | 863.6034 | 866.2044 |
| 400 | 868.8064 | | | | | | | | | |

## TABLE XV

Common Logarithms of Numbers to Four Decimal Places

| N | L. 0 | 1 | 2 | 3 | 4 | 5 | 6 | 7 | 8 | 9 | Proportional Parts 1 2 3 4 5 |
|---|------|---|---|---|---|---|---|---|---|---|---|
| 10 | 0000 | 0043 | 0086 | 0128 | 0170 | 0212 | 0253 | 0294 | 0334 | 0374 | 4 8 12 17 21 |
| 11 | 0414 | 0453 | 0492 | 0531 | 0569 | 0607 | 0645 | 0682 | 0719 | 0755 | 4 8 11 15 19 |
| 12 | 0792 | 0828 | 0864 | 0899 | 0934 | 0969 | 1004 | 1038 | 1072 | 1106 | 3 7 10 14 17 |
| 13 | 1139 | 1173 | 1206 | 1239 | 1271 | 1303 | 1335 | 1367 | 1399 | 1430 | 3 6 10 13 16 |
| 14 | 1461 | 1492 | 1523 | 1553 | 1584 | 1614 | 1644 | 1673 | 1703 | 1732 | 3 6 9 12 15 |
| 15 | 1761 | 1790 | 1818 | 1847 | 1875 | 1903 | 1931 | 1959 | 1987 | 2014 | 3 6 8 11 14 |
| 16 | 2041 | 2068 | 2095 | 2122 | 2148 | 2175 | 2201 | 2227 | 2253 | 2279 | 3 5 8 11 13 |
| 17 | 2304 | 2330 | 2355 | 2380 | 2405 | 2430 | 2455 | 2480 | 2504 | 2529 | 2 5 7 10 12 |
| 18 | 2553 | 2577 | 2601 | 2625 | 2648 | 2672 | 2695 | 2718 | 2742 | 2765 | 2 5 7 9 12 |
| 19 | 2788 | 2810 | 2833 | 2856 | 2878 | 2900 | 2923 | 2945 | 2967 | 2989 | 2 4 7 9 11 |
| 20 | 3010 | 3032 | 3054 | 3075 | 3096 | 3118 | 3139 | 3160 | 3181 | 3201 | 2 4 6 8 11 |
| 21 | 3222 | 3243 | 3263 | 3284 | 3304 | 3324 | 3345 | 3365 | 3385 | 3404 | 2 4 6 8 10 |
| 22 | 3424 | 3444 | 3464 | 3483 | 3502 | 3522 | 3541 | 3560 | 3579 | 3598 | 2 4 6 8 10 |
| 23 | 3617 | 3636 | 3655 | 3674 | 3692 | 3711 | 3729 | 3747 | 3766 | 3784 | 2 4 6 7 9 |
| 24 | 3802 | 3820 | 3838 | 3856 | 3874 | 3892 | 3909 | 3927 | 3945 | 3962 | 2 4 5 7 9 |
| 25 | 3979 | 3997 | 4014 | 4031 | 4048 | 4065 | 4082 | 4099 | 4116 | 4133 | 2 4 5 7 9 |
| 26 | 4150 | 4166 | 4183 | 4200 | 4216 | 4232 | 4249 | 4265 | 4281 | 4298 | 2 3 5 7 8 |
| 27 | 4314 | 4330 | 4346 | 4362 | 4378 | 4393 | 4409 | 4425 | 4440 | 4456 | 2 3 5 6 8 |
| 28 | 4472 | 4487 | 4502 | 4518 | 4533 | 4548 | 4564 | 4579 | 4594 | 4609 | 2 3 5 6 8 |
| 29 | 4624 | 4639 | 4654 | 4669 | 4683 | 4698 | 4713 | 4728 | 4742 | 4757 | 1 3 4 6 7 |
| 30 | 4771 | 4786 | 4800 | 4814 | 4829 | 4843 | 4857 | 4871 | 4886 | 4900 | 1 3 4 6 7 |
| 31 | 4914 | 4928 | 4942 | 4955 | 4969 | 4983 | 4997 | 5011 | 5024 | 5038 | 1 3 4 5 7 |
| 32 | 5051 | 5065 | 5079 | 5092 | 5105 | 5119 | 5132 | 5145 | 5159 | 5172 | 1 3 4 5 7 |
| 33 | 5185 | 5198 | 5211 | 5224 | 5237 | 5250 | 5263 | 5276 | 5289 | 5302 | 1 3 4 5 7 |
| 34 | 5315 | 5328 | 5340 | 5353 | 5366 | 5378 | 5391 | 5403 | 5416 | 5428 | 1 2 4 5 6 |
| 35 | 5441 | 5453 | 5465 | 5478 | 5490 | 5502 | 5514 | 5527 | 5539 | 5551 | 1 2 4 5 6 |
| 36 | 5563 | 5575 | 5587 | 5599 | 5611 | 5623 | 5635 | 5647 | 5658 | 5670 | 1 2 4 5 6 |
| 37 | 5682 | 5694 | 5705 | 5717 | 5729 | 5740 | 5752 | 5763 | 5775 | 5786 | 1 2 4 5 6 |
| 38 | 5798 | 5809 | 5821 | 5832 | 5843 | 5855 | 5866 | 5877 | 5888 | 5899 | 1 2 3 5 6 |
| 39 | 5911 | 5922 | 5933 | 5944 | 5955 | 5966 | 5977 | 5988 | 5999 | 6010 | 1 2 3 4 5 |
| 40 | 6021 | 6031 | 6042 | 6053 | 6064 | 6075 | 6085 | 6096 | 6107 | 6117 | 1 2 3 4 5 |
| 41 | 6128 | 6138 | 6149 | 6160 | 6170 | 6180 | 6191 | 6201 | 6212 | 6222 | 1 2 3 4 5 |
| 42 | 6232 | 6243 | 6253 | 6263 | 6274 | 6284 | 6294 | 6304 | 6314 | 6325 | 1 2 3 4 5 |
| 43 | 6335 | 6345 | 6355 | 6365 | 6375 | 6385 | 6395 | 6405 | 6415 | 6425 | 1 2 3 4 5 |
| 44 | 6435 | 6444 | 6454 | 6464 | 6474 | 6484 | 6493 | 6503 | 6513 | 6522 | 1 2 3 4 5 |
| 45 | 6532 | 6542 | 6551 | 6561 | 6571 | 6580 | 6590 | 6599 | 6609 | 6618 | 1 2 3 4 5 |
| 46 | 6628 | 6637 | 6646 | 6656 | 6665 | 6675 | 6684 | 6693 | 6702 | 6712 | 1 2 3 4 5 |
| 47 | 6721 | 6730 | 6739 | 6749 | 6758 | 6767 | 6776 | 6785 | 6794 | 6803 | 1 2 3 4 5 |
| 48 | 6812 | 6821 | 6830 | 6839 | 6848 | 6857 | 6866 | 6875 | 6884 | 6893 | 1 2 3 4 5 |
| 49 | 6902 | 6911 | 6920 | 6928 | 6937 | 6946 | 6955 | 6964 | 6972 | 6981 | 1 2 3 4 4 |
| 50 | 6990 | 6998 | 7007 | 7016 | 7024 | 7033 | 7042 | 7050 | 7059 | 7067 | 1 2 3 3 4 |
| 51 | 7076 | 7084 | 7093 | 7101 | 7110 | 7118 | 7126 | 7135 | 7143 | 7152 | 1 2 3 3 4 |
| 52 | 7160 | 7168 | 7177 | 7185 | 7193 | 7202 | 7210 | 7218 | 7226 | 7235 | 1 2 3 3 4 |
| 53 | 7243 | 7251 | 7259 | 7267 | 7275 | 7284 | 7292 | 7300 | 7308 | 7316 | 1 2 2 3 4 |
| 54 | 7324 | 7332 | 7340 | 7348 | 7356 | 7364 | 7372 | 7380 | 7388 | 7396 | 1 2 2 3 4 |
| N | L. 0 | 1 | 2 | 3 | 4 | 5 | 6 | 7 | 8 | 9 | 1 2 3 4 5 |

**TABLE XV** *(continued)*

| N | L. 0 | 1 | 2 | 3 | 4 | 5 | 6 | 7 | 8 | 9 | Proportional parts 1 | 2 | 3 | 4 | 5 |
|---|---|---|---|---|---|---|---|---|---|---|---|---|---|---|---|
| 55 | 7404 | 7412 | 7419 | 7427 | 7435 | 7443 | 7451 | 7459 | 7466 | 7474 | 1 | 2 | 2 | 3 | 4 |
| 56 | 7482 | 7490 | 7497 | 7505 | 7513 | 7520 | 7528 | 7536 | 7543 | 7551 | 1 | 2 | 2 | 3 | 4 |
| 57 | 7559 | 7566 | 7574 | 7582 | 7589 | 7597 | 7604 | 7612 | 7619 | 7627 | 1 | 1 | 2 | 3 | 4 |
| 58 | 7634 | 7642 | 7649 | 7657 | 7664 | 7672 | 7679 | 7686 | 7694 | 7701 | 1 | 1 | 2 | 3 | 4 |
| 59 | 7709 | 7716 | 7723 | 7731 | 7738 | 7745 | 7752 | 7760 | 7767 | 7774 | 1 | 1 | 2 | 3 | 4 |
| 60 | 7782 | 7789 | 7796 | 7803 | 7810 | 7818 | 7825 | 7832 | 7839 | 7846 | 1 | 1 | 2 | 3 | 4 |
| 61 | 7853 | 7860 | 7868 | 7875 | 7882 | 7889 | 7896 | 7903 | 7910 | 7917 | 1 | 1 | 2 | 3 | 3 |
| 62 | 7924 | 7931 | 7938 | 7945 | 7952 | 7959 | 7966 | 7973 | 7980 | 7987 | 1 | 1 | 2 | 3 | 3 |
| 63 | 7993 | 8000 | 8007 | 8014 | 8021 | 8028 | 8035 | 8041 | 8048 | 8055 | 1 | 1 | 2 | 3 | 3 |
| 64 | 8062 | 8069 | 8075 | 8082 | 8089 | 8096 | 8102 | 8109 | 8116 | 8122 | 1 | 1 | 2 | 3 | 3 |
| 65 | 8129 | 8136 | 8142 | 8149 | 8156 | 8162 | 8169 | 8176 | 8182 | 8189 | 1 | 1 | 2 | 3 | 3 |
| 66 | 8195 | 8202 | 8209 | 8215 | 8222 | 8228 | 8235 | 8241 | 8248 | 8254 | 1 | 1 | 2 | 3 | 3 |
| 67 | 8261 | 8267 | 8274 | 8280 | 8287 | 8293 | 8299 | 8306 | 8312 | 8319 | 1 | 1 | 2 | 3 | 3 |
| 68 | 8325 | 8331 | 8338 | 8344 | 8351 | 8357 | 8363 | 8370 | 8376 | 8382 | 1 | 1 | 2 | 3 | 3 |
| 69 | 8388 | 8395 | 8401 | 8407 | 8414 | 8420 | 8426 | 8432 | 8439 | 8445 | 1 | 1 | 2 | 3 | 3 |
| 70 | 8451 | 8457 | 8463 | 8470 | 8476 | 8482 | 8488 | 8494 | 8500 | 8506 | 1 | 1 | 2 | 3 | 3 |
| 71 | 8513 | 8519 | 8525 | 8531 | 8537 | 8543 | 8549 | 8555 | 8561 | 8567 | 1 | 1 | 2 | 3 | 3 |
| 72 | 8573 | 8579 | 8585 | 8591 | 8597 | 8603 | 8609 | 8615 | 8621 | 8627 | 1 | 1 | 2 | 3 | 3 |
| 73 | 8633 | 8639 | 8645 | 8651 | 8657 | 8663 | 8669 | 8675 | 8681 | 8686 | 1 | 1 | 2 | 2 | 3 |
| 74 | 8692 | 8698 | 8704 | 8710 | 8716 | 8722 | 8727 | 8733 | 8739 | 8745 | 1 | 1 | 2 | 2 | 3 |
| 75 | 8751 | 8756 | 8762 | 8768 | 8774 | 8779 | 8785 | 8791 | 8797 | 8802 | 1 | 1 | 2 | 2 | 3 |
| 76 | 8808 | 8814 | 8820 | 8825 | 8831 | 8837 | 8842 | 8848 | 8854 | 8859 | 1 | 1 | 2 | 2 | 3 |
| 77 | 8865 | 8871 | 8876 | 8882 | 8887 | 8893 | 8899 | 8904 | 8910 | 8915 | 1 | 1 | 2 | 2 | 3 |
| 78 | 8921 | 8927 | 8932 | 8938 | 8943 | 8949 | 8954 | 8960 | 8965 | 8971 | 1 | 1 | 2 | 2 | 3 |
| 79 | 8976 | 8982 | 8987 | 8993 | 8998 | 9004 | 9009 | 9015 | 9020 | 9025 | 1 | 1 | 2 | 2 | 3 |
| 80 | 9031 | 9036 | 9042 | 9047 | 9053 | 9058 | 9063 | 9069 | 9074 | 9079 | 1 | 1 | 2 | 2 | 3 |
| 81 | 9085 | 9090 | 9096 | 9101 | 9106 | 9112 | 9117 | 9122 | 9128 | 9133 | 1 | 1 | 2 | 2 | 3 |
| 82 | 9138 | 9143 | 9149 | 9154 | 9159 | 9165 | 9170 | 9175 | 9180 | 9186 | 1 | 1 | 2 | 2 | 3 |
| 83 | 9191 | 9196 | 9201 | 9206 | 9212 | 9217 | 9222 | 9227 | 9232 | 9238 | 1 | 1 | 2 | 2 | 3 |
| 84 | 9243 | 9248 | 9253 | 9258 | 9263 | 9269 | 9274 | 9279 | 9284 | 9289 | 1 | 1 | 2 | 2 | 3 |
| 85 | 9294 | 9299 | 9304 | 9309 | 9315 | 9320 | 9325 | 9330 | 9335 | 9340 | 1 | 1 | 2 | 2 | 3 |
| 86 | 9345 | 9350 | 9355 | 9360 | 9365 | 9370 | 9375 | 9380 | 9385 | 9390 | 1 | 1 | 2 | 2 | 3 |
| 87 | 9395 | 9400 | 9405 | 9410 | 9415 | 9420 | 9425 | 9430 | 9435 | 9440 | 1 | 1 | 2 | 2 | 3 |
| 88 | 9445 | 9450 | 9455 | 9460 | 9465 | 9469 | 9474 | 9479 | 9484 | 9489 | 0 | 1 | 1 | 2 | 2 |
| 89 | 9494 | 9499 | 9504 | 9509 | 9513 | 9518 | 9523 | 9528 | 9533 | 9538 | 0 | 1 | 1 | 2 | 2 |
| 90 | 9542 | 9547 | 9552 | 9557 | 9562 | 9566 | 9571 | 9576 | 9581 | 9586 | 0 | 1 | 1 | 2 | 2 |
| 91 | 9590 | 9595 | 9600 | 9605 | 9609 | 9614 | 9619 | 9624 | 9628 | 9633 | 0 | 1 | 1 | 2 | 2 |
| 92 | 9638 | 9643 | 9647 | 9652 | 9657 | 9661 | 9666 | 9671 | 9675 | 9680 | 0 | 1 | 1 | 2 | 2 |
| 93 | 9685 | 9689 | 9694 | 9699 | 9703 | 9708 | 9713 | 9717 | 9722 | 9727 | 0 | 1 | 1 | 2 | 2 |
| 94 | 9731 | 9736 | 9741 | 9745 | 9750 | 9754 | 9759 | 9763 | 9768 | 9773 | 0 | 1 | 1 | 2 | 2 |
| 95 | 9777 | 9782 | 9786 | 9791 | 9795 | 9800 | 9805 | 9809 | 9814 | 9818 | 0 | 1 | 1 | 2 | 2 |
| 96 | 9823 | 9827 | 9832 | 9836 | 9841 | 9845 | 9850 | 9854 | 9859 | 9863 | 0 | 1 | 1 | 2 | 2 |
| 97 | 9868 | 9872 | 9877 | 9881 | 9886 | 9890 | 9894 | 9899 | 9903 | 9908 | 0 | 1 | 1 | 2 | 2 |
| 98 | 9912 | 9917 | 9921 | 9926 | 9930 | 9934 | 9939 | 9943 | 9948 | 9952 | 0 | 1 | 1 | 2 | 2 |
| 99 | 9956 | 9961 | 9965 | 9969 | 9974 | 9978 | 9983 | 9987 | 9991 | 9996 | 0 | 1 | 1 | 2 | 2 |
| N | L. 0 | 1 | 2 | 3 | 4 | 5 | 6 | 7 | 8 | 9 | 1 | 2 | 3 | 4 | 5 |

# ANSWERS TO ODD-NUMBERED PROBLEMS

## CHAPTER 2, P. 19

**2.1** For lowest class in each, suggested possible answers are

(a) .0001–.0010 in., .00005–.00105 in., .00055 in., .0010 in. (avoids a negative boundary).

(b) 30–34 mph, 29.5–34.5 mph, 32 mph, 5 mph.

(c) 54,000–56,900 psi, 53,950–56,950 psi, 55,450 psi, 3000 psi.

(d) 15–19, 15–20, 17.5, 5 years.

(e) 0–9, 0–9, 4.5, 10.

(f) 45–49%, 44.5–49.5%, 47%, 5%.

(g) 1, 0–1, .5, 1 oz.

(h) 31.58–31.59, 31.575–31.595, 31.585, .02 fl oz.

(i) 110–119, 110–119, 114.5, 10.

(j) .00–.005, .00–.005, .0025, .0010 (Remember: data are discrete.).

**2.3** Range is 94.3 to 100.0. Suggest classes 94.0–94.4, etc. (so that all 94. ?'s go into one of just two classes, etc.).

**2.5** Range is 227.0 to 293.1. Suggest 225.0–229.9, etc., giving 14 classes.

**2.7** .001 ft.   **2.9** 55.3.   **2.11** .1236 in.   **2.13** $-.70(.001$ in.).   **2.15** 32.3(.0001 in.).

**2.17** 12 (discrete data).   **2.19** 75.5 psi.   **2.21** 3.166 in.   **2.23** 3.1705 fl oz.

## CHAPTER 3, P. 46

**3.1** (a) $\bar{y} = .19240$ in.,   median $= .1925$ in.,   $s = .000,464$ in.,   $R = .0012$ in.

(c) $\bar{y} = 152.7$,   median $= 156$,   $s = 6.66$,   $R = 12$.

(e) $\bar{y} = 10.357$,   median $= 10.37$,   $s = .0709$,   $R = .14$ lumens/watt.

(g) $\bar{y} = .0145$, median $= .014$,   $s = .00596$,   $R = .020$ proportion present.

(i) $\bar{y} = 6.674$,   median $= 6.674$,   $s = .00274$,   $R = .007 \ 10^{-8}$ cm$^2$/g sec$^2$.

(k) $\bar{y} = 23.2$,   median $= 23$,   $s = 4.72$,   $R = 18$ sec.

**3.3**  Tables **2.11**:  $\bar{y} = 28.15$, $s = 5.48$.     **2.13**:  $\bar{y} = 65.45$, $s = 3.00$;
                                                        $\bar{y} = 65.50$, $s = 3.05$.
        **2.15**:  $\bar{y} = 31.31$, $s = 16.63$.     **2.17**:  $\bar{y} = 67,814$, $s = 5137$.
        **2.19**:  $\bar{y} = 115.2$, $s = 20.9$.     **2.21**:  $\bar{y} = .0565$, $s = .00799$.

**3.9**  $s^2 = (n - 1)s_1{}^2 + (m - 1)s_2{}^2 + nm(\bar{y}_1 - \bar{y}_2)^2/(n + m)$. Last term could be in two: $n(\bar{y}_1 - \bar{y})^2 + m(\bar{y}_2 - \bar{y})^2$, $\bar{y}$ as in Problem 3.8.

## CHAPTER 4, P. 84

**4.1**  161,700, 4950, 358,800, 1.

**4.3**  $C(12, 3) = 220$   $C(10, 1)/C(12, 3) = 1/22$.

**4.5**  $3 \cdot 4 \cdot 4 = 48$,  1/48,  $1 \cdot 2 \cdot 3/48 = 1/8$.

**4.7**  136,080, 60,480, 20,160.

**4.11**  $A$, $A$, $A$, $W$, $\varnothing$, $A$.

**4.13**  One is $(A \cap C) \cup (B \cap C)$.   One is $(A \cup C) \cap (B \cup C)$.

**4.15**  21/45, 21/45, 3/45  All possible samples of two each.  45.

**4.17**  .729, .243, .027, .001  One space is of $20 \cdot 20 \cdot 20/6$ outcomes.

**4.19**  $A_5$,  $A_3$,  $P(A_4) - P(A_1)$,  $1 - P(A_4)$.

**4.21**  3/4, 1/16.

**4.23**  (a)  Neither defect, open circuit only, timer off only, both defects.
    (b)  One possibility $N(A \cap B) = 1$  $N(A \cap \bar{B}) = 29$  $N(\bar{A} \cap B) = 4$
$N(\bar{A} \cap \bar{B}) = 116$.

**4.25**  $P = .6(24)/\pi\,50^2 = .00183$.

**4.27**  $P(\text{alarm}) = 1 - .04^2 = .9984$, three give reliability $= .999936$  Yes

**4.29**  Assume independence $P(b > s) = P(s > b) = .5$.

**4.31**  Corner of cube: $.5^3/6 = $ probability.

**4.33**  $P(7 \cup 11) = 8/36$  $P(2 \cup 3 \cup 12) = 4/36$.

**4.35**  $C(13, 3)/C(52, 3) = 11/850$.

**4.37**  Four times answer to 4.35 $= 22/425$.

**4.39**  $4^5 \cdot 10/C(52, 5) = 10,240/2,598,960$.

**4.41**  $13C(4, 3)12C(4, 2)/C(52, 5) = 3744/2,598,960$.

**4.43**  $4C(13, 7)C(39, 6)/C(52, 13)$.

**4.45**  Tests independent. Use $P(0 \text{ impure}) = 142,880/161,700 = .8836$, $P(1) = .1128$, $P(2) = .0036$, $P(3) = .0000$, and $P(\text{pass} \mid 0) = .99^3 = .9703$, $P(\text{pass} \mid 1) = .05(.99)^2 = .0490$, $P(\text{pass} \mid 2) = (.05)^2 .99 = .0025$, then multiply respectively, and add. $P(\text{pass}) = .8629$.

**4.47**  Use  $P(2g \mid .01) = .99^2$  $P(2g \mid .10) = .9^2$  $P(2g \mid 1) = 0$.  $P(.01 \mid 2g) = .9237$, $P(.10 \mid 2g) = .0763$,   $P(1 \mid 2g) = 0$.   $P(.01 \mid 2d) = .0082$,   $P(.10 \mid 2d) = .0819$, $P(1 \mid 2d) = .9099$.

**4.49**  $P(\text{1st weak}) = 1/1960$  $P(1 \text{ part weak}) = 1/196$  Random choices.

## CHAPTER 5, P. 126

**5.1**  .130,  .346,  .346,  .154,  .026.

**5.3**  .737,  .228,  .032,  .003.

**5.5**  .335,  .402,  .201,  .054,  .009,  .001.

**5.7**  .741,  .222,  .033,  .003.

**5.9**  .951,  .048,  .001.

**5.11** .223, .335, .251, .126, .047, .014, .004, .001.
**5.13** .467, .467, .067.
**5.15** .486, .399, .105, .010.
**5.17** .337, .401, .199, .054, .009, .001.
**5.19** .200, .160, .128, .102, etc.
**5.21** .100, .090, .081, .073, etc.
**5.23** .040, .064, .077, .082, .082, .079, etc. (slow decrease).
**5.25** .001, .003, .005, .007, etc.
**5.29** .1560, .0759, .2682, .6779.
**5.31** .0416, .5681, .0798, .5589.
**5.33** .0491, .0148, .1311, .9481.
**5.35** .1607, .4628, .5372, .7983.
**5.37** .1377, .4457, .0839, .8541.
**5.39** .1064, .1131, $U = 3$ (lowest meeting condition).
**5.41** .3165, .9942, .0434.
**5.43** .3407, .9464, .0536.
**5.45** .0530, .9325, .9855.
**5.47** $\mu = 4.5$, $\sigma = 2.87$, $\alpha_3 = 0$, $\alpha_4 = 1.776$.
**5.49** 2, 9.
**5.51** 17, 1.
**5.53** $U = 11$, $L$ not available, p(0) = .01547.
**5.55** p(0) = .006738, p(1) = .033690, etc.
**5.57** (a) .001, $ijk$, each 0–9, event $i = j = k = 3$. .010.
   (b) .000,001, certain.
**5.59** .04239.
**5.61** Use Poisson with $\mu = 7$, $y = 6$.
**5.63** Use Poisson with $\mu = 16$, $y = 3$.
**5.65** Use binomial with $\phi = .50$, $n = 200$.
**5.67** Use Poisson with $\mu = .15$.
**5.69** P(0 in 20 | 7 in 30) = .000,059.
**5.71** .9489 Independence questionable, average may vary.
**5.73** P(11 in 12 | $\phi = .5$) = .00293, P(11 or 12 in 12 | $\phi = .5$) = .00317. Either is rare enough to throw doubt on equivalence of $A$ and $B$.
**5.75** .964. Binomial.
**5.77** C(13, 7)C(39, 6)/C(52, 13), 4C(13, 7) C(39, 6)/C(52, 13).
**5.79** $P(y) = 1 - (1 - \phi)^y$.
**5.81** Write out a few terms, factor out common terms, recognize series.

## CHAPTER 6, P. 161

**6.1** 4.14% 195,360–208,640 psi.
**6.3** 98.16% 4.109, 8.217 oz.
**6.5** $u = 3.117$ Gamma: .84%, normal: .09%.
**6.7** Gamma: 11.89%, 19.26%, normal: 13.57%, 20.81%.
**6.9** 5.16%, 1.90% versus observed 2.71%, 1.00%.
   $u$ carried to three decimal places and $\Phi(u)$ interpolated in 6.11–6.15:
**6.11** 1 4 6 8 17 29 34 42 34 32 14 7 6 5
   1.2 2.3 5.4 11.0 19.0 27.8 35.1 37.8 34.7 27.3 18.4 10.5 5.2 3.3.

**6.13**  1   5   13   14   26   28   38   42   32   20   14   13   2   2
        2.6 4.2   8.9 16.0 24.8 33.1 38.0 37.7 32.1 23.5 14.9   8.1 3.8 2.3.
**6.15**  1   0   5   11   13   33   21   8   4   3   1
        .3 1.3   4.5 10.9 18.8 23.2 20.3 12.7 5.7 1.8   .5.
**6.17**  .8181.
**6.19**  .6534.
**6.21**  .0162,   .0089.
**6.23**  .9636 by gamma distribution,   $\alpha_3 = .3$.
**6.25**  .9642.
**6.27**  124.
**6.29**  .0026.
**6.31**  .0173 cm;   30 cm;   .0346 cm.   Yes,   $\alpha_3 = 0$,   $\alpha_4 = 3 + (1.8 - 3)/4 = 2.7$.
**6.33**  $\mu \pm 5\sigma$,   $\mu \pm 10\sigma$.   Normal:   $\mu \pm 2.054\sigma$,   $\mu \pm 2.576\sigma$.
**6.35**  $F(y) = 1 - \exp(-y/\mu)$.   Median $= .693\mu$;   Mean $= \mu$;   Mode $= 0$.

## CHAPTER 7, P. 184

**7.7**  Binomial:   $\phi = .1$,   $n = 20$.   Binomial:   $\phi = .1$,   $n = 50$.
**7.9**  Combined proportions:   p(0) $= .209$   p(1) $= .258$.
p(2) $= .204$   p(3) $= .142$, etc., versus   Poisson $\mu = 2$:
p(0) $= .135$   p(1) $= .271$   p(2) $= .271$   p(3) $= .180$, etc.
**7.11**  $E(\chi^2) = \nu$,   $\sigma_{\chi^2} = \sqrt{2\nu}$.

## CHAPTER 8, P. 218

**8.1**  Two points:   $\mu = .10566$ in.,   $P_a = .50$;   $\mu = .1045$ in.,   $P_a = .9376$ from $u_U = 0$
and 1.535 $u_L$ is negligible in both.
**8.3**  Using (8.8) $u = 19.3$ versus 2.326, significant. Using (7.4)   $t = u = 3.56$   versus
$\pm 2.576$,   significant.
**8.5**  $t_4 = 3.54$ versus 3.747 significant. $u = 3.32$ versus $\pm 2.326$ significant. No. Could
let $\mu$ be between $72.5 \pm 11.4$ and still be $3\sigma$ from limits.
**8.7**  $s = .707$   $t_4 = 4.74$ versus 1.533. Yes. Probably.
**8.9**  $t_2 = 5.26$ versus 6.965 not significant. Random sample, normal population.
**8.11**  $k_1 = 451.753$,   $k_2 = 458.517$.   $\alpha = .338$   Too large.
**8.13**  Use $\sigma_{\bar{y}} = .0005$ in. and $\mu$, say, .099 in.,   .100 in.,   .101 in., etc., using symmetry.
**8.15**  $t_4 = -1.12$ versus $\pm 4.604$ not significant.   $\chi_4^2 = 23.0$ versus 13.277 significant.
**8.17**  $n = 9$, 22, respectively, with critical $k$ .1044,   .1106(°C)$^2$ respectively. Approve
if $s^2$ below.
**8.19**  $s^2 < .7616$ accept. P(acc | $\sigma = 1.007$) $= .10$. Reject second shipment.
**8.21**  Use Poisson tables P(3 or less | $\mu = 12$) $= .00229$ versus .02. Further improve-
ment.
**8.23**  Normal $u = (116.5 - 129.5)/8.05 = -1.61$ versus $\pm 1.96$ not significant.
**8.25**  Critical region: 8 or more. P(acc | $\mu = 5$) $= .86663$   P(acc | $\mu = 8$) $= .45296$.
**8.27**  Critical region 3 or less and 18 or more. P(3 or less | $\mu = 6$) $= .15120$   P(18 or
more | $\mu = 13$) $= .10954$.
**8.29**  Probability of any one sample may be small or zero, no matter how compatible.

## CHAPTER 9, P. 247

**9.1** $F_{2,2} = 1.31$ versus $F_{2,2,.975} = 39.0$; not significant. (9.6): $t_4 = 2.02$ versus 2.776; not significant.

**9.3** $F_{4,4} = 3.14$ versus $F_{4,4,.995} = 23.15$; not significant. $t_8 = 15.8$ versus 3.355; significant.

**9.5** $F_{19,16} = 3.30$ versus 2.719; significant. Use (9.8), (9.9): $t_{30} = 2.594$ versus 2.042; significant.

**9.7** Use (9.2): $u = -.67$ versus $\pm 1.96$; not significant. Use (9.7): $u = -3.34$ versus $\pm 1.96$; significant.

**9.9** $F = 1.011$; not significant. Use (9.7): $u = .94$ versus $\pm 1.96$; not significant.

**9.11** $F_{6,5} = 15.86$ versus $F_{6,5,.995} = 14.51$; significant.

**9.13** Independent samples $F_{4,4} = 1.05$, $t_8 = -.53$; not significant. Matched pairs $t_4 = -7.48$ versus $\pm 4.604$; significant. Small consistent difference in widely varying pairs.

**9.15** $t_{17} = -4.81$.

**9.17** Using (9.11): $u = -3.63$ versus $\pm 2.576$; significant.

**9.19** Using (9.11): $u = -1.964$ versus $\pm 1.960$; significant. Worth trying (9.13).

**9.21** Using (9.12): $u = 3.58$ versus $\pm 1.96$ or 1.645; significant.

**9.23** Using (9.15): $u = 1.36$ and $u = 5.06$ versus 2.326. Second test significant.

**9.25** Using (9.14): $u = 9.43$ versus $\pm 3.3$; significant. Using (9.17) $u = 2.73$ versus $\pm 3.3$; not significant.

## CHAPTER 10, P. 278

**10.1** $\mu$: 6.6750, 6.6814 $\sigma$: .00260, .00808 $10^{-8}$ cm$^3$/g sec$^2$.

**10.3** $\mu$: 3.862, 4.708 $\sigma$: .252, .989 units.

**10.5** $\mu$: 3303.06, 3312.62 $\sigma$: 3.25, 11.88 cal/g.

**10.7** By (10.14) or (10.16) $\mu$: .7363, .7397 $\sigma$: .01773, .02007%.

**10.9** By (10.27) ($s_1$, $s_2$ not significantly different), $\mu_1 - \mu_2$: 60, 772 microfarads.

**10.11** By (10.27) gain $\mu_2 - \mu_1$: 1.91, 4.49%; .90 confidence. 1.06, 5.34; .99 confidence.

**10.13** By (10.27) $\mu_1 - \mu_2$: $-.604$, $-.456$ 1b. (Higher after ferroxing.)

**10.15** By (10.25) $\mu_1 - \mu_2 = -1.045$, $-.355$.

**10.17** By (10.30) $\sigma_1/\sigma_2$: 1.71, 2.31.

**10.19** By (10.30), interpolating $\sigma_1/\sigma_2$: .418, .841 : By (9.8) $\nu = 39.0$. By (10.28) $\mu_1 - \mu_2$: 390, 1038.

**10.21** By (10.31) $\mu_d$: $+1.13$, $+4.87$.

**10.23** $\mu_d$: 1.716, 2.908%. Wider.

**10.25** $\mu_1 - \mu_2$: $-3.84$, $-1.76$.

**10.27** $\mu_1 - \mu_2$: $-3.01$, $+7.59$; .90 confidence. $-6.71$, $+11.29$; .99 confidence. $\sigma_1/\sigma_2$: 1.93, 6.36.

**10.29** Point estimate $\hat{\phi} = .050$. By (10.34), (10.35) $\phi$: .0333, .0739. By [4] $\phi$: .0326, .0729.

**10.31** By (10.34), (10.35) $\phi$: .119, .159. Could use $p \pm 1.645\sqrt{p(1-p)/n}$ for approximation.

**10.33** $\hat{\phi} = .12$ By (10.34), (10.35) $\phi$: .080, .174.

**10.35** By (10.34), (10.35) $\phi$: .226, .341 By [4] $\phi$: .225, .340. Appear nonhomogeneous.

**10.37** By (10.39), (10.40)   $\mu$: 27.0, 47.9 for 29 lengths;   .93, 1.65 for one length.

**10.39** $\phi_1 - \phi_2$ : .144, .358.

**10.41** $u = 8.01$ versus $\pm 2.576$ significant.   $\phi_1 - \phi_2$ : .0448, .0674.

**10.43** 400 ft² unit: $\mu_1 - \mu_2$ : $800 - 480 \pm 1.96 \sqrt{800 + 480} = 250$,   390 1 ft² unit: $\mu_1 - \mu_2$ : .625, .975.

**10.45** For two subassemblies $\mu_1 - \mu_2$ : 7.8, 26.2.   For just one, $\mu_1 - \mu_2$ : 3.9, 13.1.

**10.47** Per subassembly $\mu_1 - \mu_2$ : 32.6, 71.8.

## CHAPTER 11, P. 313

**11.1** $\hat{y} = 2.1442 - .00076x$.   $\beta_1 = 0: t_3 = -62.2$ versus $\pm 3.182$;   significant! Limits $\beta_1$ : $-.000,789$, $-.000,731$.   $\mu_{Y.300}$ : 1.9152, 1.9172.   $\sigma_\varepsilon$ : .00060, .00282.

**11.3** $\hat{y} = 3.80 - .0549(x - 1.25)$.   $\rho = 0$ : $t = -1.506$ versus $\pm 2.776$;   not significant. $\beta_1$ : $-.156$, $+.046$ (note, they contain $\beta = 0$)   $\mu_Y$ : 3.714, 3.886.

**11.5** 200.12, 92.116, 70.80, .1065, .9139, .7905, .6249.   $\beta_1 = 0$ versus $\beta_1 > 0: t = 6.19$ versus 2.500.   $\hat{y}_{187} = 90.72$.   Expected error: $s_{\hat{y}} = .2905$.

**11.7** $\hat{y} = 27.23 - .1618x$   $\beta_1 = 0$ : $t_8 = -3.59$ versus $\pm 2.306$.   Same for $\rho = 0$. $\beta_1$ : $-.246$, $-.078$.

**11.9** $\hat{y} = 14.41 + .01219x$, .8939, .7990   $\beta = 0: t_7 = 5.27$ versus $\pm 3.499$   $\beta_1$ : .0078, .0166   $\mu_{Y.1600}$ : 33.17, 34.67.

**11.11** $b_0 = -.0014$, $b_1 = .8277$, $s_\varepsilon = .07403$, $r = .98524$, $\rho = 0$   $t_{48} = 39.9$ versus 2.407, $\beta_1 = 1: t_{48} = -8.30$ versus $\pm 2.685$.   $\beta_1$ : .7859, .8694, $\mu_{.80}$ : .6607 $\pm$ .0301,     .   $\mu_{1.30}$ : 1.0746 $\pm$ .0211,   $\mu_{2.20}$ : 1.8195 $\pm$ .0425.   Depends on $X$, but average drop $1.315 - 1.087 = .228$.

**11.13** $\beta_1 = 0$   $t = -11.25$   versus   $\pm 2.620$.   $y = -.085427x + 6.4777$.   $-.1001$, $-.3564$.   75.83 for $y = 0(90°)$. Since $s_y = .111°$, $6s_y = .666°$, and with same variation in $x$ at proper level, $y$ can meet $90 \pm .333°$.

**11.15** $\beta_1 = 0$ :   $t_{47} = 10.00$   versus   $\pm 2.012$.   $\beta_1 = 1$ :   $t_{47} = -1.08$   versus   2.012, $\beta_1$ : .721, 1.084, $\mu_{107}$ : 100.04 $\pm$ 1.19,   $\mu_{113}$ : 105.46 $\pm$ .52,   $\mu_{121}$ :   112.68 $\pm$ 1.56. $r^2 = .6804$.

**11.17** Hypotheses $\rho = 0$, $\beta_1 = 0$ equivalent. Use (11.4), (11.27), and (11.34).

**11.21** Minimize $\Sigma e_i^2 = \Sigma (y_i - bx_i)^2$ relative to $b$.

**11.23** No.   Exponential does not fit much better. $\log \hat{y} = -.18285 + .01751v$.

**11.25** $\log \hat{y} = 3.7795 - .6796 \log x$ or $\hat{y} = 6019x^{-.6796}$.   Fits well.

**11.27** $\log \hat{y} = -.4201 + 1.4929 \log x$   $\hat{y} = .3801x^{1.4929}$.   Exponent is 1.5.

**11.29** $\hat{y} = 122.811 + .89667x'$   $\hat{y} = 123.621 + .89667x' - .12165x'^2$.

**11.31** Treat $\log Y$ as $y$ and $(1/T)$ as $x$ and fit a straight line, giving $b_0$ , $b_1$ .

**11.33** $an + b \Sigma x + c \Sigma (1/x) = \Sigma y$,
$a \Sigma x + b \Sigma x^2 + cn = \Sigma yx$,
$a \Sigma (1/x) + bn + c \Sigma (1/x^2) = \Sigma y/x$.

## CHAPTER 12, P. 365

**12.1** MS(sample) = 7960   MS(error) = 63.28 in $10^8$ mm²   $F_{6.14} = 125.8$ versus 2.863; significant.

**12.3** MS(operator) = 19.6   MS(lamp) = 547.0   MS(error) = .35   $F_{1.4}$(operator) = 56.0 versus 21.20,   $F_{4.4}$(lamp) = 390.7 versus 15.98.   Note that the $t^2$'s from Problem 9.13 are $F_{1.4}$'s here.

**12.5**  SS(treatment) = 30,817   SS(error) = 7,533   $F_{1,4}$ = 4.091 versus 7.709.   Take $\sqrt{F}$ for $t$.

**12.7**  Since $\sigma_\alpha{}^2 = [(-4)^2 + 1^2 + 3^2 + 0^2]/3$,   EMS($A$) = 2.94 + 5(26/3) = 46.27.

**12.9**  $s_{\bar{y}} = \sqrt{63.28/3}$ = 4.593   Range of 7: 4.83(4.593) = 22.2, etc.  Groups as follows:
$\bar{y}_4\ \bar{y}_7\ \bar{y}_1$   $\bar{y}_5\ \bar{y}_3$   $\bar{y}_6\ \bar{y}_2$

**12.11**  After coding to nearest .1 (up, if a .05): SS(company) = 9.651  SS(error) = .7867 $F_{4,11}$ = 33.7 versus 5.70 (interpolated).

**12.13**  SS(packing) = 965.98  SS(error) = 1374.97  $F_{5,15}$ = 2.11; not significant. Fixed.

**12.15**  $q$ = .2839 versus $q_{.05}$ = .2825 interpolated; significant at .05.

**12.17**  $q$ = 2.5(253,812)/(1374.97)$^2$ = .3356   $q_{.05}$ = .396 interpolated.

**12.19**  SS(inspector) = .0136  SS(ring) = 1.9736  SS(error) = .0544  $F_{4,16}$(inspector) = 1.00.   No evidence at all of inspector bias.  $\hat{\sigma}_\varepsilon$ = .058.

**12.21**  SS(time) = 85.9  SS(door) = 984.53  SS(error) = 56.27  $F_{5,20}$ = 6.11 versus 2.711  $F_{4,20}$ = 87.48 versus 2.866.   Both significant.  Only doors 1 and 2 not distinct.

**12.23**  SS(piece) = 117.8  SS(position) = 2.7  SS(error) = 5.7  $F_{3,12}$ = 82.7 versus 3.490  $F_{4,12}$ = 1.42 versus 3.259.   Orthogonal polynomials: SS(piece, linear) = 116.64 on 1 df $F_{1,12}$ = 245.6 versus 4.747.   Remainder not significant, hence linear trend.

**12.25**

| Source | SS | df | MS | F | $F_{.05}$ |
|---|---|---|---|---|---|
| Machines | 8.511 | 2 | 4.256 | 185.7 | 3.885 |
| Coatings | 7.055 | 3 | 2.352 | 102.6 | 3.490 |
| Interaction | .619 | 6 | .1032 | 4.50 | 2.996 |
| Error | .275 | 12 | .02292 | | |

**12.27**  SS(place) = 550.067   SS(time) = 38.667   SS(error) = 29.933   $F_{2,18}$ = 165.4 versus 3.57   $F_{9,18}$ = 2.583 versus 2.47   SS(drive vs. operator) = 4.05   SS(center vs. other) = 546.017.

**12.29**  SS(temperature) = 4.083   SS(catalyst) = 5.167   SS(interaction) = 68.167 SS(error) = 17.5 [find SS(error) by $(\Sigma R^2)/2$]   $F$(temperature) = 1.40 versus 5.987 $F$(catalyst) = .89 versus 5.143  $F$(interaction) = 11.68 versus 5.143.  Illustrate by graph.

## CHAPTER 13, P. 401

**13.1**  $b_1{}^*$ = .55546  $b_2{}^*$ = −.00564  $R_{0.12}^2$ = .31361  By (13.21)  $F_{2,31}$ = 7.89 versus 5.68.   No gain in adding $X_2$.

**13.3**  (a)  −.2120, −.7223   (b)  .6659   (c)  $\hat{y}$ = −.3533$x_2$ − .3032$x_3$ + 23.38, using $(n-1)s_0^2 = \Sigma\,(y - \bar{y})^2$ in (13.19)  $s_{0.23}$ = 1.183   (d)  $F_{2,97}$ = 96.7 versus 4.86 (e)  $F_{1,97}$ = 11.67 versus 6.94.

**13.5**  (a)  .7083, .2624, $y$ = 348.2 + 1.489$x_1$ + .6884$x_2$, .5008, 8.92.   (b)  $F_{2,176}$ = 88.3; significant   (c)  $F_{1,176}$ = 23.4; significant.

**13.7**  (a)  $b_1{}^*$ = .5828, $b_2{}^*$ = .4861, $b_3{}^*$ = .3039, $b_4{}^*$ = −.3701  $R_{0.1234}$ = .5757 (b)  $F_{4,115}$ = 39.0 versus 3.49; significant   (c)  $\hat{x}_0$ = −182 + 13.26$x_1$ + 19.77$x_2$ + 84.41$x_3$ − 9.902$x_4$  $s_{0.1234}$ = .3037   (d)  $F_{1,117}$ = 47.0, $F_{1,116}$ = 9.70, $F_{1,115}$ = 33.5. All significant versus critical $F$'s of about 6.87.

**13.9** (a)

| 1 | 2 | 3 | 4 | 5 | 0 |
|---|---|---|---|---|---|
| 1.00000 | −.17600 | −.42456 | +.38712 | −.21924 | +.16884 |
| | 1.00000 | +.26356 | +.13389 | +.20448 | −.56054 |
| | | 1.00000 | −.87641 | +.78077 | −.53234 |
| | | | 1.00000 | −.70099 | +.28508 |
| | | | | 1.00000 | −.52747 |
| | | | | | 1.00000 |

(b) $R^2_{0.12345} = .51314$, $s_{0.12345} = 2.1496$
$x_0 = 14.917 - 1.577x_1 - 1.276x_2 - 1.694x_3 - .5783x_4 - 41.14x_5$
(c) $b_1{}^* = -.0565$, $b_2{}^* = -.3722$, $b_3{}^* = -.4263$, $b_4{}^* = -.2135$, $b_5{}^* = -.2806$.
They are quite small relative to their standard errors.
(d) $F_{5,18} = 3.794$ versus $F_{.05} = 3.17$ Significant with all variables used.
(e) If relatively narrow, the $r$'s are decreased.
**13.11** $R^2_{0,12} = .33346$ from $r_{01} = .57659$, $r_{02} = .03175$. $s_{0.12} = 8.142$ $F_{2,6} = 1.50$
versus $F_{.05} = 5.14$ not significant.

## CHAPTER 14, P. 424

**14.3** $\chi_5{}^2 = 20.7$, lumping top class. $\chi^2_{5,.99} = 15.086$. Not binomial data.
**14.5** $\chi_4{}^2 = 2.9$ versus $\chi_{4,.95} = 9.488$. Accept normality hypothesis.
**14.9** $\chi_1{}^2 = 12.2$ using (14.14).
**14.11** $\chi_1{}^2 = 1.1$ using (14.14).
**14.13** $\chi_2{}^2 = .5$.
**14.15** $\chi_2{}^2 = 22.0$ versus $\chi^2_{2,.99} = 9.21$. Significant difference between shifts.
**14.17** Hypothesis: no difference versus a difference, then P(all $\alpha$'s same sign) $= 2(.5)^5 =$
.0625. Cannot reject at $\alpha = .05$ no matter what outcome. If alternate hypothesis is
operator 1 less than operator 2, could reject null hypothesis.
**14.19** After—before: 14 +'s, 1 −, 3 0's. Neglect last. Hypothesis: no difference
versus after greater than before. P(0 or 1 | $\phi = .50$, $n = 15$) $= .00049$. Significant.
**14.21** $\chi^2_{11} = 7.4$ versus $\chi^2_{11,.95} = 19.675$ Population could well be normal.
**14.23** $\chi^2_{11} = 9.8$ versus $\chi^2_{11,.95} = 19.675$ Population could well be normal.

# Index

## A

*A,* 40
$A_i$, factors, 323
$a_i$, 43-45
Accept hypothesis, 188-190
Acceptance region, 189, 195
Acton, F. S., 313
Algebra of expectations, 93, 94
Algebra of sets, 54
Alpha, $\alpha$, 189
Alpha moments, $\alpha_3$, $\alpha_4$, 93
Alpha, $\alpha$, risks, 189, 195, 203
Analysis of variance, 322-363
  expected mean squares, 356-359
  factors, types of, 348
  models, 323, 355
    fixed, 349, 355, 356
    mixed, 357
    random, 357
  one-factor, 323-331, 352
  sums of squares, 325, 351
    between samples, 325, 351
    error, 325, 352
    interaction, 351
    total, 327, 352-354, 356, 357
    use of ranges, 329, 350
    within sample, 325, 352
  tests, 328, 352
    two-factor, 349-362
    tests, 353, 354, 356, 357
    unreplicated design, 358-360
  unequal sample sizes 329
Anderson, R. L., 363
Anova, 360 (*see* Analysis of variance)
Antle, C. E., 161
*A posteriori* probability, 80
*A priori* distribution of parameter, 213
*A priori* probability, 80

Arithmetic mean, 12, 33-35
  efficient calculation of, 38-42
  frequency data, 41, 42
  properties of, 34
Array, 35
Aspin, A. A., 246
Assumption, 194
Average, 12 *(see also* Arithmetic mean,
  Median)
Averages, 33-35

## B

$b_0$, 289, 372
$b_1$, 289, 372
$b_i$, 372
Bain, L. J., 161
Baker, G. A., 312
Bancroft, T. A., 217
Barany, J. W., 364
Bartholomew, D. J., 246
Bartlett, M. S., 312
Bartlett's test, 345
Bayes' probabilities, 80-82
Bayes' rule, 82
Beckman, M. J., 161
Bennett, C. A., 364
Berkson, J., 313
Berretoni, J. N., 161
Betas, $\beta_0$, $\beta_1$, $\beta_i$
  $\beta_0$, intercept, population, 289
  $\beta_1$, slope, population, 289
  $\beta_i$, 372
*Beta distribution, 148-150*
  *density function, 149*
  *mean, 149*
  *moments, 149*
  *standard deviation, 149*

*Beta function, 149*
*Beta, β,* risk, 190, 195, 203
Bien, D. D., 217
Bimodal data, 13, 35
Binomial distribution, 95ff
  approximations, 103
  calculation of, 101
  conditions of applicability, 103
  derivation, 96
  distribution function, 96
  mean, 100
  moments, 99, 100
  probability function, 96
  standard deviation, 100
  tables of, 101, 454
  test on fit, 411, 412
  test on parameter, 207-209
    two parameters, 234-236, 420
Binomial expansion, 69
Bivariate data grouped, 300-302
Bobkoski, F., 161
Box, G. E. P., 313
Brandt, A. E., 364
Brownlee, K. A., 246
Brugger, R. M., 275
Brumbaugh, M. A., 247, 424
Brunk, H. D., 126
Burke, C. J., 424
Burr, I. W., 126, 160, 161, 183, 217, 364

**C**

*c,* 40
$c_4$, 176, 437
$c_5$, 176, 437
Cameron, J. M., 217
Cancellation, heavy, 3, 39, 326
Carlson, F. D., 313
Carver, H. C., 160, 217, 424
Central limit theorem, 153, 154
Central tendency, 33
Chen, H. J., 424
Chi-square distribution, 175, 203, 261, 262
  goodness of fit, 408
  mean, 175, 186
  moments, 175
  relation to gamma distribution, 175
  standard deviation, 175, 186
  table, 432-435
  variances; use for, 175, 345

Chi-square tests, 407-422
  contingency tables, 417-422
  for cell frequencies, 408-410
  goodness of fit, 407-416
Cislak, P. J., 161
Clark, R. E., 161
Class boundaries, 6
Class frequency, 6
Class interval, 7
Class limits, 6
Class mark, 6
Classes, numerical, 5
Clopper, C. J., 278
Cochran, W. G., 247, 364, 423
Coding, 40, 44, 45
Coefficient of determination, $r^2$, 299
Colton, T., 217
Combinations, 67-69
Combining frequencies 410
Complement of set or event, 51
Completely randomized design, 349
  unreplicated, 360
Conditional probability, 62
Confidence coefficient, 1-α, 259ff
Contingency tables, 417-422
  exact method, 2x2, 421
Continuity correction, 140, 147, 208, 210,
    235, 237, 420
Continuous data, 8
Confidence limits for parameters, 257-273,
    297, 298
  attribute data, 268-271
  binomial data, one, 268-271
    two samples, 272, 273
  general concept, 257
  geometrical argument, 258
  interpretation, 257, 258
  linear correlation data, 295
    error variance, 297, 298
    intercept, 297
    mean of y's, 297
    regression line value, 298
    slope, 296
  means, one, 259-261
    two, difference, 264, 265
    paired differences, 267
  Poisson data, 270, 273
  standard deviation, one, 261-263
    ratio of two, 266
  summary of cases, 276, 277
  variances, one, 261, 262
    ratio of two, 266

Coplin, E. C., 364
Correlated data, 286-305, 372-400
Correlation, 298-300
  applications, 310, 311
  coefficient, 299
Craig, C. C., 161
Cramer, H., 161, 183
Critical region, 189, 195, 196, 203, 226
Crow, E. L., 278, 312
Cumulative frequency, 10
  graph, 10-12
Cumulative function, 92
Cureton, E. E., 278
Curve shape, 13, 93, 95


**D**


Darwin, J. H., 247
Data continuous, 8
  discrete, 8
  measurement, 8
  normal 44, 45
Davies, O. L., 217, 364
Decision rule, 190, 193
Decision theory, 212ff
Defective, 130
Defects, 103
Degrees of freedom, 171, 172, 175
  analysis of variance, one-factor, 327, 328
    two-factor, 352, 353
  goodness of fit test, 408, 410
  linear model, 293, 294
    multiple, 384-387
Delta, $\delta$, 201
Deming, W. E., 183
Density function, 132
Detre, K., 247
Deviation, mean, 35-38
  standard, *see* Standard deviation
Deviations, 16, 36
Differences, significance of, 223-246 (*see* Significance of differences)
Differential effects, $\alpha_i$, 323, 355
Discrete data, 8
Discrete population, 91ff
Distribution, frequency, 4
Distribution function, 92, 133
  properties of, 133, 134
Distribution, joint, of mean and standard deviation, 177

Distribution systems, 151, 152
  general system, 152
  Pearson, 151
Distributions, *see* below
  beta, 148-150
  binomial, 95-103
  chi-square, 175
  exponential, 142-144
  F, 179
  gamma, 144-148
  geometric, 120-122
  hypergeometric, 111-119
  negative binomial, 123-124
  non-central, 182
  normal, 135-141
  Poisson, 103-110
  rectangular, 141, 142
  systems, 151, 152, 455
  $t$, 171
  uniform, 119, 120
  Weibull, 150, 151
Dixon, W. J., 217
Doolittle solution, 380-386
Dorsey, C. L., 312
Draper, N. R., 312, 400
Duncan, D. B., 363
Dunn, O. J., 364
Dunnett, C. W., 363
Dwyer, P. S., 400


**E**


$E$, 92
Empty set, 51
Engerer, E. W., 364
Epsilon, 242, 285, 292, 323, 355, 372
Equally likely events, 57
Error, of first kind, 194
  of second kind, 195
Error, random, 242, 285, 292, 323, 355, 372
  linear model, estimate of, 293
  pure, 398
Estimate of parameter, 254, *see also* Estimation of parameters
Estimation, of parameters, 253-273 (*see also,* Confidence limits)
  cases, 256, 257
  estimators, 254-257
    biased, 254, 255
    consistent, 256

efficient, 255
  maximum likelihood, 256
  sufficient, 256
  unbiased, 254
interval, 257-273
point, 253-257
Estimator of parameter, 254
Eta, multiplicative error, 309
Events, 50
  countably infinite, 58
  dependent, 59-62
  discrete spaces, 58
  disjoint, 53
  equal, 52
  equally likely, 57
  independent, 59-62
  mutually exclusive, 53
  uncountably infinite, 77
  unequal, 52
Ewing, S. P., 246
Expectations, algebra of, 93, 94
Expected value, 92
  of a function of a random variable, 92
  of a random variable, 92
Exponential distribution, 142, 143
  density function, 142
  distribution function, 143
  mean, 143
  moments, 143
  relation to geometric distribution, 143
  reliability, use in, 143
  standard deviation, 143


**F**

$f$, 41
$F_i$, 408
$f_i$, 408
$F$ distribution, 179
  confidence limits by, 266
  density function, 179
  mean, 179
  table, 438-441
  tests by, 225
  variance, 179
Factorials, 67
Factorials, logs of, 456, 457
Factors, 348
  categorical, 348
  fixed, 348

nested, 348
numerical, 348
random, 348
Fairfield, J. H., 247
Ferguson, J. H. A., 247
Ferris, C. D., 217
Finney, D. J., 161
Fisher, R. A., 246, 424
Fitting equation (*see* Least squares fitting)
Flinching, 13
Foster, L. A., 364
Fox, M., 247
Fraction defective, 8
Franklin, N. L., 364
Freeman, H. A., 246
Frequency, 6
Frequency class, 6
  nomenclature, 6
Frequency, cumulative, 10
Frequency distribution, 4
Frequency graph, 8-12
Frequency polygon, 9
Frequency table, 5
Frequency tabulation, 5, 7
Fry, T. C., 125


**G**

Gamma distribution, 144-148
  density function, 144
  discrete distributions approximated by,
      146-148
  exponential, relation to, 145
  mean, 145
  moments, 145
  normal, relation to, 143
  standard deviation, 145
  tables, 146
  test of, 414
Gamma function, 144
Gardner, R. S., 278
Generating samples, 124
Geometric distribution, 120ff
  definition, 120
  mean, 120
  moments  120
  probability function, 120
  standard deviation, 120
  use in reliability, 121
Geometric mean, 345

Ghering, L. G., 424
Glover, J. W., 160, 424
Goldfield, S. M., 313
Goodness of fit tests, 407–416
  non-chi-square tests, 416
  test, 408
Gosset, W. S., 171
Graphical representation of frequency, 8
  by computer, 9
  cumulative, 11
Graybill, F. A., 278
Grouped bivariate data, 300–302
Grubbs, F. E., 217
Guard, R. T., 364
Guenther, W. C., 218, 278
Gurland, J., 247

**H**

$H_0$, 188–191
Hadley, G., 126
Haight, F. A., 161
Half-life, 143
Harter, H. L., 161, 218, 363, 364, 446
Hartley, H. O., 364
Haug, H., 312
Herrera, L., 275
Hicks, C. R., 363
Hinchen, J. D., 313
Histogram, 8
Hoefs, R. H., 364
Holstein, J. E., 126
Homogeneity of variances, 345–347
  Bartlett's test, 345
  Q test, 345–347
Homoscedasticity, 292
  test of, 345
Houseman, E. E., 363
Hypergeometric distribution, 111ff
  approximations to, 116–117
  conditions of applicability, 118
  derivation, 111
  mean, 115
  moments, 115
  probability function, 111
  standard deviation, 115
  symmetries, 114
  tables, 113
Hypotheses, tests of, 187ff, 295–298
  attribute data, 207–212

bimomial parameter, one, 207–209
  two, 234–236
linear models, 295–298
  error variance, 297, 298
  intercept, 297
  mean of $y$'s, 297
  regression line value, 298
  slope, 296
matched pair data, 239–242
mean, $\sigma$ known, 196, 197
  $\sigma$ unknown, 198–201
means, two, 228–233
multiple, 396
Poisson parameter, one, 209, 211
  two, 236–238
summary of elements of, 194–196
variabilities, one, 202–206
  two, 225–228
Hypothesis, 187ff
  alternate, 188, 194
  composite, 188, 194
  null, 188, 194
  simple, 194
  *(see also* Hypotheses, tests of)

**I**

Interaction, 323
Intercept, $b_0$, 289
  calculation by coding, 292, 301
  distribution of, 293
  estimate of, 297
  formula for, 290, multiple, 376
  interval estimate of, 297
  tests on, 297
Interpolation, 147, 173
Interpretation of confidence interval, 258
Interpretations of hypothesis tests, 197, 198
  acceptance, 197
  rejection, 198
Intersection of events, 53
Interval estimation, 257–273 *(see* Confidence limits)
Inverse, calculation of, 397

**J**

Joint distribution of mean and standard
  deviation, 177
Johnson, N. L., 161, 247

**K**

*K,* 111
*k,* 408, 410
*k*-statistic, 45
Kendall, M. G., 125, 246
Kashyap, V. C., 312
Kerrich, J. E., 313
Kitagawa, T., 125, 424
Kittrell, J. R., 313
Klett, G. W., 275
Kramer, C. Y., 364
Kurtosis, 44


**L**

*L,* 191
Lack of fit, 338, 399
Laws, 308
  exponential, 308
  power, 308
Least squares fitting, 287-290, 306, 307,
    374-376
  after a transformation, 307-309
  multiple regression, 374-376
Levels of factors, 323
Lewis, D., 424
Li, J. C. R., 183
Lieberman, G. J., 125, 217
Linearity, test of, 338
Lipow, M., 161
Lloyd, F. R., 400
Lloyd, D. K., 161
Logan, K. H., 246
Logarithms, 458, 459
Loss, 213
Lumping frequencies  410


**M**

*M,* 119
$m_i$, 43-44
$m_i'$, 43
Machine generation of samples, 169
Mainland, D., 275, 424
Massey, F. J., 217, 424
Matrix, 388ff
  approach to multiple regression, 391-398

determinant of a square, 391
identity, 391
inverse of a square, 391
product, 389
properties, 389-391
sum, 389
transpose, 390
vector, 389
Mayhood, R. F., 247
McCall, J. T., 364
McCullough, R. S., 247
Mean, 33-35
  efficient calculation, 38-42
  frequency data, 41, 42
Mean, arithmetic, 12, 33-35
  efficient calculation of, 38-42
  frequency data, 41, 42
  weighted, 34
Mean, geometric, 345
Mean, population, 92-95, 135
Mean, sample, 12, 33-35
  mean of, 153, 157-159
  moments of, 154, 159
  standard deviation of, 153, 157-159
Mean deviation, 35-38
Measurement data, 8
Measures of curve shape, 42-45, 93, 135
Measures of variability, 35ff
Median, 12, 35
  properties of, 35
Metcalf, C. T., 312, 363
Method of attributes, 8
Method of variables, 8
Mezaki, R., 313
Midvalue, 6
Miller, R. G., Jr., 364
Mode, 12, 35
  properties of, 35
Models
    analysis of variance
    one factor, 323, 329
    two factors, 355, 356
    fixed, 355, 356
    mixed, 357
    random, 356
  factorial, 362
  Greco-Latin squares, 362
  Latin square, 362
  linear regression, 292
  matched pairs, 241
  multiple regression, 372

Moments, sample, 43-45, 134, 135
  central, 43-45, 135
  fourth, 44, 45, 135
  standardized, 43-45, 135
  third, 43-45, 135
Molina, E. C., 125, 217
Moments, population, 93, 135
Monte Carlo method, 169
Moulton, F. R., 312
Mu, for population mean, 91, 104, 119, 142
  for moments, 93
Multiple contrasts, 341-344
  interpretation of risk, 344
  multiple range test, 341-344
  Newman-Keuls test, 341-344
Multiple regression, 372-400
  coefficient of determination, 379
  confidence limits, 396
  data, 374
  error variance, 379
  fitting, 374-378
  goodness of fit, 379
  matrix approach, 391-398
  model, 372
    adequacy, 398
  multiple correlation coefficient, 379
  normal equations, 376-378
  selection of variables, 396
  significance tests, 384
  solution of normal equations, 380-384
  summary, 397, 398
  tests of coefficients, 395
  variance-covariance matrix, 395

**N**

*N*, 111
*n*, 34, 95
Natrella, M. G., 218, 278
Negative, binomial distribution, 122-124
  binomial, relation to, 122
  definition, 122
  mean, 123
  moments, 123
  probability function, 122
  standard deviation, 122
  use in reliability, 122
Newman-Keuls multiple range test, 341-344
Nicholson, W. L., 161
NID, 292

Niemann, L. J. H., 183
Noel, R. H., 247, 424
Non-central distribution, 182
Normal curve, 135-141
  approximation of probabilities, 139-141
  density function, 135
  distribution function, 137
  general density function, 138
  moments, 138
  probabilities, use in, 139-141
  properties of, 136
  sketching, 138
  test of, 413
Normal distributuion, (*see* Normal curve)
Normal equations, 376-378
  matrix, 393, 394
Normality, test of, 410, 412-414
Null set, 51
Nu, degrees of freedom, 172, 175, 179,
  408, 410, 419

**O**

O C curve, 191, 196, 201, 206, 211, 212,
  442-445
Olds, E. G., 313
Operating characteristic curve, 191, 196,
  201, 206, 211, 212, 442-445
Orthogonal contrasts, 331-334
  orthogonal property, 332
  sums of squares, 334
  tests of, 334
Orthogonal polynomials, 334-341
  equation fitted, 338
  sums of squares, 338, 339
  tables of coefficients, 336, 340
  test of linearity, 338
  tests of significance, 338, 339
Ostle, B., 363
Outcomes, 49
Owen, D. B., 125, 217, 363, 424, 432, 436,
  438, 442, 450

**P**

*p*, 95
$P_a$, 191
*P(A)*, 55
*P(Y)*, 92

*p(Y)*, 92
Pachares, J., 278
Parameter, 91, 194
Pearson, E. S., 218, 278, 364
Pearson, K., 151, 160, 423
Permutations, 66, 67, 69
Phi, lower case Greek, 96, 120, 122, 135, 408
Plait, A., 161
Point, in probability space, 49
Poisson distribution, 103ff
  approximation, use in, 109
  calculation, 105
  conditions of applicability, 110
  contagious tendency, 416
  derivation, 106 107
  limit of binomial, 109, 110
  mean, 105
  moments, 105
  postulates, 104
  probability function, 104
  standard deviation, 105
  tables, 452, 453
  test of, 415
  test on one mean, 209, 210
    two means, 236-239
Population, 16-18
  continuous, 131
  curve shape, 93, 95, 135
  discrete, 90
  mean, 92, 135
  "normal" for experiments, 184
  standard deviation, 93, 135
  variance, 93, 135
Posterior probabilities, 80-82
Probabilities, 66-80
  acceptance, 191
  calculation of discrete, 66-77
  card, 88
  continuous, 77-80
  discrete, 66-77
  examples of, 69-77
  posterior, 80-82
  rejection, 191
Probability density function, 132
Probability function, 92
  discrete random variable, 92
Probability of event, 55
  defining properties, 55
  discrete spaces, 57

interpretation of, 82, 83
laws of, 56, 61, 62

**Q**

$Q_\alpha$, 342, 446, 447
Q test, 345-347, 449
Quandt, R. E., 313

**R**

*R*, 36
*r*, 299
Random digits, 18
  table of, 450, 451
Random error, 242, 285
  linear model, estimate of, 293, 295
  one-factor analysis of variance, 327
Random numbers, 409
  table of, 450, 451
Random sampling, 18, 57, 124, 167, 168
Random variables, 83, 166
Range, 12, 35
Rectangular distribution, 141, 142
  density function, 141
  distribution function, 141
  mean, 142
  moments, 142
  standard deviation, 142
Region, acceptance, 189
  critical, 189
Regression, 285-310 (*see also* Correlation, Intercept, Multiple Regression, Slope)
  applications, 310, 311
  equation, 293, 294, 374
  intrinsically non-linear, 310
  linear, 285-305
    intrinsically, 307-309
  multiple, 372-400
    two independent variables, 385-387
  non-linear, 306-310
  polynomial, 306, 334-341
  test of linearity, 338
  transformation, use of, 307-309
Reject hypothesis, 188-190
Replication, 349
Richardson, C. H., 313
Rider, P. R., 424

Rietz, H. L., 218
Risks, 189, 195
  alpha, 189, 195
  beta, 190, 195
Robertson, W. H., 125
Romig, H. G., 125
Root-mean-square, 346
Roy, L. K., 161


**S**

*s*, 37
$s^2$, 37
Salvosa, L. R., 160, 217
Sample, 16-19
Sample size, 34
  finding for mean, 192, 193, 195, 197
  finding for standard deviation, 204, 205
  finding for two means, 241, 242
Sample space, 49
  countably infinite, 49
  finite, 49
  uncountably infinite, 50
Sample statistics, 33-45
  standard deviation, 37
  variance, 37
Sampling, 18
  randomly, 18
  with replacement, 18
  without replacement, 18
Sampling distributuons
  binomial samples, 181
  chi-square distributuon, 175
    sum of two chi-squares, 182
  differences, two means, 177-179
  *F* distribution, 179, 180
  joint for *y* and *s,* 177
  large samples, 180
  mean, one, $\sigma$ known, 170
            $\sigma$ unknown, 171
  means, two, $\sigma$'s known, 177, 188
            $\sigma$'s unknown, 178, 179
  non-central distributions, 182, 201
  non-normal populations, 174, 176, 198
  Poisson, 181
  standard deviation, 176
    two, 179, 180
  sums of two means, 177-179
  *t* distribution, 171

non-central, 182
  variance, one, 174, 175
    two, 179-181
Sandiford, P. J., 126
Satterthwaite, F. E., 247, 278
Scatter diagram, 286, 287
Scheffe, H., 247
Seefeldt, W. B., 364
Seeger, P., 364
Shakun, M. F., 313
Shapiro, S. S., 424
Shewell, C. T., 313
Shewhart, W. A., 161, 400
Sigma, capital Greek, 34
  lower case Greek, 91
Set of outcomes, 51
  Borel, 54
Sign test, 422
Significance of differences, 223-246
  binomial data, 234-236
  matched pair data, 239-241
  means, 228-233
    known $\sigma$'s, 228-230
    unknown but equal $\sigma$'s, 230-232
    unknown, unequal $\sigma$'s, 232, 233
  Poisson data, 236-239
  summary of tests, 244, 245
  variances, 225-228
Significance tests, 187-211, 223-246
  one parameter, summary, 215, 216
  two parameters, summary, 244, 245
Skewness, 42-45
Slope, $b_1$, 289
  calculation by coding, 292, 301
  direct proportion case, 304
  distribution of, 293
  formula for, 290
  interval estimation of, 296
  tests on, 296
Smith, H., 312, 400
Snedecor, G. W., 364
Sobel, E., 313
Standard deviation, population, 93, 135,
  table of distribution constants, 437
Standard deviation, sample, 35, 37-42
  calculation of, 38-42
  distribution of, 176, 437
    curve shape, 437
    mean, 176, 437
    standard deviation, 176, 437

table of constants, 437
frequency data, 41
tests of, 202, 203-206, 448
Statistic, efficient, 189 (*see* Estimation of parameters)
Statistic, sample, 33-45, 195
distribution of 166ff, 195
Steel, R. G. D., 364
Stevens, W. L., 364
Stuart, A., 246
Studentized range, 342, 446, 447
Student's distribution, 171 (*see also* Samping distributions *t,* and *t* distribution)
Summarization of data, 33-45
Summation, 34
laws of, 34
Sums, 152, 153
mean of, 153
moments, 159, 160
standard deviation of, 153
Sutcliffe, M. I., 275
Srivastava, A. B. L., 247
Sukhatme, B. V., 247
Systems of distributions, 151, 152
General system, 152
Pearson, 151


**T**

*t,* 171
$T_i$, 325
*t* distribution
density function, 172
difference of means, 172
distributuon function, 173
linear model, 294, 295, 300, 304
mean, 172
mean test, 199
moments, 172
standard deviation, 172
tables of, 173, 436
Tate, R. F., 275
Tchebycheff's inequality, 155
Tchebycheff's theorem, 155
Testing type of distribution, 407, 410-416
Tests of hypotheses, 187-211, 223-246
one parameter, summary 215, 216 (*see* Hypotheses, tests of)
Tews, G., 218

Theoretical frequency distributions, test of, 409-416
Thoman, D. R., 161
Thomas, P. O., 218
Tukey, J. W., 247


**U**

U, 191, 238, 273
*u,* 135, 138, 139, 147, 153, 171, 181, 189, 227, 229, 234-237, 262, 378
Uniform distribution, 119, 120, 141
mean, 119
moments, 119
probability function, 119
standard deviation, 119
test of, 409
Union, of events 52


**V**

*v,* 40, 378
$\bar{v}$, 40
Vahle, H., 218
Van Valen, L., 247
Variable, random, 83
dependent, 285, 372
independent, 285, 372
input, 372
output, 372
Variability measures, 35ff
Variance, sample, 37
distribution of, 174, 175
linear model, 293
distribution of, 293
error, 293
population, 93
laws for, 94
test of, 202, 203, 206
Variances, homogeneity of, 345-347
Vector, 388, 389
Venn diagrams, 53-54


**W**

Wallace, D. L., 364
Wasan, M. T., 161

Watson, C. C., 313
Watson, G. S., 313
Weaver, C. L., 217
Weaver, W. R., 364
Weibull, W., 161
Weibull distribution, 150, 151
    density function, 150
    distribution function, 150
    fitting, 151
    mean, 150
    moments, 150
    standard deviation, 150
    use in life testing, 150
Weighted mean, 34
Weintraub, S., 125
Weir, J. B. deV., 247
Welch, B. L., 246, 247
Weyl, P. R., 278, 363

White, C., 247
Whitin, T. M., 126
Wilk, M. B., 424
Williams, C. M., 125
Williams, F., 126

**X**

$X$, 285
$x$, 285
Xi, 336

**Y**

$Y$, 92
$y$, 34, 95
$\bar{y}$, 34
Youden, W. J., 313